Robert N. Hammer

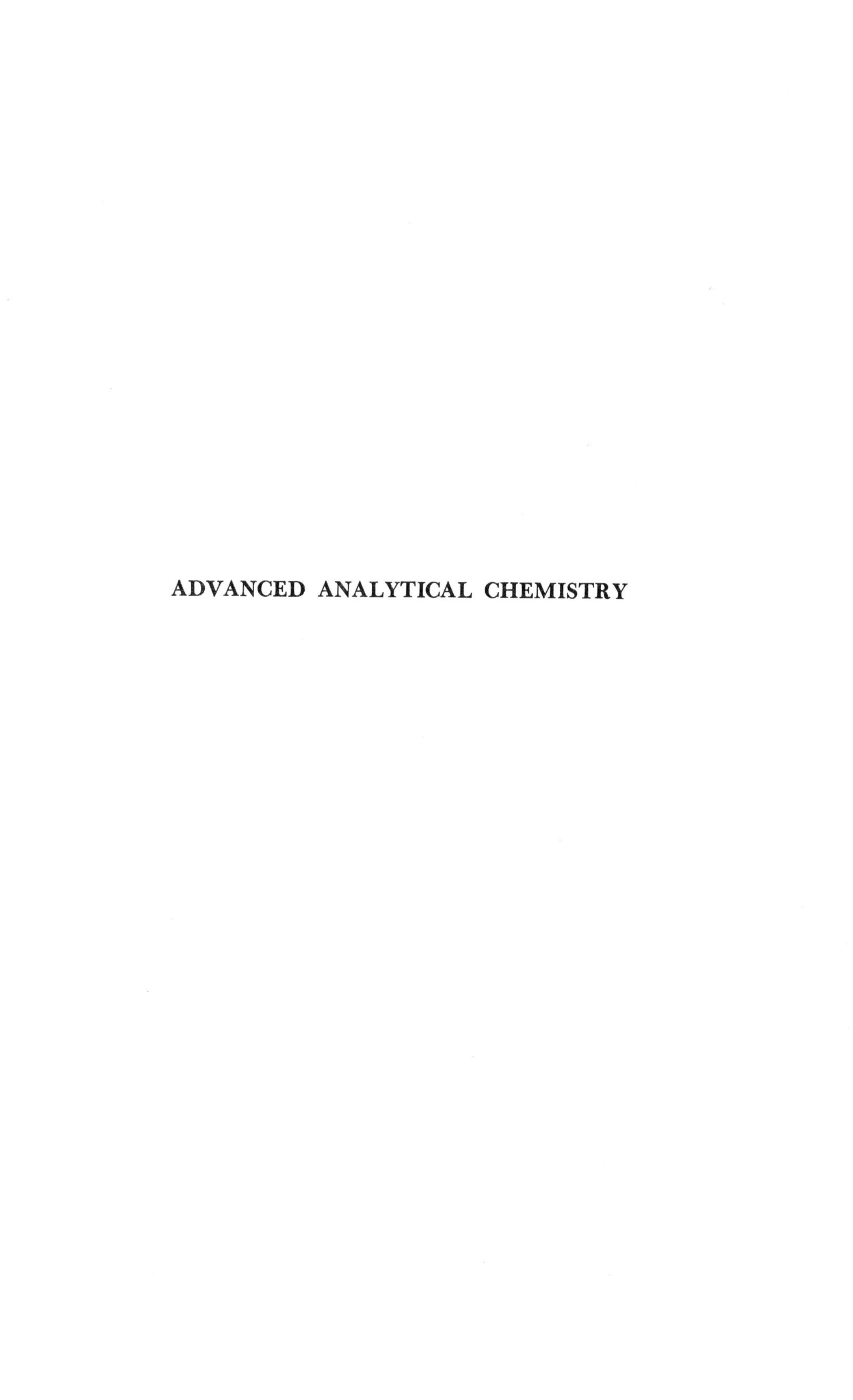

ADVANCED ANALYTICAL CHEMISTRY

Advanced Analytical Chemistry

by

LOUIS MEITES
Associate Professor of Analytical Chemistry
Polytechnic Institute of Brooklyn

and

HENRY C. THOMAS
Professor of Chemistry
University of North Carolina

with a chapter on
The Absorption of Infrared Radiation
by

ROBERT P. BAUMAN
Assistant Professor of Physical Chemistry
Polytechnic Institute of Brooklyn

McGraw-Hill Book Company, Inc.
NEW YORK TORONTO LONDON
1958

Library of Congress Catalog Card Number 57-13339

THE MAPLE PRESS COMPANY, YORK, PA.

PREFACE

As recently as thirty years ago, analytical chemistry was a rather quiescent field of scientific endeavor in which few if any surprises might have been expected by a casual observer. Volumetric and gravimetric analysis, which had been spurred on by the widespread interest in geochemical analyses during the latter half of the nineteenth century, had reached a point at which new reagents and new procedures were appearing only slowly. The brilliant work of T. W. Richards and G. P. Baxter and their students at Harvard on the determination of the atomic weights of the elements, of E. F. Smith and his students at the University of Pennsylvania on electroanalysis, and of many other great analytical chemists as well, had reached its peak. It would certainly not have been difficult to imagine that analytical chemistry was about to relinquish its status as an equal partner of the other branches of chemical science and to assume a dull and unfruitful position of subservience.

Reasonable though such a belief might have appeared, the facts have been quite different. During these thirty years analytical chemistry has undergone a revolution which has again thrust it into the very foreground of the advances in chemical science. Many factors have contributed to this renaissance, but it would unquestionably have been quite impossible had it not been for the growing perception of the ways in which the newer instrumental techniques could be turned to advantage by the analytical chemist. It is the theory and practice of these techniques on which the present proud status of analytical chemistry so largely depends.

One result of this development has been a rapid increase in both the number and the popularity of college and university courses dealing with these instrumental techniques. Though in different institutions these courses are designated by a variety of names—"instrumental analysis," "advanced quantitative analysis," and several others—they all share the common aim of introducing the student to the instruments and instrumental techniques which are so important in modern analytical chemistry.

The field of instrumental analysis may be broadly divided into three parts. One deals with the instruments themselves and includes their design, construction, and operation; the second consists of the practical ways in which the instruments can be used in the laboratory to analyze unknowns or to secure other information of value to the analytical chemist; and the third is the study of the theoretical foundations of the instrumental procedures, the general types of information that can be secured by the use of each, and their comparative advantages and limitations. It is the last of these which is most strongly emphasized in the present text.

We have adopted—and have discussed in some detail in Chapter 1—a broad definition of analytical chemistry. It is our feeling that any more restricted definition would have been unjust to both the present state of the field and the far-reaching interests of today's analytical chemists. In keeping with this definition, we have laid particular emphasis on the use of the instrumental techniques in securing data of physicochemical importance rather than on their use in the routine analysis of unknowns. This emphasis is perhaps most clearly visible in our choice of illustrative experiments, concerning which more will be said in a later paragraph.

Our definition of analytical chemistry has also led us to subordinate the instruments themselves to the theory which underlies their use. This is partly due to the rapid strides being made by the designers and manufacturers of scientific apparatus, which make it distinctly possible that any instrument now on the market will have become somewhat out of date in a year or two. It is also partly due to the financial problems inherent in such courses—problems which it would be presumptuous to describe to any instructor who has struggled with them—as well as to the fact that no two institutions are likely to have quite the same complement of equipment available, the possible variety being so wide that the book would have grown out of all proportion had we attempted to cope with it. But it is primarily due to our conviction that the theoretical foundation of an instrumental technique is a subject whose intellectual content renders it fit for inclusion in a college curriculum, while knob-turning and dial-reading *per se* is not, being best served by the manuals of operation provided by the manufacturers of the instruments.

Furthermore, no attempt has been made to discuss every one of the techniques of instrumental analysis. Courses in this subject are generally one-semester courses, and we believe that an attempt to describe twenty or thirty types of instrumental measurements in one semester is inherently unlikely to impart any real appreciation of their scopes and limitations to the student. This is especially true of those fields—such as mass spectrometry, electron and X-ray diffraction, Raman spectroscopy, nuclear magnetic resonance, and so on—which involve the use of equipment so intricate and costly that the student is almost necessarily barred from gaining any practical experience with it. Although infrared spectrophotometry certainly falls into this category, at least so far as most institutions are concerned at the present time, it is both so closely related to the subject of Chapter 8 and so important in its own right that it could not justifiably have been excluded. It has been discussed by Professor Robert P. Bauman in Chapter 9.

We have, then, chosen to discuss only a relatively small number of instrumental methods. We believe these to be the ones of most widespread utility throughout all of the branches of chemistry, and the ones which are most essential to an understanding of the principles of even those techniques which we have chosen to omit.

This decision has permitted an unusually high degree of correlation of the several chapters, as well as what we feel to be a unique unification of the laboratory work associated with the course. Here, in an effort to encourage the student to compare the various techniques with respect to both the complexity of the experimental manipulations and the nature, accuracy, and reliability of the data they provide, we have undertaken a radical departure from previous practice in the teaching of the subject. After some years of experience with laboratory programs of this nature, we are satisfied that this aim, which is surely one of the most important facets of the education of an analytical chemist, is exceptionally well served by thus studying a single simple system from a number of different angles by a number of different techniques.

This limitation has had one further substantial advantage. It has allowed the primary responsibility for the preparation of each chapter to be assumed by an author who has had extensive and continuing experience with the application of the technique under discussion in the course of his own research. In this way we hope to have achieved a thoroughly reliable and authoritative treatment of each technique, which in so extensive a field seems impossible to accomplish in any other way.

In this connection it is a pleasure to acknowledge the very substantial contribution made by Dr. Charles Merritt, Jr., to the preparation of Chapter 8 as well as to the design of a number of the experiments.

With a few notable exceptions such as Section 8-3, we have deliberately refrained from choosing material from the field of organic chemistry to illustrate the points being discussed. This was certainly not done with a view to furthering the impression, which is already all too prevalent in student circles, that analytical chemistry and organic chemistry are as East and West in Kipling's poem: the impression is not only wrong, but positively ludicrous. On the contrary, it reflects the fact that the behavior of a complex organic molecule is nearly always so much more complicated than that of a simple inorganic ion that one can scarcely hope to understand the ways in which organic substances behave when investigated by these techniques without first having a good grasp of the more straightforward phenomena encountered in dealing with simpler substances. We may perhaps be excused for taking this opportunity to record our belief that the lack of any course dealing with the partnership of organic chemistry and modern analytical chemistry is probably the most serious single fault of the customary curriculum in chemistry.

This book is a result of our experience with two different courses in advanced analytical chemistry. One of these is an undergraduate course offered in the second half of the junior year. This follows a term of classical quantitative inorganic analysis, a full year of organic chemistry, and one term of physical chemistry; it is accompanied by the second term of physical chemistry; and it may be followed in the senior year by an elective program of research in analytical chemistry which forms the basis of a senior thesis. The other is a one-term graduate course which, according to the needs and interests of the student, may serve either as a substitute for or as an introduction to more specialized courses in such fields as absorption spectroscopy, radiochemical analysis, and electroanalytical chemistry. The dearth of textbooks suitable for either of these courses has led us to attempt to satisfy the needs of both, in so far as this could be done in a single book. We have therefore included a substantial amount of material in excess of the needs of an undergraduate course, and at a number of places have suggested topics whose further discussion by the instructor will both interest and benefit the graduate student.

Everyone engaged in teaching such courses as these is well aware of the numerous problems which they present. We urgently solicit the comments, opinions, and criticisms of every colleague who shares these problems with us.

Louis Meites
Henry C. Thomas

CONTENTS

CHAPTER 1

THE NATURE OF ANALYTICAL CHEMISTRY

Few students beginning a course in advanced analytical chemistry have any very accurate knowledge of the responsibilities and interests of the analytical chemist or of the methods and tools with which he pursues these interests. Before examining techniques and instruments in detail, therefore, it is wise to try to describe the functions of the analytical chemist and his relation to the field of chemistry today.

Contrary to the impression which the student has probably derived from his course in elementary quantitative analysis, the analytical chemist is neither primarily nor even largely engaged in carrying out analyses, any more than the organic chemist spends his time in preparing aniline from nitrobenzene. The actual performance of an analysis, according to a detailed set of directions, is entrusted to an analyst. The analyst's responsibility is simply to execute the instructions of the analytical chemist with as much care and skill as he can command.

Analytical procedures do not, however, spring out of thin air. It is the acquisition and correlation of the fundamental knowledge which underlies such procedures, the application of this knowledge to the development of a method which will give a suitably accurate answer while consuming a minimum amount of the analyst's time, and the improvement of previously proposed analytical methods which constitute the responsibilities of the analytical chemist.

Let us look, for example, at a typical analytical procedure, such as the Mohr titration of chloride with silver nitrate using potassium chromate as the indicator, from the viewpoint of the analytical chemist. In order to tell whether this will work at all, we would have to know the solubility products of silver chloride and silver chromate. To find the lowest and highest pH values at which good enough results can still be obtained, we would have to investigate the dissociation constants of chromic acid and the solubility products of silver oxide and silver carbonate. We would certainly want to know something about the extent to which silver chloride adsorbs its own ions, because any considerable difference between the adsorbabilities of silver and chloride ions near the equivalence point would change the shape of the titration curve, perhaps appreciably. If the chloride solution happens to contain a reducing agent, such as stannous ion, we would have to inquire whether the chromate might not be reduced and the titration ruined, and in order to answer the question we might have to study the kinetics of the reaction between stannous and chromate ions, the standard potentials of the two half-reactions, the activity coefficients of all the ions involved, and so on. Or suppose that the sample contained only a very small amount of chloride, or that in addition to the chloride it contained a large amount of copper sulfate. In either of these cases we would be reluctant to recommend the use of the Mohr method, and so we would seek another way of locating the end point. We might then begin to think about a potentiometric titration, which would raise a whole new set of problems.

Now many of these inquiries might seem to the student to be more properly the concern of the physical chemist, but the distinction between physical and analytical chemistry is not so easy to draw as the usual undergraduate courses in these subjects might make one think. When a physical chemist measures the dissociation constant of an acid, he usually does it for the sake of the theoretical conclusions he can draw from the results of his measurements. An analytical chemist might make the very same measurements, but he would probably do it for the sake of the practical use to which he expected to put them (or suspected that someone else might some day want to put them). Although the distinction between the analytical and the organic chemist is also vanishing in much the same way, as witnessed by the steadily increasing use of organic reagents in analytical procedures and of such primarily analytical tools as the polarograph in the study of the structure and properties of organic compounds, the habit of thinking of synthesis and analysis in terms of "never the twain shall meet" makes it all too easy to consider organic and analytical chemistry as two branches of science with not very much in common. Largely as a result of this delusion, the chemical world has today no more pressing need than for people well trained in just these two fields.

At this point we may define analytical chemistry as that branch of chemical research which has the development and improvement of practical analytical procedures as its goal. Perhaps the most important word in this definition is the word "research." A constant assault on new problems—the discovery of new facts and their assimilation into an ordered pattern of thought, the development of new procedures and the improvement of old ones for reaching known goals, and the discernment of new problems whose solutions will repay the effort expended—is the very lifeblood of analytical chemistry as we have defined it. Though it goes without saying that the analytical chemist must have an intimate knowledge of the theory and practice of each of the techniques he employs, it is well to keep in mind the fact that these techniques are important, not for their own sakes, but only as means by which the desired information can be secured. More important than anything else to a research worker is the ability to reason clearly: to interpret accurately the data secured in one experiment and relate it to other available information, and to see what must logically be the next step in his inquiry.

We are now confronted with the problem of how to go about studying analytical chemistry. In just the same way that the hammer and the saw and the plane are the tools of the carpenter, the buret and the balance, the potentiometer, the conductance bridge, and the spectrophotometer are the tools of the analytical chemist. But no apprentice ever became a carpenter by even the most painstaking examination of a hammer and a saw and a plane. So the student of analytical chemistry must begin by using the analytical chemist's tools as the analytical chemist himself would use them, that is, in solving a problem which might be of interest to the analytical chemist.

Now of course it would not be possible to start one's education in analytical chemistry with a very difficult or complicated problem, any more than an apprentice carpenter would start making Duncan Phyfe furniture as soon as he reports to work. This book includes a number of simple problems, each of which might be of analytical importance, and approaches each of them from several different angles. Thus, directions are given for the determination of the solubility product of cupric oxalate by a number of independent methods. (We might suppose that cupric oxalate had been proposed as a primary standard in iodometry, or as a substance which might be precipitated and weighed by an analyst trying to determine oxalate in the presence of, say, tartrate. Or we might suspect that oxalate in such a mixture could be determined by an appropriate kind of titration with a standard copper solution.) In following these directions, the student should strive to gain familiarity with the instrument he is using, and especially with the principles underlying the operation of the instrument. After completing two or more such experiments, he should examine the results with a view to assessing their relative merits, the errors in each, and the most probable value of the quantity he has measured. No step in the experiment approaches in importance this final careful scrutiny of the data secured.

So far as the details of operating the various instruments are concerned, it is an unfortunate fact that no educational institution can afford to buy enough instruments to provide a comprehensive survey of all the tools which the practicing analytical chemist may use. A mass spectrometer, which may cost \$20,000 to \$40,000, is an excellent investment for an oil company which can use it for analyses that would otherwise be nearly or even totally impossible and save the salaries of several technicians who would be needed to carry out the long and tedious manipulations required by older techniques. But the few colleges which can afford to buy a mass spectrometer at all cannot afford to use it for student instruction, and so it would have been unrealistic to include it here. On the same principle we suggest the use of a Beckman Model B spectrophotometer or a Bausch and Lomb Spectronic 20 "colorimeter" instead of the much more popular, accurate, and expensive Model DU, and an inexpensive homemade polarograph is recommended instead of a commercial instrument. In no case has any compromise involving the essentials of the method been made (as would have been the case, in the authors' opinion, if a filter photometer had been recommended instead of a spectrophotometer), and any student who later encounters a more elaborate instrument can confidently be referred to the manufacturer's instruction book for information on how to turn the knobs. It is left to the instructor to supply such information for the particular instruments available, for to have attempted to do so here would have entailed sacrificing much of the far more important theoretical material.

It is assumed that the student has a working knowledge of elementary physical chemistry. Since so much of the subject matter of modern analytical chemistry is inextricably intertwined with physical chemistry, an adequate discussion of the former is scarcely possible without constant reference to the latter. No attempt will be made to define terms or develop concepts which may reasonably be supposed to be part of the content of courses in physics and physical chemistry.

CHAPTER 2

IONIC EQUILIBRIA IN SOLUTIONS

2-1. Introduction. In this text we shall discuss some of the instrumental techniques of the analytical chemist. To illustrate the applications of these techniques, we have devised a number of experiments dealing with ionic equilibria in solutions. This represents a considerable departure from tradition in the teaching of advanced analytical chemistry, which is normally concerned with the purely analytical use of such techniques, and it therefore seems justifiable to include a brief description of the reasons for our choice.

It may first of all be pointed out that the study of these equilibria constitutes a significant fraction of the work of the modern analytical chemist. Few practical methods of either separation or analysis could be devised rationally without a knowledge of the equilibrium and rate constants which govern them. The study of these equilibria is also of great importance in developing the systematic chemistry of the elements, which is a subject whose significance can hardly be overestimated.

Moreover, this approach makes it possible to coordinate the laboratory work accompanying the course to an extent which would be impossible under any other regime. Carrying out a polarographic determination of, say, lead in a zinc-base alloy may give the student some facility in the manipulation of a polarograph and some insight into the theoretical principles of polarography. But no combination of experiments of this type can give him a broad perspective of the field of instrumental analysis, or much (if any) insight into the reasoning which dictates the use of a polarographic method in preference to a conductometric or spectrophotometric one for conducting such an analysis. When, on the other hand, several independent techniques are applied to the study of the same problem, it becomes a simple matter to compare the significance and reliability of the experimental results.

In the present chapter we shall discuss some of the principles of ionic equilibria in solutions and some of the experimental methods by which these equilibria may be studied.

2-2. The Study of Ionic Equilibria. In its broad outline, any procedure for studying an equilibrium mixture consists of only three steps.

These are the preparation of the solution, its analysis, and the mathematical interpretation of the experimental results. Of the first of these nothing will be said here; enough examples appear in the experiments included at the end of the text to illustrate the principles involved.

We may distinguish two kinds of procedures for the analysis of a solution. These we shall term—rather loosely, to be sure—"destructive" and "nondestructive," referring to the effect of the analysis on the equilibrium. For example, the analysis of an acetic acid solution by titration with standard sodium hydroxide is a destructive analysis, because the addition of sodium hydroxide changes the position of the equilibrium between the undissociated acid and its ions. On the other hand, a measurement of the pH of the same solution would be a nondestructive analysis for hydrogen ion, because the act of measuring the pH does not displace the dissociation equilibrium. It is characteristic of destructive analyses that they give, not the concentration of a single species participating in an equilibrium, but the sum of at least two concentrations. The titration of an acetic acid solution with sodium hydroxide thus gives the sum of the concentrations of undissociated acetic acid and of hydrogen ion in equilibrium with it; this sum is often called the "analytical concentration" of acetic acid. Most nondestructive measurements, on the contrary, give the concentration of a single species, although this is not invariably the case (*e.g.*, with conductance measurements).

It is convenient to introduce at this point a notation which will appear frequently throughout this book. We may illustrate it by considering a solution prepared by adding exactly 1 mole of acetic acid to enough water to yield exactly 1 liter of solution. The analytical concentration of acetic acid is evidently 1 mole per liter, either by synthesis or by titration with standard base. However, the solution contains 0.004 M hydrogen ion from the dissociation of the acid, and hence the concentration of the *undissociated* acetic acid is actually only 0.996 M. To remove this ambiguity, which is often very confusing, we shall use the convention suggested by Swift for expressing analytical concentrations. According to this convention, the above solution is 1 F (formal) in acetic acid: it contains one formula weight (60.05 g.) of CH_3COOH per liter. At the same time it contains 0.996 M acetic acid, 0.004 M hydrogen ion, and 0.004 M acetate ion. The formality of a solution is a convenient expression of its over-all composition as found by destructive analysis.

It should hardly need to be mentioned that a solution prepared by dissolving a mole of sodium chloride in water and diluting to a liter is 1 F (*not* 1 M) in sodium chloride; at the same time it is 1 M in sodium ion and 1 M in chloride ion because sodium chloride is a strong electrolyte. Students sometimes feel that "everyone knows" what a "1 M sodium chloride" solution is; this amounts to substituting slovenliness for accuracy, intention for meaning, and mind reading for scientific communication.

In addition to these two kinds of analytical methods, the chemist has available three mathematical principles which pertain to solutions at equilibrium:

1. All relevant equilibrium constant expressions must be obeyed.
2. The solution must contain just as many positive as negative charges (the so-called "electroneutrality rule").
3. No matter of any kind is either created or destroyed during the approach to equilibrium (the familiar law of conservation of mass).

The first and third of these need no further explanation. The power of the second may be illustrated by a very simple example: the composition of a solution prepared by adding 10^{-8} mole of hydrogen chloride to a liter of pure water. If we naïvely take the hydrogen ion concentration as equal to the analytical concentration of acid added, we compute pH = 8, which is absurd. Invoking the electroneutrality rule, we write $[H^+] = [Cl^-] + [OH^-]$. As chloride ion cannot participate in any equilibrium with the other substances present, its concentration must be 10^{-8} M. At the same time (from rule 1 above) $[OH^-] = K_w/[H^+]$, so that

$$[H^+] = 10^{-8} + \frac{10^{-14}}{[H^+]}$$

whence $[H^+] = 1.05 \times 10^{-7}$, the correct answer. (At this point the student would do well to take the time to convince himself that the electroneutrality-rule statement for a solution of oxalic acid in water is $[H^+] = [HC_2O_4^-] + 2[C_2O_4^{=}] + [OH^-]$.)

We shall now proceed to discuss a number of successively more complicated equilibrium situations. In each we shall assume that one has available a saturated solution of the salt under discussion and wishes to determine the solubility product of the salt. Other types of equilibria will be discussed in subsequent chapters.

2-3. The Calcium–Sulfate System. The equilibrium between solid calcium sulfate and its saturated solution, because only one reaction is involved,[1] is an example of the simplest kind of equilibrium situation that can be encountered. This reaction

$$CaSO_4 = Ca^{++} + SO_4^{=}$$

has as its equilibrium constant the solubility product of calcium sulfate

$$K = [Ca^{++}][SO_4^{=}]$$

[1] Although the reaction $SO_4^{=} + H_2O = HSO_4^- + OH^-$ does proceed to a small extent, the dissociation constant of bisulfate ion, *ca.* 10^{-2}, is so large that the concentration of bisulfate is always negligibly small except in strongly acidic solutions.

Inspection of this equation suggests an obvious approach to the measurement of K. We might simply analyze the saturated solution for both calcium and sulfate, and then compute the product of their concentrations. (It is worth noting that in this instance there would be no question of a choice between destructive and nondestructive analyses, for there are no equilibria which can be displaced.)

However, before setting out to conduct the analyses, it would be wise to temper experimental zeal with chemical caution. The determination of calcium in the saturated solution [by titration with a standard solution of disodium dihydrogen ethylenediaminetetraacetate, $Na_2H_2(OOCCH_2)_2$-$NCH_2CH_2N(CH_2COO)_2$] is a simple, straightforward, and accurate procedure. The determination of sulfate would be best done either gravimetrically by weighing barium sulfate, amperometrically (*cf.* Chap. 6) by titration with standard lead nitrate, or conductometrically (*cf.* Chap. 5) by titration with standard barium chloride or nitrate. In any of these methods one would expect serious errors due to coprecipitation of calcium. Hence, instead of spending the time required to carry out a sulfate determination of doubtful validity, it would probably be better to determine the calcium concentration alone. From the electroneutrality rule we have, since the concentrations of hydrogen and hydroxyl ions are equal,

$$[Ca^{++}] = [SO_4^{=}]$$

whence

$$K = [Ca^{++}]^2$$

so that measurement of the calcium concentration alone serves to yield the desired result.

2-4. The Silver–Acetate System. This system is considerably more complicated than the one just discussed, because in addition to the solubility equilibrium described by the equation

$$AgOAc = Ag^+ + OAc^- \qquad K = [Ag^+][OAc^-]$$

we must take into account the hydrolysis of acetate ion according to

$$OAc^- + H_2O = HOAc + OH^- \qquad K_h = \frac{[HOAc][OH^-]}{[OAc^-]} = \frac{K_w}{K_a}$$

where K_a is the dissociation constant of acetic acid. From the electroneutrality rule we have

$$[Ag^+] + [H^+] = [OAc^-] + [OH^-]$$

from which, incidentally, we could drop the concentration of hydrogen ion since it is very small compared to any of the other terms (the solubility of silver acetate is about 0.04 F, and the hydrolysis necessarily renders the solution alkaline). In addition, from the law of conservation of mass we have

$$[Ag^+] = [OAc^-] + [HOAc]$$

since one acetate ion entered the solution together with each silver ion.

Now it is a simple matter to determine the concentration of silver ion directly [*e.g.*, by titration with standard thiocyanate, either potentiometrically (Chap. 3) or by the Volhard method, or by direct potentiometry with a silver indicator electrode (Chap. 3)]. In contrast to the calcium–sulfate system, we cannot take $[Ag^+] = [OAc^-]$, for the silver ion concentration is necessarily larger than the acetate concentration. Moreover, for experimental reasons, it is desirable to avoid the unpleasant necessity of measuring the acetate ion concentration, which is not at all easy to do.

We may therefore proceed to substitute calculation for experiment in the following fashion. Two alternatives are available.

1. From the electroneutrality-rule expression given above

$$[Ag^+] + [H^+] = [OAc^-] + [OH^-]$$

it is clear that, knowing $[Ag^+]$ and the pH of the saturated solution (from which both $[H^+]$ and $[OH^-]$ can readily be calculated), we could compute $[OAc^-]$ without further ado.

2. Or we might start from the previously deduced equation

$$[Ag^+] = [OAc^-] + [HOAc]$$

and, employing the dissociation constant of acetic acid, write

$$[Ag^+] = [OAc^-]\left(1 + \frac{[H^+]}{K_a}\right)$$

Again, knowing $[Ag^+]$ and the pH of the saturated solution, we could compute the acetate ion concentration. In this case, however, we would also have to know or measure the dissociation constant of acetic acid.

Evidently these two approaches are entirely equivalent experimentally; both substitute a relatively simple pH measurement for an acetate ion determination which would be anything but simple. The student should pause to contemplate the effect which a trace of acetic acid, present as an impurity in the silver acetate, would have on the value of K calculated by each of these methods. It may be remarked that, whereas $[Ag^+]$ could be determined by either a destructive or a nondestructive method, no additional information of any value would then be secured by a destructive analysis for acetate.

2-5. The Silver–Iodide System. This is an example of a system in which the salt may dissolve in two ways:

$$AgI = Ag^+ + I^- \qquad K_1 = [Ag^+][I^-]$$

and $2AgI = Ag^+ + AgI_2^-$. The second of these equilibria is most conveniently described in terms of the reaction

$$AgI_2^- = Ag^+ + 2I^- \qquad K_2 = \frac{[Ag^+][I^-]^2}{[AgI_2^-]}$$

Applying the electroneutrality rule gives

$$[Ag^+] = [I^-] + [AgI_2^-]$$

(hydrogen and hydroxyl ions are omitted because neither reacts with silver or iodide, and so their concentrations will be equal). This, together with the equation for K_2, gives

$$[Ag^+] = [I^-] + \frac{[Ag^+][I^-]^2}{K_2}$$

As in the silver–acetate system, we have two alternatives to consider. This time, however, they are by no means equivalent. They are:

1. Direct measurement of $[Ag^+]$ and $[I^-]$. If some nondestructive method can be found for the determination of each of these, K_1 can be calculated immediately. For example, $[Ag^+]$ could be determined by direct potentiometry with a silver wire electrode, and $[I^-]$ by direct potentiometry with the indicator electrode $Pt|I_2(s), I^-$ (*cf.* Chap. 3). Or, in principle, one might determine $[Ag^+]$ and the total analytical concentration of iodide (equal to $[I^-] + 2[AgI_2^-]$), and calculate the free iodide concentration from these values and the electroneutrality-rule expression. In practice, it is improbable that an analytical determination of total iodide would be satisfactory because of its extremely low concentration. The student should try to determine the feasibility of attacking the entirely similar case of silver chloride by either of these methods.

2. Direct determination of one of these two concentrations and the determination of K_2. The preceding equation gives

$$[Ag^+]\left(1 - \frac{[I^-]^2}{K_2}\right) = [I^-]$$

from which either concentration can be calculated with the value of K_2 and the other concentration. By such a method the solution of the silver chloride problem would become, if not simple, at least reasonably straightforward.

This naturally raises the question of how K_2 can be determined. Suppose that $[Ag^+]$ in the saturated silver iodide solution has been determined as suggested, and that the complex ion formation now has to be taken into account. Quite unfortunately but also quite generally, all that will be known is that the salt is appreciably soluble in solutions containing an excess of its anion; usually even the formula of the complex ion is unknown. One approaches this problem by constructing the equilibrium (in this case)

$$AgI + nI^- = AgI_{n+1}^{-n}$$

This is done by preparing several solutions of known iodide concentrations considerably in excess of that in a pure saturated silver iodide solution. This serves to repress the solubility of silver iodide according to the equation

$$AgI = Ag^+ + I^-$$

to such an extent that the total concentration of silver in the solution is virtually equal to $[AgI_{n+1}^{-n}]$ alone. After these mixtures have stood for a long enough time to ensure that equilibrium has been attained, they are analyzed for their total silver contents. As these can generally be made to fall within an analytically accessible range by selection of appropriate iodide (or other anion) concentrations, this is not often a matter of much difficulty. Numerous examples may be found in the experiments accompanying later chapters.

When the results of these analyses are available, their dependence on $[I^-]$ is examined. In this case it might be found that the analytical concentration of silver was proportional to the iodide concentration over a fairly wide range; this would establish n as 1, and would give a value for $[AgI_2^-]/[I^-]$. A little algebra shows that this ratio corresponds to K_1/K_2.

The direct determination of complex dissociation constants by potentiometric and polarographic measurements is discussed in Chaps. 3 and 6, and methods for their determination by spectrophotometric means are discussed in Chap. 8.

Many actual systems are much more complicated than the three which have been discussed. For example, the cupric–oxalate system involves all of the following equilibria:

$$\begin{aligned}
CuC_2O_4 &= Cu^{++} + C_2O_4^{=} \\
Cu(C_2O_4)_2^{=} &= Cu^{++} + 2C_2O_4^{=} \\
Cu^{++} + H_2O &= Cu(OH)^+ + H^+ \\
C_2O_4^{=} + H_2O &= HC_2O_4^- + OH^- \\
HC_2O_4^- + H_2O &= H_2C_2O_4 + OH^-
\end{aligned}$$

and all five of these must be studied in the elucidation of the system. However, the exercise of a little ingenuity will always serve to reveal how the principles outlined in the foregoing paragraphs may be applied to the study of any combination of equilibria.

2-6. Activity Coefficients and Related Matters. The equilibrium constant of a reaction is a relationship among the activities of the reactants and products which is obeyed whenever thermodynamic equilibrium is attained. For example, for the equilibrium $HOAc = H^+ + OAc^-$

$$K = \frac{a_{H^+}a_{OAc^-}}{a_{HOAc}}$$

For many purposes it is convenient to express the activity of each species as the product of its concentration and a factor known as the activity coefficient; when the concentration is expressed in moles per liter, the activity coefficient is given the symbol f and is called the rational activity coefficient. Thus $a_{H^+} = f_{H^+}[H^+]$, and

$$K = \frac{f_{H^+}f_{OAc^-}}{f_{HOAc}} \frac{[H^+][OAc^-]}{[HOAc]}$$
$$= \frac{f_{H^+}f_{OAc^-}}{f_{HOAc}} K'$$

The quantity K is known as the "activity," "true," or "thermodynamic" equilibrium constant, and it is independent of the ionic strength. However, K', which is called the "concentration" or "formal" equilibrium constant, is not truly constant, because the activity coefficients in the foregoing expression vary with ionic strength, and so K' must also vary to preserve the constancy of K. This fact has prompted physical chemists to concentrate their attention on K rather than on K', which is equivalent to determining values of K' at zero ionic strength, where all of the activity coefficients are equal to unity. It is for this reason that the dissociation constants, solubility products, standard potentials, and other equilibrium data, in handbooks or in the original literature, almost always pertain to infinitely dilute solutions. Applying these data to practical problems—to problems dealing with real solutions having finite ionic strengths—often involves certain difficulties which are worth discussing.

For example, from the standard potentials

$$Fe(CN)_6^{-3} + e = Fe(CN)_6^{-4} \qquad E^0 = +0.36 \text{ v.}$$
$$I_3^- + 2e = 3I^- \qquad E^0 = +0.54 \text{ v.}$$

one calculates (*cf.* Sec. 3-3) that the equilibrium constant of the reaction

$$I_3^- + 2Fe(CN)_6^{-4} = 3I^- + 2Fe(CN)_6^{-3}$$

is 1.2×10^6. This indicates that the reaction goes fairly completely to the right. However, this is true only at zero ionic strength. In solutions of unit ionic strength, the equilibrium constant is approximately twelve orders of magnitude smaller than the value just calculated. Hence it is possible to determine ferricyanide volumetrically by adding excess iodide and titrating with thiosulfate, even though such a procedure could hardly be very successful in the face of the unfavorable equilibrium which prevails at zero ionic strength.

Of course it is evident that this difficulty could be avoided if we knew the activity coefficient, at the particular ionic strength in question, of each of the ions entering into an equilibrium. The activity coefficient of a single ion is, however, a quantity incapable of exact definition within the realm of classical thermodynamics. Activity coefficients are generally found from measurements of the potentials of galvanic cells, *e.g.*, the cell Pt, $H_2(g)|HCl(c)$, $AgCl(s)|Ag$. This is a so-called "cell without liquid junction" (Secs. 3-4 and 3-5): the composition of the solution is the same throughout the cell. From suitable measurements with cells of this type we could secure the "mean ionic activity coefficient"[1] of the hydrogen and chloride ions in an HCl solution of any desired concentration, but not the activity coefficient of either ion separately. On paper one might design a cell with liquid junction whose potential depended on the activity of one ion alone, but the exact interpretation of the data in the face of the unknown liquid-junction potential would be impossible.

In addition, the solutions with which the analyst must work, and which the analytical chemist must therefore consider, are frequently so complex that activity coefficient data are wholly lacking.

It therefore appears that the analytical chemist is often unable either to make direct use of the available data pertaining to infinitely dilute solutions or to correct them to apply to the particular conditions under which he must work. In this quandary he adopts either one of two courses. Although neither of these is wholly satisfactory, both give results which are more useful than the values pertaining to zero ionic strength.

[1] If we specify that c is much greater than the molar solubility of AgCl, and if we assume that the deviation of hydrogen gas from ideality is negligible, the e.m.f. of this cell at 25°C. is given by

$$E_{\text{cell}} = E^0_{\text{AgCl,Ag}} - 0.05915 \log c^2 - 0.05915 \log f_{\text{H}^+}f_{\text{Cl}^-} + 0.05915 \log \sqrt{p_{\text{H}_2}}$$

In this equation E_{cell}, c, and p_{H_2} are experimentally measurable quantities. By a suitable extrapolation of the measured potentials to zero ionic strength (where all activity coefficients are unity) we may eliminate the next to the last term and so find $E^0_{\text{AgCl,Ag}}$, but there is no way of separating the two activity coefficients from each other and so of determining their individual values. To emphasize this impossibility it is customary to employ this combination of activity coefficients as the definition of the so-called "mean ionic activity coefficient" $f_\pm$,

$$f_{\text{H}^+}f_{\text{Cl}^-} = f_\pm^2$$

The single-ion activity coefficient problem is discussed very lucidly and thoroughly in R. G. Bates, "Electrometric pH Determinations," John Wiley & Sons, Inc., New York, 1954.

Undoubtedly the better of these alternatives consists of selecting a single finite ionic strength at which to carry out an entire study. If, for example, we tried to apply the thermodynamic dissociation constants of oxalic acid to a study of the copper–oxalate equilibria mentioned earlier, we would find sharp inconsistencies in the data. We might, however, study these equilibria by measuring the solubility of cupric oxalate in 1 F sodium perchlorate ($\mu = 1$), the dissociation constant of the complex in 0.33 F potassium oxalate ($\mu = 1$), the dissociation constants of oxalic acid in solutions having ionic strengths of 1, and so on. From these data we could secure a completely consistent and reliable picture of the system at an ionic strength of 1, and we might take this as at least a fairly accurate representation of its behavior at ionic strengths not too different from 1.

The other approach is a much simpler one, and one which has no justification whatever other than expedience. It consists simply of neglecting activity coefficients completely. For example, we might study the solubility of lead thiocyanate by determining the concentrations of lead ion present at equilibrium in solutions containing 0.1, 0.2, and 0.5 M thiocyanate. Doubtless we would find that the "constant" $K = [Pb^{++}][SCN^-]^2$ varied appreciably over this interval because of the changes in the activity coefficients. We would also find, however, that our data would be quite as useful in, say, developing a method for the gravimetric determination of lead as $Pb(SCN)_2$ or for the amperometric titration of thiocyanate with a standard lead nitrate solution as if we knew the thermodynamic solubility product of lead thiocyanate to three or four significant figures. Similarly, the bare fact that the solubility of hydrous ferric oxide in 10 F sodium hydroxide is of the order of 10^{-5} F may of itself cast a great deal of light on a proposed method for the separation of iron and aluminum. One of the hardest, and yet one of the most important, things that the beginning research worker must learn is to temper the accuracy he demands of his experiments to the purpose to which his results will be put.

In this text we shall generally adopt the first of these approaches, although we shall occasionally use the second as well. In our discussions we shall make more use of "formal" than of "standard" potentials and other equilibrium constant data. These terms may be defined very briefly as follows. The standard potential of the $Fe(CN)_6^{-3}$, $Fe(CN)_6^{-4}$ couple is the limit toward which the potential of the couple tends when the ferrocyanide and ferricyanide concentrations are equal and the ionic strength approaches zero. The formal potential of this couple is the potential of the couple at equal ferrocyanide and ferricyanide concentrations at some defined finite value of the ionic strength. Thus we may speak of the formal potential of the couple at $\mu = 0.1$ or at $\mu = 1.0$.

Often it is desirable to specify the composition of the solution even more completely than this. For example, the formal potential of the cupric-cuprous couple is quite different in 3 F potassium iodide than in 3 F potassium nitrate; the formation of a strong cuprous iodide complex in the former medium greatly favors the tendency for +2 copper to be reduced. In such a case we would refer to the formal potential of the cupric-cuprous couple in 3 F potassium iodide rather than to the formal potential of the couple at $\mu = 3$. In the same way we might speak of the formal solubility product of silver chloride at $\mu = 2.5$, or of the formal dissociation constant of lactic acid at $\mu = 0.5$.

PROBLEMS

The answers to the odd-numbered problems are given on page 402.

1. What is the pH of a solution prepared by dissolving 5×10^{-7} mole of hydrogen chloride in a liter of pure water?

2. If the pH of the solution is observed with equipment sensitive to ± 0.02 pH unit, what weight of sodium hydroxide must be added to a liter of pure water to produce a barely noticeable change of pH?

3. The dissociation constant of acetic acid is 1.75×10^{-5}. What will be the pH of a 10^{-3} F solution of acetic acid?

4. How many moles of acetic acid would have to be added to a liter of pure water to give a solution of pH 6.0?

5. The solubility of silver chloride in water at 10°C. is 0.89 mg. per l. What is its solubility product at this temperature?

6. The solubility of barium sulfate in water at 25°C. is 0.25 mg. per l., and that of radium sulfate is 0.002 mg. per l. If a mixture of barium and radium sulfates is shaken with water at 25°C. until equilibrium is reached, what will be the concentration of radium ion in the solution?

7. One liter of water at 25°C. will dissolve 2.62 g. of silver benzoate (C_6H_5COOAg). The dissociation constant of benzoic acid at this temperature is 6.3×10^{-6}. What is the pH of the resulting solution?

8. From the data in Problem 7 calculate the solubility product of silver benzoate at 25°C.

9. At 25°C. the solubility products of silver chloride and silver bromide are 1.8×10^{-10} and 4.0×10^{-13}, respectively, and the dissociation constant of the $Ag(NH_3)_2^+$ ion is 7×10^{-8}. If the dissociation constant of ammonia is 1.8×10^{-5}, calculate the concentration of each species present in a solution prepared by shaking 0.01 F ammonia with excess silver chloride and silver bromide.

10. What fraction of the bromide originally present in a solution containing 0.01 M bromide and 0.1 M chloride will be precipitated if just enough silver nitrate is added to initiate the precipitation of silver chloride? Assume that the volume of the solution is unchanged by the addition of the silver nitrate.

11. The solubility of silver chloride in 0.036 F sodium chloride is 1.9×10^{-6} F; in 0.5 F sodium chloride it is 2.8×10^{-5} F. What is the formula of the complex ion which predominates in these solutions, and what is its dissociation constant?

12. With the aid of the answer to Problem 11 and the solubility product of silver chloride given in Problem 9, calculate the solubility of silver chloride in 0.0092 F sodium chloride, and compare your result with the experimental value, 9.1×10^{-7} F.

13. One liter of water at 20°C. dissolves 0.14 mg. of silver sulfide. Taking into account the hydrolysis of sulfide ion to HS^- (but not the further hydrolysis of HS^- to H_2S), calculate the solubility product of silver sulfide. The second dissociation constant of hydrogen sulfide is 3.5×10^{-13}.

14. The pH of a saturated solution of manganous sulfide in pure water is 7.17, and the first and second dissociation constants of hydrogen sulfide are 9×10^{-8} and 3.5×10^{-13}, respectively. What is the solubility product of manganous sulfide?

15. The solubility of silver phosphate, Ag_3PO_4, in pure water is 6.9 mg. per l. at 25°C. The dissociation constants of phosphoric acid are 1.1×10^{-2}, 7.5×10^{-8}, and 4.8×10^{-13}. What is the pH of a saturated silver phosphate solution?

16. What would be the pH of a buffer of which 1 l. would dissolve 30 mg. of silver phosphate?

17. The solubility product of silver carbonate is 6.15×10^{-12}, and that of silver chloride is 1.8×10^{-10}. The acid dissociation constants of carbon dioxide are 4.3×10^{-7} and 4.4×10^{-11}. What will be the concentration of each ion present in a solution prepared by shaking 0.001 mole of silver carbonate with a liter of 0.001 *F* hydrochloric acid until equilibrium is reached?

18. Using the data given in Problem 17, compute the concentration of a sodium carbonate solution which will just completely metathesize 0.001 mole of silver chloride to silver carbonate when the silver chloride is shaken with 1000 ml. of the sodium carbonate solution.

CHAPTER 3

POTENTIOMETRIC MEASUREMENTS

3-1. Introduction. Potentiometric measurements in analytical chemistry are of two kinds: those in which the potential of a galvanic cell is measured and used to calculate the concentration of one constituent of a solution, and those in which the variation of a potential during the addition of a standard solution of a reagent is used to find the end point of a titration. Procedures of the first kind are termed "direct potentiometry"; a titration in which the end point is found from potentiometric data is called a "potentiometric titration." Potentiometric titrations are, of course, destructive in the sense explained in the preceding chapter, while direct potentiometry is nondestructive.

Potentiometric titrations have two important advantages over ordinary volumetric methods. First, they make it possible to locate the end point accurately even when the solution is deeply colored or when a suitable indicator cannot be found. Second, they are capable of considerably better accuracy, because they eliminate indicator errors and the uncertainties involved in judging the color change of an indicator.

Primarily because its accuracy is limited by factors which will be discussed later, direct potentiometry finds but one important application in routine analytical work. This application—the electrometric measurement of pH—is, however, of immense importance, and we shall consider it separately in Chap. 4.

3-2. Sources of Potential in a Galvanic Cell. Consider the cell comprised of a copper electrode immersed in a copper sulfate solution and a zinc electrode immersed in a zinc sulfate solution, the two solutions being separated by a porous diaphragm which permits the flow of electricity but prevents mechanical mixing (Fig. 3-1). We may describe this cell schematically by writing

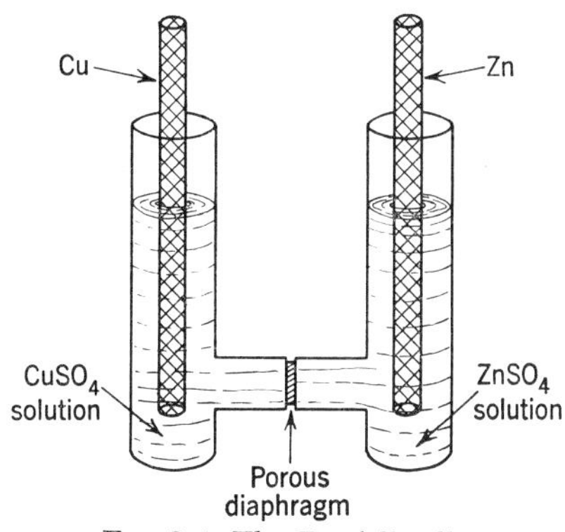

FIG. 3-1. The Daniell cell.

$$Cu|CuSO_4(c_1)||ZnSO_4(c_2)|Zn$$

The single vertical line indicates a locus of electrical potential difference across a phase boundary; the double vertical line indicates a locus of electrical potential difference across the interface between two solutions having different compositions. The former is called an "electrode potential"; the latter is called a "liquid-junction potential." We shall discuss these sources of potential in subsequent sections. The concentrations c_1 and c_2 are generally expressed in moles (or formula weights) per liter, and the partial pressure of a gas involved in a cell reaction may be expressed in either atmospheres or millimeters of mercury [*e.g.*, as in the cell Pt, H_2(0.5 atm.)|HCl(2 F)|Cl_2(650 mm.), Pt].

By convention, the electrode which appears on the left-hand side of a schematic representation of a galvanic cell is always considered to be the electrode at which oxidation occurs (*i.e.*, the anode), and the right-hand electrode is considered to be the electrode at which reduction occurs (the cathode). The half-reactions which occur in the Daniell cell as written above are therefore

$$\text{At the Cu electrode:} \qquad \mathrm{Cu} = \mathrm{Cu^{++}} + 2e$$
$$\text{At the Zn electrode:} \quad \mathrm{Zn^{++}} + 2e = \mathrm{Zn}$$

The cell reaction is then simply the sum of the two half-reactions:

$$\mathrm{Cu} + \mathrm{Zn^{++}} = \mathrm{Cu^{++}} + \mathrm{Zn}$$

If this reaction is spontaneous, then:

1. When the cell is short-circuited, electrons will be liberated at the left-hand electrode (the anode) and consumed at the right-hand electrode (the cathode), and will therefore flow from left to right through the external circuit; we shall consider this to be a positive current.
2. The potential of the cell is given a positive sign.
3. The left-hand electrode is arbitrarily considered to be the negative electrode of the cell.

In the Daniell cell as we have written it, and with c_1 and c_2 equal to 1 F, the cell reaction is not spontaneous. Therefore the cell potential is negative and the copper electrode is the positive electrode of the cell. The potential of the cell is given by the equation

$$E_{\mathrm{cell}} = E_{\mathrm{Zn^{++},Zn}} - E_{\mathrm{Cu^{++},Cu}} + E_j \tag{3-1}$$

where E_j is the liquid-junction potential. Note the use of the subscripts in specifying the direction in which the half-reactions proceed.

The electrode potentials in Eq. (3-1) are given by the Nernst equation, which for the copper electrode is

$$E_{\mathrm{Cu^{++},Cu}} = E^0_{\mathrm{Cu^{++},Cu}} - \frac{RT}{nF_y} \ln \frac{a_{\mathrm{Cu}}}{a_{\mathrm{Cu^{++}}}} \tag{3-2}$$

In this equation $E^0_{Cu^{++},Cu}$ is the "standard potential" of the $Cu^{++}|Cu$ electrode. This is defined as the potential of the cell

$$\text{Pt, } H_2(1 \text{ atm.})|H^+(a = 1), Cu^{++}(a = 1)|Cu$$

This is tantamount to defining the potential of the electrode

$$\text{Pt, } H_2(1 \text{ atm.})|H^+(a = 1)$$

as zero. This electrode, the so-called "normal hydrogen electrode" or N.H.E., is arbitrarily assigned a potential of 0.0000 volt at all temperatures. The quantity R in Eq. (3-2) is the gas constant (8.316 v.-coulombs per °C.), T is the absolute temperature (298.2° at 25°C.), n is the number of electrons appearing in the half-reaction, and F_y is the faraday (96493 coulombs). The symbols $a_{Cu^{++}}$ and a_{Cu} represent the activities of cupric ion and of metallic copper, respectively. The activity of massive solid copper is taken as unity, by definition.[1] We could use the Debye-Hückel equation, together with certain extrathermodynamic assumptions relating to the effective diameter of the cupric ion, to calculate the activity coefficient of cupric ion if the solution were sufficiently simple and sufficiently dilute, but the information required for such a calculation is so rarely available that we shall henceforth generally take activities and concentrations to be equal. Introducing these various numerical values into Eq. (3-2) gives, at 25°C.,

$$E_{Cu^{++},Cu} = E^0_{Cu^{++},Cu} + \frac{0.05915}{2} \log [Cu^{++}] \tag{3-3}$$

From tables of standard potentials, of which those given by Latimer[2] are the most comprehensive and reliable, one finds $E^0_{Cu^{++},Cu} = +0.337$ v.†

[1] However, see Sec. 7-2.

[2] "The Oxidation States of the Elements and Their Potentials in Aqueous Solutions," Prentice-Hall, Inc., Englewood Cliffs, N.J., 1952, pp. 340–348.

† In these pages we have described the conventions established by the International Union of Pure and Applied Chemistry at its 1953 meeting in Stockholm. It is to be hoped that these will be universally adopted in place of the conflicting "American" and "European" sign conventions which have plagued electrochemists for many years.

It is an integral part of the Stockholm conventions that the terms "electrode potential" and "standard potential" are reserved for half-reactions in which reduction occurs. Thus, the standard potential of a copper electrode is the potential of the cell Pt, H_2(1 atm.)$|H^+(a = 1)$, $Cu^{++}(a = 1)|Cu$ in which the half-reaction occurring at the copper electrode is $Cu^{++} + 2e = Cu$. This standard potential is given the symbol $E^0_{Cu^{++},Cu}$ to emphasize the direction of the half-reaction.

However, in Latimer's compilation the half-reactions are written as oxidation

(Note that the sign of this value indicates that the reaction

$$H_2 + Cu^{++} = 2H^+ + Cu$$

is spontaneous when all of the substances involved are present at unit activity.)

In the same way, the potential of the zinc electrode is given by

$$E_{Zn^{++},Zn} = E^0_{Zn^{++},Zn} + \frac{0.05915}{2} \log [Zn^{++}] \tag{3-4}$$

With the aid of a table, which gives $E^0_{Zn^{++},Zn} = -0.763$ v., we thus secure

$$E_{Zn^{++},Zn} = -0.763 + \frac{0.05915}{2} \log [Zn^{++}] \tag{3-5}$$

Combining Eqs. (3-1), (3-3), and (3-5) then finally gives the potential of the Daniell cell as (neglecting E_j)

$$E_{cell} = -1.100 - \frac{0.05915}{2} \log \frac{[Cu^{++}]}{[Zn^{++}]} \tag{3-6}$$

If the electrodes are short-circuited and the cell reaction is allowed to proceed, the concentration of zinc ion will increase and that of cupric ion will decrease. Both of these changes will cause the cell potential to decrease numerically, and at equilibrium this will become zero. Then

$$\log \frac{[Cu^{++}]}{[Zn^{++}]} = -37.2$$

whence K, the equilibrium constant of the cell reaction, is found to be $10^{-37.2}$, or 6.3×10^{-38}.

Sign conventions, being of necessity highly arbitrary, are unfortunately difficult to learn and remember. In Lingane's words, "the conscientious use of a pencil and paper is the only method known to man for mastering thermodynamic calculations in general, and 'cell problems' in particular." The student is advised to have recourse to the problems at the end of this chapter.

3-3. Applications of Standard Potential Data. We have seen how the equilibrium constant of the reaction between copper and zinc ion can be

---

reactions. For example, the cupric ion–copper couple appears there in the form

$$Cu = Cu^{++} + 2e \qquad E^0 = -0.337 \text{ v.}$$

where we might use the symbol $E^0_{Cu,Cu^{++}}$ for the quantity tabulated. This value is related to the standard potential of the copper electrode, $E^0_{Cu^{++},Cu}$, as defined above, by the equation

$$E^0_{Cu^{++},Cu} = -E^0_{Cu,Cu^{++}}$$

so that the standard potential of the copper electrode is +0.337 v.

calculated from the standard potentials of the two half-reactions involved. This can also be done with other kinds of reactions, whether they look like redox reactions or not. For example, the solution of silver chloride in water according to the equation $AgCl = Ag^+ + Cl^-$, a process whose equilibrium constant is the solubility product of silver chloride, may be considered as the sum of the two half-reactions

$$\begin{aligned} AgCl + e &= Ag + Cl^- \\ Ag &= Ag^+ + e \end{aligned}$$

These reactions will occur in the cell

$$Ag|Ag^+(c_1)||Cl^-(c_2), AgCl(s)|Ag$$

whose potential is given by

$$\begin{aligned} E_{\text{cell}} &= E_{AgCl,Ag} + E_j - E_{Ag^+,Ag} \\ &= E^0_{AgCl,Ag} - E^0_{Ag^+,Ag} - 0.05915 \log [Ag^+][Cl^-] + E_j \end{aligned} \qquad (3\text{-}7)$$

The values of the standard potentials appearing in this equation are $E^0_{Ag^+,Ag} = +0.7991$ v., and $E^0_{AgCl,Ag} = +0.2222$ v. If the liquid-junction potential in Eq. (3-7) be neglected (it can be reduced to a very small value by the use of an appropriate "salt bridge"), then at equilibrium $E_{\text{cell}} = 0$ and

$$\begin{aligned} \log K_{AgCl} &= \frac{E^0_{AgCl,Ag} - E^0_{Ag^+,Ag}}{0.05915} \\ &= \frac{0.2222 - 0.7991}{0.05915} = -\frac{0.5769}{0.05915} = -9.753 \end{aligned}$$

whence

$$K_{AgCl} = 1.77 \times 10^{-10}$$

In a similar fashion we may write the equation for the dissociation of water, $H_2O = H^+ + OH^-$, as the sum of the half-reactions

$$\begin{aligned} 2H_2O &= O_2 + 4H^+ + 4e & E^0_{O_2,H_2O} &= +1.229 \text{ v.} \\ O_2 + 2H_2O + 4e &= 4OH^- & E^0_{O_2,OH^-} &= +0.401 \text{ v.} \end{aligned}$$

The potential of the corresponding cell, Pt, $O_2(p_1)|H^+(c_1), OH^-(c_2)|O_2(p_1)$, Pt, is

$$E_{\text{cell}} = E^0_{O_2,OH^-} - E^0_{O_2,H_2O} - \frac{0.05915}{4} \log [H^+]^4[OH^-]^4 \qquad (3\text{-}8)$$

Again setting $E_{\text{cell}} = 0$ as the condition for equilibrium, this gives

$$\log [H^+][OH^-] = -\frac{0.828}{0.05915} = -14.00$$

In general, it is possible to find from standard-potential data the equilibrium constant of any reaction which can be expressed as the sum of two half-reactions whose standard potentials are known. For this reason, data on solubility and other equilibria are often expressed as standard-potential data. For example, new information on the solubility product of lead oxalate might well be combined with the standard potential of the half-reaction $Pb^{++} + 2e = Pb$, so that the quantity reported would be the standard potential of the half-reaction

$$PbC_2O_4 + 2e = Pb + C_2O_4^{=}$$

This is no less useful than reporting $K_{PbC_2O_4}$ directly and has the virtue of permitting all equilibrium data to be collected in one table.

It is worth noting that, by definition, standard potentials are strictly applicable only to infinitely dilute solutions, in which all activity coefficients are equal to unity. This, incidentally, is the reason why concentrations have been written in the equations appearing in this section. It follows that any equilibrium constant calculated from standard-potential data must be a "true" or "thermodynamic" equilibrium constant which is always true as far as activities are concerned, but which can be used for the exact calculation of concentrations only at infinite dilution. For example, the solubility product of silver chloride calculated above is the ion-*activity* product of silver chloride. We could use it to calculate the *concentrations* of silver and chloride ions in equilibrium with, say, 0.0001 F potassium chloride (whose ionic strength is so low that to all intents and purposes it is infinitely dilute). But if we wanted to find the concentrations of silver and chloride ions in, say, 0.1 F potassium nitrate saturated with silver chloride, we would have to write the ion-activity product in the form

$$f_{Ag^+}[Ag^+]f_{Cl^-}[Cl^-] = f_{\pm}^2[Ag^+][Cl^-] = 1.77 \times 10^{-10}$$

and look up (or make some assumption about) a value of the mean ionic activity coefficient of silver chloride at an ionic strength of 0.1.

The converse of this, applied to the lead oxalate example given above, means that we should be careful to avoid reporting a value for the standard potential of the half-reaction $PbC_2O_4 + 2e = Pb + C_2O_4^{=}$ on the basis of a value of the concentration product $[Pb^{++}][C_2O_4^{=}]$ which was obtained by studying solutions whose ionic strengths are too high to justify setting activities equal to concentrations.

3-4. The Liquid-junction Potential. The liquid-junction potential is that component of the potential of a galvanic cell which arises at the boundary of two dissimilar solutions; it results from the fact that the various ions concerned tend to diffuse across the boundary at unequal rates. Liquid-junction potentials may be classified according to the kind of difference in composition existing across the boundary. Although for certain physicochemical purposes it is possible to construct a cell "without liquid junction" (which consists simply of two dissimilar electrodes in contact with the same solution), the great majority of practical e.m.f. measurements are made with cells which do have liquid junctions. Examples of cells "without liquid junction" can be found in Secs. 2-6 and 3-6.

The simplest kind of liquid junction is that between two solutions containing the same electrolyte at different concentrations, such as HCl(0.1 F)||HCl(0.01 F) or KCl(sat'd.)||KCl(0.1 F). Let us consider the first of these. It is obvious that both hydrogen and chloride ions will tend to diffuse across the boundary from left to right (*i.e.*, from the more concentrated to the more dilute solution). However, the hydrogen ion diffuses about five times as rapidly as the chloride ion, and therefore the right-hand side of the boundary rapidly acquires a net positive charge. Naturally this accumulation of positive charge on the right-hand side of the boundary tends to retard the diffusion of hydrogen ion and accelerate that of chloride ion by virtue of the electrostatic forces set up, and so a steady state is reached in which the potential difference between the two solutions has a constant value. It should be noted that, to conform with the conventions stated in Sec. 3-2, whereby a cell whose negative electrode is on the left-hand side is assigned a positive potential, the sign of the liquid-junction potential across the boundary [−HCl(0.1 F)|| HCl(0.01 F) +] must be positive.

Of course it is immediately evident that the magnitude of the liquid-junction potential depends largely on the relative mobilities of the ions present at the boundary. The potential across the junction HCl(0.1 F)|| HCl(0.01 F) is relatively high (+40 mv.) inasmuch as the hydrogen and chloride ions move at quite disproportionate rates. On the other hand, the liquid-junction potential is much smaller when the ions move at nearly the same speed. For potassium chloride, whose ions have almost identical mobilities (the transference numbers of potassium and chloride ions are 0.491 and 0.509, respectively), the liquid-junction potential is much smaller. Thus E_j is only about −1.0 mv. for the junction KCl(0.1 F)||KCl(0.01 F); it is negative because chloride ion moves slightly faster than potassium ion, so that the right-hand side of the boundary acquires a negative charge.

A more complicated kind of liquid junction is that formed between two solutions which contain a common ion at the same concentrations, such as KCl(0.1 F)||HCl(0.1 F). There is virtually no tendency for the common ion to diffuse across the boundary, because its concentration is the same on both sides, and so we have to consider only the effect of the motion of hydrogen and potassium ions. Of these, the hydrogen ion moves much more rapidly, and therefore the left-hand side of the junction acquires a net positive charge and E_j is negative.

The still more complex junction between two solutions having no ion in common [such as $NaNO_3(0.1\ F)\|KCl(0.05\ F)$] involves the simultaneous diffusion of four or more ions, and attempts to calculate E_j in such instances have as yet been unsuccessful.

3-5. "Elimination" of the Liquid-junction Potential. Some idea of the magnitude of the liquid-junction potential is afforded by Table 3-1,

TABLE 3-1. LIQUID-JUNCTION POTENTIALS AT 25°C.*

Boundary	E_j, *mv.*	*Boundary*	E_j, *mv.*
$LiCl(0.1\ F)\|KCl(0.1\ F)$	− 8.9	$KCl(0.1\ F)\|KCl(3.5\ F)$	+ 0.6
$NaCl(0.1\ F)\|KCl(0.1\ F)$	− 6.4	$NaCl(0.1\ F)\|KCl(3.5\ F)$	− 0.2
$NH_4Cl(0.1\ F)\|KCl(0.1\ F)$	+ 2.2	$NaCl(1\ F)\|KCl(3.5\ F)$	− 1.9
$NaOH(0.1\ F)\|KCl(0.1\ F)$	−18.9	$NaOH(0.1\ F)\|KCl(3.5\ F)$	− 2.1
$NaOH(1\ F)\|KCl(0.1\ F)$	−45	$NaOH(1\ F)\|KCl(3.5\ F)$	−10.5
$KOH(1\ F)\|KCl(0.1\ F)$	−34	$KOH(1\ F)\|KCl(3.5\ F)$	− 8.6
$HCl(0.1\ F)\|KCl(0.1\ F)$	+27	$HCl(0.1\ F)\|KCl(3.5\ F)$	+ 3.1
$H_2SO_4(0.05\ F)\|KCl(0.1\ F)$	+25	$H_2SO_4(0.05\ F)\|KCl(3.5\ F)$	+ 4

* The data are taken from G. Milazzo, "Elektrochemie," Springer-Verlag, Vienna, 1952; *cf.* R. G. Bates, "Electrometric pH Determinations," John Wiley & Sons, Inc., New York, 1954, p. 41.

which gives the approximate values of E_j across some typical boundaries. From these data we may draw three important conclusions. Firstly, the liquid-junction potential across a boundary at which only neutral salts (KCl, $NaCl$, NH_4Cl, etc.) are involved is less than when a strong acid or base is present on one side of the boundary. Secondly, the liquid-junction potential is invariably decreased by increasing the concentration of potassium chloride on one side of the boundary. Thirdly, if the concentration of potassium chloride on one side of the boundary is kept constant, an increase in the concentration of the solution on the other side of the boundary always results in an increase of the liquid-junction potential.

The first of these simply reflects the fact that both hydrogen and hydroxyl ions have unusually high mobilities and therefore diffuse across the boundary more rapidly than any other ion present. The second conclusion follows from the fact that potassium and chloride ions are nearly equally mobile; when they are present in large excess, virtually all of the current is carried across the boundary by their motion (which, since the cations and anions move at practically the same rate, contributes very little to the liquid-junction potential), and very little of the current is carried by the ions of the other salt. Finally, increasing the concentration of the other salt increases the fraction of the current carried by its ions; the higher the concentrations of these ions and the greater the disparity between their mobilities, the greater the liquid-junction potential becomes.

We have gone into these considerations in some detail because virtually every cell which the analytical chemist uses involves a salt bridge and consequently a liquid-junction potential. Because the nature of the junction is almost always too complex to permit a direct evaluation of E_j, it is customarily neglected; one might say that ignorance is the mother of necessity. One must remember that the quantity being neglected is frequently many times larger than the probable error of the actual instrumental measurement, and that it does no good to measure the potential of a cell to 0.1 or 0.01 mv. when the cell includes a liquid junction whose potential may amount to 10 or 20 mv. It is this fact far more than any other which limits the accuracy attainable in analyses by direct potentiometry.

Most textbooks of physical chemistry imply that the liquid-junction potential is eliminated or decreased to a negligible value by the use of a saturated potassium chloride salt bridge. However, this is true *only* when the concentrations of the other ions present at the boundary are very much smaller than 4.2 F (the solubility of potassium chloride at 25°C.); as is shown by Table 3-1, the liquid-junction potential at the interface between nearly saturated potassium chloride and an 0.1 or 1 F solution of another substance may be anything but negligible. We use salt bridges of saturated potassium chloride, *not to eliminate the liquid-junction potential or render it negligible*, but because no other way is known of consistently securing smaller liquid-junction potentials.

3-6. How Standard Potentials Are Measured. Let us suppose that we wished to determine the standard potential of the $Cu^{++}|Cu$ electrode. We might do this by measuring the e.m.f. of the cell $Cu|CuCl_2(c)$, $AgCl(s)|Ag$; this is a cell "without liquid junction." Its potential is given by (assuming that c is very much larger than the molar solubility of AgCl in the solution)

$$\begin{aligned} E_{\text{cell}} &= E_{\text{AgCl,Ag}} - E_{\text{Cu}^{++},\text{Cu}} \\ &= E^0_{\text{AgCl Ag}} - 0.05915 \log f_{\text{Cl}^-}[\text{Cl}^-] - E^0_{\text{Cu}^{++},\text{Cu}} + \frac{0.05915}{2} \log \frac{1}{f_{\text{Cu}^{++}}[\text{Cu}^{++}]} \\ &= E^0_{\text{AgCl,Ag}} - E^0_{\text{Cu}^{++},\text{Cu}} - \frac{0.05915}{2} \log (4c^3) - \frac{0.05915}{2} \log f_{\text{Cu}^{++}} f^2_{\text{Cl}^-} \qquad (3\text{-}9) \end{aligned}$$

Of the quantities appearing on the right-hand side of Eq. (3-9), $E^0_{\text{AgCl,Ag}}$ is known [from entirely similar measurements of the potential of the cell Pt, $H_2|HCl(c)$, $AgCl(s)|Ag$], and c is known, leaving $E^0_{\text{Cu}^{++},\text{Cu}}$ and $f_{\text{Cu}^{++}} f^2_{\text{Cl}^-}$ as unknowns. Since it is known that the activity coefficients become equal to unity at zero ionic strength, the cell potential is measured at a number of values of c; then the corresponding values of $E_{\text{cell}} - E^0_{\text{AgCl,Ag}} + (0.05915/2) \log (4c^3)$ are plotted against some suitable function of c and extrapolated to $c = 0$. Both empirically and from the Debye-Hückel equation it is found that straight lines are secured in such extrapolations when the square root of the ionic strength is chosen for the abscissa scale. This is desirable because a straight line can be more accurately and conveniently extrapolated than any other function. The value of the extrapolated quantity at $\mu = 0$ is, of course, numerically equal to but of the opposite sign from the desired $E^0_{Cu^{++},Cu}$. Quite evidently, once this value is known, one can use the actual values of E_{cell} at finite (*i.e.*, nonzero) ionic strengths to calculate the corresponding values of $f_{Cu^{++}}f^2_{Cl^-}$. From these values, in turn, one can calculate the "mean ionic activity coefficient," denoted by the symbol $f_\pm$ and defined by the equation

$$f_\pm = \sqrt[3]{f_{Cu^{++}}f^2_{Cl^-}}$$

or, in general, for a salt A_mB_n

$$f_\pm = \sqrt[m+n]{f_A{}^m f_B{}^n} \tag{3-10}$$

3-7. Standard Potentials *vs.* Formal Potentials. Strictly speaking, the standard potential of a half-reaction provides an accurate description of the chemistry of the corresponding couple only at infinite dilution. For example, one finds listed in tables the standard potential of the half-reaction $Sn^{++++} + 2e = Sn^{++}$. However, stannic tin cannot exist in solution as a simple ion, and hydrous stannous oxide precipitates from solutions of +2 tin unless a considerable excess of acid is present. Undoubtedly any attempt to use the reported value of this standard potential to describe the action of stannous chloride solutions as reducing agents could only lead to error and confusion.

More useful information for most practical purposes is provided by the "formal potential" of a half-reaction. As Lingane wrote, "the formal potential includes in one fell swoop all effects resulting from variation of activity coefficients with ionic strength, complexation, acid-base dissociation, liquid-junction potentials, etc., and thus is a kind of 'practical standard potential.'"[1] In the example mentioned, one might measure the formal potential of the stannic–stannous couple in 1 *F* HCl: this would be simply the potential of the cell, in which *c* is much smaller than 1 *F*,

$$\text{Pt, } H_2(1 \text{ atm.})|\text{HCl}(1\ F), \text{Sn(IV)}(c), \text{Sn(II)}(c)|\text{Pt}$$

[Note the use of "Sn(IV)" to symbolize all of the species containing +4 tin which may be present.] An illustration of the practical utility of the formal potential was given in Sec. 2-6. Formal potentials are usually given the symbol $E^{0\prime}$ so that they can be differentiated from the corresponding standard potentials. Thus, the formal potential of the

[1] J. J. Lingane, "Electroanalytical Chemistry," Interscience Publishers, Inc., New York, 1953, pp. 41–42.

stannic-stannous couple in 1 *F* HCl would be represented by the symbol $E^{0'}_{\mathrm{Sn(IV),Sn(II);1}F\mathrm{HCl}}$.

3-8. Types of Indicator Electrodes. In ordinary analytical practice one is interested in changes of the potential of only one of the two electrodes which comprise the cell. For example, in the potentiometric titration of ferrous iron with dichromate, one merely wants to locate the point at which the potential of the solution is changing most rapidly. A platinum wire electrode immersed in the solution serves to detect changes in this potential and transmit them to the external measuring circuit. An electrode which performs this function is called an "indicator" electrode. In addition, a second electrode is needed so that there will be a complete electrical circuit through the cell. If, as is the case in routine analytical work, the only datum of interest is the volume of reagent required to reach the equivalence point, one can use a "bimetallic electrode system" (see below). This consists of a platinum indicator electrode, together with an electrode made of a different material (such as tungsten) whose potential is almost but not quite independent of the composition of the solution. The rate of change of the potential difference between these two electrodes with respect to the volume of reagent added reaches a maximum at or very close to the equivalence point of the titration.

However, the use of a bimetallic electrode system often results in a considerably distorted titration curve, and so in all exploratory work it is customary to use as the second electrode a "reference" electrode whose potential is known and is truly independent of the composition of the solution being titrated.

The indicator electrode in a potentiometric system has the function of providing a potential which depends in a predictable fashion on the concentrations of the substances taking part in the half-reaction which is of interest. For the purposes of direct potentiometry, this virtually amounts to requiring thermodynamic reversibility, which is to say conformity with the Nernst equation; otherwise we should have to establish empirically the relationship between solution composition and indicator-electrode potential, which is likely to be tedious as well as unreliable. Potentiometric titrations, on the other hand, can be (and often are) made with electrodes which do not wholly satisfy the Nernst equation, because in a titration only the shape of the curve is of interest, and the actual potentials are of no concern. For the purpose of selecting an indicator electrode, we may classify half-reactions into a number of categories, depending on whether the substances participating in them are all present in the liquid phase (homogeneous reactions) or also include solids or gases (heterogeneous reactions), and on whether the half-reactions are reversible or irreversible.

Half-reactions such as

$$Fe^{+++} + e = Fe^{++}, \qquad Ce^{IV} + e = Ce^{III}, \qquad I_3^- + 2e = 3I^-$$

and many others, are both homogeneous and reversible. Consequently, the potential of an inert metal electrode such as gold or platinum (the latter is almost invariably used in practical work) immersed in a solution containing these substances is accurately expressed by the Nernst equation. The definition of an inert metal here is not quite so simple as might be thought at first glance, for there are many metals which do not react with solutions chemically but which nonetheless cannot be used as indicator electrodes. Among these are tungsten, tantalum, and numerous alloys (including stainless steel); these always have thin oxide films on their surfaces which prevent the free interchange of electrons between metal and solution. Hence the potential of a tungsten electrode in, say, a ferrous–ferric solution is almost constant and independent of the ferric: ferrous concentration ratio. For this reason a tungsten electrode can be used as a sort of reference electrode in potentiometric redox titrations, together with a platinum wire as the indicator electrode. Such "bimetallic electrode systems" were extensively investigated around 1930, and details concerning their analytical applications may be found in the original literature. Now that calomel reference electrodes can be secured commercially (from manufacturers of glass-electrode pH meters), however, bimetallic electrode systems have lost the advantage of ease of construction which they possessed many years ago, and so they are used less and less frequently as time goes on.

Platinum wire indicator electrodes are usually prepared by winding a length of the wire around a small glass tube to form a helix, then removing the glass and connecting one end of the platinum, by means of a wire connector, to a copper or nickel wire leading to the electrical circuit. Increasing the area of platinum exposed to the solution by winding it into a helix not only decreases the length of time required to reach a steady potential, but also virtually eliminates the possibility of securing erratic results due to insufficient stirring. It is wise to clean the electrode immediately before use by immersing it briefly in concentrated nitric acid, rinsing it with water, then heating it to bright redness in the oxidizing cone of a Meker or Bunsen burner flame. This serves to remove or render harmless any impurities which may be present on the surface of the wire.

Platinum electrodes are unsuitable for work with solutions containing powerful reducing agents such as chromous, titanous, and vanadous ions. This is because the reduction of hydrogen ion or water,

$$H^+ + Cr^{++} = \tfrac{1}{2}H_2 + Cr^{+++}$$
$$H_2O + Cr^{++} = \tfrac{1}{2}H_2 + OH^- + Cr^{+++}$$

is catalyzed at a platinum surface, and so the composition of the solution at the surface of the electrode (and with it the observed potential) changes as the measurements are being made. In such cases one may use a mercury electrode, which does not have this catalytic effect (which is to say that the overpotential[1] associated with the deposition of hydrogen gas on a mercury surface is very high). The mercury is contained in a little cup immersed in the solution, and is connected to the electrical circuit by a wire of some metal which is insoluble in mercury (such as platinum, nickel, or tungsten) and which does not come in contact with the solution. Solutions of such strong reducing agents are still very difficult to work with even when mercury indicator electrodes are used, because the ease with which they react with atmospheric oxygen requires special precautions to ensure the exclusion of air.

Unlike the reversible half-reactions mentioned, which can be made to proceed in either direction by an infinitesimal change in the potential of an electrode, many more or less irreversible half-reactions are known. Some of these are

$$SO_4^= + 2H^+ + 2e = SO_3^= + H_2O$$
$$Cr_2O_7^= + 14H^+ + 6e = 2Cr^{+++} + 7H_2O$$
$$MnO_4^- + 8H^+ + 5e = Mn^{++} + 4H_2O$$
$$S_2O_8^= + 2e = 2SO_4^=$$
$$S_4O_6^= + 2e = 2S_2O_3^=$$

A reversible half-reaction is one in which the interchange of electrons among the electrode and the oxidized and reduced forms of the redox couple involved proceeds virtually instantaneously, so that equilibrium is very rapidly established and maintained at the electrode surface. In an irreversible half-reaction, on the other hand, either the reduction or the oxidation (or both) involves a slow step, so that the observed potential depends not only on the activities of the oxidized and reduced species but also on the factors which influence the rate of the slow step. In the sulfate–sulfite couple, for example, the oxidation of sulfite ion proceeds fairly rapidly though not instantaneously, but the reduction of sulfate is very slow. The potential of a platinum indicator electrode in a sulfate–sulfite solution therefore does vary with the sulfite activity, but is completely independent of the activity of sulfate. It may be remarked that the standard potentials of these irreversible half-reactions are

[1] *Cf.* Sec. 7-2.

generally calculated from calorimetric data instead of being found potentiometrically.

Direct potentiometry with such irreversible systems is out of the question, for the measured potential cannot be related unambiguously to the composition of the solution. On the other hand, potentiometric titrations are still feasible, provided that a reagent can be found which reacts with the solution at a reasonable rate and in a stoichiometric fashion. Thus, sulfite can be titrated potentiometrically with iodine or with permanganate.

Because of the existence of these irreversible systems, it is essential during the course of any measurement of a standard or formal potential to conduct special experiments designed to test the reversibility of the half-reaction being studied. This can be done by systematically varying the concentration of each of the substances taking part in the half-reaction while holding all of the other concentrations constant. Then, if the half-reaction is

$$aA + bB + ne = yY + zZ$$

we should find that, subject to the errors involved in neglecting liquid-junction potentials and activity coefficients,

$$\frac{\Delta E}{\Delta \log [A]} = + \frac{0.05915}{n} a$$
$$\frac{\Delta E}{\Delta \log [Y]} = - \frac{0.05915}{n} y$$

and so on. Unfortunately, this is often rather tedious experimentally, but still it cannot be omitted in any but the most slovenly work. The student will find another method of studying the reversibility of an electrode reaction described in Sec. 6-12.

Heterogeneous reactions involving one substance in the gas phase and having any practical importance are very few in number; only the hydrogen–hydrogen ion couple is worth citing here. A cell of the type

$$\text{Pt (platinized), } H_2(g)|H^+(c)||\text{reference electrode}$$

is of course the standard of reference for pH measurements. It is discussed further in Chap. 4.

When one or more of the substances participating in a half-reaction is a solid, many cases of great practical importance arise. The simplest possible instance of this is a half-reaction involving a metal and its ion, such as

$$Ag^+ + e = Ag$$
$$Cu^{++} + 2e = Cu$$

According to the Nernst equation, the potential of a silver wire immersed in a solution containing silver ion is

$$E_{Ag^+,Ag} = E^0_{Ag^+,Ag} - 0.05915 \log \frac{a_{Ag}}{a_{Ag^+}}$$

By convention, the activity of pure massive solid silver is taken as unity, and so the measured potential is a function of the silver ion activity alone. Accordingly, this electrode system can be used either for the measurement of a silver ion activity by direct potentiometry, or for following the change of the silver ion activity during a titration with, say, a chloride solution. An electrode of this kind, for which the electrode reaction involves only a metal and its ions, is known as an *electrode of the first order.*

One might perhaps expect that any metal could be used as an electrode of the first order to follow the concentration of its own ions, but this is very far from being the case. We may, in fact, divide the various metals into two classes: (1) those which are soft, ductile, rather freely soluble in mercury, and which function as electrodes of the first order, and (2) those which are hard, brittle, and only very slightly soluble in mercury, and which do not serve as reproducible electrodes. The first group includes such elements as silver, lead, copper, cadmium, and mercury; the second includes such elements as iron, cobalt, nickel, chromium, molybdenum, and tungsten. The explanation of this difference in behavior rests upon the very large change in internal energy which must result from any strain or deformation of the crystal structure of a hard metal and which is reflected in its electrode potential. Since two samples of the metal would naturally be most unlikely to be exactly equally strained, it is virtually impossible to secure data sufficiently reproducible for any purpose whatever. We shall see in Chap. 6 that the ions of these hard metals are invariably irreversibly reduced at a dropping mercury cathode.

The preparation of an electrode of the first order is quite analogous to that of a platinum electrode. The wire is wound into a helix in exactly the same way, but then it is cleaned by being dipped briefly into fairly concentrated nitric acid, and washed thoroughly with water. This should be repeated before each use to remove any foreign matter.

Electrodes of such heavy metals as lead and copper are particularly prone to the formation of films of basic (*i.e.*, hydroxy–) compounds. With lead, for example, this generates at the electrode surface a new equilibrium such as

$$Pb(OH)(NO_3) + H^+ + 2e = Pb + H_2O + NO_3^-$$

Accordingly, the electrode will assume an obviously undesirable response to pH and nitrate concentration. To avoid the formation of these adherent films, it is often advantageous to amalgamate the surface of the electrode very lightly by immersing the freshly cleaned wire in a dilute acidified solution of mercurous nitrate, then washing very thoroughly. This forms a saturated lead (or copper, etc.) amalgam. This treatment does not alter the activity of the electrode metal, because the chemical potentials of the solid metal and of the saturated amalgam (which is in equilibrium with it) are necessarily the same. But it does eliminate the effect of any slight distortion or strain of the crystal structure of the metal, and hence gives rise to more accurate and reproducible results.

Suppose, however, that one wished to follow the concentration of chloride in a solution. An electrode of the first order [Pt, $Cl_2(g)|Cl^-$] would be rather unpleasant to work with, and, in addition, platinum is attacked chemically by chlorine. In such cases one uses an *electrode of the second order*, which involves the ion in question, a slightly soluble metallic salt of that ion, and an electrode of the corresponding metal. Electrodes of the second order which are responsive to the chloride activity include, for example, $Ag|AgCl(s), Cl^-$ or $Hg|Hg_2Cl_2(s), Cl^-$. As above, the potential of the silver electrode is given by

$$E = E^0_{Ag^+,Ag} + 0.05915 \log a_{Ag^+} \qquad (3\text{-}11)$$

where the numerical value of $E^0_{Ag^+,Ag}$ is +0.7991 v. if the standard potential of the hydrogen ion–hydrogen gas couple is arbitrarily defined as zero. However, the activity of silver ion in the presence of solid silver chloride depends on the activity of chloride:

$$a_{Ag^+} = \frac{K_{AgCl}}{a_{Cl^-}}$$

Accordingly, the potential of the silver electrode becomes

$$E = E^0_{Ag^+,Ag} + 0.05915 \log K_{AgCl} - 0.05915 \log a_{Cl^-} \qquad (3\text{-}12)$$

The sum of the first two terms on the right-hand side of this equation is, of course, the potential of the silver chloride–silver electrode at unit activity of chloride, *i.e.*, the standard potential of the half-reaction $AgCl + e = Ag + Cl^-$.

We may consider an indicator electrode of the second order as being composed of two parts: an electrode of the first order, reversible to the concentration of an ion, and a salt whose solubility equilibrium relates the concentration of that ion to the concentration of another ion. From our discussion of electrodes of the first order it follows that no electrode of the second order involving a hard metal could be expected to function reversibly. However, since electrodes of the second order are used only to follow anion concentrations, this is not of much importance; it is easy to select a metal which gives a reversible electrode of the first order and forms a slightly soluble salt with the anion being determined.

The preparation of an electrode of the second order is similar in every detail to that of an electrode of the first order. It is only necessary to add an excess of the intermediate salt (*e.g.*, silver chloride in the preceding example) and to allow enough time to ensure that equilibrium is attained.

It hardly seems necessary to mention that in an electrode of the second order the solubility of the intermediate salt should be negligibly small compared to the concentration of the anion which is being determined. Obviously, if the reverse were true the total anion concentration indicated by the electrode potential would be practically equal to that arising from the solubility of the intermediate salt alone, and would therefore swamp the quantity in which one is interested.

Though on paper one can write an electrode of the second order which is responsive to the concentration of a cation [*e.g.*, Pt, $Cl_2(g)|AgCl(s)$, Ag^+], in practice such electrodes are used only to measure the concentrations of anions. In order, therefore, to be able to apply potentiometric measurements to cations for which electrodes of the first order are unavailable (such as calcium ion), one has to resort to an *electrode of the third order*, which involves two slightly soluble salts having a common anion. For example, the electrodes of the third order $Pb|PbC_2O_4(s)$, $CaC_2O_4(s)$, Ca^{++} or $Hg|Hg_2C_2O_4(s)$, $CaC_2O_4(s)$, Ca^{++} are reversible to the calcium ion activity, and might be employed for the direct potentiometric measurement of an unknown calcium ion activity. In the first of these, the potential of the electrode depends on the activity of lead ion:

$$E = E^0_{\mathrm{Pb^{++},Pb}} - \frac{0.05915}{2} \log \frac{1}{a_{\mathrm{Pb^{++}}}}$$

However, the activity of lead ion is dependent on the activity of oxalate:

$$a_{\mathrm{Pb^{++}}} = \frac{K_{\mathrm{PbC_2O_4}}}{a_{\mathrm{C_2O_4^{=}}}}$$

and the activity of oxalate ion is in turn dependent on that of calcium ion:

$$a_{\mathrm{C_2O_4^{=}}} = \frac{K_{\mathrm{CaC_2O_4}}}{a_{\mathrm{Ca^{++}}}}$$

so that the potential of the electrode is finally given by

$$E = E^0_{\mathrm{Pb^{++},Pb}} + \frac{0.05915}{2} \log \frac{K_{\mathrm{PbC_2O_4}}}{K_{\mathrm{CaC_2O_4}}} + \frac{0.05915}{2} \log a_{\mathrm{Ca^{++}}}$$

Because solubility equilibrium in such a complicated system is usually only slowly attained, such electrodes are characteristically rather sluggish and intractable. Against this, however, has to be set the fact that, to choose an example of some biochemical importance, the activity of calcium ion, and the equilibria involving calcium ion, in blood serum would probably be even more difficult to study in any other way.

3-9. Reference Electrodes. The ideal reference electrode would possess all of the following characteristics:

1. Ease of preparation from readily available materials.
2. Rapid attainment of an accurately reproducible potential.
3. Invariance of potential over long periods of time.
4. Negligible thermal hysteresis (*i.e.*, its potential should respond rapidly to changes of temperature and should return promptly to its initial value when the electrode is warmed and then cooled to its original temperature).
5. Low polarizability (*i.e.*, it should be possible to pass moderate currents through it for short periods of time without significantly altering the potential; this is largely dependent on the area of the interface between the metal and the solution).

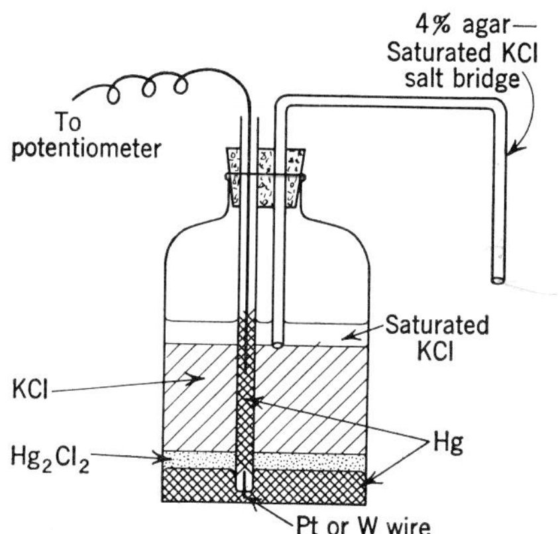

FIG. 3-2. Saturated calomel reference electrode.

Though many reference electrodes have been proposed and studied, one which satisfies all of these criteria has not yet been found. Probably the best, and certainly the most widely used, are the silver–silver chloride–potassium chloride and the mercury–mercurous chloride–potassium chloride ("calomel") electrodes. Their potentials are given by equations of the form

$$E = E^0 - 0.05915 \log a_{Cl^-}$$

and hence depend on the chloride ion activity. Electrodes unsaturated with respect to potassium chloride, such as the confusingly named "0.1 *N* calomel electrode" [$Hg|Hg_2Cl_2(s)$, $KCl(0.1\ N)$], which is 0.1 *N* (*i.e.*, 0.1 *F*) in potassium chloride and not in calomel, have somewhat lower temperature coefficients of potential than the corresponding "saturated" electrodes. For this reason they are widely used in very precise electrochemical work. However, this property is not of much practical importance to the analytical chemist, and the saturated electrodes are so much easier to prepare and store that they are almost universally used.

A saturated calomel electrode suitable for most work is shown in Fig. 3-2. It is prepared by placing a layer of pure mercury about 5 mm.

deep in a 2-oz. wide-mouth bottle, adding a few grams of pure mercurous chloride and a large excess of potassium chloride, and then nearly filling the bottle with saturated potassium chloride solution. The neck of the bottle is covered with a very thin film of silicone grease (to prevent "creeping" of potassium chloride), and a clean two-hole neoprene stopper is inserted. (Rubber stoppers always contain some sulfur, which may cause the slow formation of mercuric sulfide and considerably affect the potential of the electrode.) Into one hole of the stopper is inserted a glass tube through which a short length of platinum or tungsten wire is sealed as shown in the figure; this is then partly filled with mercury into which is dipped a copper (or, better, nickel) wire leading to the potentiometer. Into the other hole is inserted a 6- or 8-mm. i.d. glass tube filled with a 4 per cent agar gel saturated with potassium chloride. This is made by warming 4 g. of pure agar with 100 ml. of water to about 90°C. on a steam or water bath until the agar is entirely dissolved, then adding 35 g. of potassium chloride. The glass tube, previously bent to the shape shown, is then filled by suction through a rubber tube, with care to avoid trapping air bubbles (which would greatly increase the electrical resistance of the bridge), then cooled and inserted in place. To keep the agar from drying out, the tip of the bridge must always be placed in a saturated potassium chloride solution when not in use.

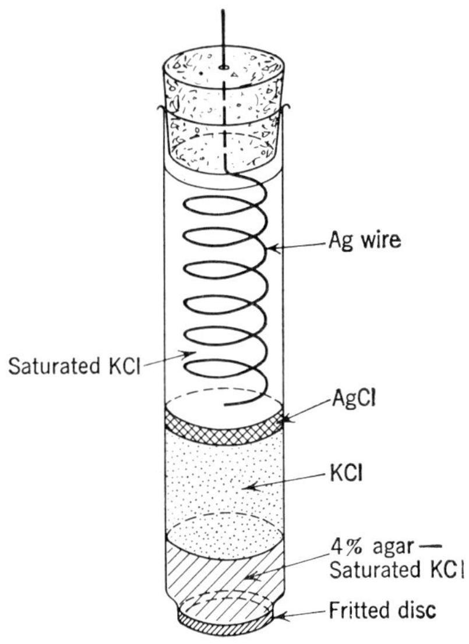

Fig. 3-3. Silver–silver chloride–saturated potassium chloride reference electrode.

The electrode must be prepared several days before it is to be used so that a steady potential will be attained. Though the potential of this electrode (which is generally abbreviated S.C.E.) is somewhat affected by air, owing to the formation of basic mercury chlorides, the effect is not large enough to be important for most purposes.

In exactly the same fashion one can prepare a mercury–mercurous sulfate–saturated potassium sulfate reference electrode for use with solutions which react with chloride ion. The bridge should then contain 4 per cent agar and 0.5 F potassium sulfate.

Though a silver–silver chloride electrode can be prepared in the same type of vessel, the authors recommend the design shown in Fig. 3-3.

One end of a sealing tube (Corning No. 39570) containing a 10-mm. medium-porosity fritted disc is ground off flush with the disc; then a 5-mm. layer of 4 per cent agar-saturated potassium chloride gel is placed above the disc. When this has set, about 10 g. of potassium chloride is added, followed by about 10 ml. of saturated potassium chloride solution and a few drops of saturated silver nitrate (or a few tenths of a gram of solid silver chloride). A tightly wound helix of stout silver wire, etched with 1:1 nitric acid and thoroughly washed with water, is inserted. The end of this wire is allowed to project through a one-hole neoprene stopper, and the protruding end is connected to a wire leading to the potentiometer by means of a wire connector.

For many purposes it is quite feasible to use a "fibre-type" saturated calomel electrode such as is sold for use with most commercial pH meters. In such electrodes the salt bridge consists of a small porous fiber through which saturated potassium chloride slowly flows from the interior of the electrode. Compared with the electrodes just described, a fibre-type electrode has a much smaller current-carrying capacity because of the relatively small area of mercury exposed. This means that the passage of an appreciable current (*e.g.*, by depressing the potentiometer key when the circuit is greatly unbalanced) may temporarily alter the potential of the reference electrode slightly. Of occasional importance is the fact that the resistance of such an electrode is comparatively high, about 2500 ohms or so. This is immaterial in potentiometric work, but it rules out the use of such an electrode as a reference electrode in polarographic work involving finite currents.

3-10. Electrical Apparatus for Potentiometric Measurements. It might perhaps be thought that the e.m.f. of a galvanic cell could be measured by connecting an ordinary d-c voltmeter across the electrodes and observing the reading. When this is done, the voltmeter reading is always smaller numerically than the expected potential of the cell. As time goes on it drifts slowly and irregularly toward zero, and indeed becomes equal to zero if the circuit is left connected for a long time. This is due to the fact that the voltmeter has a finite resistance and permits the flow of an electrical current around the circuit. Consequently a chemical reaction takes place in the cell, and the composition of the solution at the surface of each electrode will change. With the cell $Ag|AgCl(s), KCl(s)||Fe^{++}(c_1), Fe^{+++}(c_2)|Pt$, for example, the flow of current through the cell will result in a decrease of the chloride ion concentration at the surface of the silver electrode, and also in an increase of c_1 and a decrease of c_2 at the surface of the platinum electrode. These changes will cause the cell potential to tend toward zero; the cell is said to have become "polarized" (*cf.* Sec. 6-5).

Accordingly it is possible to measure the potential of a cell accurately only if no appreciable current is allowed to flow through it. For purposes for which a continuous indication of the potential is of value and in which a small error due to polarization is of no importance, a high-resistance vacuum-tube voltmeter can be used. The best results, however, are secured by the use of a "potentiometer," which is essentially an instrument in which the unknown potential of the cell is opposed by a variable known potential; when the latter is made equal to the potential of the cell, no current flows through the cell circuit. This point of balance is detected by a galvanometer in series with the cell.

In Chap. 7 we shall need to distinguish between "voltage" and "potential" in discussing the phenomena which occur in electrolytic cells. These terms are often the source of much confusion, because they are expressed in the same units (volts) and designated by the same symbol (E). The student will be saved much harassment if he remembers that a measurement with a *volt*meter permits the flow of an electrical current, whereas a measurement with a *potentio*meter does not. Voltage and potential are related by the equation

$$\text{Potential} = \text{voltage} + iR$$

where R is the resistance of the circuit (in ohms) and i is the current (in amperes) represented by the flow of electrons through the external circuit from the left-hand electrode to the right-hand electrode.

According to this statement, whose brevity cloaks a good deal of complexity, i would be positive if the cell

$$\text{Pt}|\text{Fe(II)}(0.1\,F),\ \text{Fe(III)}(0.1\,F),\ H_2SO_4(1\,F)\|\text{Ce(III)}(0.1\,F),\ \text{Ce(IV)}(0.1\,F),\ H_2SO_4(1\,F)|\text{Pt}$$

were connected to a voltmeter, for in this cell, whose potential is about +0.7 v., the reaction

$$\text{Fe(II)} + \text{Ce(IV)} = \text{Fe(III)} + \text{Ce(III)}$$

will proceed spontaneously from left to right as written as the cell discharges, so electrons will flow from left to right around the external circuit, and the voltage will be less positive than the potential. On the other hand, if the cell

$$\text{Pt}|\text{Ce(III)}(0.1\,F),\ \text{Ce(IV)}(0.1\,F),\ H_2SO_4(1\,F)\|\text{Fe(II)}(0.1\,F),\ \text{Fe(III)}(0.1\,F),\ H_2SO_4(1\,F)|\text{Pt}$$

whose potential is −0.7 v., is connected to a voltmeter, the spontaneous occurrence of the same chemical reaction will force electrons to flow from right to left around the external circuit. Hence i must be taken as negative, and so the voltage will be less negative than the cell potential. In either of these cases the voltage is numerically *smaller* than the potential;

the difference between them, of course, is dissipated as iR drop in the voltmeter and in the cell itself.

In polarography (Chap. 6) and in most other electrolytic techniques (Chap. 7), it is necessary to force a reaction in the nonspontaneous direction by applying an external source of e.m.f. to a cell. Suppose that we wished to bring about the nonspontaneous reaction

$$\text{Fe(III)} + \text{Ce(III)} = \text{Fe(II)} + \text{Ce(IV)}$$

in the first of the above cells by reducing ferric iron at the left-hand electrode and oxidizing cerium(III) at the right-hand electrode. This would involve withdrawing electrons from the right-hand electrode and supplying them to the left-hand electrode, and therefore i would be negative. Consequently the voltage which would have to be applied across the two electrodes must be more positive than the potential of the cell. To bring about the same nonspontaneous reaction in the other cell, it would be necessary to apply a voltage more negative than the potential of the cell, which is −0.7 v.

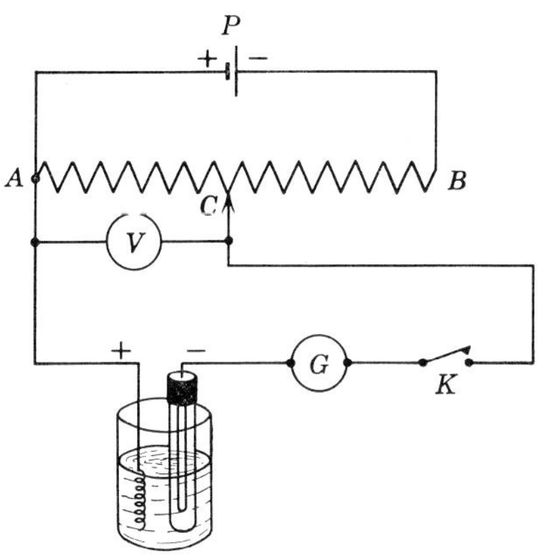

FIG. 3-4. Simple potentiometer employing a nonlinear voltage divider and a voltmeter as the measuring element.

These considerations are generally summarized by saying that potential is numerically greater than voltage for the spontaneous discharge of an electrochemical cell, but that voltage is numerically greater than potential when a nonspontaneous reaction is forced to occur by electrolysis.

A very simple type of potentiometer is shown in Fig. 3-4. Here P is a battery whose voltage is at least equal to the potential of the cell being measured. (As the latter almost never exceeds 1.5 v., P may be merely a single dry cell.) The "bridge" or voltage divider AB need not be exactly linear, although it should be at least approximately so in order that the sensitivity of adjustment will be nearly independent of the magnitude of the potential being measured.

In operation, the position of the slider C is adjusted until no deflection of the galvanometer G is observed when the tapping key K is momentarily depressed. Then the voltage drop imposed across AC by the battery [which is equal to $(R_{AC}/R_{AB})E_P$, where E_P is the e.m.f. of the battery] and indicated by the voltmeter V is equal to the potential of the cell. Thus the instrument is direct-reading, and its over-all sensitivity and accuracy are determined primarily by the characteristics of the voltmeter. An accuracy of ± 0.01 v. is not difficult to secure, and this is adequate for most potentiometric titrations, although an error of this magnitude would almost always be intolerable in direct potentiometry.

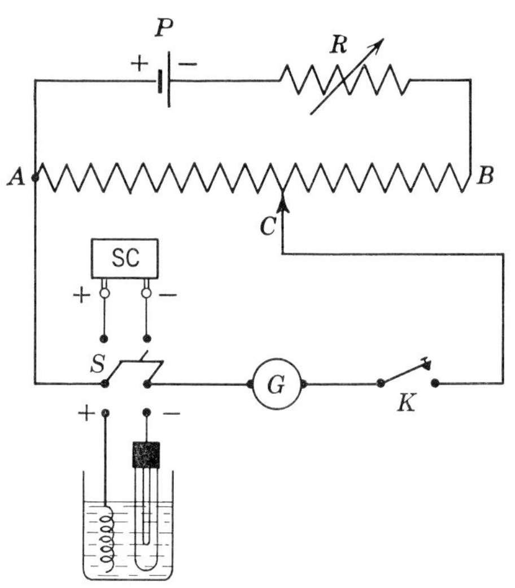

FIG. 3-5. Circuit diagram of a potentiometer employing a linear voltage divider.

A more elaborate and accurate instrument can be constructed according to the scheme shown in Fig. 3-5. Here *P*, *G*, and *K* have the same functions as in the circuit of Fig. 3-4. The bridge *AB*, however, must now be accurately linear (*i.e.*, the resistance between *A* and *C* must be a linear function of the distance between them) and must be provided with a scale which gives the ratio of the distances between *AC* and *AB* (equal to the corresponding ratio of resistances, R_{AC}/R_{AB}). *R* is most conveniently a four-dial resistance box whose setting can be varied from 0.1 to 1000 ohms.

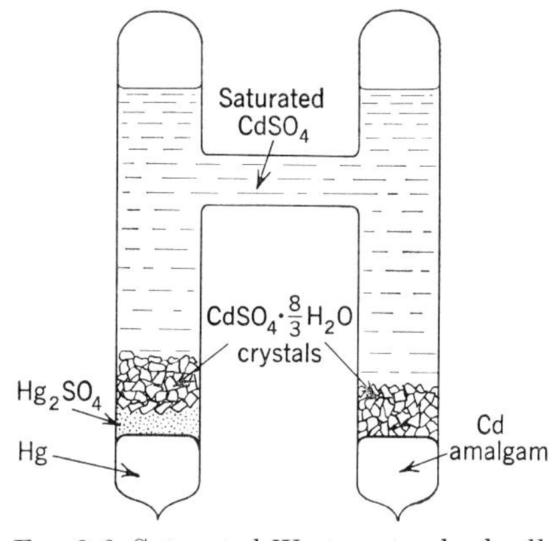

FIG. 3-6. Saturated Weston standard cell.

The standard cell *SC* may be a "saturated Weston cell," such as is shown in Fig. 3-6. At any given temperature the e.m.f. of this cell is constant and is determined by the standard potentials of the reactions

$$Cd^{++} + x\mathrm{Hg} + 2e = \mathrm{Cd(Hg)}_x \qquad \text{(saturated amalgam)}$$

and

$$Hg_2SO_4 + 2e = 2\mathrm{Hg} + SO_4^{=}$$

and by the solubilities of cadmium and mercurous sulfates. Its e.m.f. at 20°C. is taken as 1.018300 v., and it changes with temperature between 0 and 40°C. according to the equation

$$E = 1.018300 - 0.000,040,6(t - 20) - 0.000,000,95(t - 20)^2 + 1 \times 10^{-9}(t - 20)^3$$

where t is the temperature in °C. The change of the e.m.f. with temperature depends primarily on the changing solubility of cadmium sulfate and the effects of the resulting changes in the cadmium and sulfate ion concentrations on the potentials of the two electrodes.

In most laboratory work it is customary to use an "unsaturated" Weston cell. This differs from the cell shown in Fig. 3-6 in that no excess of cadmium sulfate is present; the electrolyte is saturated with cadmium sulfate at 4°C. and is therefore unsaturated at room temperature. The e.m.f. of this cell is about 1.019 v. at 20°C., and its temperature coefficient (−0.01 mv. per degree) is only about one-fourth that of the saturated Weston cell.

The sensitivity of the galvanometer G should depend on the resistance of the cell and on the accuracy of measurement which one wants to achieve. If an accuracy of ±1 mv. is desired (and in virtually all analytical work this is quite sufficient), and if the circuit resistance is 1000 ohms, which is a typical value, it follows from Ohm's law that the galvanometer should be able to detect a current of 1 microamp. This could be accomplished with a galvanometer whose sensitivity is 1 microamp. per mm. (*i.e.*, whose pointer deflects 1 mm. for each microampere of current), though in practice one would probably select a rather more sensitive galvanometer to secure larger and more conveniently observable deflections. The galvanometer should be damped by means of a variable resistance connected across its terminals (*e.g.*, as shown in Fig. 3-8). When one starts to measure an unknown potential, this damping resistance should be very small in order to prevent a large off-balance current from passing through the galvanometer. As the point of balance is more and more closely approached, the damping resistance should be increased up to its maximum value, thus permitting the experimenter to take advantage of the highest possible galvanometer sensitivity in making the final adjustment of the potentiometer.

The potentiometer scale is "standardized" by setting the slider C at a point which corresponds to the e.m.f. of the standard cell. Then, with the standard cell connected into the circuit and the resistance box set at some arbitrary value, the key K is depressed momentarily, and the direction in which the pointer of the galvanometer deflects is noted. The setting of R is then changed considerably, and K is again depressed for an instant. If the galvanometer pointer now deflects in the opposite direction, intermediate values of R are tried until a setting is found at which the galvanometer does not deflect when K is depressed. At this point the voltage across AB is equal to $E_{SC} \times (R_{AB}/R_{AC})$. For example, if AB has 2000 divisions and C is set at 1019 divisions (assuming the e.m.f. of the standard cell to be 1.019 v.), each division on AB will correspond to 0.001 v., and potentials up to 2 v. can be measured.

As the voltage developed by the battery P changes slowly with time, it is advisable to repeat this standardization at intervals of from 15 to 30 min. It is convenient to leave a new battery connected into the

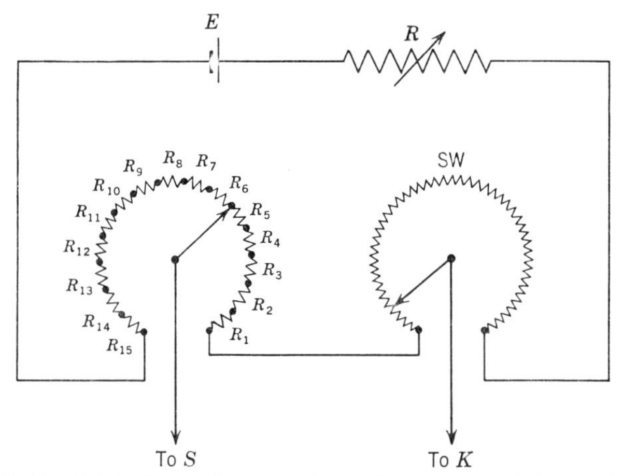

FIG. 3-7. Simplified circuit diagram of a typical commercial potentiometer.

circuit for a few days to discharge it partially, because the e.m.f. of a dry cell or storage battery is much more stable when it is partially discharged than when it is fresh. Thereafter, however, the battery should always be disconnected whenever the circuit is not in use.

In the interest of securing greater accuracy with a slide wire of limited length, commercial potentiometers usually replace the simple bridge arrangement shown in Fig. 3-5 with the arrangement illustrated by Fig. 3-7.

In this diagram R_1, R_2, . . . , R_{15} are fixed resistors mounted on a rotary switch, and SW is a slide wire; the total resistance of SW is equal to that of each of the fixed resistors. When the total voltage across the switch and slide wire has been adjusted to 1.6 v. by the method just described, the voltage across each fixed resistor, as well as the voltage across SW, is exactly 0.1 v. Thus SW permits interpolation between the 0.1-v. settings secured by manipulating the rotary switch. It is obvious that an accuracy of ± 1 mv. can be secured even if SW is linear (or can be read) only to ± 1 per cent, whereas the specifications which must be satisfied by the bridge in Fig. 3-5 are much more stringent.

The circuit of Fig. 3-5, modified according to Fig. 3-7, is essentially the circuit employed in the Leeds and Northrup students' potentiometer, the necessary external connections to which are shown in Fig. 3-8. When a connection is made to the binding post marked "0.01," an end coil is introduced in series with the slide wire and rotary switch; this changes the range of the instrument to 0 to 0.016 v. instead of the usual 0 to 1.6 v. Such small voltages are of considerable practical importance in the measurement of temperature using thermocouples. The arrangement of the

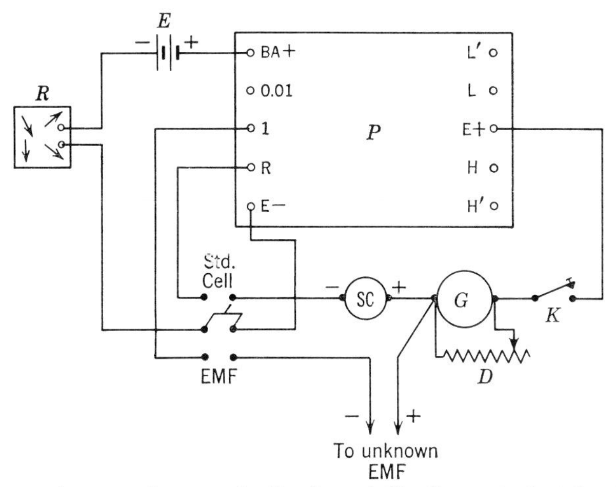

FIG. 3-8. External connections to the Leeds and Northrup students' potentiometer.

switch in Fig. 3-8 is recommended solely for convenience in standardizing the instrument when the 0.01 range is in use.

The posts marked *H*, *H*′, *L*, and *L*′ serve to permit the use of the slide wire alone in a Wheatstone bridge (Chap. 5); they are not used in potentiometric work.

Several commercially available potentiometers are so constructed that a precision of ±0.01 mv. can easily be secured. However, this great sensitivity is quite useless in analytical work. Nor do we recommend the use of a glass electrode pH meter for potentiometric measurements, for such an instrument, though certainly very convenient for the purpose, is in fact so convenient that the student is generally encouraged to subordinate any attempt to understand the principles involved in favor of a more or less mechanical repetition of the required switch turnings and button pushings.

3-11. Techniques of Direct Potentiometry. Experimentally, the performance of an analysis by direct potentiometry is extremely simple. It involves merely the selection and preparation of a suitable indicator electrode (Sec. 3-8), the immersion of this electrode and a reference electrode

(Figs. 3-2 and 3-3) in the solution which is to be analyzed, and the measurement of the resulting potential. In measuring the potential it is desirable to make observations at intervals of a few minutes to ensure that the value recorded actually corresponds to a true equilibrium potential. This is especially important when the solution is poorly poised ("poising" is the redox analog of "buffering" in pH measurements), for electrodes are very often slow to attain their final potentials when immersed in such solutions.

Unfortunately, the interpretation of the measured potential is often a matter of considerable difficulty. We have already referred to the fact that the existence of the liquid-junction potential severely limits the accuracy which may be attained. The potential which is actually measured by the potentiometer is given by

$$E_{\text{cell}} = E_{ref} + E_j - E_{ind} \tag{3-13}$$

where E_{ind} and E_{ref} are the potentials of the indicator and reference electrodes, respectively. Taking E_{ref} as fixed by the nature of the reference electrode, we have, for an indicator electrode of the first order involving a metal M and its univalent ion M^+ [*cf.* Eq. (3-11)]

$$E_{\text{cell}} = E_{ref} + E_j - E^0_{M^+,M} - 0.05915 \log a_{M^+} \tag{3-14}$$

or

$$a_{M^+} = \text{antilog} \frac{1}{0.05915} (E_{\text{cell}} + E^0_{M^+,M} - E_j - E_{ref}) \tag{3-15}$$

Now an error of only ± 1 mv. in the term within parentheses corresponds to an error of $\pm$ antilog $1/59$, or ± 4 per cent, in a_{M^+}. With a divalent ion the situation is even worse; an error of ± 1 mv. in the term in parentheses represents an error of ± 8 per cent in a_{M^+}. As the uncertainty in E_j alone is often several times as large as this, it is easy to see why direct potentiometry is not more widely used in analytical practice. An obvious further drawback is that analyses are usually carried out for the purpose of determining concentrations rather than activities. Even if the liquid-junction potential were completely eliminated and if the cell potential were measured so precisely as to render the error in a_{M^+} negligible, the activity-coefficient data required to permit the calculation of the concentration of the ion from its activity would only very rarely be available.

In order to minimize the effects of any uncertainties or variations in E_j and E_{ref}, as well as to permit the interpretation of the result in terms of the concentration, rather than the activity, of the ion being determined, it is customary to adopt the following procedure. A "blank" solution is prepared which has, as nearly as possible, the same ionic strength, pH, and concentration of each individual ion (especially those which may form complexes with the ion being determined) as the solution which is to be analyzed. Portions of this "blank" solution are then treated with known concentrations of the ion to be determined, and the corresponding potentials are then plotted against these concentrations. The range of known concentrations should be wide enough to include the potential obtained with the unknown solution, for extrapolation involves dangers which should be evident. After the potential secured with the unknown solution is measured, the corresponding concentration can easily be read off the graph. This amounts to determining a sort of "formal potential" which includes not only activity effects, liquid-junction potentials, and the like (*cf.* Sec. 3-7), but also any deviation of the potential of the particular reference electrode used from the expected value (due, perhaps, to impurities in the chemicals used in preparing the reference electrode).

Both the precision and the accuracy of the data secured by direct potentiometry are intimately related to the poising capacity of the redox couple being studied. For example, a silver electrode immersed in a 1 *F* silver nitrate solution (or in a 1 *F* sodium chloride solution containing an excess of solid silver chloride) attains a constant potential almost immediately. This potential is practically unaffected by stirring, by the passage of moderate currents, or by the occurrence of any small amount of chemical reaction in the solution.

Suppose, on the other hand, that a silver electrode is placed in a sodium perchlorate solution which contains no silver ion at all. Though at the instant at which the electrode is placed in the solution the silver ion concentration is zero, this situation cannot persist. According to the Nernst equation, a silver electrode immersed in a solution entirely free from dissolved silver ion would have an infinitely negative potential. That is, the metallic silver would react with any reducible constituent of the solution whatever, for example

$$4Ag + O_2 + 2H_2O = 4Ag^+ + 4OH^-$$

Even if dissolved oxygen were completely removed, silver ion could still be produced by the reaction

$$2Ag + 2H_2O = 2Ag^+ + H_2 + 2OH^-$$

Of course it would be a simple matter to predict the equilibrium concentration of silver ion formed by either of these reactions. However, in view of the fact that they proceed very slowly, it would be most unreasonable to suppose that the composition of the solution, say, five minutes after immersing the silver electrode, would actually correspond to the equilibrium value thus calculated. On the contrary, the actual concentration of silver ion then present would also depend on such factors as the rate constants of all of the possible reactions, the area and roughness of the silver electrode, the temperature, the rate of stirring, and so on.

In this situation it would evidently be futile to attempt to calculate the potential of the silver electrode from *a priori* considerations.

Of course the concentration of silver ion produced could be determined by a measurement of the electrode potential, but here a practical difficulty arises. The reactions just mentioned produce silver ion right at the surface of the electrode and it is the concentration there, not in the body of the solution, which determines the value of the electrode potential. Now if the solution is stirred, this trace of silver ion will be distributed throughout the solution; this decreases the concentration of silver ion at the electrode surface and so changes the potential. If enough time is allowed, these processes will continue until the solution becomes saturated with silver oxide formed by the reaction

$$2Ag^+ + 2OH^- = Ag_2O + H_2O$$

Thereupon the electrode potential will become constant over a long period of time.

These considerations are summarized by saying simply that a metal electrode in a very dilute solution of its ions acquires an "oxygen electrode function," *i.e.*, that its potential will depend partly or wholly on the concentration of oxygen (or, indeed, of any other oxidizing agent) in the solution. [The student will find it instructive to speculate on the "inertness" of a platinum indicator electrode in solutions of oxidizing agents, such as cerium(IV), whose potentials are comparable to those of the PtO, Pt and PtO_2, Pt couples.] It therefore follows that direct potentiometry cannot be applied in very poorly poised solutions. The exact lower limit in any specific instance must be established experimentally, but as a very rough guide it may be stated that trouble will generally be encountered if the concentration of any species entering into the expression for the electrode potential is less than about 10^{-5} to 10^{-6} F.

3-12. Potentiometric Titrations. A potentiometric titration is essentially a titration whose end point is found from data on the effect of the volume of titrant on the e.m.f. of a galvanic cell. Since the position of the titration curve with respect to the volume axis is independent of the value of E_j or of any knowledge of E^0 or E_{ref}, and since the location of the end point need not involve the personal errors which enter into a decision concerning the color change of an indicator, potentiometric titration constitutes one of the most accurate analytical methods available.

In the succeeding sections of this chapter we shall see that the equivalence point either exactly or virtually coincides with the inflection point of a plot of the cell potential (or a quantity related to it, such as the pH, pAg, etc.) against the volume of reagent added. The characteristic shape of a titration curve near the equivalence point is shown in Fig. 3-9*a*. We may use either of two methods of finding the equivalence point from such a curve: we may calculate or measure the e.m.f. at the equivalence

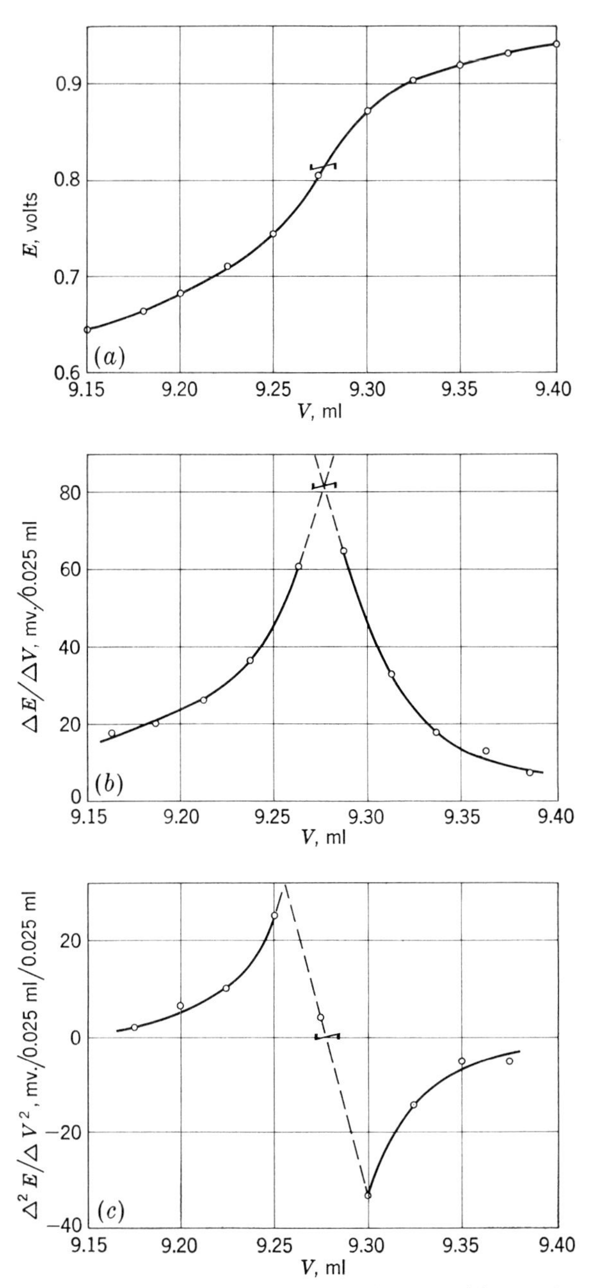

FIG. 3-9. Titration curves for the titration of Table 3-2: (*a*) experimental titration curve, (*b*) first derivative curve, and (*c*) second derivative curve.

point and find the volume of reagent required to produce that e.m.f., or we may employ some method designed to locate the inflection point itself.

The first of these procedures, titration to the equivalence point potential, naturally demands that this value be known. In view of the customary high slope of the e.m.f.-volume curve near the equivalence point, knowledge of the equivalence point potential to even ± 0.02 v. generally suffices. If the formal potentials of the half-reactions at the indicator electrode are known, the equivalence point potential can easily be calculated by the methods outlined in this and the following sections. Otherwise it can be found directly from potentiometric data as explained below. This procedure is employed in the operation of some autotitrators (Sec. 3-17), and also in a commonly used method for locating the end point in a "coulometric titration" (Sec. 7-11).

The alternative procedure, location of the inflection point, obviates the necessity of finding or calculating the equivalence point potential (and thus any error resulting from the selection of an incorrect value for this potential). It requires the measurement of a number of points on the e.m.f.-volume curve in the vicinity of the equivalence point. Some typical data secured in an actual case are shown in the first two columns of Table 3-2. From these data we could locate the inflection point either graphically or analytically.

The graphical approach is illustrated by Fig. 3-9*a*. It involves merely locating by inspection the point at which the curve has its maximum slope.[1] This technique suffers from two disadvantages: it takes some time to plot the titration curve, and the amount of personal judgment inherent in the processes of drawing the curve through the experimental points and locating the point of maximum slope may introduce a considerable error into the result. Constructing a plot of $\Delta E/\Delta V$ against V, as illustrated in Fig. 3-9*b*, is mentioned in virtually every elementary textbook as a method of locating the end point, but in actual fact it is never done in analytical practice. It is both much more tedious and appreciably less precise than the method illustrated by the third and fourth columns of Table 3-2.

This so-called "analytical" method of locating the end point depends on the fact that the second derivative of the curve $(\Delta^2E/\Delta V^2)$ is zero at the point where the slope of the curve $(\Delta E/\Delta V)$ is a maximum. This is, of course, simply a mathematical definition of an inflection point. Figure 3-9 shows, for the data of Table 3-2, the relationships among the titration curve itself and its first and second derivatives. A little thought

[1] The literature contains descriptions of more complicated, and doubtless more accurate, graphical methods of finding the inflection point, and some of these are still finding practical use. They all, however, involve the tedium of plotting the titration curve, and to our knowledge none equals the accuracy of the analytical method described below.

will reveal the way in which the calculations shown in Table 3-2 were made, and comparison with Fig. 3-9 will serve to clarify the principles involved.

It will be seen that, in securing the data shown in Table 3-2, a number of equal increments of the ceric sulfate were added. The use of equal

TABLE 3-2. "ANALYTICAL" LOCATION OF THE END POINT OF A POTENTIOMETRIC TITRATION*

Volume of ceric sulfate, ml.	E.m.f., volts	$\Delta E/\Delta V$ (mv./0.025 ml.)	$\Delta^2 E/\Delta V^2$ (mv./0.025 ml./0.025 ml.)
9.150	0.644		
		18	
9.175	0.662		2
		20	
9.200	0.682		6
		26	
9.225	0.708		10
		36	
9.250	0.744		25
		61	
9.275	0.805		4
		65	
9.300	0.870		−33
		32	
9.325	0.902		−14
		18	
9.350	0.920		− 6
		12	
9.375	0.932		− 4
		8	
9.400	0.940		

$$\text{Volume at end point} = 9.275 + 0.025\left(\frac{4}{4+33}\right)$$
$$= 9.278 \text{ ml.}$$
$$\text{Potential at end point} = 0.805 + 0.065\left(\frac{4}{4+33}\right)$$
$$= 0.812 \text{ v.}$$

* The data were obtained during the titration of a phosphoric acid solution of ferrous iron with standard ceric sulfate. A platinum wire indicator electrode and a saturated calomel reference electrode were used.

increments naturally saves some time in making the calculations, but it is by no means necessary. The use of 0.025-ml. aliquots of the ceric sulfate was simply a matter of convenience, for the buret used in the original experiment happened to be graduated in 0.025-ml. divisions.

Some little care must be taken in experiments of this kind to ensure that the increments of reagent added are neither too large nor too small.

From the data of Table 3-2 one can calculate that the use of 0.075-ml. increments, beginning at 9.15 ml., would lead to an error of −0.016 ml. in the position of the end point. On the other hand, data taken at intervals too closely spaced give a disproportionate weight to the normal error of measurement. The proper increment to add in any practical case can be judged only from experience.

It need hardly be mentioned that such small increments of reagent as this are not added throughout the titration. One very soon learns how to judge, from the rate at which the measured potential is changing, whether a large volume of reagent can be added without overstepping the end point, or whether the end point is so close that the addition of the reagent in small increments should be begun. When one is interested only in the location of the end point, one merely observes the potential occasionally during a titration to gain some idea of the distance to the end point, but there is no need to record the values until the end point is very near.

3-13. Theory of Potentiometric Precipitation Titrations. In this section we shall discuss the calculations of the titration curves for the titrations of silver with chloride and of a halide mixture with silver.

We shall assume first that 10.00 ml. of 0.1000 F silver nitrate is to be titrated with 0.1000 F sodium chloride. The apparatus with which this might be done is shown in Fig. 3-10. Note the manner in which the

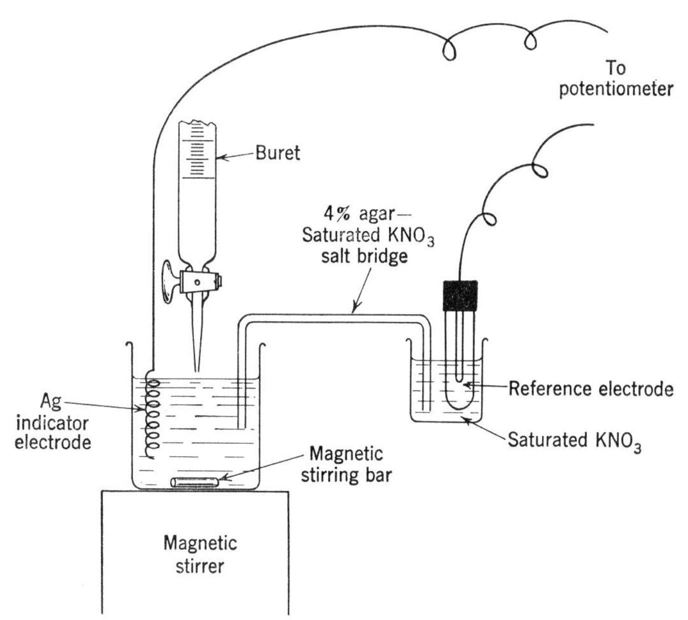

Fig. 3-10. Apparatus for potentiometric titration of silver nitrate with sodium chloride, using a silver indicator electrode.

solution being titrated is protected against contamination by chloride ion from the reference electrode.

At the start of the titration, the potential of the silver electrode is

$$E_{ind} = 0.799 + 0.059 \log 0.1 = 0.740 \text{ v.}$$

vs. the normal hydrogen electrode (N.H.E.), that is, the potential of the cell Pt, H_2(1 atm.)|$H^+(a = 1)$, $Ag^+(a = 0.1)$|Ag is +0.740 v. If an S.C.E. is used as the reference electrode, its potential will be included in the measured potential of the cell, and from Table B of the Appendix we find $E_{ref} = +0.246$ v. *vs.* N.H.E. Then, by Eq. (3-13), the measured potential—neglecting activity coefficients as well as E_j—will be

$$E_{\text{cell}} = 0.246 - 0.740 = -0.494 \text{ v.}$$

The cell referred to is, of course, Ag|Ag^+(0.1 *M*)||KCl(*s*), Hg_2Cl_2(*s*)|Hg. Its potential is *negative*, and therefore, according to the sign conventions outlined in Sec. 3-2, the silver electrode is the *positive* electrode of the cell and will have to be connected to the binding post marked + on the potentiometer. This is the same thing as saying that the potential of the silver indicator electrode is 0.494 v. more positive than the potential of the S.C.E. It is customary to write that the potential of the indicator electrode is +0.494 v. *vs.* S.C.E., and this terminology will be used henceforth.

Now suppose that some volume, V ml., of 0.1000 F sodium chloride is added, such that $0 < V < 10$. This will contain $0.1V$ millimole of chloride, which will precipitate $0.1V$ millimole of silver chloride, and so the concentration of silver ion remaining will be simply $(1 - 0.1V)/(10 + V)$. The numerator here gives the number of millimoles of silver remaining dissolved; the denominator gives the volume of the solution in milliliters. In just the same way as we did for the initial point, we can then calculate the potential of the silver electrode in a solution containing this concentration of silver ion. Thus we can define the titration curve until the equivalence point has almost been reached.

Very near the equivalence point, however, we must take into account the silver ions resulting from the solubility of silver chloride. In actuality, the total number of millimoles of silver ion in the solution is always the sum of two quantities: the number of millimoles of silver "in excess" (*i.e.*, as yet untitrated), and the number of millimoles of silver chloride dissolved in the solution. When the concentration of silver ion is large, the solubility of silver chloride is comparatively small and may safely be ignored, but as one nears the equivalence point it becomes more and more essential to use the complete equation

$$[Ag^+] = \frac{1 - 0.1V}{10 + V} + [Cl^-] \tag{3-16}$$

Since the solution is saturated with silver chloride the last term here is, of course, equal to $K_{AgCl}/[Ag^+]$. Table 3-3 illustrates the magnitude of the error which may be made by neglecting this correction.

TABLE 3-3. POTENTIOMETRIC TITRATION OF 0.1 M Ag^+ WITH 0.1 M Cl^-*

V_{Cl^-}, ml.	$[Ag^+]$, calculated by		E_{ind}, v., *vs.* S.C.E.	$\Delta E/\Delta V$, mv./ml.	$\Delta^2 E/\Delta V^2$, mv./ml.^{2}
	Eqs. (3-16) and (3-17)	Neglecting AgCl solubility			
0	0.1000	0.1000	+0.494		
				− 5	
1	0.0818	0.0818	+0.489		
				− 6	
5	0.0333	0.0333	+0.466		
				− 12	
9	5.3×10^{-3}	5.3×10^{-3}	+0.418		− 22
				− 67	
9.9	5.03×10^{-4}	5.03×10^{-4}	+0.358		− 1147
				− 635	
9.99	5.33×10^{-5}	5.00×10^{-5}	+0.300		− 54500
				−3333	
9.999	1.61×10^{-5}	5.00×10^{-6}	+0.270		−333300
				−5000	
10.00	1.34×10^{-5}	0	+0.265		0
				−5000	
10.001	1.12×10^{-5}	3.6×10^{-5}	+0.260		+311200
				−3444	
10.01	3.38×10^{-6}	3.6×10^{-6}	+0.229		+ 56790
				− 633	
10.1	3.62×10^{-7}	3.62×10^{-7}	+0.172		+ 1149
				− 64	
11	3.78×10^{-8}	....	+0.114		+ 12
				− 6	
20	5.4×10^{-9}	....	+0.064		

* See text for details. $K_{AgCl} = 1.8 \times 10^{-10}$.

At the equivalence point we have simply $[Ag^+] = [Cl^-] = \sqrt{K_{AgCl}}$. Then and later, the total chloride concentration is given by

$$[Cl^-] = \frac{0.1V - 1}{10 + V} + \frac{K_{AgCl}}{[Cl^-]} \tag{3-17}$$

completely analogous to Eq. (3-16). This equation was used to calculate the silver ion concentrations after the equivalence point shown in Table 3-3. The student will profit by checking some of these calculations.[1]

[1] For the sake of simplicity we have neglected the complex ions, such as $AgCl_2^-$, which exert an appreciable influence on the solubility of silver chloride even at chloride concentrations of 0.03 M.

In an actual titration, of course, the measured values will deviate slightly from those we have given here because the activity coefficients are not unity, as we have assumed them to be, and also because of the existence of the liquid-junction potential. Neither of these effects, however, will alter the volume of chloride used to reach the end point. As Table 3-3 shows, the titration curve is almost completely symmetrical (the small deviations from perfect symmetry result partly from rounding off the values of E to the nearest millivolt and partly from the steadily increasing volume of the solution), and the analytically determined end point coincides exactly with the equivalence point.

The manner in which the value of K_{AgCl} could be found from the experimental data in the first and fourth columns of Table 3-3 should be obvious. From the first few values of the measured potential one can secure, by using Eq. (3-14), a value of $(E_{ref} + E_j - E^0_{ind})$, since the concentration of silver ion is accurately known (in terms of its original concentration and the fraction of the equivalent amount of chloride added). This serves as a "calibration" of the apparatus in which any deviation of E_{ref} from the expected value, as well as the value of E_j, is taken into account. Then, from the values of E_{cell} some distance past the equivalence point, one can readily calculate K. It may be stressed that the potentials near the equivalence point are not used in such calculations; not only are the results excessively affected by small errors in the volume measurements (a reflection of the poor poising of the solutions), but the potentials themselves are much more difficult to measure precisely in this region.

Values of K thus secured are formal solubility products, not thermodynamic ones; they refer to the solubility of the substance at the particular ionic strength used in the titration. To minimize apparent variations in K due to changes in the ionic strength, it is convenient to use a reagent whose ionic strength is approximately the same as that of the solution being titrated. For example, the ionic strength of the solution in Table 3-3 is initially 0.1, decreases only to 0.05 at the equivalence point, and then slowly increases toward its initial value as excess chloride is added. The variation of K over this twofold range of ionic strength would be almost too small to detect.

The student is left to work out for himself the theory of the titration of chloride (or any other halide) with silver. In doing so he should bear in mind the fact that the potential of the silver electrode will be hopelessly undefined until enough silver ion has been added to saturate the solution with the silver halide, for the reasons explained in Sec. 3-11.

A rather unusual kind of titration curve is that for a mixture of halides with silver ion. Suppose, for example, that 10.00 ml. of a solution containing 0.1000 M bromide and 0.1000 M chloride is titrated with 0.1000 M

TABLE 3-4. POTENTIOMETRIC TITRATION OF 0.1 M Br^- − 0.1 M Cl^- WITH 0.1 M Ag^+*

V_{Ag^+}, ml.	$[Br^-]$	$[Cl^-]$	$[Ag^+]$	E_{ind}, v., *vs.* S.C.E.	$\Delta E/\Delta V$, mv./ml.	$\Delta^2 E/\Delta V^2$, mv./ml.^{2}
0	0.1	0.1	(0)	Undefined		
5×10^{-10}	0.1	(0.1)	5×10^{-12}	−0.115		
					3	
1	0.0818	(0.091)	6.1×10^{-12}	−0.112		
					6	
5	0.0333	(0.0667)	1.50×10^{-11}	−0.087		
					12	
9	5.3×10^{-3}	(0.0526)	9.44×10^{-11}	−0.038		22
					67	
9.9	5.03×10^{-4}	(0.0503)	9.94×10^{-10}	+0.022		606
					367	
9.99	(1.39×10^{-4})	0.05002	3.6×10^{-9}	+0.055		− 7414
					0	
9.999	(1.39×10^{-4})	0.0500	3.6×10^{-9}	+0.055		0
					0	
10.00	(1.39×10^{-4})	0.0500	3.6×10^{-9}	+0.055		4
					4	
11	(1.19×10^{-4})	0.0429	4.2×10^{-9}	+0.059		
					5	
15	(5.55×10^{-5})	0.0200	9.0×10^{-9}	+0.079		
					11	
19	(9.6×10^{-6})	3.45×10^{-3}	5.21×10^{-8}	+0.124		22
					66	
19.9	(9.3×10^{-7})	3.34×10^{-4}	5.39×10^{-7}	+0.183		1147
					634	
19.99	(1.06×10^{-7})	3.80×10^{-5}	4.74×10^{-6}	+0.240		43320
					2800	
20	(3.73×10^{-8})	1.34×10^{-5}	1.34×10^{-5}	+0.266		0
					2800	
20.01	(1.32×10^{-8})	(4.74×10^{-6})	3.80×10^{-5}	+0.292		−43320
					634	
20.1	(1.51×10^{-9})	(5.42×10^{-7})	3.32×10^{-4}	+0.349		− 1151
					64	
21	(1.55×10^{-10})	(5.57×10^{-8})	3.23×10^{-3}	+0.407		− 12
					6	
30	(2.0×10^{-11})	(7.2×10^{-9})	2.50×10^{-2}	+0.458		

* See text for details. Values in parentheses are not needed for calculating E_{ind}.

silver. The solubility products are taken as $K_{AgBr} = 5.0 \times 10^{-13}$ and $K_{AgCl} = 1.8 \times 10^{-10}$. The data for the titration curve are given in Table 3-4.

What is worthy of note here is the shape of the curve near the bromide equivalence point. As silver bromide precipitates, the bromide concentration decreases and the silver ion concentration increases. When the latter becomes just large enough to satisfy the solubility product of silver

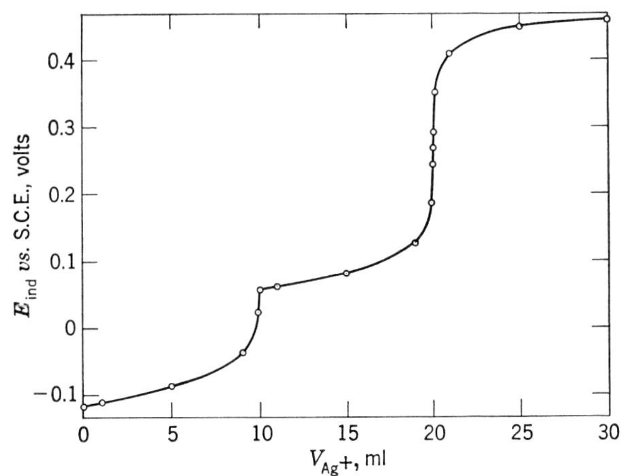

FIG. 3-11. Calculated titration curve for the potentiometric titration of 10.00 ml. of 0.1000 *M* bromide–0.1000 *M* chloride, using a silver indicator electrode.

chloride, we have

$$[Br^-] = \frac{K_{AgBr}}{[Ag^+]} = \frac{K_{AgBr}}{K_{AgCl}}[Cl^-] = 1.39 \times 10^{-4}$$

Employing the appropriate analogue of Eq. (3-16), we find that the volume of silver solution required to decrease the bromide concentration to this value is given by V in the equation

$$\frac{1 - 0.1V}{10 + V} = 1.39 \times 10^{-4} - \frac{5 \times 10^{-13}}{1.39 \times 10^{-4}}$$

The second term on the right is negligibly small and can be dropped without further ado. Solving the resulting equation gives $V = 9.972$ ml. When once this volume has been reached, a small further addition of silver merely precipitates a small fraction of the chloride, and so the rate of change of the silver ion concentration drops abruptly to virtually zero. The titration curve is plotted in Fig. 3-11, which shows the characteristic discontinuity which occurs at the point where silver chloride begins to precipitate.

The end point of the bromide titration has to be found by extrapolating the steep portion of the curve preceding the discontinuity and the nearly horizontal portion of the curve following the discontinuity; the end point is then the point at which the extrapolated lines intersect. The procedure evidently has an inherent error of -0.28 per cent when the bromide and chloride concentrations are equal; this increases as the proportion of chloride in the mixture increases. The error is larger if the end point is found by the analytical method, for this method cannot cope with a discontinuity in the titration curve.

In practice, neither the titration error nor the shape of the curve is quite identical with what is predicted by theory. This is because silver bromide and silver chloride form solid solutions, so that an appreciable fraction of the chloride is precipitated before the first end point. Consequently too much silver is consumed in the first part of the titration, and too little silver is consumed in the second part, which means that the amount of bromide found is too high while that of chloride is too low. The relative errors involved in determinations of bromide and chloride at equal concentrations are usually about ± 3 per cent. However, if the bromide/chloride ratio is considerably different from 1, the relative error in the determination of the one present in the smaller amount will naturally be far greater.

Moreover, the fact that solid solutions of the halides are formed means that the precipitation of silver chloride, instead of beginning abruptly, becomes relatively more and more important as the first end point is approached. This tends to "smear out" the discontinuity, so that the curve in this region actually becomes somewhat rounded.

When a solution containing roughly equal concentrations of iodide, bromide, and chloride is titrated with silver nitrate, the coprecipitation of silver bromide and silver chloride during the first part of the titration leads to an error of $+3$ to 5 per cent in the amount of iodide found. The quantity of bromide calculated from the distance between the first and second end points is usually very nearly correct, because the amount of silver chloride coprecipitated with the silver bromide is almost equal to the amount of silver bromide previously lost by coprecipitation with silver iodide. Finally, the quantity of chloride calculated from the distance between the second and third end points is about 3 to 5 per cent low because of the incorporation of this amount of chloride in the precipitate already present at the second end point. Of course the coprecipitation phenomena do not affect the *total* amount of silver nitrate used to titrate the three halides.

The best results are achieved in these titrations when a fairly high concentration (0.1 to 0.5 F) of a salt such as barium nitrate is present in the solution. This greatly decreases the extent of the coprecipitation and makes it possible to secure results which are accurate to ± 0.25 per cent or so.

3-14. Theory of Potentiometric Redox Titrations. Let us suppose that 10.00 ml. of 0.1000 M ferrous solution is to be titrated with 0.1000 F ceric sulfate, and let us suppose that both solutions are 1 F in sulfuric acid. The half-reactions involved, together with their formal potentials in 1 F sulfuric acid (Appendix, Table A), are

$$Fe^{+++} + e = Fe^{++} \qquad E^{0\prime} = +0.68 \text{ v.}$$
$$Ce(SO_4)_3^{=} + e = Ce^{+++} + 3SO_4^{=} \qquad E^{0\prime} = +1.44 \text{ v.}$$

Assume further that the reference electrode is an S.C.E. and that the indicator electrode is an inert metal such as platinum.[1]

In the original solution we might naïvely take $[Fe^{++}] = 0.1$ and $[Fe^{+++}] = 0$. However, a solution of this composition would be impossible to prepare. Having an infinitely negative potential, it would be an infinitely strong reducing agent and hence would react with oxygen, hydrogen ion, or some other oxidizing agent to give a trace of ferric ion. What the resulting concentration of ferric ion would be, however, is something we cannot calculate. It would certainly depend on the length of time during which this reaction was allowed to proceed. Of course it is a familiar fact that ferrous solutions are slowly oxidized on standing when exposed to the air. In view of these considerations, we must regard the potential of the original solution as undefined. A completely analogous situation was discussed in Sec. 3-11.

When a small amount of ceric sulfate has been added, however, we can safely neglect the trace of ferric ion present as an impurity or formed by an extraneous reaction. Letting V be the volume of ceric sulfate added, the concentration of ferrous ion will be

$$[Fe^{++}] = \frac{1 - 0.1V}{10 + V}$$

while the concentration of ferric ion will be

$$[Fe^{+++}] = \frac{0.1V}{10 + V}$$

If we let f be the fraction of the iron which is equivalent to V ml. of ceric sulfate, then

$$f = \frac{V}{10}$$

and

$$\frac{[Fe^{++}]}{[Fe^{+++}]} = \frac{1 - f}{f}$$

[1] One might assume that silver might be used because it appears in neither of the half-reactions, but this would be a mistake. Silver is a fairly strong reducing agent and would react to some extent with the oxidizing constituents of the solution. Thereafter its potential would reflect, not the true redox potential of the solution, but merely the concentration of silver ion which had been produced.

Consequently the potential of the platinum electrode will be

$$E_{Fe^{+++},Fe^{++}} = 0.68 - 0.059 \log \frac{1-f}{f} \qquad (3\text{-}18)$$

After the addition of, say, 0.1 ml. of the ceric solution, f is 0.01 (*i.e.*, 1 per cent of the iron will have been oxidized) and $E_{Fe^{+++},Fe^{++}}$ will be $+0.56_2$ v. The potential of the indicator electrode, E_{ind}, will therefore be $+0.31_6$ v. *vs.* S.C.E. (*cf.* the third paragraph of Sec. 3-13). In this way we can calculate the potentials given in Table 3-5 up to (but obviously not including) $f = 1$.

At the equivalence point, by definition, we shall have added just as many, say x, ceric ions as we originally had ferrous ions. Suppose that y ceric ions react with ferrous ions; then we shall have in the solution, when equilibrium is reached, $x - y$ ceric ions, the same number of ferrous ions, y cerous ions, and y ferric ions. Therefore

$$[Ce(SO_4)_3^{=}] = [Fe^{++}]$$

and

$$[Ce^{+++}] = [Fe^{+++}]$$

We have two expressions for the potential of the solution:

$$E_{Ce} = 1.44 - 0.059 \log \frac{[Ce^{+++}]}{[Ce(SO_4)_3^{=}]}$$

(the concentration of sulfate ion does not appear in the log term because it is implicitly included in the value of $E^{0\prime}$), and

$$E_{Fe} = 0.68 - 0.059 \log \frac{[Fe^{++}]}{[Fe^{+++}]}$$

The student should convince himself that equilibrium in this reaction is defined by the statement that

$$E_{Ce} = E_{Fe}$$

Combining the two equations, then, gives *at equilibrium*

$$0.059 \log \frac{[Fe^{+++}][Ce^{+++}]}{[Fe^{++}][Ce(SO_4)_3^{=}]} = 1.44 - 0.68 = 0.76$$

whence

$$\frac{[Fe^{+++}][Ce^{+++}]}{[Fe^{++}][Ce(SO_4)_3^{=}]} = 10^{0.76/0.059} = 10^{12.85} = 7 \times 10^{12}$$

Obviously this is the equilibrium constant of the reaction in 1 F sulfuric acid. At the equivalence point, then,

$$\frac{[Fe^{+++}]}{[Fe^{++}]} = \frac{[Ce^{+++}]}{[Ce(SO_4)_3^{=}]} = \sqrt{K} = 2.6 \times 10^6$$

This gives

$$E_{Fe} = 0.68 + 0.059 \log (2.6 \times 10^6) = +1.06 \text{ v.}$$

or

$$E_{ind} = 1.06 - 0.246 = 0.81_4 \text{ v. } vs. \text{ S.C.E.}$$

Note that the potential at the equivalence point is the mean of the formal potentials of the half-reactions. This is true whenever the half-reactions involve equal numbers of electrons. Moreover, when this is true, the

TABLE 3-5. POTENTIOMETRIC TITRATION OF Fe^{++} WITH $Ce(SO_4)_3^{=}$*

$V_{Ce(SO_4)_3^=}$, ml.	E_{ind}, v., *vs.* S.C.E.	$\Delta E/\Delta V$, mv./ml.	$\Delta^2E/\Delta V^2$, mv./ml.^{2}
0	Undefined		
0.1	$+0.31_6$		
		69	
1.0	$+0.37_8$		
		14	
5.0	$+0.43_4$		
		14	
9.0	$+0.49_0$		22.5
		69	
9.9	$+0.55_2$		1186
		656	
9.99	$+0.61_1$		27480
		2030	
10.00	$+0.81_4$		0
		2030	
10.01	$+1.01_7$		−27480
		656	
10.1	$+1.07_6$		− 1192
		66	
11	$+1.13_5$		− 119
		7	
20	$+1.19_4$		− 0.4
		1.4	
40	$+1.22_2$		

* See text for details.

titration curve is symmetrical around the equivalence point. (When n_1 is not equal to n_2 in the half-reactions

$$\text{Ox}_1 + n_1e = \text{Red}_1 \qquad E^{0\prime} = E_1^{0\prime}$$
$$\text{Ox}_2 + n_2e = \text{Red}_2 \qquad E^{0\prime} = E_2^{0\prime}$$

the potential at the equivalence point will be

$$E = \frac{n_1E_1^{0\prime} + n_2E_2^{0\prime}}{n_1 + n_2} \tag{3-19}$$

and the titration curve will be markedly asymmetrical around the equivalence point.)

Beyond the equivalence point we can neglect the ceric ion remaining in the solution by virtue of the incompleteness of the titration reaction. Taking the concentration of ceric ion, then, as $0.1(V - 10)/(10 + V)$ for values of V greater than 10, and the concentration of cerous ion as $1/(10 + V)$ (because the 1 millimole of ferrous ion originally present will produce 1 millimole of cerous ion), the potential at any point beyond the equivalence point is most conveniently given by

$$E = 1.44 + 0.059 \log (0.1V - 1)$$

Table 3-5 shows the values calculated from these equations.

Whether or not a potentiometric titration is practicable depends primarily on the speed of the reaction and the difference between the formal potentials of the two half-reactions. Often it is necessary to heat the solution or to add a catalyst to ensure that an equilibrium potential is attained reasonably soon after the addition of each portion of reagent, and when the reaction is very slow it is easy to secure misleading results if this is not done. Incidentally, it should be pointed out that nearly every reaction proceeds quite slowly in the vicinity of the equivalence point, so that in that region it is generally necessary to wait a minute or two between the addition of a portion of the reagent and the measurement of the potential.

Ordinarily the change of potential around the equivalence point will be large enough for the titration to yield accurate results only when the formal potentials of the two half-reactions involved differ by at least 0.3 v. Actually, this figure is considerably dependent on the values of n_1 and n_2; the larger these are, the sharper will be the change of potential around the equivalence point for a given difference of formal potentials. It is essential to use formal, and not standard, potentials in applying this criterion. For example, the standard potentials of the half-reactions

$$H_3AsO_4 + 2H^+ + 2e = HAsO_2 + 2H_2O \qquad E^0 = +0.559 \text{ v.}$$
$$I_3^- + 2e = 3I^- \qquad E^0 = +0.536 \text{ v.}$$

differ by only 20 mv., and from this one would conclude that the titration of +3 arsenic with iodine is not feasible. This is true in the presence of 1 M hydrogen ion, but the formal potential of the arsenate–arsenite half-reaction becomes 59 mv. less positive for each unit increase in the pH of the solution, so that the titration is quite feasible at any pH above about 5.

The formal potential of a half-reaction is most readily determined with the aid of data such as those in Table 3-6, which is self-explanatory.

TABLE 3-6. FORMAL POTENTIAL OF THE FERRIC–FERROUS COUPLE IN 2 F H_3PO_4*

V_{KMnO_4}, ml.	E_{ind}, v., *vs.* Ag|AgCl(s), KCl(s)	$0.05915 \log \frac{25.93 - V}{V}$	$E_{ind} + 0.05915 \log \frac{25.93 - V}{V}$ $= E^{0'}_{Fe^{+++},Fe^{++}}$, v., *vs.* Ag|AgCl($s$), KCl($s$)
0	+0.1860		
2.0	+0.2070	+0.0636	(+0.2706)
4.0	+0.2205	+0.0435	+0.2640
6.0	+0.2310	+0.0303	+0.2613
8.0	+0.2400	+0.0208	+0.2608
10.0	+0.2480	+0.0120	+0.2600
12.0	+0.2560	+0.0039	+0.2599
14.0	+0.2640	−0.0041	+0.2599
16.0	+0.2720	−0.0122	+0.2598
18.0	+0.2810	−0.0210	+0.2600
20.0	+0.2920	−0.0312	+0.2608
22.5	+0.3120	−0.0483	+0.2637
24.0	+0.3300	−0.0648	+0.2652
25.0	+0.3530	−0.0846	(+0.2684)

The end point was found at 25.93 ml. of $KMnO_4$. Average $E^{0'}_{Fe^{+++},Fe^{++}} = +0.2614 \pm 0.0016$ v. *vs.* Ag|AgCl(s), KCl(s); $E_{ref} = +0.201$ v. *vs.* N.H.E.; $E^{0'}_{Fe^{+++},Fe^{++}} = +0.462_4$ v. *vs.* N.H.E.

* 1.005 g. of $Fe(NH_4)_2(SO_4)_2 \cdot 12H_2O$ was dissolved in 500 ml. of 2 F H_3PO_4 and titrated with 0.02 F $KMnO_4$. A bright platinum indicator electrode was used, and the reference electrode was a silver–silver chloride–saturated potassium chloride electrode. The data were secured by A. Acuna and G. L. Jacobs at the Polytechnic Institute of Brooklyn.

3-15. Differential Potentiometric Titrations. Suppose that a dilute hydrochloric acid solution containing both stannous and ferrous ions is titrated with permanganate. Being the stronger of the two reducing agents present, the stannous ion will be oxidized first; then, as its concentration decreases and the potential becomes more and more positive, the oxidation of ferrous ion will begin. When this titration is carried out visually, only one end point can conveniently be detected. That is the one corresponding to the oxidation of both the stannous and ferrous ions, and the familiar pink color of excess permanganate appears only when the equivalence point of the iron oxidation has been passed.

Such a titration is evidently incapable of distinguishing or differentiating between the two stages of the oxidation; all that can be determined is the volume of permanganate equivalent to the total reducing power of the solution. By potentiometric methods, however, it is easily possible to locate both of the equivalence points. A titration in which this is done is called a "differential" titration[1] because it serves to differentiate between two reducing (or oxidizing) agents present in the same solution. Differential titrations are, of course, very closely related to the simpler titrations considered in the preceding section of this chapter, and in fact they present only one problem of any substantial difficulty from a theoretical point of view. This has to do with the potentials of the indicator electrode at the equivalence points of the titration.

If the titration described in the first sentence of this section were carried out potentiometrically, it would be convenient to consider the titration curve as being divided into three sections. In the first, stannous ion would be oxidized, and the potential of an inert electrode immersed in the solution could be calculated most conveniently from the expression

$$E = E^{0\prime}_{\mathrm{Sn(IV),Sn(II)}} - \frac{0.059}{2} \log \frac{[\mathrm{Sn(II)}]}{[\mathrm{Sn(IV)}]}$$

Following the equivalence point of this oxidation, ferrous ion would be oxidized in the second portion of the curve, and there it would be simplest to calculate the electrode potential from the equation

$$E = E^{0\prime}_{\mathrm{Fe^{+++},Fe^{++}}} - 0.059 \log \frac{[\mathrm{Fe^{++}}]}{[\mathrm{Fe^{+++}}]}$$

After the second equivalence point had been passed, the electrode potential could be calculated from the volume of the solution, the amount of manganous ion formed during the two oxidation steps, and the known excess of permanganate present, using the equation

$$E = E^{0\prime}_{\mathrm{MnO_4^-,Mn^{++}}} - \frac{0.059}{5} \log \frac{[\mathrm{Mn^{++}}]}{[\mathrm{MnO_4^-}][\mathrm{H^+}]^8}$$

Once the first equivalence point has been passed, the titration curve becomes essentially identical with that for an ordinary titration of ferrous ion with permanganate. Nor is there any difficulty associated with the portion of the titration during which stannous ion is being oxidized. However, the potential at the first equivalence point is obviously not equal to the potential at the equivalence point of the direct titration of stannous ion with permanganate. Given the formal potentials in 1 F

[1] The term "differential potentiometric titration" has also been used in the literature to denote a titration in which one measures the difference in indicator electrode potential produced by the addition of a fixed aliquot of reagent, and locates the equivalence point from the variation of this difference with the total volume of reagent which has been added. This amounts to securing the $\Delta E/\Delta V$ *vs.* V curve directly; since $\Delta E/\Delta V$ is the derivative of the titration curve, such a procedure is probably better described as a "derivative potentiometric titration."

hydrochloric acid

$$SnCl_6^{=} + 2e = Sn^{++} + 6Cl^- \qquad E^{0\prime} = +0.14 \text{ v.}$$
$$Fe^{+++} + e = Fe^{++} \qquad E^{0\prime} = +0.70 \text{ v.}$$
$$MnO_4^- + 8H^+ + 5e = Mn^{++} + 4H_2O \qquad E^{0\prime} = +1.45 \text{ v.}$$

we can calculate with the aid of Eq. (3-19) that the potential at the equivalence point of the latter titration would be +1.07 v. At this potential all but about 0.0001 per cent of the iron would be in the ferric state. This, however, is impossible, for at the first equivalence point of the titration of the stannous–ferrous mixture only enough permanganate has been added to oxidize the stannous ion without reacting with any of the ferrous ion.

Assume that the original solution contained 0.01 M stannous ion and 0.1 M ferrous ion. At the first equivalence point, then, it will contain (neglecting the effect of dilution by the added permanganate) 0.01 F stannic tin and 0.1 M ferrous ion. In such a solution, the reaction

$$SnCl_6^{=} + 2Fe^{++} = Sn^{++} + 2Fe^{+++} + 6Cl^-$$

would produce stannous and ferric ions:

$$[Fe^{+++}] = 2[Sn^{++}]$$

since two ferric ions are formed for every stannous ion. By the method outlined in Sec. 3-3, we can easily calculate the equilibrium constant of this reaction from the formal potentials of the two corresponding half-reactions. This gives (dropping $[Cl^-]$ because it is 1 M)

$$\frac{[Fe^{+++}]^2[Sn^{++}]}{[Fe^{++}]^2[SnCl_6^{=}]} = 1.2 \times 10^{-19}$$

Now, knowing $[Fe^{++}]$, $[SnCl_6^{=}]$, and the relationship between $[Fe^{+++}]$ and $[Sn^{++}]$, we can calculate the concentration of either of the latter ions present at the equivalence point. For example, we find that the concentration of ferric ion is

$$[Fe^{+++}] = 2.9 \times 10^{-8}$$

and from this and the known concentration of ferrous ion we calculate

$$E = 0.70 - 0.059 \log \frac{0.1}{2.9 \times 10^{-8}} = +0.31 \text{ v.}$$

Though this is not very far from the potential at the equivalence point of an ordinary titration of stannous ion with ferric ion, which by Eq. (3-19) would be +0.33 v., the two are not quite identical, contrary to what is stated in certain other discussions of potentiometric titrations which the reader may consult. The difference between the two situations is this: in the direct titration just mentioned, the concentrations of the stannic and ferrous ions produced by the titration reaction and present at the equivalence point would necessarily be related by the equation

$$2[Sn(IV)] = [Fe^{++}]$$

In the differential titration, on the other hand, the ratio between these concentrations is determined, not by the stoichiometry of the stannous–ferric reaction, but by the stannous/ferrous ratio in the original solution.

In exactly the same way it can be shown that the potential of the indicator electrode at the equivalence point of the titration of the ferrous ion is not quite identical with that at the equivalence point of an ordinary titration of ferrous ion with permanganate in the absence of stannous ion. The detailed treatment is left to the student.

The use of a differential titration for the determination of two reducing agents present in the same solution needs no further comment. However, it is worth pointing out that the use of a differential titration for the determination of a single constituent of a solution may often simplify the preparation of the solution enormously. The student may have encountered in his elementary course in quantitative analysis a method for the determination of iron in an ore which involves the following scheme: the ore is dissolved in hydrochloric acid and the resulting solution is treated with a small but definite excess of stannous chloride. The mixture is then cooled and a large excess of mercuric chloride is added to oxidize the excess stannous ion; then the solution is titrated with dichromate, ceric sulfate, or (after adding Zimmermann-Reinhardt solution) potassium permanganate. Unless a small precipitate of mercurous chloride is formed when the mercuric chloride is added, accurate results are impossible to secure, and even under optimum conditions the mercurous chloride is attacked to some extent by the dichromate or other oxidizing agent.

This determination could be made much more easily and rapidly by means of a differential titration. Starting with the stannous–ferrous mixture resulting from the reduction step, one could simply carry out a potentiometric titration with one of the reagents mentioned above. Two end points would be found: the first would correspond to the oxidation of the excess stannous ion, and the second to the oxidation of the ferrous ion. The volume of reagent equivalent to the iron content of the sample would then be simply the difference between the volumes of reagent consumed at the second and at the first end points. Not only would this procedure eliminate the small error due to the reaction of the oxidizing agent with the suspended mercurous chloride (by eliminating the mercurous chloride), but it would also make it unnecessary to ensure that only a very small excess of stannous chloride was added in the reduction step. Moreover, although in the volumetric procedure the solution must be cooled between the reduction step and the addition of the mercuric chloride, the potentiometric titration could be carried out in the hot solution.

3-16. Titrations of Solutions Containing an Excess of the Reduced Form of the Reagent. In Chap. 7 we shall discuss the potentiometric location of the equivalence point of a "coulometric titration." Since the theory involved in such a measurement is very closely allied to what has been set forth in the preceding section, it seems advisable to discuss it briefly here.

In effect, such a "titration" of ferrous ion by ceric ion would involve the addition of known amounts of ceric ion to a solution which initially contained a very low concentration of ferrous ion and a much higher one of cerous ion. The presence of the large excess of cerous ion considerably modifies the titration curve calculated in Sec. 3-14, as will be seen from the following discussion.

Let us suppose that a solution containing 10^{-6} *M* ferrous ion and 0.1 *M* cerous ion in 1 *F* sulfuric acid is titrated with ceric ion under such conditions that the volume of the solution does not change during the titration; these concentrations and conditions are realistic ones for the practical performance of coulometric titrations, as the student will see from Sec. 7-11.

At the equivalence point of such a titration, the concentrations of ferrous and ceric ions will be equal, in accordance with the considerations outlined in Sec. 3-14. Let us say that

$$[Fe^{++}] = [Ce(SO_4)_3^{=}] = c$$

From the conservation equation, then,

$$[Fe^{+++}] = 10^{-6} - c$$

The initial concentration of cerous ion is so large that the small additional amount formed during the titration can be neglected; that is, the concentration of cerous ion remains essentially constant at 0.1 *M*. So, with the aid of the equilibrium constant for this reaction calculated in Sec. 3-14, we have

$$\frac{(10^{-6} - c)(0.1)}{c^2} = 7 \times 10^{12}$$

The student should note the similarity between this situation and the one discussed in Sec. 3-15. Solving this equation gives

$$c = 1.2 \times 10^{-10}$$

whence, by the Nernst equation, since $10^{-6} - c$ is virtually equal to 10^{-6},

$$E = 0.68 - 0.059 \log \frac{1.2 \times 10^{-10}}{10^{-6}} = 0.91 \text{ v. } vs. \text{ N.H.E.}$$
$$= 0.66_4 \text{ v. } vs. \text{ S.C.E.}$$

which is 150 mv. less positive than the value which would have corresponded to the equivalence point if no cerous ion had been present in the original solution. To arrive at the latter value, we should have to add excess ceric ion until

$$E_{\text{Ce}} = 1.06 = 1.44 - 0.059 \log \frac{0.1}{[\text{Ce}(\text{SO}_4)_3^{=}]}$$

The concentration of excess ceric ion would then be

$$[\text{Ce}(\text{SO}_4)_3^{=}] = \frac{0.1}{\text{antilog } (0.38/0.059)} = 3.6 \times 10^{-8} \, M$$

This corresponds to a 3.6 per cent excess of the reagent. It is obvious that the end point potential in such a titration would have to be adjusted to correspond to the ratio of the ferrous and cerous concentrations in the initial solution.

In an ordinary titration of ferrous iron with ceric sulfate, the observed potential when a 100 per cent excess of the reagent had been added would be equal to the formal potential, 1.44 v. *vs.* N.H.E., of the ceric–cerous couple. In the titration just described, however, the potential at that point would be only

$$E_{\text{Ce}} = 1.44 - 0.059 \log \frac{0.1}{10^{-6}} = 1.14 \text{ v.}$$

because of the large excess of cerous ion present. This corresponds to a potential of $1.14 - 0.246 = 0.89_4$ v. *vs.* S.C.E., which in the normal titration (Table 3-5) is attained after the addition of much less than 0.1 per cent excess ceric sulfate. The presence of the excess cerous ion therefore considerably lowers the potentials attained beyond the equivalence point, and so tends to decrease both the accuracy and the precision with which the end point can be located. In the practical execution of a coulometric titration it is necessary to have an excess of cerous ion in the initial solution. However, it is clear that this is disadvantageous rather than otherwise from the point of view of the potentiometry involved, and that—especially when the formal potentials of the two half-reactions are more nearly equal than in our example—considerable judgment may need to be exercised in providing an excess of the reagent precursor (in this case, cerous ion) large enough to allow 100 per cent current efficiency to be attained, but yet not so large that the end point cannot be detected with the desired accuracy.

3-17. Autotitrators. A number of instruments which perform potentiometric titrations automatically have been designed in the last few years. These are so useful in many fields of research, in addition to their time-saving function in routine analysis, as to deserve a brief mention here.

The instruments thus far described are based on three fairly different principles. In the earliest models, the reagent was added to the solution at a known constant rate by a motor-driven syringe, and the potential was recorded on the chart of a recording potentiometer. This operates in such a way that at any instant the horizontal deflection of a pen is proportional to the cell potential; meanwhile a strip of paper is drawn vertically past the pen at a constant rate, so that the instrument automatically plots potential against time (*i.e.*, against the volume of reagent). This arrangement would work satisfactorily only if the titration reaction were practically instantaneous and if the indicator electrode responded immediately to changes in the composition of the solution. As these conditions are not usually fulfilled (especially near the equivalence point), instruments of this kind have been superseded by newer types.

One of these employs a circuit which automatically stops the addition of reagent (by closing a solenoid in a buret valve, disconnecting the power input of a motor driving a syringe, or some other method) when the cell potential reaches some predetermined value, normally the potential expected at the equivalence point. The solution is meanwhile stirred by a magnetic stirrer, and the tip of the buret or syringe and the indicator electrode are positioned in such a way that the incoming reagent is stirred past the indicator electrode. This causes the solution around the electrode always to be at a slightly more advanced stage of the titration than the solution as a whole. Consequently the addition of reagent is stopped before the true equivalence point is reached. As the stirring continues, the solution becomes uniform and the electrode potential changes accordingly. In most instruments, this causes more reagent to be added, and so on until the equivalence point potential is finally attained. Instruments of this kind are capable of an accuracy and precision quite as good as can be secured by a trained technician, and of course the time which the operator must expend on each titration is very much less.

In the newest kind of autotitrator, the reagent is again added at a known constant rate, but the cell potential is fed into an electronic circuit which essentially differentiates it twice. When the output of this circuit (which is proportional to $\Delta^2E/\Delta V^2$) becomes zero after passing through a large value, the addition of reagent is automatically stopped. Like the direct-recording instruments, this will give erroneous results unless the titration reaction is very rapid. However, it has the advantage that the equivalence point potential need not be known.

Apart from their obvious uses in routine analyses, autotitrators can be used to maintain the pH, redox potential, chloride concentration, etc., of a solution constant while a chemical reaction is taking place. This might be of great utility in, for example, the study of corrosion phenomena occurring in very poorly buffered or poised media such as natural waters. With the aid of a device for plotting the volume of reagent used against time, autotitrators can also be used for following the rate of a reaction.

PROBLEMS

In solving the following problems, assume that the temperature is 25°C. and (unless otherwise directed) that all activity coefficients are equal to unity. The answers to the starred problems are given on page 402.

For each of the reactions given in Problems 1 to 10:

a. Separate the reaction into its component half-reactions.
b. Write the schematic representation of a cell in which the reaction will occur in the direction written.
c. Compute the standard potential of the cell, using the data given in Table A of the Appendix.
d. Assign the appropriate sign to each of the two electrodes, assuming that all substances are present at unit activity.
e. Calculate the equilibrium constant of the reaction.

***1.** $Ag + Fe^{+++} = Ag^{+} + Fe^{++}$
2. $2MnO_4^{-} + 16H^{+} + 10Cl^{-} = 2Mn^{++} + 8H_2O + 5Cl_2$
3. $2Cu^{++} + 5I^{-} = Cu_2I_2 + I_3^{-}$
4. $VO_2^{+} + V^{+++} = 2VO^{++}$
5. $AgBr = Ag^{+} + Br^{-}$
***6.** $AgCl + I^{-} = AgI + Cl^{-}$
7. $NiO_2 + Cd + 2H_2O = Ni(OH)_2 + Cd(OH)_2$
8. $2UO_2^{+} + 4H^{+} = UO_2^{++} + U^{++++} + 2H_2O$
9. $Fe + 2Fe^{+++} = 3Fe^{++}$
10. $2MnO_4^{-} + 3Mn^{++} + 4OH^{-} = 5MnO_2 + 2H_2O$

For each of the cells given in Problems 11 to 18:

a. Write the equations for the half-reactions occurring at the two electrodes and for the cell reaction.
b. Compute the potential of the cell.

***11.** Pt, H_2(0.5 atm.)|HCl(0.1 *F*)|Cl_2(0.5 atm.), C
12. Pt, H_2(0.5 atm.)|NaOH(2.5 *F*)|O_2(2 atm.), Pt
13. Pb|$PbSO_4$(*s*), K_2SO_4(0.1 *F*)||K_2SO_4(0.2 *F*), $PbSO_4$(*s*)|Pb
14. Ag|AgCl(*s*), KCl(0.01 *F*)||KI(0.01 *F*), AgI(*s*)|Ag
15. Pt|Fe^{++}(0.01 *M*), Fe^{+++}(0.1 *M*), $HClO_4$(0.1 *F*)||$HClO_4$(0.1 *F*), VO_2^{+}(0.005 *M*), VO^{++}(0.01 *M*)|Pt
***16.** Ag|Ag_2SO_4(*s*), K_2SO_4(*s*), $PbSO_4$(*s*)|Pb
17. Pt|$SO_3^{=}$(0.1 *M*), $SO_4^{=}$(1 *M*), $S_2O_8^{=}$(0.001 *M*)|Pt

18. $Pt|H_3AsO_4(0.001\ F),\ HAsO_2(0.01\ F),\ HClO_4(1\ F)||HClO_4(1\ F),\ I_2(10^{-5}\ M),\ IO_3^-(0.1\ M)|Pt$

For each of the following sets of half-reactions, compute the standard potential of the third from the data given for the first and second.

*__19.__ $Cu^{++} + 2e = Cu \quad E^0 = +0.337$ v.
$Cu^+ + e = Cu \quad E^0 = +0.521$ v.
$Cu^{++} + e = Cu^+ \quad E^0 = ?$

20. $Cr^{+++} + 3e = Cr \quad E^0 = -0.74$ v.
$Cr^{+++} + e = Cr^{++} \quad E^0 = -0.43$ v.
$Cr^{++} + 2e = Cr \quad E^0 = ?$

21. $ClO_2^- + H_2O + 2e = ClO^- + 2OH^- \quad E^0 = +0.66$ v.
$ClO^- + H_2O + 2e = Cl^- + 2OH^- \quad E^0 = +0.89$ v.
$ClO_2^- + 2H_2O + 4e = Cl^- + 4OH^- \quad E^0 = ?$

22. $MnO_4^- + e = MnO_4^= \quad E^0 = +0.564$ v.
$MnO_4^= + 2H_2O + 2e = MnO_2 + 4OH^- \quad E^0 = +0.60$ v.
$MnO_4^- + 2H_2O + 3e = MnO_2 + 4OH^- \quad E^0 = ?$

23. The solubility product of silver azidodithiocarbonate, $AgSCSN_3$, is 1.0×10^{-16}. Calculate the potential of the cell $Ag|AgSCSN_3(s),\ KSCSN_3(0.01\ F),\ KOH(0.1\ F)|H_2$-(700 mm.), Pt.

*__24.__ The standard potential of the half-reaction $Pb^{++} + 2e = Pb$ is known to be -0.126 v. When a solution initially containing 0.001 F perchloric acid and 0.0001 F stannous perchlorate is shaken with excess metallic lead, the solution is found to contain $3.15 \times 10^{-5}\ M$ lead ion at equilibrium. What is the standard potential of the half-reaction $Sn^{++} + 2e = Sn$? (Assume that metallic lead reacts with only stannous ion under these conditions.)

25. The solubility product of lead oxalate is 8.3×10^{-12}. What is the standard potential of the half-reaction $PbC_2O_4 + 2e = Pb + C_2O_4^=$?

*__26.__ A chemist wishes to determine copper in the presence of arsenic by an iodometric titration. If his sample contains 10 mole per cent of copper and 1 mole per cent of arsenic, if the solution at the end point of the titration will contain 0.1 M iodide and $1 \times 10^{-6}\ M$ triiodide, and if the relative error in the copper determination must not exceed 0.1 per cent, what is the minimum pH at which the titration can be carried out? The relevant standard potentials are

$$I_3^- + 2e = 3I^- \qquad E^0 = +0.536 \text{ v.}$$
$$H_3AsO_4 + 2H^+ + 2e = HAsO_2 + 2H_2O \qquad E^0 = +0.559 \text{ v.}$$

27. With the aid of the following formal potentials in 1 F hydrochloric acid:

$$SnCl_6^= + 2e = Sn^{++} + 6Cl^- \qquad E^{0\prime} = +0.14 \text{ v.}$$
$$Fe^{+++} + e = Fe^{++} \qquad E^{0\prime} = +0.70 \text{ v.}$$

calculate the potential at the equivalence point of the titration of 0.01 M Fe^{+++} with 0.01 M Sn^{++} in 1 F HCl. What fraction of the ferric ion will remain unreduced at the equivalence point?

*__28.__ The reaction which occurs when a 1 F hydrochloric acid solution containing vanadyl ion is passed through a silver reductor is described by the equations

$$VO^{++} + 2H^+ + e = V^{+++} + H_2O \qquad E^{0\prime} = +0.34 \text{ v.}$$
$$AgCl + e = Ag + Cl^- \qquad E^{0\prime} = +0.22 \text{ v.}$$

What fraction of the vanadyl ion initially present in a 0.01 M solution of vanadyl ion in 1 F hydrochloric acid will remain unreduced when the solution is shaken with excess metallic silver until equilibrium is reached?

29. Exactly 50 ml. of a 0.00100 *F* solution of potassium ferrocyanide in 0.01 *F* hydrochloric acid was titrated with 0.0100 *F* ceric sulfate, and the course of the titration was followed potentiometrically, using a platinum wire indicator electrode and a saturated calomel reference electrode. When 1.50 ml. of the ceric solution had been added, the potential of the platinum electrode was +0.200 v. *vs.* S.C.E. What is the formal potential of the ferricyanide–ferrocyanide couple in 0.01 *F* hydrochloric acid?

***30.** A strongly alkaline solution of sodium perrhenate, $NaReO_4$, was titrated with a solution containing an unknown concentration of a weak reducing agent whose identity is immaterial. The titration was followed potentiometrically, using a platinum indicator electrode and a saturated calomel reference electrode. After 0.50 ml. of the reducing agent had been added, the potential of the indicator electrode was +0.254 v. *vs.* S.C.E.; when 6.00 ml. of the reducing agent had been added, this potential was +0.165 v. *vs.* S.C.E. No precipitate was observed during the titration. The end point occurred at 9.40 ml. of the reducing agent. To what oxidation state was the rhenium reduced, and what was the formal potential of this half-reaction in the medium used for the titration?

31. The potential of the cell

$$\text{Pt, } H_2(1 \text{ atm.})|HCl(c), \ AgCl(s)|Ag$$

is 0.5727 v. when c is 0.001129 *F* and 0.6353 v. when c is 3.288×10^{-4} *F*. Find the standard potential of the cell, and calculate the mean ionic activity coefficient of 0.001129 *F* hydrochloric acid.

32. Silver chloride is appreciably soluble in concentrated chloride solutions because of the formation of complex ions such as $AgCl_2^-$, $AgCl_3^=$, etc. What difference between the titration curves of 0.1 *M* chloride with 0.1 *M* silver and of 5 *M* chloride with 5 *M* silver would result from this fact? How would these curves differ if no chloro complexes were formed?

33. Why is it possible to secure more precise results in the potentiometric titrations of halides when silver nitrate is used as the reagent than when mercurous nitrate is used?

***34.** A solution containing 0.10 *M* chromic ion and 0.001 *M* vanadic ion, V^{+++}, in 0.1 *F* hydrochloric acid is to be analyzed for its vanadium content by a potentiometric titration with standard chromous chloride.

a. What will be the potential (*vs.* N.H.E.) of an inert metal indicator electrode at the equivalence point?

b. What percentage error would result from setting an autotitrator to stop adding reagent at the point where this potential was −0.33 v. (the mean of the two standard potentials)?

CHAPTER 4

THE MEASUREMENT OF pH

4-1. Introduction. Hydrogen ions[1] are involved in many different kinds of reactions, both inorganic and organic, and both the equilibrium positions of these reactions and the rates at which they proceed are therefore pH-dependent. Consequently the measurement of pH is perhaps the most frequently performed of all physicochemical measurements, and this is the justification for discussing it separately in the present chapter.

For many years the "degree of acidity" of a solution (as distinguished from its "total acidity" as found by an alkalimetric titration) could be measured only by the use of indicators. Originally only very rough distinctions, such as between "acidic toward litmus" and "alkaline toward litmus," could be drawn in this way. As time went on, however, more indicators were discovered and arranged in a series, and so, with the aid

[1] With the exception of one section in this chapter, we have chosen to use the term "hydrogen ion" throughout this book in preference to "hydronium ion" or "oxonium ion." This usage is not intended to suggest that the unhydrated proton exists as such in aqueous solutions, for that is unquestionably not the case. However, it is equally certainly true that each proton is bound to, not one, but several molecules of water; the best available evidence indicates that the average hydration number of the proton is approximately 5, so that we should actually write $H_{11}O_5^+$ instead of H_3O^+. Then, to be consistent, we would have to write $Li(H_2O)_5^+$ instead of Li^+, $Na(H_2O)_4^+$ instead of Na^+, and so on. Otherwise we would be implying that there is some important qualitative difference between the hydration energies of hydrogen ion on the one hand and all other cations and anions on the other, and we know of no grounds on which any such implication can be based.

Moreover, writing H_3O^+, $H_{11}O_5^+$, $H(H_2O)_n^+$, or any other similar formula for the hydrated proton seems to us to lay special emphasis on its hydration as opposed to its solvation in other media. What the hydrated proton (H_3O^+ for simplicity) is to aqueous solutions, the ammoniated proton (NH_4^+ for simplicity) is to solutions in liquid ammonia, and the solvated proton ($C_2H_5OH_2^+$ for simplicity) is to solutions in absolute ethanol.

On the basis of these considerations, we have decided to represent the proton simply as H^+ (except in Sec. 4-6 to emphasize the fact that the pH-responsive characteristics of a glass electrode depend on the presence of both protons and water molecules). The student must realize that this representation carries with it no more implication concerning the actual solvation of the proton than do the equally oversimplified formulas Li^+ and Na^+.

of several indicators, it became possible to differentiate more and more finely between the "degrees of acidity" of various solutions. Far from being outmoded, the use of indicators is today very important in making quick pH measurements, and with the aid of any of several simple techniques it is an easy matter to measure the pH of a solution to ± 0.1 pH unit.

Later on it was found that the pH of a solution could be measured potentiometrically. Hildebrand introduced the use of the "hydrogen electrode" for this purpose in 1912, and for about 20 years thereafter much effort was expended on the development of indicator electrodes which would respond to changes in the hydrogen ion concentration. Of these, the "quinhydrone electrode" is still used fairly widely, and the antimony electrode finds considerable use in industrial process control.

However, most pH measurements today are made with a "glass electrode pH meter." This ubiquitous instrument was made possible by the discovery, in 1906, that the potential developed across a glass membrane surrounded by a solution is related to the acidity of that solution. This potential is extremely difficult to measure by ordinary potentiometric methods, for the resistance of a glass membrane thick enough to be reasonably sturdy may amount to tens or even hundreds of megohms. One pH unit corresponds to 0.059 v. (at 25°C.). If one wanted to make measurements to ± 0.01 pH unit with a 60-megohm glass electrode, the galvanometer would have to be sensitive enough to show a measurable deflection with a current of only 1×10^{-11} ampere flowing through its coil. (This tiny current will produce a deflection of only 1 mm. on the most sensitive galvanometer commercially available today.)

Consequently the use of the glass electrode in pH measurements really became practicable only when the science of electronics had progressed to the point at which it was possible to design a stable and reliable high-gain d-c amplifier. The extent to which these instruments are used nowadays in the practical measurement of pH is a compelling tribute to the place of electronics in the chemical laboratory.

4-2. The Colorimetric Measurement of pH. Let us assume that one has an acid whose dissociation constant is accurately known, and an indicator for which K_a (or K_w/K_b) is fairly close to that of the acid. If we prepare a series of solutions containing known ratios of the acid and its salt, the hydrogen ion concentration of each can be calculated from the equation

$$[H^+] = \frac{[HX]}{[X^-]} K_a \tag{4-1}$$

Now if a small amount of the indicator is added to each solution, the ratio of the colored forms of the indicator will in turn be determined by

(assuming that the indicator is an acid)

$$\frac{[\text{color}]_1}{[\text{color}]_2} = \frac{[\text{Ind}^-]}{[\text{HInd}]} = \frac{K_{\text{HInd}}}{[\text{H}^+]} \tag{4-2}$$

so that

$$\frac{[\text{color}]_1}{[\text{color}]_2} = \frac{K_{\text{HInd}}}{K_a} \frac{[\text{X}^-]}{[\text{HX}]} \tag{4-3}$$

The estimation of the pH of an unknown buffer solution is then experimentally very simple. It consists merely of adding the same total concentration of indicator to the unknown solution, and matching the color of the resulting solution with that of one of the known buffers. If the color of the indicator in the unknown buffer is intermediate between the colors of two of the known solutions, the pH of the unknown is estimated by a sort of visual interpolation.

Although this is apparently a perfectly straightforward procedure, it actually contains several pitfalls. One of these involves the fact that the color of a solution of an indicator frequently depends upon the ionic strength of the solution as well as its pH. This is so because, at a constant activity of hydrogen ion, we actually have

$$\frac{[\text{Ind}^-]f_{\text{Ind}^-}}{[\text{HInd}]f_{\text{HInd}}} = \frac{K_{\text{HInd}}}{a_{\text{H}^+}}$$

instead of the simpler Eq. (4-2). What is seen by the eye (or by a colorimeter, *cf.* Chap. 8) is the ratio of *concentrations* of the two colored forms of the indicator. Or, by the use of a spectrophotometer, we can measure the *concentration* of one of the two colored forms by observing the absorbance of the solution at a wavelength at which one absorbs light but the other does not; then, if we know the total *concentration* of the indicator in the solution ($= [\text{Ind}^-] + [\text{HInd}]$), we can calculate the *concentration* of the other colored form (or we could measure it directly at another wavelength). All of these are substantially equivalent, however, because all give *concentrations* rather than *activities*.

In the range of ionic strengths in which most measurements are made (0.01 to 1 M), the activity coefficient of the indicator anion varies considerably with ionic strength, whereas that of the uncharged indicator acid is usually much closer to 1 and changes very little with ionic strength. The result is that the ratio of concentrations of the anion and the uncharged molecule varies with ionic strength even though the pH is kept constant. However, when Eq. (4-2) is used to interpret the color of the solution, only the effect of pH on the ratio of the two colored forms is taken into account, and so an error will result from neglecting the variation of the activity coefficients. This phenomenon is known as the "salt error" of the indicator, and its magnitude in some typical cases is illustrated by the data in Table 4-1.

About the only rational way to deal with the salt error would involve the use of buffered comparison solutions whose ionic strengths closely approximated the ionic strength of the unknown solution. Because this is not very practical—and also because one would need to know the formal dissociation constant of the acid in the reference buffer at that ionic strength (and such data are unfortunately not very numerous)—the practical accuracy of the determination of pH with indicators is considerably poorer than the attainable precision.

TABLE 4-1. FORMAL DISSOCIATION CONSTANTS (AT 20°C.) OF VARIOUS INDICATORS*

Indicator	pK_{HInd} at ionic strength of				
	0	0.01	0.05	0.1	0.5
Thymol blue (acid range)	1.65	....	1.65	1.65	1.65
Methyl orange	3.46	3.46	3.46	3.46	3.46
Bromphenol blue	4.10	4.06	4.00	3.85	3.75
Bromcresol purple	6.40	6.28	6.21	6.12	5.9
Bromthymol blue	7.30	7.19	7.13	7.10	6.9
Thymol blue (alkaline range)	9.20	9.01	8.95	8.90	

* Taken from I. M. Kolthoff and H. A. Laitinen, "pH and Electrotitrations," John Wiley & Sons, Inc., New York, 1941, p. 41.

A second complication arises with very poorly buffered solutions. Unless the solution being tested and the indicator solution have exactly equal pH values, it is evident that the color of a mixture of the two will not correspond to the pH of the solution being examined. Suppose that the indicator solution has a somewhat lower pH than the unknown. Then the color of a mixture of these solutions will change more and more toward the acidic color of the indicator as more of the indicator is added. In this case a little base is added to the indicator solution and the experiment is repeated with a fresh portion of the unknown. The true pH of the unknown solution is eventually found from the color of a mixture which does not change when more indicator is added. The solution of the indicator then has the same pH as the unknown, and the two solutions are said to be "isohydric."

In attempts to use indicators in measuring the pH of biological fluids, a third source of error has been identified. This is the "protein error," which is caused by the presence of protein molecules or other large organic molecules which form large aggregates called "micelles." Bromthymol blue assumes an intermediate color in a neutral phosphate buffer. If a small amount of a surface-active material such as sodium dodecylsulfonate ($C_{12}H_{25}SO_3Na$) is added to such a mixture, the dodecylsulfonate ions aggregate into micelles: the sulfonate groups protrude into the solution, and the hydrocarbon moieties cluster together in the interiors of the micelles. This gives the interior of the micelle the properties of an organic solvent having much more affinity for the neutral indicator molecule than for its charged ion. Hence the uncharged molecules of indicator are selectively extracted into the interior of the micelle. This displaces the equilibrium in the aqueous "phase," and the color changes sharply.

Of these three sources of error, the one which results from insufficient buffering of the unknown solution is probably the most important. It is this which constitutes the chief limitation on the use of the "indicator papers" or "pH papers" which are so popular for rapid and approximate pH measurements. The pH of distilled water found by using a pH paper may differ by a unit or more from that found by more refined techniques.

Of course it need hardly be mentioned that the colorimetric determination of pH is possible only if the solution is nearly or completely colorless.

4-3. The Hydrogen Electrode. The potential of the cell

$$\mathrm{Pt, H_2}(g)|\mathrm{H^+}||\mathrm{KCl}(s), \mathrm{Hg_2Cl_2}(s)|\mathrm{Hg}$$

is given (at 25°C.) by the equation

$$E_{\text{cell}} = E_{ref} + E_j - E^0_{\mathrm{H^+,H_2}} + \frac{0.05915}{2} \log \frac{p_{\mathrm{H_2}}}{a^2_{\mathrm{H^+}}} \tag{4-4}$$

where $a_{\mathrm{H^+}}$ is the activity of hydrogen ion and E_{ref} is the potential of the saturated calomel reference electrode (+0.246 v. *vs.* N.H.E.). By definition, the standard potential of the hydrogen electrode is zero at all temperatures, and so if we define the pH by the equation[1]

$$\mathrm{pH} = -\log a_{\mathrm{H^+}} \tag{4-5}$$

Eq. (4-4) reduces to

$$E_{\text{cell}} = 0.05915\ \mathrm{pH} + (E_{ref} + E_j) + 0.05915 \log \sqrt{p_{\mathrm{H_2}}}$$

In practice, the two terms within parentheses are combined into a single experimentally determined constant, ϵ, and we write

$$\mathrm{pH} = \frac{E_{\text{cell}} - \epsilon}{0.05915} - \log \sqrt{p_{\mathrm{H_2}}} \tag{4-6}$$

[1] This is a grossly oversimplified definition, largely because of the difficulty of assigning a concrete meaning to the activity coefficient of a single ion. Though we cannot go more deeply into the definition of the pH scale here, the student will profit from a study of the thorough and lucid exposition of the subject in R. G. Bates, "Electrometric pH Determinations," John Wiley & Sons, Inc., New York, 1954, pp. 16–33.

TABLE 4-2. pH VALUES OF STANDARD BUFFER SOLUTIONS

Composition of solution	pH at			
	20°C.	25°C.	30°C.	50°C.
0.1 *F* HCl	....	1.09	....	....
0.05 *F* $KHC_2O_4 \cdot H_2C_2O_4 \cdot 2H_2O$	1.68	....	1.69	1.71
$KHC_4H_4O_6$, saturated at 25°C	....	3.56	3.55	3.55
0.05 *F* $KHC_8H_4O_4$	4.00	4.01	4.01	4.06
0.025 *F* KH_2PO_4, 0.025 *F* Na_2HPO_4	6.88	6.86	6.85	6.83
0.05 *F* $Na_2B_4O_7 \cdot 10H_2O$	9.22	9.18	9.14	9.01

The value of ϵ is determined by measuring the potential of the cell with a buffer solution of defined pH, preferably one of the ones in Table 4-2. The pH values of these reference buffers have been established by painstaking work at the National Bureau of Standards in Washington, and it is to them that the practical scale of pH values is referred.

In measuring ϵ, it is advisable to use the standard buffer whose pH most closely approximates that of the unknown solution. This is because of the relatively high contributions of hydrogen and hydroxyl ions to the liquid-junction potential (Table 3-1, p. 24). A check with a second buffer takes very little additional time and ensures that the apparatus is functioning properly; if measurements with two solutions of known pH do not give the same value of ϵ, something is obviously wrong.

It should be noted that this procedure not only minimizes the uncertainty in E_j but also eliminates the effect of any deviation of the potential of the reference electrode from the expected value. Once ϵ has been determined, the determination of pH involves only a simple potentiometric measurement.

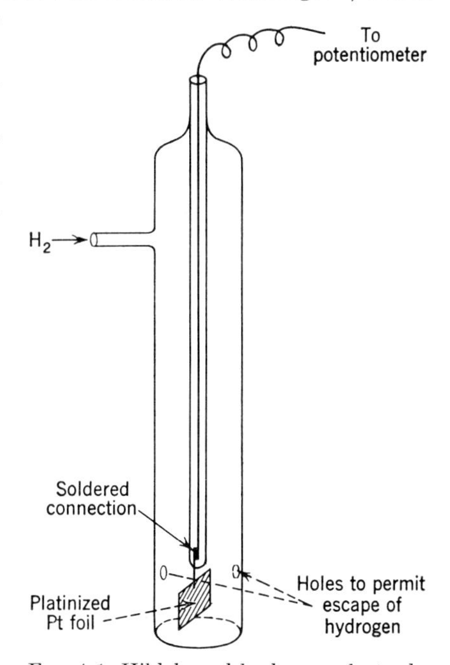

FIG. 4-1. Hildebrand hydrogen electrode.

Figure 4-1 shows the construction of the hydrogen electrode devised by Hildebrand. It consists of a square of thin platinum foil in an inverted glass bell. Provisions are made for connecting the platinum foil to a wire leading to the potentiometer, and also for bubbling hydrogen over the surface of the platinum. This ensures that the solution at the surface of the electrode is saturated with hydrogen at a known pressure, which is equal to the atmospheric pressure at the time of the experiment minus the vapor pressure of water at the temperature of the solution. We may note in passing that, if the measurements are being made at 25°C. and at sea level, an error of 10 mm. in the partial pressure of hydrogen would change the measured pH by only about $\log \sqrt{725/735} = 0.003$ unit. To produce this error in the partial pressure of hydrogen would require an error of 10 mm. in the barometric pressure or of 7°C. in the temperature of the solution (the effect of the latter is due to the variation with temperature of the vapor pressure of water). Therefore it is unnecessary in ordinary work to measure either of these quantities with great accuracy.

A bright platinum wire or foil does not behave as a reversible hydrogen electrode, which is to say that its potential does not depend on the pressure of hydrogen gas and on the activity of hydrogen ion in the manner predicted by the Nernst equation. This is the reason for using a "platinized" platinum electrode, which, because of its greater surface area and increased ability to adsorb hydrogen, is able to catalyze the half-reaction and provide a reproducible and meaningful potential.

A platinum electrode is platinized, *i.e.*, covered with a thin layer of very finely divided metallic platinum, by the following procedure. A platinizing solution is prepared by dissolving 1 g. of platinum in aqua regia, evaporating to near dryness, adding about 25 ml. of concentrated hydrochloric acid and repeating the evaporation (to remove nitrate), and finally diluting to 100 ml. with water. A trace (0.1 mg.) of lead acetate added to the platinizing solution increases the rate at which the platinum black will be deposited. The thoroughly cleaned electrode to be platinized is now immersed in this solution and connected to the negative terminal of a 3-v. d-c power supply; a piece of scrap platinum wire attached to the positive terminal is used as the anode. The current is allowed to flow for about 10 seconds; then the electrode is removed from the solution and examined. If necessary, the electrolysis is allowed to continue until enough platinum has been deposited to produce a dull grayish surface. Too thick a coat is harmful rather than beneficial.

The platinized electrode should then be used to electrolyze a dilute solution of sulfuric acid, being made alternately the cathode and the anode at intervals of 15 seconds or so for several minutes. Again a piece of scrap platinum wire is used as the other electrode. This treatment removes traces of chloroplatinate, chlorine, etc., which may have been occluded in the platinum black.

An electrode once platinized should always be stored in distilled water or in dilute sulfuric acid. The catalytic activity of the coating of "platinum black" disappears rapidly if the electrode is allowed to dry out.

Hydrogen gas may be secured from a Kipp generator charged with very pure zinc and hydrochloric acid. The evolution of hydrogen by pure zinc is slow, and may be speeded up by the addition of a little cupric chloride to the acid. The gas is washed successively with silver nitrate (to remove arsine, AsH_3, which "poisons" the electrode very rapidly), dilute sodium hydroxide, dilute sulfuric acid, and finally with water. The first three of these may be dispensed with if compressed tank hydrogen is used. In accurate work some provision must be made for the removal of oxygen, which alters the potential of the electrode. This may be done by passing the hydrogen through an acidified chromous chloride solution or through a heated tube filled with copper turnings or platinized asbestos.

The hydrogen electrode cannot be used in solutions containing oxidizing or reducing agents, because these substances affect its potential so that the measured value cannot be interpreted. Proteins and other materials which are adsorbed on the surface of the platinum cause sluggish and erratic behavior. Nonetheless, the hydrogen electrode is the standard to which all pH measurements are referred.

4-4. The Quinhydrone Electrode. Many of the experimental difficulties associated with the hydrogen electrode are circumvented by the use of the quinhydrone electrode. Quinhydrone is an equimolecular compound of *p*-benzoquinone (Q) and *p*-benzohydroquinone (H_2Q); these are often, though rather loosely, referred to simply as quinone and hydroquinone.

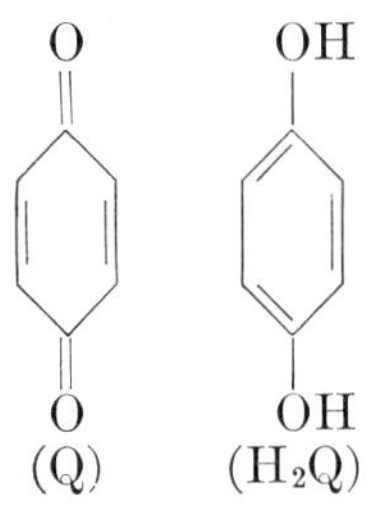

The redox equilibrium between these two compounds is expressed by the half-reaction

$$Q + 2H^+ + 2e = H_2Q$$

and so the potential of an inert metal electrode (such as a bright platinum wire) in a solution containing these substances is given by

$$E = E^0_{Q,H_2Q} - \frac{0.05915}{2} \log \frac{a_{H_2Q}}{a_Q a^2_{H^+}}$$

or, in view of Eq. (4-5), by

$$E = E^0_{Q,H_2Q} - \frac{0.05915}{2} \log \frac{a_{H_2Q}}{a_Q} - 0.05915 \text{ pH} \qquad (4\text{-}7)$$

In a solution prepared from quinhydrone, the *concentrations* (and therefore, to a good approximation at least in solutions of low ionic strength, the activities also) of quinone and hydroquinone are equal, and so

$$E = E^0_{Q,H_2Q} - 0.05915 \text{ pH}$$

The potential of the cell[1]

$$\text{Pt}|\text{quinhydrone}(s),\ H^+||\text{KCl}(s),\ Hg_2Cl_2(s)|\text{Hg}$$

is therefore given by

$$\begin{aligned} E_{\text{cell}} &= E_{ref} + E_j - E^0_{Q,H_2Q} + 0.05915 \text{ pH} \\ &= \epsilon + 0.05915 \text{ pH} \end{aligned}$$

so that

$$\text{pH} = \frac{E_{\text{cell}} - \epsilon}{0.05915} \qquad (4\text{-}8)$$

which may be compared with Eq. (4-6). It is evident that the determination of ϵ with the aid of standard buffers will be made in just the same way as in measurements with the hydrogen electrode.

The potential of the quinhydrone electrode, like that of the hydrogen electrode, is affected by oxidizing and reducing agents. This is because such substances change the ratio a_{H_2Q}/a_Q in Eq. (4-7). The quinhydrone electrode is also subject to a salt error similar to, though much smaller than, that described in Sec. 4-2, for the activity coefficients of quinone and hydroquinone do not vary in quite the same way as the ionic strength is changed. In most routine pH measurements, however, these limitations are not serious.

The one really severe limitation on the use of the quinhydrone electrode is the fact that it is subject to quite large errors in alkaline solutions. This reflects two phenomena. One is the dissociation of hydroquinone, which naturally decreases the activity of the undissociated hydroquinone,

$$H_2Q = H^+ + HQ^-$$

[1] The student may wonder why the presence of solid quinhydrone is specified, since, according to what has just been said, the measured potential ought to be independent of the actual concentration of quinhydrone. This is true, but in practice it is found that more time is required for this potential to be reached if the solution is not saturated (or nearly so) with quinhydrone. Moreover, a very dilute solution of quinhydrone would be considerably more poorly poised, and hence the observed potential would be more sensitive to traces of oxidizing or reducing impurities.

As hydroquinone is a very weak acid ($K_a = 10^{-10}$), its dissociation in acidic solutions is negligible, and becomes of importance only around pH 8 to 9. In addition, hydroquinone is spontaneously air-oxidized in solutions whose pH is above about 8,

$$2H_2Q + O_2 = 2Q + 2H_2O$$

Both of these effects decrease the activity of the hydroquinone relative to that of the quinone, and hence the quinhydrone electrode is of little or no use at pH values greater than 8 or 9. However, the error does not appreciably affect the location of the equivalence point of an acid-base titration performed with a quinhydrone electrode, provided that this does not occur at a pH higher than about 9. Even though the potential does not accurately represent the pH of the solution, in other words, the location of the sharp rise at the equivalence point is substantially unchanged.

If, for some reason, it were absolutely necessary to extend the range of the quinhydrone electrode above pH 9, this could be done by saturating the solution with both quinhydrone and *either* quinone (the so-called "quino-quinhydrone electrode") *or* hydroquinone (the "hydroquinhydrone electrode"), and at the same time bubbling a stream of hydrogen or nitrogen into the solution to remove dissolved air.

4-5. The Antimony Electrode. This is representative of a number of metal–metal oxide electrodes which have been found to be more or less reversible toward hydrogen ion. The potential of a metallic antimony electrode immersed in a solution saturated with hydrous antimony(III) oxide, Sb_2O_3, may be considered to be determined by the equations

$$Sb_2O_3 + 6H^+ + 6e = 2Sb + 3H_2O$$

and $$E = E^0_{Sb_2O_3,Sb} - \frac{0.05915}{6} \log \frac{1}{a^6_{H^+}} = E^0_{Sb_2O_3,Sb} - 0.05915 \text{ pH}$$

The antimony electrode has the advantages of extreme simplicity and ruggedness, and it therefore finds a good deal of use in industrial plants which must monitor or control the pH of flowing liquids or suspensions. However, its useful range is limited by the fact that the oxide is amphoteric:

$$Sb_2O_3 + 2OH^- = 2SbO_2^- + H_2O$$
$$Sb_2O_3 + 2H^+ = 2SbO^+ + H_2O$$

and hence the electrode cannot be used in solutions which are either strongly acidic or strongly alkaline. Moreover, unless great care is taken to prepare the thermodynamically stable (cubic) crystalline modification of the oxide, the potential of the electrode does not vary in quite the theoretical manner as the pH is changed. Finally, the potential of an antimony electrode is considerably affected by traces of tartrate, citrate, and other organic materials which form soluble complexes with antimony.

Other metal–metal oxide electrodes which have been used for pH measurements are the $Hg|HgO, OH^-$ and the $Ag|Ag_2O, OH^-$ electrodes. These may be of occasional use in work with strongly alkaline solutions, since neither mercuric oxide nor silver oxide is amphoteric to any significant extent. However, the potentials of both are greatly altered by the presence of halide ions and many other ions which form stable complexes or slightly soluble mercury or silver salts.

Unlike the hydrogen electrode, which is an electrode of the first order (*cf.* Sec. 3-8), and whose potential depends on the activity of *hydrogen* ion, these metal–metal oxide electrodes are actually electrodes of the second order whose potentials depend on the activity of *hydroxyl* ion. Consequently they may yield pH values which differ slightly from those obtained from measurements with the hydrogen electrode unless the change of K_w with temperature is taken into account.

4-6. The Glass Electrode. The glass electrode is today by far the most popular of the hydrogen ion–responsive electrodes. In large part this is due to the ease with which a pH meter reading can be made. Moreover, the glass electrode is virtually unique in that its potential is unaffected by oxidizing and reducing agents. Nonetheless, a measurement with a glass electrode pH meter must be carefully made and intelligently interpreted. In Perley's words,[1] "The attitude that the pH number read on a meter using a glass electrode is certainly correct is rather unfortunate. There are definite limitations with the glass electrode system as with the other pH-responsive systems. These limitations may be less than with other systems, but it is important to recognize the facts."

The glass electrode consists essentially of a silver–silver chloride electrode immersed in a dilute hydrochloric acid solution contained in a glass bulb. The complete cell is

$$Ag|AgCl(s),\ HCl|glass|unknown\ solution||reference\ electrode$$

The silver–silver chloride electrode serves merely to make an electrical connection to the inner wall of the glass bulb. The concentration of hydrochloric acid is immaterial, and a buffered solution containing sodium or potassium chloride is often used instead. Indeed, it has been asserted that the entire silver–silver chloride electrode could just as well be replaced by a metal film deposited on the inner wall of the glass.[2] This

[1] G. A. Perley, Electrometric pH Measurements, in D. F. Boltz (ed.), "Selected Topics in Modern Instrumental Analysis," Prentice-Hall, Inc., Englewood Cliffs, N.J., 1952.

[2] M. R. Thompson, *Bur. Standards J. Research*, **9**, 833 (1932).

is not done in commercial glass electrodes, partly because of the difficulty of securely attaching the electrical lead wire to such a film.

In this cell there are five sources of potential:

1. The potential of the silver–silver chloride electrode
2. The potential of the reference electrode (usually an S.C.E.)
3. The liquid-junction potential at the boundary between the unknown solution and the reference electrode
4. The potential across the phase boundary between the glass and the hydrochloric acid solution inside the bulb
5. The potential across the phase boundary between the glass and the unknown solution

At any given temperature, the first, second, and fourth of these are obviously constant. The liquid-junction potential has been discussed in Sec. 3-4. The variation of the cell potential with the composition of the unknown solution is therefore due primarily to changes in the potential across the glass–solution interface.

Although it is a well-established experimental fact that, subject to the restrictions to be discussed below, this potential varies with the pH of the unknown solution according to the Nernst equation

$$E = k - \frac{2.303RT}{nF_y}\,\mathrm{pH}$$

(where k depends on the temperature and on the composition of the glass, and where n is 1), the mechanism on which this hydrogen ion response depends is not yet entirely clear. The glass itself undoubtedly consists of a network of SiO_4 tetrahedra containing interstitial alkali and alkaline earth metal ions. (A typical pH-responsive glass, the famous Corning 015, contains 21.4 per cent Na_2O, 6.4 per cent CaO, and 72.2 per cent SiO_2, the percentages being mole percentages.) The alkali metal ions at the glass–solution boundary are presumably capable of being exchanged for hydronium ions in the solution; on passing an electric current through a glass membrane, hydronium ions migrate through the glass. The activity of hydronium ion in the silicate network is different from that of hydronium ion in the solution, and it is this difference which gives rise to the observed potential.

It is worthy of note that the potential of a glass electrode is dependent on the activity of water in the solution being examined. If this activity is not unity (for example, if the ionic strength is extremely high, or if a nonaqueous solvent is present), or if the interior of the glass is not in equilibrium with water at unit activity, the measured potential deviates from the expected value. This is why a new glass electrode must always be soaked in distilled water for a day or so before it is first used, and thereafter always kept covered with water.

This same phenomenon gives rise to what is known, rather misleadingly, as the "acid error" of the glass electrode. In strongly acidic solutions the activity of water is less than unity, and so an error is introduced. It is apparent that, as long as the glass electrode responds properly to changes in pH, the potential of the cell

$$\mathrm{Ag|AgCl}(s),\ \mathrm{HCl|glass|H^+|H_2(1\ atm.),\ Pt}$$

will be constant and independent of changes in the hydrogen ion activity. Figure 4-2 shows the deviations from this behavior observed by Dole in

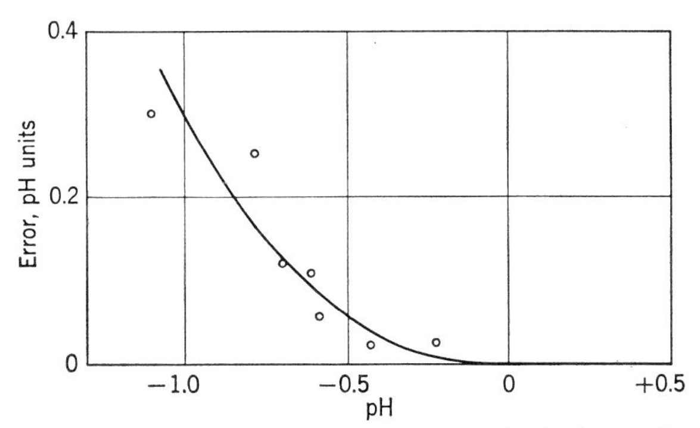

FIG. 4-2. Error of the glass electrode in hydrochloric acid solutions. In this region the pH indicated by the glass electrode is too high.

hydrochloric acid solutions. The error first becomes appreciable at a pH near zero (where the activity of hydrogen ion is unity), and reaches 0.3 pH unit at a pH of -1, which corresponds to just under 5 F hydrochloric acid. An exactly similar error would be encountered in measuring the pH of any other solution in which the activity of water was substantially smaller than in pure water. Consequently the glass electrode does not give accurate values for the pH of solutions containing high concentrations of dissolved salts (*e.g.*, saturated calcium chloride) or substantial mole fractions of nonaqueous solvents (*e.g.*, 50 per cent ethanol). In either of these cases an additional error is introduced by the large liquid-junction potential between the unknown solution and the saturated potassium chloride in the S.C.E., but these two sources of error are entirely independent of each other. The latter is inherent in potentiometric pH measurements and has nothing to do with the indicator electrode used, but the former is characteristic of the glass electrode alone and could be eliminated by using a hydrogen electrode in place of the glass electrode.

The name "acid error" for this error of the glass electrode is unfortunate, because of the implication that the error is found only in strongly acidic solutions. It would be much more accurate to call it a "water activity error."

The "alkaline error" of the glass electrode is usually considerably larger and more important than the water activity error. The alkaline error is due to the fact that other cations than hydronium ion are capable

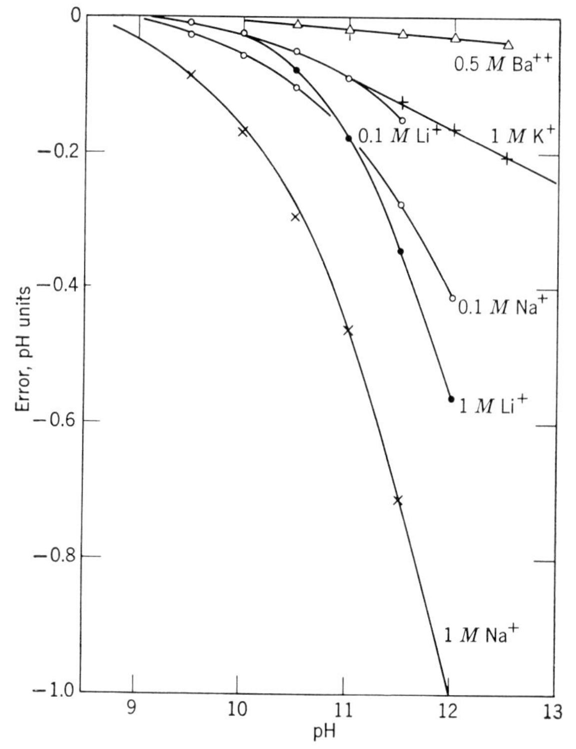

FIG. 4-3. Error of the Corning 015 glass electrode in strongly alkaline solutions containing various cations. In this region the indicated pH is too low.

of carrying current through the glass membrane. The error introduced in this way is negligible as long as the pH is lower than about 9, but in solutions more alkaline than this it is not surprising that ions such as sodium, potassium, lithium, and barium can begin to compete with the hydronium ion when they are present at a concentration which may be 10^9 or more times as great. The alkaline error for a typical soda-lime glass such as Corning 015 is shown in Fig. 4-3. These data are also due to Dole, who carried out much of the research on the theory of the glass electrode.

It is quite evident that sodium ion is the worst offender; this is particularly unfortunate because sodium ion is present in such a large fraction of the solutions with which the chemist has to deal. In recent years other glasses have been developed by replacing most or all of the Na_2O in the original sodium glasses with Li_2O. Perley found that a lithium glass electrode containing 28 per cent Li_2O, 2 per cent Cs_2O, 4 per cent BaO, 3 per cent La_2O_3, and 63 per cent SiO_2 had an error of only −0.12 pH unit in a 2 *M* sodium ion solution at pH 12.8. The error of a Corning 015 glass electrode is almost −1.5 pH unit under these conditions. Lithium glass electrodes are now available commercially for measurements in strongly alkaline solutions.

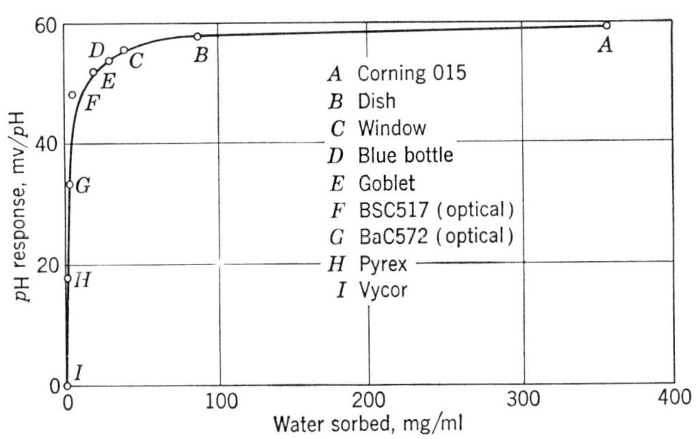

FIG. 4-4. Hygroscopicities and pH responses of nine glasses (*reproduced by permission from R. G. Bates, "Electrometric pH Determinations," John Wiley & Sons, Inc., New York*, 1954).

Glass electrodes are also likely to give erroneous results when used with very poorly buffered solutions which are nearly neutral. Glass (especially the soft glass from which glass electrodes must be made, for reasons to be explained in the following paragraphs) is slightly soluble in water, and the resulting solution is alkaline. Consequently the layer of solution at the glass–liquid interface, whose composition determines the potential of the electrode, will have a slightly higher pH than the bulk of the solution being tested. The resulting error can be somewhat decreased by vigorous stirring. However, it must be noted that some time may be required to attain an equilibrium potential with such solutions, as is true in any potentiometric measurement in a poorly poised solution, and that the solubility of the glass always results in some increase in the pH of the solution being examined.

It might be thought that this error could be decreased by using an electrode made out of a less soluble glass, such as Pyrex. The data shown in Fig. 4-4 show why this is impossible. The quantity plotted on the abscissa scale is not the solubility of the glass, which would be very difficult to determine experimentally, but the quantity of water taken up by a milliliter of the finely powdered dried glass on exposure to an atmosphere of known relative humidity. For these very similar glass–water systems, solubility and sorption are more or less equivalent measures of the affinity between water and glass.

Figure 4-4 shows that Vycor glass (which is made by allowing an ordinary glass to stand in an acid solution until practically all of the constituents of the glass except silica are leached out, then washing, drying, and firing), which has virtually no affinity for water, also displays practically no pH response when made into a glass electrode. At the other extreme, Corning 015 glass, which does display a theoretical pH response ($\Delta E/\Delta \text{pH} = 59.15$ mv. at 25°C.), has the highest affinity for water (and consequently the highest solubility) of any of the glasses tested. The smoothness of the curve through the data secured with these glasses shows that there is, in fact, an intimate relationship between the affinity of a glass for water and its suitability for use as a pH-responsive electrode.

This phenomenon is also closely related to the anomalous behavior of a glass electrode after drying or long immersion in a nonaqueous solution. If a glass electrode is dried in an oven, it loses most or even all of its pH-responsive property, which can be regained only after long soaking in water. An electrode which has been thoroughly "seasoned" by immersion in water can easily be used to measure the "pH" of a solution containing, say, 95 per cent ethanol (although the large liquid-junction potential between such a solution and the saturated aqueous potassium chloride in the calomel reference electrode makes it impossible to attach any precise quantitative significance to the number which results from the measurement). This is because the inside of the glass membrane is still in virtual equilibrium with pure water. However, if the electrode were allowed to stand in such a solution for a long period of time, much of the water would be leached out of the glass, and the potential of the electrode would become practically independent of the composition of the solution.

4-7. The pH Meter. A pH meter is an electronic instrument which operates in either of two ways:

1. The glass electrode is used in an ordinary potentiometric circuit, and the off-balance currents are electronically amplified so that a milliammeter can be used to detect the point of balance.

2. The potential of the glass electrode cell is impressed across a high resistance, and the current flowing through this resistance is amplified and then passed through a milliammeter whose deflection indicates the pH directly.

The first kind of instrument is typified by the Beckman Model G pH meter. With this instrument a slide wire is adjusted until no meter deflection is observed; since the slide wire is calibrated in pH units, the pH may be read directly from its scale. The second kind is typified by the Leeds and Northrup 7664 pH indicator, in which a needle moves across a scale graduated in pH units to a position corresponding to the pH of the solution. The former is characterized by a greater precision, and the latter by a greater ease and rapidity, of measurement. Many pH meters are available today which are line-operated, *i.e.*, powered from a 115-v. a-c line, and most of these are of the direct-reading type. The Beckman Model G pH meter, however, is entirely d-c operated, and derives its power from dry cells contained within the instrument.

No matter what kind of pH meter is used, its scale must be calibrated at each time of use to read the known pH of some standard buffer. This is necessary because different glass and calomel electrodes may vary slightly, and because the potential of any one glass electrode may shift from time to time. The procedure is entirely analogous to the measurement of ϵ with a hydrogen or quinhydrone electrode. The glass and calomel electrodes are rinsed thoroughly and immersed in a solution of known pH (Table 4-2). In accordance with the manufacturer's instructions, the meter is then adjusted to read the pH of the known buffer. Sometimes it is desirable to make a first rough adjustment with one portion of the standard buffer, then to replace this with a fresh portion and make the final adjustment; this ensures that the electrodes and sample beaker are free from any trace of material which may affect the equilibrium. At this point the instrument is ready for the measurement of unknown pH values. In a long series of measurements, it is advisable to check the pH of the standard buffer occasionally.

Most pH meters have either three or four adjustments. One sets the zero position of the milliammeter needle to compensate for any residual grid or plate current in the electronic circuit. The second is a "temperature compensator" which varies the instrument's "definition" of a pH unit; this should change from, *e.g.*, 54.1 mv. at 0°C. to 66.0 mv. at 60°C. (The temperature compensator does *not* compensate for actual changes in the pH of a standard or unknown buffer with temperature, which may be appreciable, as Table 4-2 shows.) The third is a comparison of the e.m.f. across the slide wire (or some other portion of the circuit) with the e.m.f. of an internal standard cell; this is often omitted from line-operated pH meters which depend on the linearity of the meter deflections. The fourth is the adjustment by which the meter is made to indicate the known pH of the standard buffer.

Even after these adjustments have been made, and after all of the errors peculiar to the glass electrode have been accounted for, it must be remembered that the instrument cannot differentiate between a potential arising at the surface of the glass electrode and a liquid-junction potential. It is therefore imperative to keep in mind the factors outlined in Sec. 3-4.

4-8. Strong Acid–Strong Base Titrations. Let us suppose that 10.00 ml. of an 0.1000 F solution of a strong monobasic acid (*e.g.*, HCl, HNO_3, or $HClO_4$) is titrated with an 0.1000 F solution of a strong monoacid base such as NaOH or KOH. Taking activities equal to concentrations, the pH of the initial solution is 1.00. After, say, 1.00 ml. of base has been added, the concentration of hydrogen ion is

$$[H^+] = \frac{(10.00 \times 0.1000) - (1.00 \times 0.1000)}{10.00 + 1.00} = 0.0818$$

The numerator here is the difference between the number of millimoles of acid initially present and the number of millimoles of acid which have been neutralized by the addition of base, and the denominator is the total volume of the solution in milliliters. In general

$$[H^+] = \frac{V_aC_a - V_bC_b}{V_a + V_b} \tag{4-9}$$

where the meanings of the symbols are self-evident.

Equation (4-9) suffices for the calculation of the hydrogen ion concentration, and therefore of the pH, until a pH of about 6 is reached. To all intents and purposes, this covers the entire titration curve up to the equivalence point unless the acid is extremely dilute. Very near the equivalence point, however, it is necessary to take into account the hydrogen ions produced by the dissociation of water; these are equal in number to the hydroxyl ions, and since there is no other source of hydroxyl ion we may write Eq. (4-9) in the expanded form

$$[H^+] = \frac{V_aC_a - V_bC_b}{V_a + V_b} + [OH^-] \tag{4-10}$$

which, since $[OH^-] = K_w/[H^+]$, may be solved by applying the well-known algebraic equation for the solution of a quadratic. At the equivalence point, of course, $V_aC_a = V_bC_b$ by definition, and so $[H^+] = \sqrt{K_w}$. Beyond the equivalence point, by the same reasoning, we have

$$[OH^-] = \frac{V_bC_b - V_aC_a}{V_a + V_b} + [H^+] \tag{4-11}$$

completely analogous to Eq. (4-10).

Differentiation of Eq. (4-10) or (4-11) shows that the slope of the titration curve, expressed as $d\text{pH}/dV_b$, increases continuously from the start of the titration to a maximum at the equivalence point, then decreases continuously as more and more excess base is added. Regardless of the conditions of the experiment, the pH at the equivalence point is necessarily equal to 7.00, provided, of course, that acidic or basic impurities such as carbon dioxide and ammonia are rigorously excluded. This requirement is a very sensitive criterion of the accuracy of a titration; when the interpolated pH is equal to 7.0 ± 0.2 at the point where the second derivative of the titration curve is zero, one may have considerable confidence in the reliability of the result. If the original data are treated by the method illustrated by Table 3-2, this criterion is a very easy one to apply.

4-9. Weak Acid–Strong Base Titrations. In this section we shall consider the phenomena observed when a solution of a weak monobasic acid is titrated with a strong base. Representing the acid by HX, the equilibrium constant of the reaction

$$HX = H^+ + X^-$$

is

$$K_a = \frac{[H^+][X^-]}{[HX]} \tag{4-12}$$

At the start of the titration, hydrogen ion is produced by two reactions. One of these is the dissociation of the acid; the other is the dissociation of water. A little thought shows that the total number of hydrogen ions in the solution is equal to the sum of the numbers of OH^- and X^- ions, or

$$[H^+] = [OH^-] + [X^-]$$

so that

$$[X^-] = [H^+] - \frac{K_w}{[H^+]}$$

If the formal concentration of the acid is C_a moles per liter, the actual concentration of HX when the dissociation equilibrium has been established is equal to $C_a - [X^-]$. Making the appropriate substitutions into Eq. (4-12) gives

$$\begin{aligned} K_a &= \frac{[H^+]([H^+] - K_w/[H^+])}{C_a - ([H^+] - K_w/[H^+])} \\ &= \frac{[H^+]^3 - K_w[H^+]}{C_a[H^+] - [H^+]^2 + K_w} \end{aligned} \tag{4-13}$$

Fortunately it is usually possible to simplify Eq. (4-13) considerably. If $[H^+]$ is equal to or greater than about 10^{-6} M, then $[H^+]^2 \gg K_w$, and the terms involving K_w can be neglected, giving

$$K_a = \frac{[H^+]^3}{C_a[H^+] - [H^+]^2} = \frac{[H^+]^2}{C_a - [H^+]} \tag{4-14}$$

If in addition the acid is sufficiently concentrated that $C_a \gg [H^+]$, we may neglect the second term in the denominator and thus secure the familiar approximation

$$[H^+] = \sqrt{C_a K_a} \tag{4-15}$$

Except in the virtually unheard-of case in which $K_a < 10^{-16}$, this simple equation is sufficiently accurate whenever C_a is greater than about $100K_a$. Like most of the other approximate equations we shall deduce, it should not be used for the calculation of K_a or $[H^+]$ when the acid is both fairly strong and quite dilute.

After the addition of V_b ml. of base (such that the equivalence point is not yet reached), the number of millimoles of X^- produced by the titration reaction

$$HX + OH^- = H_2O + X^-$$

will be essentially equal to V_bC_b, the number of millimoles of base added (unless the acid is so weak that this reaction is substantially incomplete even in the presence of excess free acid. The problems which then arise are of some interest, but here we must be content to note that the titration curve has scarcely any inflection near the equivalence point and is therefore analytically useless). To this we must add the quantity of X^- produced by the dissociation of the as yet unneutralized HX, so that

$$[X^-] = \frac{V_bC_b}{V_a + V_b} + [H^+] \tag{4-16}$$

It is convenient to introduce here the quantity f, which is defined as the fraction of the equivalent volume of base added and which is given by the equation

$$f = \frac{V_bC_b}{V_aC_a}$$

With the aid of this definition we may rewrite Eq. (4-16) as

$$[X^-] = fC_a \frac{V_a}{V_a + V_b} + [H^+]$$

In an entirely similar fashion we secure

$$[HX] = (1 - f)C_a \frac{V_a}{V_a + V_b} - [H^+]$$

Combining these with Eq. (4-12) gives

$$K_a = [H^+] \left(\frac{fC_a \dfrac{V_a}{V_a + V_b} + [H^+]}{(1 - f)C_a \dfrac{V_a}{V_a + V_b} - [H^+]} \right) \tag{4-17}$$

Provided that f is not very close to either 0 or 1, and also that C_a is larger than about $100K_a$, the first terms of the numerator and denominator will be so much larger than $[H^+]$ that the latter can be neglected, and in that case

$$K_a = [H^+]\left(\frac{f}{1-f}\right) \tag{4-18}$$

The restrictions on the validity of this equation instead of the exact Eq. (4-17) should be evident. Incidentally, the form of Eq. (4-17) is such that the calculation of K_a from an experimentally measured pH value is not at all difficult, although the reverse calculation (which involves the solution of a quadratic equation) is much more formidable.

Equation (4-18) may be written as

$$\text{pH} = \text{p}K_a + \log\frac{f}{1-f}$$

whence, by differentiation

$$\frac{d\text{pH}}{df} = \frac{0.4343}{f(1-f)} \tag{4-19}$$

Consequently the slope of the titration curve approaches a maximum value both as f approaches zero (at the beginning of the titration) and as f approaches 1 (at the equivalence point). The sharp rise in pH at the start of the titration is characteristic of a weak acid, and it becomes more and more pronounced as K_a decreases. This and other important features of such titration curves are illustrated by Fig. 4-5.

It is apparent from both Fig. 4-5 and Eq. (4-19) that the slope of the titration curve is least when $f = 0.5$, at the mid-point of the titration.

At the equivalence point, we see from the equation for the titration reaction that just as many hydroxyl ions as HX molecules will remain unreacted. To this must be added the number of hydroxyl ions produced by the dissociation of water, giving

$$[OH^-] = [HX] + [H^+] \tag{4-20}$$

The concentration of X^- will be

$$[X^-] = C_a\left(\frac{V_a}{V_a + V_b}\right) - [HX] = C_s - [HX]$$

Substituting these relationships into Eq. (4-12) gives finally

$$K_a = \frac{[H^+]^3 + C_s[H^+]^2 - K_w[H^+]}{K_w - [H^+]^2} \tag{4-21}$$

As these solutions are obviously alkaline, by virtue of Eq. (4-20), it is generally possible to neglect $[H^+]^2$ in comparison with K_w in the denominator. In addition, if C_s is large compared with $[H^+]$, we can neglect the first term of the numerator in comparison with the second, and finally, if $C_s[H^+]$ is large compared with K_w, the third term of the numerator may

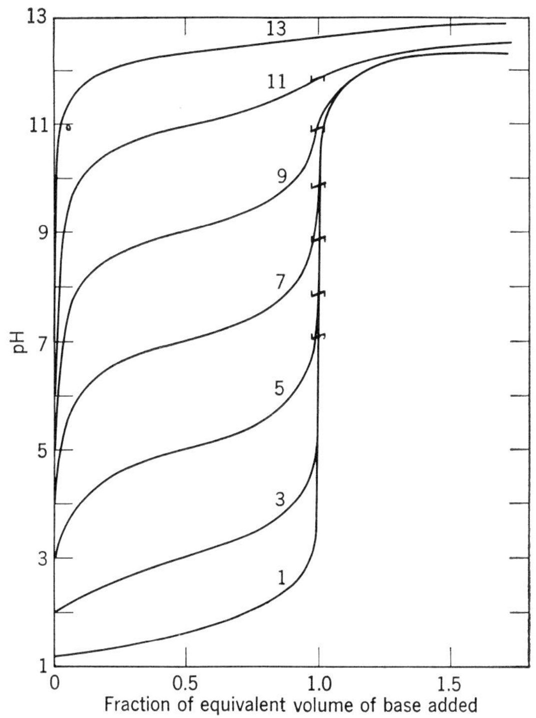

FIG. 4-5. Calculated titration curves for the potentiometric titrations of 0.1 F solutions of various monobasic weak acids with 0.1 F sodium hydroxide. The number beside each curve is the value of pK_a for the corresponding acid.

be omitted as well, leaving only

$$[H^+] = \sqrt{\frac{K_w K_a}{C_s}} \tag{4-22}$$

Beyond the equivalence point, we can neglect the trace of hydroxyl ion coming from the dissociation of water and write

$$[OH^-] = \frac{V_b^* C_b}{V_a + V_b} + [HX] = \frac{K_w}{[H^+]}$$

where V_b^* is the volume of *excess* base added. This is evidently defined by the equation

$$V_b^* = V_b - \frac{V_a C_a}{C_b}$$

Since, again,

$$[X^-] = C_s - [HX]$$

we obtain, after combining these equations with Eq. (4-12),

$$C_s[H^+]^2(V_a + V_b) - K_w(V_a + V_b)(K_a + [H^+]) + V_b^*C_b[H^+](K_a + [H^+]) = 0 \quad (4\text{-}23)$$

Because the solutions are usually quite strongly alkaline, we can generally neglect $[H^+]$ in comparison with K_a. If in addition $C_s[H^+]^2$ is small compared with K_wK_a, Eq. (4-23) can be simplified to

$$[H^+] = \frac{K_w(V_a + V_b)}{V_b^*C_b} \quad (4\text{-}24)$$

This is the same result that would be secured from Eq. (4-11) by neglecting the second term on the right, which takes into account the hydroxyl ion derived from the dissociation of water. We could have derived Eq. (4-24) directly, though without any concrete understanding of its limitations, if we had chosen to neglect the hydroxyl ion arising from the hydrolysis of X^-. In general we may use this approximation whenever the acid being titrated is moderately strong (say, $K_a > 10^{-7}$) and whenever points only very slightly beyond the equivalence point need not be considered.

4-10. Calculation of K_a from Titration Data. Data secured in titrations of the kind we have just discussed are frequently used for the calculation of the dissociation constant of the acid being titrated. Not only is a knowledge of K_a of importance in the study of equilibria in which a weak acid takes part, but in addition the technique of the calculation is representative of many kinds of treatment of experimental data. For this reason we shall present here an analysis of some actual data secured[1] when 0.3500 g. of a certain weak monobasic acid was dissolved in exactly 100 ml. of water and titrated with 0.0530 N (0.0265 F) barium hydroxide.

Some of the original data are given in Table 4-3. They indicate that the equivalence point was reached when 40.77 ml. (2.16 meq.) of barium hydroxide had been added. From this it may be calculated that the equivalent (or molecular) weight of the acid is 162, and that the original solution of the acid was 0.0216 F.

At the start of the titration the pH was 2.40, which corresponds to a hydrogen ion activity of 4×10^{-3} M. By the simple Eq. (4-15), we calculate

$$K_a = \frac{[H^+]^2}{C_a} = 7.4 \times 10^{-4}$$

[1] By A. Acuna, J. J. Chap, and G. L. Jacobs at the Polytechnic Institute of Brooklyn.

We note, however, that $[H^+]$ is not negligible compared with C_a, but is nearly one-fifth of C_a. Therefore we must use the more accurate Eq. (4-14)

$$K_a = \frac{[H^+]^2}{C_a - [H^+]} = 9.1 \times 10^{-4}$$

Equation (4-13) gives the same result, as would be expected, for the dissociation of water in so acidic a solution furnishes only an insignificant number of additional hydrogen ions.

TABLE 4-3. TITRATION OF A WEAK MONOBASIC ACID WITH BARIUM HYDROXIDE*

Ml. of $Ba(OH)_2$	pH	*Ml. of* $Ba(OH)_2$	pH
0	2.40	39.50	4.35
5.00	2.50	40.00	4.65
10.00	2.60	40.50	5.15
15.00	2.70	41.00	8.10
20.00	2.85	41.50	9.00
25.00	3.03	42.50	9.80
30.00	3.30	45.00	10.30
35.00	3.60	50.00	10.70

* See text for experimental details.

After the addition of 10 ml. of barium hydroxide, f was $10/40.77 = 0.246$, and from this and the measured pH, Eq. (4-18) yields

$$K_a = (2.51 \times 10^{-3}) \left(\frac{0.246}{0.754}\right) = 8.2 \times 10^{-4}$$

However, $C_a = 2.16 \times 10^{-2}$, which is only about 25 times as large as K_a, and so the exact Eq. (4-17) should be employed. This gives

$$K_a = (2.51 \times 10^{-3})(0.596) = 1.50 \times 10^{-3}$$

After 30 ml. of barium hydroxide had been added, f was $30/40.77 = 0.736$. Equation (4-18) then gives

$$K_a = (5.0 \times 10^{-4}) \left(\frac{0.736}{0.264}\right) = 1.40 \times 10^{-3}$$

whereas Eq. (4-17) yields

$$K_a = (5.0 \times 10^{-4})(3.27) = 1.64 \times 10^{-3}$$

Note that the values of K_a calculated from these last two points by Eq. (4-17) agree very well (± 5 per cent), whereas those obtained by using the approximate Eq. (4-18) agree to only ± 20 per cent. Both the exact and approximate equations give low results at the initial point; this is, at least in part, because the experimental error of measuring the pH exerts a greater influence on the value of K_a at this point than at the later ones. This in turn simply reflects the fact that the "measurement" of f with a buret is subject to considerably less error than the measurement of $[H^+]$ with a pH meter. It may be pointed out that an error of only ± 0.01 pH unit (which corresponds at 25°C. to ± 0.6 mv. in the measurement of the potential) represents an error of very nearly 2.4 per cent in $[H^+]$.

On careful examination of these data it appears that the measured pH of the initial solution is about 0.15 pH unit too high. This could arise partly from the error in measurement, and partly from the presence of a small amount of a basic impurity (*e.g.*, NaX) in the weak acid used. Such an impurity may have a very large effect on the initial pH, but it has very little influence on the pH values of the well-buffered solutions secured after some base has been added.

The pH of the virtually completely unbuffered solution obtained at the equivalence point is extremely sensitive to small amounts of impurities. Moreover, the calculation of K_a from the interpolated pH at the equivalence point again involves the square of the hydrogen ion activity, with the consequent doubling of the experimental error. Worse still, the interpolation necessary to find the equivalence point pH may itself give a somewhat erroneous value, because the titration curve is not exactly a straight line between the points employed in the interpolation. It is therefore no surprise to find that the calculation of K_a from the equivalence point pH gives results which are usually in very poor agreement with those calculated from the preceding points.

From the data in Table 4-3 the interpolated pH at the equivalence point is found to be 6.76. This is evidently an impossible value for the pH at the equivalence point of a weak acid–strong base titration. The student should endeavor to find a possible explanation for the error.

4-11. Weak Base–Strong Acid Titrations. In the interest of saving space, we shall not discuss titrations of this class in detail. The development of the theory involved is completely analogous to that of the theory of the weak acid–strong base titration, and will present no stumbling blocks to the student who is thoroughly conversant with the material presented in Sec. 4-9.

4-12. Titrations of Dibasic Acids. It is convenient to differentiate weak dibasic acids into two rather arbitrary groups: those for which K_1 is greater than about $10^6 K_2$, and those for which K_1 is less than $10^6 K_2$. The ratio K_1/K_2 has a definite physical significance: it is the equilibrium constant of the reaction

$$H_2X + X^{=} = 2HX^- \qquad K = \frac{K_1}{K_2}$$

When K is large, it is impossible for substantial concentrations of H_2X

and $X^=$ to be present in the same solution. This means that we can almost always neglect the concentration of one of them, and consider without appreciable error that a solution contains either H_2X and HX^- alone, or HX^- and $X^=$ alone. On the other hand, if K is small (and any value of K which is less than about 10^6 is considered "small" for this purpose), we shall often have to take the concentrations of all three of these species into account in the calculations.

The corollary of this is that a sharp and analytically useful end point will be observed in the alkalimetric titration of a dibasic weak acid at the equivalence point of the reaction

$$H_2X + OH^- = HX^- + H_2O$$

when and only when K_1/K_2 is large. Otherwise the slope of the titration curve around this equivalence point will be small, and if K_1/K_2 is less than about 100 it may even be impossible to find a point of inflection on the titration curve. For example, only one end point appears on the titration curve for citric acid ($K_1 = 8 \times 10^{-4}$, $K_2 = 2 \times 10^{-5}$, $K_3 = 4 \times 10^{-6}$) with sodium hydroxide; it corresponds to the third equivalence point

$$H_3C_6H_5O_7 + 3OH^- = C_6H_5O_7^{\equiv} + 3H_2O$$

The alkalimetric titration of an acid in the first group may be divided into three sections. Before the first equivalence point we can consider that the solution contains only the weak acid H_2X and its anion HX^-, and that the dissociation of HX^-, which is a much weaker acid than H_2X, is negligibly small. The theory of this portion of the titration is then identical with that of an ordinary monobasic acid as outlined in Sec. 4-9, taking K_a as equal to K_1.

Beyond the first equivalence point a similar line of reasoning leads us to consider that the solution contains only the weak acid HX^- and its anion $X^=$, and that the concentration of H_2X remaining can be neglected. During this portion of the titration, then, we may again apply the equations of Sec. 4-9, but in this case K_a must be taken as equal to K_2.

At the first equivalence point, however, we have a situation which differs from any previously considered. The solution here is essentially a solution of pure NaHX in water. Three reactions may take place:

$$HX^- = H^+ + X^= \qquad (a)$$
$$HX^- + H^+ = H_2X \qquad (b)$$
$$H_2O = H^+ + OH^- \qquad (c)$$

The total number of hydrogen ions present in the resulting solution is equal to the sum of the number of hydrogen ions produced in reaction (a) (which is equal to the number of $X^=$ ions formed at the same time) and the number of hydrogen ions formed in reaction (*c*) (which is equal to the number of OH^- ions formed), *minus* the number of hydrogen ions consumed in reaction (*b*) (which is equal to the number of molecules of H_2X produced). Solutions of sodium hydrogen carbonate, sodium hydrogen sulfide, and so on, are alkaline because reaction (*b*) for such solutions consumes more hydrogen ions than reaction (*a*) produces, and so the solution is left with an excess of hydroxyl ions. On the other hand, solutions of sodium hydrogen tartrate and sodium hydrogen oxalate are acidic because more hydrogen ions are produced by reaction (*a*) than are consumed by reaction (*b*).

In view of these considerations, the total concentration of hydrogen ion in the solution is given by the equation

$$[H^+] = [X^=] + [OH^-] - [H_2X]$$

Using the equations for K_1 and K_2 to express the concentrations of $X^=$ and H_2X in terms of $[HX^-]$ gives

$$[H^+] = K_2 \frac{[HX^-]}{[H^+]} + \frac{K_w}{[H^+]} - \frac{[H^+][HX^-]}{K_1}$$

A little algebra now produces the following equation:

$$[H^+] = \sqrt{\frac{K_1(K_2[HX^-] + K_w)}{K_1 + [HX^-]}} \qquad (4\text{-}25)$$

(Unless the solution is extremely dilute or the values of K_1 and K_2 very close together, the concentration of HX^- may be taken as C_s, the formal concentration of NaHX, without serious error. To introduce the conservation equation

$$[HX^-] = C_s - [H_2X] - [X^=]$$

would immensely complicate the algebra involved.)

If $[HX^-]$ is large compared to both K_1 and K_w/K_2, Eq. (4-25) reduces to the familiar

$$[H^+] = \sqrt{K_1K_2} \qquad (4\text{-}26)$$

The second group of acids, for which K_1 and K_2 are not greatly different, requires a slightly modified treatment because of the fact that appreciable concentrations of H_2X and $X^=$ can exist in the same solution. Without discussing the titration in detail, we shall examine a typical calculation from which the student will be able to deduce the principles involved.

Suppose that 20.00 ml. of an 0.1000 *F* solution of a weak dibasic acid for which K_1 is 1×10^{-4} and K_2 is 1×10^{-6} is titrated with 0.1000 *F* sodium hydroxide. According to Eq. (4-25) or (4-26), the pH after the addition of 20.00 ml. of the sodium hydroxide will be 5.00. However, if Eq. (4-17) or (4-18) is used to calculate the pH after the addition of 19.80 ml. of sodium hydroxide, where $f = 0.99$, the value secured is 6.00. Quite obviously the pH is not going to *decrease* as the next 0.20 ml. of base is added, and the calculation has to be made instead by the following procedure.

In this solution we have a total of 2.000 millimoles of the $X^=$ radical in all of its forms: partly as $X^=$ itself, partly as HX^-, and partly as undissociated H_2X. As the volume of the solution is 39.80 ml., we have

$$[H_2X] + [HX^-] + [X^=] = \frac{2.000}{39.80} = 0.05025$$

We also have, from the electroneutrality rule,

$$[Na^+] + [H^+] = [HX^-] + 2[X^=] + [OH^-]$$

From these equations we secure, since $[Na^+] = 1.980/39.80 = 0.04975$,

$$0.04975 + [H^+] = \frac{0.05025(K_1[H^+] + 2K_1K_2)}{[H^+]^2 + K_1[H^+] + K_1K_2} + [OH^-] \qquad (4\text{-}27)$$

In this case, as in almost all others of any practical importance, the concentration of sodium ion is much larger than either $[H^+]$ or $[OH^-]$, so that we may safely drop the second term on each side of the equation. After rearrangement we then secure

$$0.99[H^+]^2 - 0.01K_1[H^+] - 1.01K_1K_2 = 0$$

Solving this, and neglecting the term $10^{-4}K_1^2$ in comparison with $4.00K_1K_2$, gives

$$[H^+] = \frac{0.01K_1 + 2.00\sqrt{K_1K_2}}{1.98} = 1.06 \times 10^{-5}$$

or

$$\text{pH} = 4.97$$

As would be expected, the pH is slightly *less* than at the first equivalence point, where the pH is 5.00.

4-13. Acid Dissociation Constants of Metal Ions. An application of pH measurements which is rather highly specialized, but which is of considerable utility in the study of equilibria involving metal ions, is to the determination of such equilibrium constants as that for the reaction

$$Cu(H_2O)_4^{++} + H_2O = Cu(OH)(H_2O)_3^+ + H_3O^+$$

In spite of what was said in Sec. 4-10 about the inadvisability of calculating K_a from the pH of the original solution before the addition of base, in this instance no other procedure is available. On titrating a solution of cupric ion (or of almost any other heavy metal ion) with base, some of the hydrous oxide will be precipitated by the further reaction

$$Cu(H_2O)_4^{++} + 2OH^- = Cu(OH)_2(H_2O)_2 + 2H_2O$$

in the presence of the local excess of base at the point where the drops of reagent enter the solution. The resulting hydrous oxide is so very finely divided, and its adsorptive power is therefore so great, that very little significance can be attached to the measured pH of the suspension. The situation is moreover greatly complicated by the formation of various basic salts under such conditions, and also by the polymerization phenomena observed with many heavy metal ions.[1]

An example of this kind of measurement is provided by the following treatment of the data of Table 4-4. These values were secured by Jones

TABLE 4-4. THE ACID DISSOCIATION CONSTANT OF VANADYL ION

$VOSO_4$ concentration, F	pH	$[H^+]$, M	$K \left(= \frac{[H^+]^2}{C - [H^+]} \right)$
0.050	3.305	$4.9_5 \times 10^{-4}$	5.0×10^{-6}
0.020	3.51	3.1×10^{-4}	4.8×10^{-6}
0.010	3.685	$2.0_6 \times 10^{-4}$	4.4×10^{-6}
0.005	3.845	$1.4_3 \times 10^{-4}$	4.2×10^{-6}
0.002	4.05	8.9×10^{-5}	4.2×10^{-6}
0.001	4.21	6.2×10^{-5}	4.1×10^{-6}
0.0005	4.35	$3.5_5 \times 10^{-5}$	(2.7×10^{-6})
0.0002	4.60	2.5×10^{-5}	3.6×10^{-6}
0.0001	4.70	2.0×10^{-5}	5.0×10^{-6}
			Mean: $(4.4 \pm 0.4) \times 10^{-6}$

and Ray,[2] who prepared solutions containing known concentrations of very pure vanadyl sulfate, $VOSO_4$, and measured their pH values with a glass electrode pH meter. The method of calculating K, which is the equilibrium constant of the reaction

$$VO^{++} + H_2O = VOOH^+ + H^+$$

should be evident.

In their measurements Jones and Ray used two samples of vanadyl sulfate. One had been crystallized from a solution containing excess sulfuric acid. Because it was feared that, in spite of thorough washing, these crystals might contain enough occluded sulfuric acid to cause a significant error, a portion of this material was recrystallized from pure water. Although Jones and Ray secured identical results with the

[1] L. Pokras, *J. Chem. Educ.*, **33**, 152, 223, 282 (1956).
[2] G. Jones and W. A. Ray, *J. Am. Chem. Soc.*, **66**, 1571 (1944).

original and recrystallized vanadyl sulfate, this is a precaution well worth remembering in making measurements of the same kind with other substances.

On the other hand, this is not the only way in which contamination by sulfuric acid could be detected. The question is worth going into in some detail by way of illustrating how experimental data can be manipulated so that they will yield the maximum possible amount of information. It may be remarked that beginners frequently tend to blame unexpected results on chance alone (or, what is much the same thing, on "experimental errors"). It is undeniable that some experiments seem to founder on the rocks of bad luck, but more often the failure of an experiment is due to some perfectly straightforward chemical or instrumental cause; if we take the trouble to ferret out the reason for an initial failure, our second attempt is far more likely to be successful.

To illustrate these principles we shall discuss the data shown in Table 4-5. These data were secured[1] during the course of an attempt to determine the equilibrium constant of the reaction

$$Pb(H_2O)_4^{++} + H_2O = Pb(OH)(H_2O)_3^{+} + H_3O^{+}$$

which, for the sake of convenience, we shall abbreviate as

$$Pb^{++} + H_2O = PbOH^{+} + H^{+}$$

The experimental procedure employed was that described in Expt. I-5*a*, page 424.

TABLE 4-5. pH VALUES OF LEAD NITRATE SOLUTIONS

$[Pb^{II}]$, F	$[H^+]$, M	$K = \dfrac{[H^+]^2}{C - [H^+]}$	$\dfrac{[H^+]}{C}$
0.100	2.29×10^{-3}	5.3×10^{-5}	0.023
0.0200	4.07×10^{-4}	8.5×10^{-6}	0.020
0.00400	7.4×10^{-5}	1.4×10^{-6}	0.018_5
0.00080	7.4×10^{-6}	5.5×10^{-8}	0.009_3
0.00016	1.26×10^{-6}	1.0×10^{-8}	0.007_9

Although the chemical problem here is, on the face of it, entirely similar to that of Table 4-4, it is at once apparent that some completely different phenomenon must be involved. Seven of the nine values of K calculated from the data of Table 4-4 fall within 15 per cent of the mean; but the values of K in Table 4-5 vary continuously as the concentration of lead ion is changed, and the first value is more than 5000 times as large as the last. Naturally one would never think of merely averaging

[1] By M. Gutterson and S. Karp at the Polytechnic Institute of Brooklyn.

these values of K; evidently the "constant" is not constant at all, and the average would be worse than meaningless.

In searching for an explanation of this behavior, we might begin by doubting that the equilibrium had been correctly formulated. Perhaps the mono-hydroxy complex does not exist, so that we should have written (again in the simplified form)

$$Pb^{++} + 2H_2O = Pb(OH)_2 + 2H^+$$

If we assume that the hydrous lead oxide precipitates, so that its activity is constant, we must again take K as being equal to $[H^+]^2/[Pb^{++}]$; this is no improvement. On the other hand, we might imagine (though not with very much confidence) that the hydrous oxide is a weak electrolyte and remains in solution. On the basis of this assumption we would write the equilibrium constant as $0.5[H^+]^3/[Pb^{++}]$, but it takes very little time to show that this is even less constant than the "constant" previously calculated.

At this point we might observe that the ratio of the hydrogen to lead ion concentrations is much more nearly constant than either of the above quantities. In other words, if one of these solutions is diluted to 10 times its volume, its pH increases by approximately one unit. This, however, is not the behavior characteristic of a weak acid; on the contrary, it shows that our solutions contain a strong acid. We immediately suspect the aquo-lead ion of being a strong acid, but then we see that this is not so: the value of $[H^+]/[Pb^{++}]$ would have to be very nearly 1 rather than 0.01. Therefore the solutions must contain some strong acid as an impurity. This train of thought led to the discovery that the stock lead nitrate solution employed in the experiment was one to which a small amount of perchloric acid had been added to prevent hydrolytic precipitation of the lead.

We now observe that the values tabulated in the last column of Table 4-5 are not truly constant; they decrease steadily, and to an extent considerably larger than the expected error of measurement, as C is decreased. We can imagine two reasons why a systematic change is to be expected: one is the fact that the activity coefficient of hydrogen ion increases with decreasing ionic strength in this range, and the other is the fact that the extent to which lead ion hydrolyzes will certainly increase as its concentration is decreased. Both of these effects, however, should cause $[H^+]/[Pb^{++}]$ to *increase* as C decreases, which is exactly the reverse of what is observed. There is no way of avoiding the conclusion that the water used in making these measurements contained some basic impurity, and this was in fact found to be the case.

To be sure, we have not succeeded in calculating the acidic dissociation constant of the aquo-lead ion from the data at hand. But at least we are in a position to repeat the experiment with a far greater probability of success than if we had failed to extract all of the information contained in these data. Nature yields her secrets unwillingly; to the experimenter who fails to use his data to the utmost she often yields them not at all.

4-14. Hydrolytic Titrations. Titrations of this type are performed by adding a salt of a weak acid to a solution containing a metal ion which reacts with the anion of the weak acid to give a precipitate or a complex ion. As long as an appreciable concentration of the free metal ion remains unreacted, the concentration of the anion is held to very small values. Around the equivalence point, however, there is a rapid increase in the concentration of this anion and therefore, in accordance with Eq. (4-22), a rapid increase of the pH.

For example, calcium may be determined by a titration with a standard solution of sodium palmitate, $CH_3(CH_2)_{14}COONa$. Palmitic acid is a quite weak acid, and calcium palmitate is very insoluble; the end point is readily detected by the use of phenolphthalein. Zinc can be determined by a titration with standard potassium cyanide: there is a fairly sharp break at the equivalence point of the reaction

$$Zn^{++} + 2CN^- = Zn(CN)_2$$

followed by another, much less pronounced, at the equivalence point of the reaction

$$Zn(CN)_2 + 2CN^- = Zn(CN)_4^=$$

Although hydrolytic titrations are fairly attractive in principle, they are little used in practice. Both acidic and basic impurities naturally interfere. Any other heavy metal present would be counted as calcium in the palmitate method, for all heavy metal palmitates are insoluble; by the same token there are few heavy metals which would not interfere in the titration of zinc with cyanide, either by competing with the zinc for the cyanide ions or by reacting with the hydroxyl ion resulting from the hydrolysis of the cyanide. Consequently the method is virtually limited to the analysis of pure neutral salt solutions, and even there one often finds considerable deviations from the expected stoichiometry because of the coprecipitation of either the reagent or the excess metal ion. An excellent discussion of the practical performance of hydrolytic titrations is given by Kolthoff and Stenger.[1]

PROBLEMS

The answers to the starred problems will be found on page 403.

***1.** Ten drops of 0.0015 *F* methyl orange are added to 50 ml. of a certain buffer

[1] I. M. Kolthoff and V. A. Stenger, "Volumetric Analysis," 2d ed., vol. 2, chap. 6, Interscience Publishers, Inc., New York, 1947.

solution, and the color of the mixture is found to be identical with that of a mixture of 50 ml. of water with seven drops of 0.0015 F methyl orange and six drops of an 0.00075 F solution of the sodium salt of methyl orange. What is the pH of the buffer? pK for methyl orange is 3.46.

2. What would be the pH of a solution prepared by adding 0.50 ml. of 0.0015 F methyl orange to 50 ml. of water?

3. A chemist adds 0.10 ml. of an 0.10 per cent solution of phenolphthalein (molecular weight 318) to 50.0 ml. of water and finds that he can just discern a pink color after the addition of enough sodium hydroxide to raise the pH to 9.1. If pK_a for phenolphthalein is 9.85, what concentration of the indicator anion is required to give a barely visible color?

4. Using the answer to Problem 3, calculate the volume of 0.10 per cent phenolphthalein solution that should be added to 50 ml. of a solution of an acid in a titration in which 30 ml. of base will be used to arrive at an equivalence point at pH 8.00.

***5.** A certain indicator has a molecular weight of 343 and an acid dissociation constant of 5×10^{-8}. What volume of 0.1000 F sodium hydroxide is required to neutralize 0.25 ml. of an 0.10 per cent solution of the acid form of the indicator to an exactly intermediate color?

6. Exactly 0.20 ml. of 0.002 F thymol blue is added to 25.0 ml. of an acidic buffer solution. By spectrophotometric measurements it is found that the concentration of the yellow form of the indicator in the resulting mixture is 1.0×10^{-5} M. What is the pH of the buffer?

7. Calculate the value of K_w in glacial acetic acid, which has a density of 1.0511 g./ml. and which contains 99.5 per cent acetic acid by weight.

8. The variation of the density of a sodium chloride solution with its concentration is shown by the following data:

G./l. of NaCl	*Density, g./ml.*
10	1.0053
41	1.0268
84.5	1.0559
130.3	1.0857
178.6	1.1162
229.6	1.1478

Calculate the value of K_w in each of these solutions. Do your results seem to justify the common assertion that K_w is not appreciably affected by the presence of small concentrations of electrolytes?

***9.** What volume of 0.0980 F sodium hydroxide would have to be added to 50.0 ml. of pure water to secure a solution of pH 12.60?

10. A certain solution, whose pH is 3, is known to contain either hydrochloric acid or acetic acid. How could you identify the acid without using any apparatus except a graduate, any reagent except water, or any instrument except a pH meter?

11. The potential of the cell

$$\text{Pt, } H_2(720 \text{ mm.})|\text{unknown buffer}||\text{S.C.E.}$$

is +0.303 v. What is the pH of the unknown buffer? (Neglect the liquid-junction potential.)

12. The equivalence point of a certain titration is known to occur at a potential of +0.40 v. *vs.* N.H.E. If the end point is to be taken as the point where the potential of the cell

$$\text{indicator electrode}|\text{solution}||\text{buffer, quinhydrone}(s)|\text{Pt}$$

is zero, what should be the pH of the buffer used in the quinhydrone "reference" electrode? (Neglect the liquid-junction potential.)

***13.** What percentage of the acid will remain untitrated at the end point if an 0.1000 F hydrochloric acid solution is titrated to pH 4.60 with 0.1000 F sodium hydroxide?

14. What is the smallest concentration of hydrochloric acid that can be titrated with sodium hydroxide of the same concentration if the end point will be taken at pH 6.00 and if the titration error must be smaller than 0.1 per cent?

15. An autotitrator was used for the titration of 0.0005 F hydrochloric acid with 0.0005 F sodium hydroxide. It was found that the volume of base used was 0.05 per cent low when the autotitrator was set to an end point pH of 7.00. At what pH was the titration actually being stopped?

16. What is the pH of an 0.10 F solution of trichloroacetic acid, for which $K_a = 0.20$?

***17.** An 0.100 F solution of a certain monobasic weak acid has a pH of 6.80. What is the dissociation constant of the acid?

18. Exactly 50 ml. of 0.2000 F sodium hydroxide is added to 250.0 ml. of 0.1000 F bromic acid. If the pH of the resulting solution is 1.50, what is the dissociation constant of bromic acid?

19. What weight of solid sodium hydroxide should be added to 1000 ml. of 0.200 F trichloroacetic acid ($K_a = 0.20$) to give a solution whose pH is 1.00?

20. The dissociation constant of benzoic acid (C_6H_5COOH) is 6.3×10^{-6} and its solubility in water at 25°C. is 0.025 F. A sample of commercial benzoic acid is shaken with just enough water to dissolve it completely, and the pH of the resulting solution is found to be 3.60. What percentage by weight of sodium benzoate is present if this is the only impurity?

***21.** An 0.0100 F solution of a newly discovered monobasic acid has a pH of 6.86 according to measuring equipment which is reliable to ± 0.02 pH unit. Between what limits must the dissociation constant of the acid lie?

22. A solution of acetic acid is titrated with sodium hydroxide and the end point is taken at pH 7.00. What is the percentage error of the titration?

23. What will be the pH of a mixture produced by suspending 0.0100 mole of pure benzoic acid in 100 ml. of water and adding 10.0 ml. of 0.100 F sodium hydroxide? Use the numerical data given in Problem 20.

24. What is the pH of a solution prepared by dissolving 0.01 mole of acetic acid and 10^{-5} mole of sodium acetate in a liter of water?

***25.** What is the pH of an 0.01 F solution of sodium picrate? K_a for picric acid is 0.16.

26. An 0.1 F solution of the sodium salt of a certain monobasic weak acid has a pH of 12.75. What is the dissociation constant of the acid?

27. The dissociation constants of succinic acid are $K_1 = 6.6 \times 10^{-5}$ and $K_2 = 2.8 \times 10^{-6}$. What is the pH of an 0.1 F solution of succinic acid?

28. What mole fraction of sodium hydrogen succinate would have to be present to cause a deviation of 0.05 pH unit from the value calculated in Problem 27?

***29.** A solution of an unknown acid is titrated with sodium hydroxide. Only one equivalence point can be detected on the titration curve; it occurs at 40.0 ml. of sodium hydroxide. The pH of the solution is 4.50 when 10.0 ml. of sodium hydroxide has been added, and is 6.10 when 30.0 ml. of sodium hydroxide has been added. Is the acid monobasic or dibasic?

30. Derive an exact equation for the concentration of A^- present at equilibrium in a solution containing $C_s F$ NaA. Assume that the dissociation constant of the acid HA is known.

***31.** It is proposed to investigate the behavior of a microcoulometer operating on the following principle. The electric current to be integrated is passed through two platinum electrodes immersed in a saturated solution of silver acetate, whose solubility product is 4.2×10^{-3}. At the cathode, silver ion is reduced:

$$Ag^+ + e = Ag$$

while the anode reaction is

$$2OAc^- + H_2O = \frac{1}{2}O_2 + 2HOAc + 2e$$

An excess of solid silver acetate is added to keep the electrolyte saturated at all times.

Assume that such a coulometer is constructed and that it contains exactly 10 ml. of electrolyte. K_a for acetic acid is 1.75×10^{-5}. One faraday is 9.65×10^4 coulombs.

For how long must a current of 1 microampere flow through the coulometer to change the initial pH by 0.05 unit?

CHAPTER 5

CONDUCTOMETRIC MEASUREMENTS

5-1. Introduction. The electrical conductivity of a solution is dependent upon both the nature and the concentration of every ionic species present in the solution. For example, a 1 F solution of sodium chloride is a better conductor of electricity—it has a higher conductance[1]—than an 0.1 F solution, because the more concentrated solution contains a greater number of ions per unit volume. When a voltage is applied across two electrodes immersed in one of these solutions, the dissolved ions will migrate toward the electrodes: the sodium ions toward the negative electrode, and the chloride ions toward the positive electrode. This migration of ions through the solution constitutes the flow of electric current through the solution,[2] and this current will be greater in the more concentrated solution because of the larger number of ions moving through the solution. However, the current which will flow depends not only on the number of ions which can be set in motion, but also on the rate at which these ions move. It is for this reason that 1 F hydrochloric acid has a much higher conductivity than 1 F sodium chloride; other things being equal, a hydrogen ion moves through an aqueous solution nearly seven times as rapidly as a sodium ion.

We shall see in a subsequent section of this chapter that the conductance of a solution is given by an equation of the form

$$\frac{1}{R} = k(c_A\lambda_A + c_B\lambda_B + \cdots + c_Z\lambda_Z)$$

where k is a constant of proportionality, c_i is the concentration in *equivalents per liter* of the ith ion in the solution, and λ_i is a numerical constant characteristic of that ion.

Suppose that a solution contains only two ions, which is to say that it contains only one electrolyte, which may be either strong or weak. Then

$$c_A = c_B$$

[1] "Conductance" is the reciprocal of resistance. "Conductivity" and "specific conductance" are synonymous and denote the reciprocal of "resistivity" or "specific resistance."

[2] Evidently this is quite different from the way in which a current flows through a metallic conductor, and so it is no surprise to find that electrolytic and metallic conductors behave differently in several ways.

because the concentrations are expressed in normalities. Letting c_A and c_B be represented by c, we have

$$\frac{1}{R} = kc(\lambda_A + \lambda_B)$$

It happens that the constant λ, although characteristic of the ion to which it refers, is not quite independent of concentration. However, we can circumvent this small difficulty by some such expedient as measuring the conductance of each of a number of known solutions and plotting $1/R$ against c; then we can easily find the concentration of an unknown solution from its measured conductance.

This is the principle of *direct conductometry*, which might be defined as a method of determining the concentration of a solution from the result of a single conductance measurement (excluding, of course, any calibration experiments carried out for the purpose of determining the values of λ). A little thought will reveal that direct conductometry should be applicable whenever there is some known relationship among the concentrations of all the ions present in the solution. This is most commonly the case when only two ions are present, as in solutions of sodium chloride or acetic acid or saturated silver chloride. So direct conductometry can be used to analyze a sodium chloride solution, to determine the dissociation constant of acetic acid, or to measure the solubility product of silver chloride. Sometimes direct conductometry can be useful in analyzing solutions containing three ions (*e.g.*, a solution of sodium potassium tartrate, in which $c_{K^+} = c_{Na^+} = \frac{1}{2}c_{C_4H_4O_6^=}$), but more often the concentrations of the ions present in such solutions will not necessarily be related in any simple way, and then a single conductometric measurement will not suffice for the calculation of any of the ionic concentrations.

This does not mean that conductance measurements are of no use in work with solutions containing several ions. For example, a mixture of sodium chloride and sodium nitrate could be analyzed by a combination of direct conductometry and a potentiometric titration of chloride. Moreover, conductometric measurements can be used to follow the titration of one constituent of a mixture with a chemical reagent; this is known as a *conductometric titration*. In such a titration one measures the conductance of a solution as a function of the volume of reagent added. To take a simple example, let us suppose that a hydrochloric acid solution is titrated with sodium hydroxide. Up to the equivalence point the conductance decreases because the reaction

$$HCl + NaOH = H_2O + NaCl$$

results in the replacement of the very mobile hydrogen ion with the much more slowly moving sodium ion. After the equivalence point, however, the conductance rises again as more and more excess base is added. An appropriate graphical treatment of these data will give a V-shaped titration curve consisting of two straight lines whose point of intersection corresponds to the equivalence point. It will be seen that there are many titrations which are hardly feasible by volumetric or potentiometric methods but which give very good results when performed conductometrically.

5-2. Units of Conductance. The resistance of a column of solution of uniform cross-sectional area A (cm.^{2}) between two electrodes l cm. apart is

$$R = \rho\left(\frac{l}{A}\right) \tag{5-1}$$

The proportionality constant ρ, which is called the *specific resistance*, is evidently equal to the measured resistance in a cell for which l and A are equal. Because the concentration of a solution, the viscosity of water, and the degree of ionic hydration (which affects the size of an ion and hence the rate at which it can move through the solution) all vary with temperature, ρ is also dependent on temperature. For solutions of electrolytes ρ decreases with increasing temperature,[1] and the temperature coefficient is generally of the order of -2 per cent per degree. It follows from what has been said above that ρ is also dependent on the concentration and nature of the solution.

The quantity (l/A) is known as the "cell constant." With the types of cells used in analytical practice, it is impossible to calculate the cell constant from direct measurements of the areas of the electrodes and the distance between them, primarily because the electric field is inhomogeneous and is not confined to the volume of solution bounded by the electrodes. Consequently, the cell constant has to be determined by measuring the resistance of a solution for which ρ is known. For this purpose it is advisable to use a solution whose resistance will be of about the same order of magnitude as the resistances to be encountered in later work. Table 5-1 lists the compositions and specific resistances of some

TABLE 5-1. SPECIFIC RESISTANCES OF POTASSIUM CHLORIDE SOLUTIONS AT 25°C.*

Grams of KCl *(weighed in air against brass weights) per liter of solution*	ρ, *ohm-cm.*
76.40_1	8.981_3
7.456_9	77.78
0.7440_5	709.8

* The data are calculated from the results of G. Jones and B. C. Bradshaw, *J. Am. Chem. Soc.*, **55**, 1780 (1933).

[1] Exactly the reverse is true of a metallic conductor.

solutions suitable for this purpose. It may be remarked that the cell constant need not be measured at all for conductometric titrations.

Since the conductance of a solution is the reciprocal of its resistance,

$$\frac{1}{R} = \frac{1}{\rho}\frac{A}{l} = \kappa\frac{A}{l} \tag{5-2}$$

where κ is called the *specific conductance*. The conductance itself is expressed in reciprocal ohms or mhos, and the specific conductance is expressed in reciprocal ohms per cm.

The *equivalent conductance* Λ is the specific conductance of a hypothetical solution which contains 1 gram-equivalent of solute per cm.^{3} Expressing the concentration c in gram-equivalents per 1000 cm.^{3} (which differs from its value in gram-equivalents per liter by only 0.0027 per cent, an entirely negligible amount)

$$\Lambda = 1000\frac{\kappa}{c} = 1000\frac{l}{A}\frac{1}{R}\frac{1}{c} \tag{5-3}$$

A less commonly used unit, the *molar conductance*, is defined by an equation identical with Eq. (5-3) except that the concentration is taken in gram-moles per 1000 cm.^{3}

5-3. Effect of Concentration on Equivalent Conductance. Figure 5-1*a* shows the way in which the equivalent conductance of a hydrochloric acid solution varies with the concentration of the acid. At zero concentration (infinite dilution), the extrapolated equivalent conductance Λ^0 is 426.05 mho-cm.^{2} per equivalent. At all finite concentrations, however, Λ is less than Λ^0, and it decreases continuously with increasing concentration.

This has two causes. As the concentration of the acid is increased, the number of oppositely charged particles in the vicinity of each ion increases. The electrostatic forces of attraction exerted by this shell of oppositely charged ions tend to retard the motion of the central ion through the solution, and therefore the conductance is smaller than it would be if these interionic attraction forces did not exist. This is the so-called "interionic attraction effect."

Moreover, as an ion moves through the solution, it carries water molecules with it. At any instant, therefore, we have a stream of hydrated cations moving toward one electrode and a stream of hydrated anions moving toward the other electrode. This produces the same effect as though each ion were moving through a stream of solvent flowing in the opposite direction. This second retarding influence is known as the "electrophoretic effect."

Although the mathematical treatment of these effects, which is primarily the work of Debye and Hückel and Onsager, is much too involved for discussion here, we may note that at low concentrations ($c \leq$ about $0.001\ N$) Λ is given by an equation of the form

$$\Lambda = \Lambda^0 - A\sqrt{c} \tag{5-4}$$

where A is a numerical constant which involves the temperature as well as a number of terms characteristic of both the solvent and the solute.

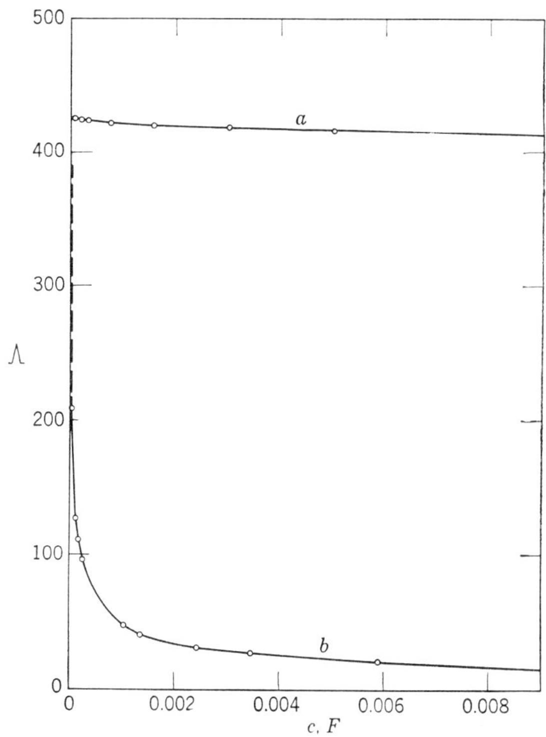

FIG. 5-1. The effects of concentration on the equivalent conductances of (*a*) hydrochloric acid and (*b*) acetic acid.

The increasingly large deviations from the simple Eq. (5-4) at concentrations much above $0.001\ N$ are due primarily to the mathematical approximations made in the derivation.

Figure 5-1*b* is a plot of some data on the equivalent conductance of acetic acid at various concentrations. The effect of concentration on Λ for a weak electrolyte is much more pronounced than that for a strong electrolyte. This is because the degree of dissociation decreases with increasing concentration, and this effect overshadows both the interionic and electrophoretic effects. In a later section we shall see how the data of Fig. 5-1*b* can be used for the calculation of the dissociation constant of acetic acid.

5-4. Equivalent Ionic Conductances. The equivalent conductance of a particular ion in a solution containing a single salt is related to the equivalent conductance of the salt by the equation

$$\lambda_i = t_i \Lambda \tag{5-5}$$

where t_i is the transference number of the ion. Since by definition

$$t_+ + t_- = 1$$

this becomes

$$\lambda_+ + \lambda_- = \Lambda \tag{5-6}$$

If the solution contains more than one salt, the total conductance is given by

$$\frac{1}{R} = \frac{A}{1000l} \sum_i c_i \lambda_i \tag{5-7}$$

The fraction of this total conductance which is due to any one ion is simply equal to the transference number of that ion in the solution:

$$t_i = \frac{c_i \lambda_i}{\sum_i c_i \lambda_i} \tag{5-8}$$

(From this, incidentally, we see that the transference number of an ion can be reduced virtually to zero by the addition of a large excess of some other salt. We shall refer to this fact in discussing the use of supporting electrolytes in polarography in Chap. 6.)

Since it is possible to evaluate transference numbers by methods entirely independent of conductometric measurements (*e.g.*, the Hittorf and moving boundary methods, which are discussed in textbooks of physical chemistry), it is a relatively simple matter to evaluate the equivalent conductance of an ion in almost any desired environment.

At infinite dilution, the value of λ for any given ion is rigorously constant and independent of the nature of the accompanying ion. Thus the equivalent conductance of chloride ion at infinite dilution, $\lambda^0_{Cl^-}$, is found to be equal to 76.34 regardless of whether it is calculated from data on hydrochloric acid, potassium chloride, or barium chloride.

Values of λ^0 are found in the following manner. From data on the equivalent conductance of some salt (say potassium chloride), the value of Λ^0 is found by plotting Λ *versus* $\sqrt{c}$ and extrapolating to $c = 0$. The transference number of chloride ion is then measured in potassium chloride solutions of various concentrations and extrapolated to give $t^0_{Cl^-}$. Then

$$\lambda^0_{Cl^-} = t^0_{Cl} - \Lambda^0_{KCl}$$

and

$$\lambda^0_{K^+} = \Lambda^0_{KCl} - \lambda^0_{Cl^-}$$

From these values and the equivalent conductances at infinite dilution of other salts containing these ions, it is a simple matter to calculate other values of λ^0. For example,

$$\lambda^0_{Na^+} = \Lambda^0_{NaCl} - \lambda^0_{Cl^-}$$

and

$$\lambda^0_{NO_3^-} = \Lambda^0_{KNO_3} - \lambda^0_{K^+}$$

and so on. Table C in the Appendix lists a number of values of λ^0 obtained in this manner.

At finite concentrations, however, the equivalent conductance of an ion is not quite independent of its environment. For example, the equivalent conductance of chloride ion at $c = 0.1$ equivalent per liter is 65.98 in hydrochloric acid, 65.79 in potassium chloride, 65.58 in sodium chloride, and 65.49 in lithium chloride. Comparison with Table C in the Appendix shows that in these four electrolytes the mobility of the cation decreases from hydrogen to lithium in the order given. Evidently the effect of cation mobility is very slight, and can be ignored for most practical purposes.

The equivalent conductance of barium ion at infinite dilution is intermediate between the values for potassium and sodium ions. However, the equivalent conductance of chloride ion in 0.1 N barium chloride is only 60.53, and so it is evident that the valence of the oppositely charged ion does exert a considerable effect. At concentrations smaller than 0.1 N, however, this effect becomes progressively smaller, and it can be neglected in all but the most precise work at concentrations below about 0.001 N.

5-5. Determination of Dissociation Constants. According to MacInnes and Shedlovsky,[1] the equivalent conductance of 0.02000 F acetic acid is 11.563 mho-cm.^{2} per g.-equiv. At infinite dilution

$$\Lambda^0_{HOAc} = \lambda^0_{H^+} + \lambda^0_{OAc^-} = 349.8 + 40.9 = 390.7$$

If we assume that λ_{H^+} and λ_{OAc^-} have the same values in 0.02 F acetic acid as at infinite dilution, then the ratio of the observed equivalent conductance to Λ^0 is the fraction α of the acetic acid which is dissociated:

$$\alpha = \frac{11.563}{390.7} = 0.02960$$

If c is the formal concentration of the acetic acid, then

$$[H^+] = [OAc^-] = \alpha c$$

$$[HOAc] = c - \alpha c$$

and so

$$K_a = \frac{[H^+][OAc^-]}{[HOAc]} = \frac{\alpha^2 c}{1 - \alpha} = 1.806 \times 10^{-5}$$

[1] D. A. MacInnes and T. Shedlovsky, *J. Am. Chem. Soc.*, **54**, 1429 (1932).

It will be seen that this value of α corresponds to approximately 0.0006 N hydrogen and acetate ions. At this concentration of hydrochloric acid Λ is 422.5 and t_{H^+} is 0.822, giving $\lambda_{H^+}^{0.0006} = 347.3$. In 0.0006 N sodium acetate Λ is 89.1 and t_{OAc^-} is 0.448, so that $\lambda_{OAc^-}^{0.0006} = 39.9$. If, therefore, we had a solution containing these concentrations of hydrogen and acetate ions, we should have an equivalent conductance of

$$\Lambda = 347.3 + 39.9 = 387.2$$

instead of 390.7 as at infinite dilution. Hence a better approximation to the value of α is

$$\alpha = \frac{11.563}{387.2} = 0.02986$$

from which a better value of K_a is 1.838×10^{-5}.

This is the *formal* dissociation constant of acetic acid at an ionic strength of 0.0006. It would have to be multiplied by the activity coefficient ratio $f_{H^+}f_{OAc^-}/f_{HOAc}$ to secure the thermodynamic dissociation constant. Alternatively, we could find the latter from a plot of the logarithm of K_a against $\sqrt{c}$; at low values of c this will be a straight line which can easily be extrapolated to infinite dilution. The value of K_a^0 thus found is 1.753×10^{-5}, which is in remarkable agreement with the value of 1.754×10^{-5} found by Harned and Ehlers[1] by potentiometric measurements.

It is not possible to determine the dissociation constants of a dibasic acid in this way, because any measured conductance could correspond to an infinite number of values of K_1 and K_2. More accurately, even assuming that the manner in which λ varies with c for each ion is exactly known, we have six unknowns—$[H^+]$, $[H_2X]$, $[HX^-]$, $[X^=]$, K_1, and K_2—and only five equations:

$$\frac{1}{R} = \left(\frac{A}{1000l}\right)([H^+]\lambda_{H^+} + [HX^-]\lambda_{HX^-} + 2[X^=]\lambda_{X^=})\dagger$$

$$[H^+] = [HX^-] + 2[X^=]$$

$$[H_2X] + [HX^-] + [X^=] = c_a$$

and the equations for K_1 and K_2.

5-6. Determination of Solubility Products. Let us consider the experimental determination of the solubility product of silver chloride. If c is the formal or normal concentration of the saturated solution, it is

[1] H. S. Harned and R. W. Ehlers, *J. Am. Chem. Soc.*, **55**, 652 (1933).

† The factor $2[X^=]$ arises from the fact that $[X^=]$ is in moles per liter, whereas the concentration of $X^=$ employed in the conductance equation must be in equivalents per liter.

evident from Eqs. (5-2) and (5-3) that

$$\Lambda = \frac{1000l}{RcA} \tag{5-9}$$

In addition, by Eq. (5-6),

$$\Lambda = \lambda_{Ag^+} + \lambda_{Cl^-}$$

so that

$$c = \frac{1000l}{RA(\lambda_{Ag^+} + \lambda_{Cl^-})} \tag{5-10}$$

The solubility of silver chloride is so slight (*ca.* 10^{-5} F) that it can be assumed without serious error that λ_{Ag^+} and λ_{Cl^-} are equal to $\lambda^0_{Ag^+}$ and $\lambda^0_{Cl^-}$, respectively, and so the concentration of a saturated silver chloride solution can be found directly from a measurement of its conductance with a cell of known cell constant. It will be remembered from Sec. 5-5 that an error of less than 2 per cent in K_a resulted from taking λ equal to λ^0 even at an ionic concentration of 6×10^{-4} N. Since greater accuracy than this is not often required, it is almost always justifiable to make this approximation whenever the ionic concentrations are less than about 0.001 N.

To determine the concentration of a solution more concentrated than 0.001 N, or in a case where no reasonably reliable values of the equivalent ionic conductances are available, one can proceed in the following fashion. Suppose, for example, that one wished to measure the solubility of the salt cadmium mandelate, $(C_6H_5CHOHCOO)_2Cd$. The product of the concentration and equivalent conductance of its saturated solution, calculated with the aid of Eq. (5-9) from the experimentally measured resistance, is equal to

$$c\Lambda_{CdM_2} = (\lambda_{Cd^{++}} + \lambda_{M^-})c$$

where M^- represents the mandelate ion, for which not even λ^0 is known If one measures the equivalent conductances of cadmium nitrate, sodium nitrate, and sodium mandelate solutions of identical normalities, one can calculate

$$\Lambda_{Cd(NO_3)_2} + \Lambda_{NaM} - \Lambda_{NaNO_3} = \lambda_{Cd^{++}} + \lambda_{M^-} = \Lambda_{CdM_2}$$

at each of a number of concentrations within a range which is thought to include the value of c for the saturated cadmium mandelate solution. If we now multiply each of the resulting values of Λ_{CdM_2} by the concentration to which it refers, we can construct a plot of $\Lambda_{CdM_2}c$ *versus* c. This will be very nearly a straight line. We can now locate the point at which the interpolated value of $\Lambda_{CdM_2}c$ is equal to the measured value of $1000l/RA$ for the saturated solution. At this point the value of c read off the graph is the normality of the saturated solution, and the remainder of the calculation is obvious.

5-7. Conductometric Acid-Base Titrations. In the titration of a strong acid with a strong base, the reaction which occurs up to the equivalence point may be written as

$$H^+ + X^- + M^+ + OH^- = H_2O + M^+ + X^-$$

The net result of this reaction is the replacement of some hydrogen ions with an equal number of M^+ ions. The equivalent conductance of hydrogen ion ($\lambda^0_{H^+} = 349.8$) is very much larger than the equivalent conductance of any other cation (*e.g.*, $\lambda^0_{Na^+} = 50.1$, $\lambda^0_{K^+} = 73.5$, $\lambda^0_{Ba^{++}} = 63.6$, etc.), and therefore the conductance of the solution drops until the equivalence point is reached. After the equivalence point, however, as excess M^+ and OH^- are added, the conductance begins to rise again.

Before the equivalence point is reached, by Eq. (5-7)

$$\frac{1}{R} = \frac{A}{1000l}(c_{H^+}\lambda_{H^+} + c_{M^+}\lambda_{M^+} + c_{X^-}\lambda_{X^-}) \tag{5-11}$$

If the initial solution has a volume of V_a ml. and a concentration of C^0_a equivalents of acid per liter, and if f is the fraction of the acid neutralized by the addition of V_b ml. of MOH, then

$$c_{H^+} = (1 - f)C^0_a\left(\frac{V_a}{V_a + V_b}\right)$$

$$c_{X^-} = C^0_a\left(\frac{V_a}{V_a + V_b}\right)$$

and

$$c_{M^+} = fC^0_a\left(\frac{V_a}{V_a + V_b}\right)$$

Therefore, up to the equivalence point

$$\frac{1}{R} = \frac{AC^0_a}{1000l}\left(\frac{V_a}{V_a + V_b}\right)[(1 - f)\lambda_{H^+} + f\lambda_{M^+} + \lambda_{X^-}] \tag{5-12}$$

or

$$\frac{1}{R} = \frac{AC^0_a}{1000l}\left(\frac{V_a}{V_a + V_b}\right)[\lambda_{H^+} + \lambda_{X^-} + (\lambda_{M^+} - \lambda_{H^+})f]$$

This has the form

$$\frac{1}{R}\left(\frac{V_a + V_b}{V_a}\right) = k_1 - k_2f \tag{5-13}$$

where the minus sign results from the fact that λ_{H^+} is greater than λ_{M^+}. Consequently a plot of the left-hand side of this equation against the volume of reagent added will be a straight line, for this volume is evidently proportional to f.

Beyond the equivalence point the concentration of hydrogen ion is so small that its contribution to the conductance is negligible. We then have three ions which can carry the current through the solution. These are M^+, X^-, and OH^-, and their concentrations are

$$c_{M^+} = \frac{V_b C_b}{V_a + V_b}$$

$$c_{X^-} = \frac{V_a C_a^0}{V_a + V_b}$$

$$c_{OH^-} = \frac{V_b C_b - V_a C_a^0}{V_a + V_b}$$

where C_b is the concentration of the base. With the aid of Eq. (5-7),

$$\frac{1}{R}\left(\frac{V_a + V_b}{V_a}\right) = \frac{A}{1000l}\left[C_a^0(\lambda_{X^-} - \lambda_{OH^-}) + \frac{C_b}{V_a}(\lambda_{M^+} + \lambda_{OH^-})V_b\right] \quad (5\text{-}14)$$

and this is of the form

$$\frac{1}{R}\left(\frac{V_a + V_b}{V_a}\right) = -k_3 + k_4 V_b \quad (5\text{-}15)$$

(since λ_{OH^-} is always larger than λ_{X^-}; *cf.* Table C of the Appendix). After the equivalence point, therefore, a plot of $(1/R)(V_a + V_b)/V_a$ *versus* the volume of reagent used will be a straight line, but the slope of this line will be positive. Hence the titration curve will be V-shaped, and the equivalence point will lie at the point of intersection of the two straight lines.

Actually, if the titration is made in such a way that the conductances are measured very close to the equivalence point, it will be found that the points deviate slightly from the extrapolated straight lines. This is because there is a small concentration of hydroxyl ion in the solution just before the equivalence point and a small concentration of hydrogen ion in the solution just after the equivalence point, and these make a contribution to the conductance which was neglected in the derivations of Eqs. (5-12) and (5-14). Consequently the data near the equivalence point will fall on a curve like the solid line in Fig. 5-2 rather than on the dashed straight lines.

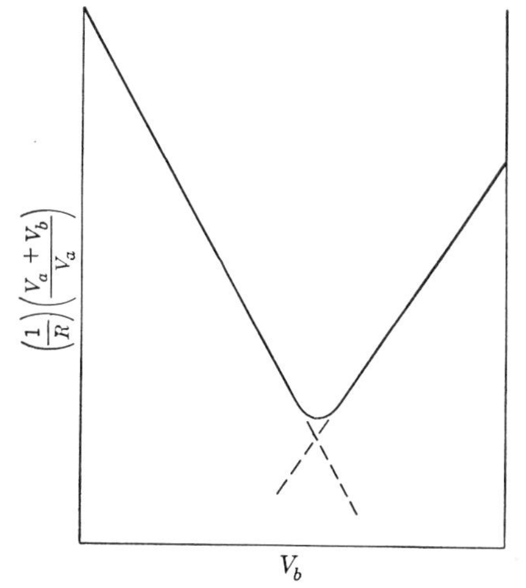

FIG. 5-2. Schematic titration curve for the conductometric titration of 0.1 *F* hydrochloric acid with 0.1 *F* sodium hydroxide. The curvature around the equivalence point is greatly exaggerated.

In a titration of a strong acid with a strong base this curvature around the equivalence point is scarcely noticeable unless both solutions are extremely dilute. This simply reflects the fact that the titration reaction

$$H^+ + OH^- = H_2O \qquad K = 10^{14}$$

is so nearly complete that it is impossible for appreciable concentrations of these ions to be present simultaneously. In general, however, we see that points near the equivalence point are of relatively little assistance in locating the end point of a conductometric titration. This is exactly the opposite of what is true of a potentiometric titration. It follows from this that the easiest and best way to locate the end point of a conductometric strong acid–strong base titration is to make several measurements sufficiently far before the end point, and several more measurements sufficiently far beyond the end point, to ensure that the curved portion of the graph is not encroached upon. Then the quantity $(1/R)(V_a + V_b)/V_a$ is plotted against V_b, and two straight lines are drawn through the experimental points and extrapolated until they meet at the end point.

Because only a comparatively small number of observations must be made in order to define the two straight-line portions of the titration curve, conductometric titrations are very rapid. Their accuracy and precision are determined by the characteristics of the instrument used for the conductance measurements, and by the angle at which the straight-line portions of the titration curve intersect. When, as in the case of a strong acid–strong base titration, this angle is quite acute, the end point is easily located within a few tenths of a per cent even if the accuracy and precision of the bridge are only about ± 1 per cent. Conductometric titrations are more accurate than analyses by direct conductometry, partly because the error of measuring a volume of reagent with a buret is considerably smaller than the error of measuring the conductance of a solution with the apparatus generally available, and also because the process of drawing the titration curve has the effect of averaging out the random experimental errors which afflict the individual measurements. As we shall see in Chap. 6, the very same reasons cause amperometric titrations to be more accurate than analyses by direct polarography.

It will be noted that, even if the factor $(V_a + V_b)/V_a$ is neglected in Eqs. (5-13) and (5-15), a plot of R itself against the volume of reagent, V_b, will consist of two lines which are curved rather than straight. This is because either of these equations becomes

$$R = \frac{1}{a + bV_b}$$

and so the slope of either branch of the titration curve is not constant, but is a function of V_b. Therefore it is essential to use the conductance and not the resistance for the construction of the titration curve.

In practical work it is desirable to keep this "dilution factor" $(V_a + V_b)/V_a$ as nearly equal to unity as possible by using a reagent which is considerably more concentrated than the solution being titrated. If C_b is greater than about $20C_a^0$, no appreciable error will result from neglecting the dilution correction altogether. As this means that V_b will generally be quite small (of the order of a few milliliters), it is advantageous to use a 10-ml. microburet graduated in 0.01- or 0.025-ml. divisions. This is also true, and for exactly the same reason, in amperometric titrations (Chap. 6).

If the strong acid is titrated with a weak base, such as ammonia, instead of with a strong base like potassium hydroxide, the portion of the curve up to the equivalence point will not be affected in any significant respect. The value of $\lambda^0_{NH_4^+}$ is almost exactly equal to that of $\lambda^0_{K^+}$, and so at identical values of f the conductance of the ammonium chloride–hydrochloric acid solution secured in the one case will be almost the same as that of the potassium chloride–hydrochloric acid solution secured in the other. However, the addition of excess ammonia to the pure ammonium chloride solution present at the equivalence point will have scarcely any effect on the conductance. Ammonia is only slightly dissociated even in pure solutions, and its dissociation is repressed by the ammonium ions formed during the neutralization. So to all intents and purposes the excess of ammonia does not give any new ions, and the conductance of the solution (after correction for dilution) remains constant after the equivalence point is reached. The titration curve is consequently ⎿-shaped. This would appear to be somewhat less conducive to accuracy than the V-shaped curve secured in the titration with strong base, but in practice it is found that the two titrations give about equally accurate results.

An entirely different kind of curve is secured when a weak acid is titrated. At the very beginning of the titration, the reaction

$$HX + M^+ + OH^- = H_2O + M^+ + X^-$$

effectively adds X^- ions to the solution. The resulting increase in the concentration of X^-, by virtue of the dissociation equilibrium

$$HX = H^+ + X^-$$

decreases the concentration of hydrogen ion. Because the equivalent conductance of hydrogen ion is extremely high, the resulting decrease of $c_{H^+}\lambda_{H^+}$ is greater than the sum of the simultaneous increases of $c_{M^+}\lambda_{M^+}$ and

$c_{X^-}\lambda_{X^-}$, and so the net conductance of the solution *decreases.* Later in the titration, however, the concentration of hydrogen ion is decreased to so low a value that its contribution to the conductance of the solution (which by now contains relatively high concentrations of M^+ and X^-)

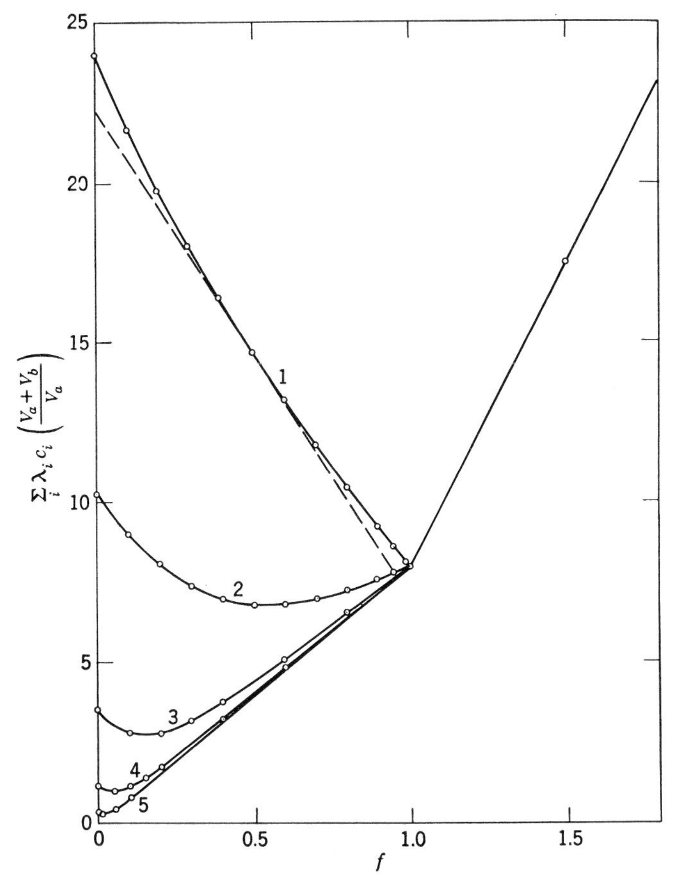

FIG. 5-3. Calculated titration curves for the conductometric titrations of 0.1 F solutions of various monobasic weak acids with sodium hydroxide. The number beside each curve is the value of pK_a for the corresponding acid. The quantity plotted on the ordinate scale is equal to $(1000l/A)(1/R)$, and was computed by assuming that the equivalent ionic conductance of the anion of each acid is 40. The dashed straight line is drawn to emphasize the curvature of the titration curve for the acid of $pK_a = 1$.

is negligibly small. Then the neutralization of a further fraction of the acid effectively only adds more M^+ and X^- ions, and so the conductance *increases.* Accordingly the conductance passes through a minimum, and then begins to increase again, before the equivalence point is reached, and the titration curve resembles one of the curves in Fig. 5-3. Beyond the equivalence point, of course, the weak acid–strong base curve becomes identical with the strong acid–strong base curve, for obvious reasons.

This minimum in the conductometric titration curve is quite characteristic of the weak acid–strong base (or weak base–strong acid) titration, just as is the rapid increase of pH near the beginning of such a titration when it is carried out potentiometrically. The exact location of the minimum depends on the values of K_a and C_a^0; the smaller the

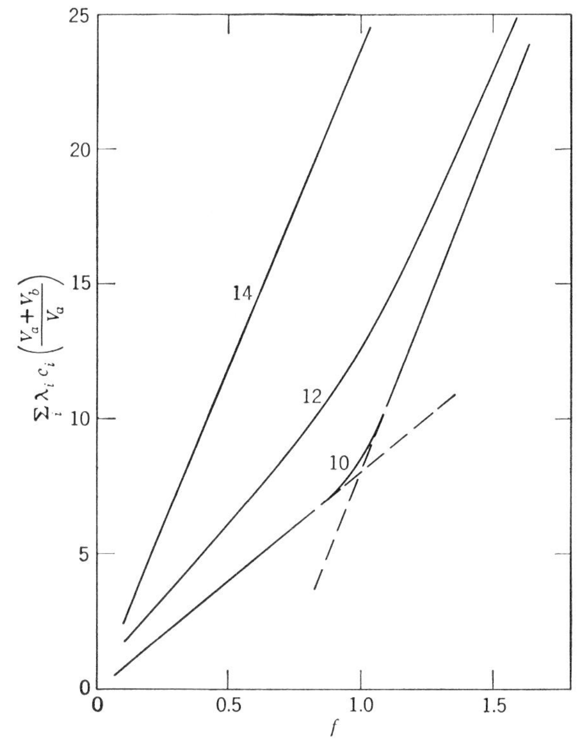

FIG. 5-4. Continuation of Fig. 5-3 for successively weaker acids.

fraction of the acid which is dissociated in the original solution, the smaller will be the fraction of the acid which has to be neutralized before the increase in $c_{M^+}\lambda_{M^+} + c_{X^-}\lambda_{X^-}$ resulting from the addition of a portion of base becomes numerically larger than the decrease in $c_{H^+}\lambda_{H^+}$. So as K_a decreases the minimum moves to smaller and smaller values of f. If $C_a^0K_a$ is less than about 10^{-7}, the minimum occurs at so small a value of f as to be virtually undetectable. In that case the conductance of the initial solution is practically equal to zero, and a plot of $(1/R)(V_a + V_b)/V_a$ *vs.* V_b is linear up to the equivalence point. In view of Eq. (5-12), the slope of this line will be equal to

$$\frac{d(1/R)(V_a + V_b)/V_a}{dV_b} = \frac{AC_b}{1000lV_a}(\lambda_{M^+} + \lambda_{X^-}) \tag{5-16}$$

since c_{X^-} now depends on f, and f in turn is related to V_b by the equation

$$f = \frac{V_bC_b}{V_aC_a^0} \tag{5-17}$$

After the equivalence point, by differentiating Eq. (5-14), the slope of the titration curve is seen to be

$$\frac{d(1/R)(V_a + V_b)/V_a}{dV_b} = \frac{AC_b}{1000lV_a}(\lambda_{M^+} + \lambda_{OH^-}) \tag{5-18}$$

Since no other anion is as mobile as hydroxyl ion, it follows from Eqs. (5-16) and (5-18) that the slope of the line following the equivalence point is always greater than the slope of the line preceding the equivalence point. This is illustrated by Figs. 5-3 and 5-4. The difference in slope is never as large as in a strong acid–strong base titration, however, and this is the chief limitation on the accuracy with which the end point of a weak acid titration can be found. Although Figs. 5-4 and 4-5 indicate that the smallest value of K_a at which good results can still be obtained is slightly less for conductometric than for potentiometric titrations (in each case the limiting factor is the degree of incompleteness of the reaction at the equivalence point), it is interesting to note that conductometric titrations cannot be used at all for acids whose dissociation constants are between about 2×10^{-3} and 0.2, despite the fact that the corresponding potentiometric titrations give excellent results.

The angle between the two titration lines can be decreased, thereby permitting somewhat better accuracy in the location of the end point, by using ammonia (or some other fairly weak base) instead of sodium hydroxide. Once the equivalence point of a titration with ammonia has been passed, the ammonium ions produced by the neutralization repress the dissociation of the excess ammonia to such an extent that the conductance remains practically constant, as illustrated by Fig. 5-5. The titration of a weak acid with ammonia gives good results when done conductometrically even if K_a is as small as 10^{-7}, even though there is almost no inflection around the equivalence point of the corresponding potentiometric titration.

Although limitations of space prevent an extended discussion of the subject here, it may be mentioned that certain mixtures of acids or bases can be titrated more accurately by the conductometric than by the potentiometric method. For example, the potentiometric determination of hydrochloric acid in the presence of acetic acid gives only a quite poorly defined end point, but when the titration is performed conducto-

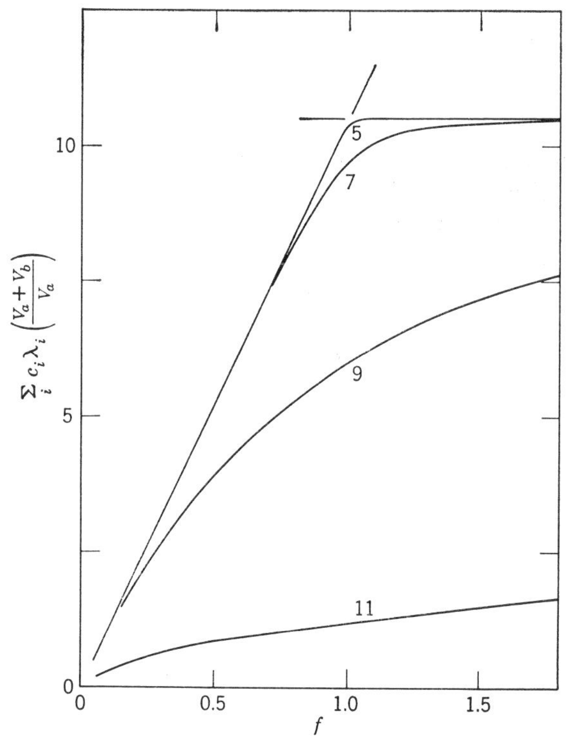

FIG. 5-5. Calculated titration curves for the conductometric titrations of 0.1 F solutions of various monobasic weak acids with ammonia. For details see the legend of Fig. 5-3.

metrically with ammonia as the reagent both acids can be determined with satisfactory accuracy.

5-8. Conductometric Precipitation Titrations. A reaction in which a precipitate is formed may be written

$$M^+ + A^- + N^+ + B^- = NA + M^+ + B^-$$

The net effect of this reaction up to the equivalence point (assuming for simplicity that the solubility of NA is negligible) is the replacement of the A^- ions in the original solution with B^- ions from the reagent. More generally, the ion being precipitated is supplanted by the ion of the reagent which has the same charge. A little reflection will show that if λ_{B^-} is less than λ_{A^-}, the conductance will decrease until the equivalence point; whereas if λ_{B^-} and λ_{A^-} are the same, the conductance will remain unchanged; and if λ_{B^-} is greater than λ_{A^-}, the conductance will increase. Beyond the equivalence point the conductance invariably increases as both N^+ and B^- ions are added in excess. So three kinds of titration curves are possible: one ∨-shaped, one _/-shaped, and one ╯-shaped. The first of these will naturally give rise to the most precise results. It follows, therefore, that in titrating silver ion ($\lambda^0 = 61.9$) with chloride it would be better to use lithium chloride ($\lambda^0_{Li^+} = 38.7$) than potassium chloride ($\lambda^0_{K^+} = 73.5$) as the reagent.

Titrations of mixtures of substances which form precipitates with the reagent (*e.g.*, of bromide and chloride with silver ion) are not feasible by conductometric methods, because the differences between the equivalent ionic conductances involved are too small to permit the accurate location of the end points.

An interesting kind of titration, but one which has had little use, is that of a metal ion with an organic reagent, such as

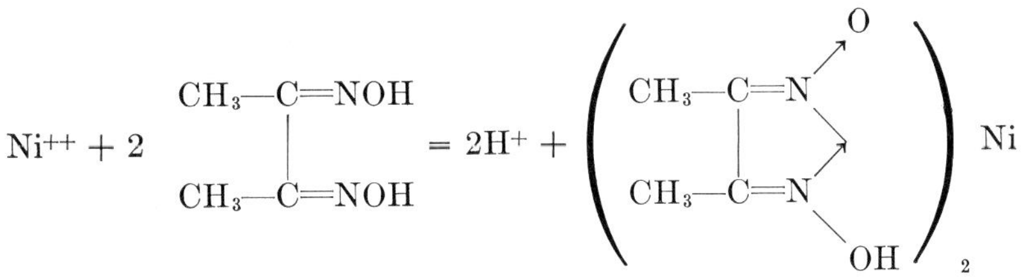

$$Ni^{++} + 2 \begin{array}{c} CH_3-C=NOH \\ | \\ CH_3-C=NOH \end{array} = 2H^+ + \left(\begin{array}{c} \qquad\quad O \\ \qquad \nearrow \\ CH_3-C=N \\ | \qquad \searrow \\ CH_3-C=N \\ \qquad \searrow \\ \qquad\quad OH \end{array} \right)_2 Ni$$

This is the familiar reaction between nickel ion and dimethylglyoxime to form the scarlet nickel–dimethylglyoxime precipitate employed in the gravimetric determination of nickel. On titrating a solution of a nickel salt with dimethylglyoxime (or, better, the related compound α-furildioxime, which has the advantage over dimethylglyoxime of being fairly soluble in water), hydrogen ion is liberated, causing the conductance to increase up to the equivalence point. As the reagent is virtually un-ionized, an excess will not affect the conductance (after correction for dilution) and a /‾-shaped titration curve will result.

5-9. Conductometric Redox Titrations. Almost all redox titrations have to be carried out in the presence of a large excess of acid or base. A typical titration is that of ferrous iron with permanganate, in which an 0.002 *N* (0.002 *F*) solution of ferrous ion in 1 *N* (0.5 *F*) sulfuric acid is titrated with 0.1 *N* (0.02 *F*) permanganate. Although the resulting reaction

$$MnO_4^- + 5Fe^{++} + 8H^+ = Mn^{++} + 5Fe^{+++} + 4H_2O$$

does consume hydrogen ion, thus decreasing the conductance of the solution up to the equivalence point, the fraction of the hydrogen ions thus removed is very small, so that the entire change in the conductance is less than 1 per cent. For this reason, even though it might be possible to perform the titration with fair accuracy if very refined equipment were available for the conductance measurements, conductometric redox titrations are scarcely ever made.

5-10. Potentiometric *vs.* Conductometric Measurements. We have seen that many kinds of measurements which can be performed potentiometrically according to Chaps. 3 and 4 can also be made conductometrically. When and why is one of these kinds of measurements selected in preference to the other?

This is a question to which there is no very clear-cut answer. If one of these techniques were invariably superior, it is obvious that the other would long ago have been relegated to the museum. In choosing between the two, one has to take into account the nature and composition of the sample, the kind and accuracy of the information desired, and the quality of the apparatus which is available. Perhaps the simplest way to describe what is involved in making such a choice is to cite a typical example.

Suppose that one wants to determine the solubility of a slightly soluble salt, such as lead thiocyanate. This might be done by measuring the conductance of a pure saturated aqueous solution of the salt and using the considerations outlined in Sec. 5-6, or it might be done by measuring the potential of a lead electrode in a solution saturated with lead thiocyanate and containing a known concentration of excess thiocyanate ion (Sec. 3-13). Or one might prepare a saturated solution of lead thiocyanate and analyze it for either lead or thiocyanate ion by a potentiometric or conductometric titration. The student will be able to discern numerous other possibilities after reading later chapters, but for the moment we shall concern ourselves with only these six.

It has to be emphasized that in selecting one of the methods just outlined, the chemist will be greatly influenced by his own experience with it under the experimental conditions he must use. If in his own hands analyses by direct potentiometry with heavy metal electrodes have proved to be difficult and unreliable, he will tend to shun them in favor of other methods. If the distilled water available in his laboratory is of such poor quality that the solvent correction (Sec. 5-11) is a large fraction of the total conductance of the saturated lead thiocyanate solution, he will prefer to use a potentiometric technique. These are things which cannot be treated by general rules; they must be left to the chemist's own judgment, and this properly reflects to a very large extent his individual experiences and abilities.

More than this, however, is involved in making the final choice. Let us suppose that the solubility of lead thiocyanate in pure water is thought to be approximately 0.01 F, and that an accuracy of ± 10 per cent in the solubility product will suffice (note that this corresponds to an accuracy of ± 3 per cent in the concentration of either ion in the saturated solution).

The expected concentrations of the ions are then $[Pb^{++}] = 0.01$ *M* and $[SCN^-] = 0.02$ *M*. These are large enough so that a titration method ought to be feasible if a suitable reagent can be found. One would not think, for example, of attempting either a potentiometric or a conductometric titration of a 10^{-8} *M* solution; either the volume of reagent used would be far too small to measure accurately, or the reagent solution would be so dilute that it would be very unlikely to be stable.

In order to carry out a lead determination by direct potentiometry with an accuracy of ± 3 per cent, we would have to secure a potential measurement which would be accurate to ± 0.5 mv. This does not seem very practical, since the liquid-junction potential alone may be at least this great, and so this possibility would probably be rejected. If we use direct conductometry, however, we can tolerate an over-all error of ± 3 per cent [compare Eq. (5-10), noting that the value of c secured there must be cubed to obtain the solubility product]. This is quite feasible even with a relatively inexpensive conductance bridge.

The potentiometric titration of lead with, say, sulfate would probably not give very good results, partly because of the difficulty of getting stable potentials with a lead electrode, but primarily because lead sulfate is so soluble ($K_{PbSO_4} = 1 \times 10^{-8}$) that the rate of change of the electrode potential near the equivalence point would not be very great (especially since a tenfold change in the concentration of lead ion will alter the electrode potential by only 30 mv. at 25°C.). But on the other hand, a potentiometric titration of the thiocyanate with silver nitrate, using a silver indicator electrode, would be capable of giving much better results, partly because silver thiocyanate is much less soluble than lead sulfate ($K_{AgSCN} = 1 \times 10^{-12}$), and also because n for the silver ion–silver electrode is only 1, so that a tenfold change in the concentration of silver ion will change the electrode potential by 59 mv. at 25°C.

The conductometric titration of this solution with a standard lithium sulfate solution ($\lambda^0_{Pb^{++}} = 73$, $\lambda^0_{Li^+} = 38.7$) would give a $\vee$-shaped titration curve. Because lead sulfate is not very insoluble, there would be a good deal of curvature in the vicinity of the equivalence point, but this merely means that the two straight lines would have to be constructed from data obtained considerably before and considerably after the equivalence point. In either of these cases the solubility of the lead sulfate would be reduced practically to zero by the excess of lead or sulfate ions. *This ability to cope with a very incomplete reaction is one of the chief advantages of conductometric* (as well as amperometric and spectrophotometric) *titrations over potentiometric titrations.* Incidentally, the titration curve could be considerably improved by the addition of 25 per cent ethanol to decrease the solubility of the lead sulfate. (It should be noted that even a tenfold reduction of the solubility of lead sulfate would not suffice to give a potentiometric titration curve with a sharp inflection at the equivalence point.)

A conductometric titration of the solution with silver nitrate ($\lambda^0_{SCN^-}$ = *ca.* 70†, $\lambda^0_{NO_3^-}$ = 71.4) would give a curve whose shape would be less favorable to the accurate location of the end point than that secured in the lead-sulfate titration, even though silver thiocyanate is much less soluble than lead sulfate. An additional difficulty here would arise from the adsorption of ions onto the colloidal particles of the silver thiocyanate, which would cause the conductance to deviate from the expected values. This may be remedied by the addition of 0.05 to 0.1 per cent gelatin, which serves to coagulate the precipitated silver thiocyanate.

In this example, purely chemical considerations lead to the conclusion that the desired information could be best secured by one of three methods: direct conductometry, potentiometric titration with silver nitrate, or conductometric titration with lithium sulfate, and the chemist would have to choose among these. It hardly needs to be mentioned that a conscientious worker would apply at least two of them as a check on each other.

By the same reasoning processes, if the required accuracy had been much better than a few per cent, we should have decided that, of the methods mentioned, only the potentiometric titration would be likely to give sufficiently accurate results. The development of this train of thought is left to the student as an instructive exercise.

The most important restriction on the use of conductometric measurements is that *the solution must not contain large concentrations of foreign ions.* Although the concentration of, say, a pure 10^{-4} *F* copper sulfate solution can be found with considerable accuracy by direct conductometry, this would be difficult even with the most precise apparatus if 0.01 *F* sulfuric acid were present as well, even if the concentration of sulfuric acid were very accurately known. The determination of 0.01 *F* chloride in a 1 *F* sodium nitrate solution by a potentiometric titration with silver nitrate would be a relatively easy matter; conductometrically it would be all but impossible. It is this, more than anything else, which limits the practical applications of conductometric methods.

On the other hand, when sufficiently precise apparatus for conductance measurements is available, such as the Leeds and Northrup Company's Jones and Josephs bridge (which permits measurements to ±0.001 per cent), the accuracy of a conductometric titration can easily equal or surpass that of most other common methods even in the presence of appreciable concentrations of foreign salts.

† Apparently the exact value of this quantity has never been measured. However, it is probably not much different from the values of λ^0 for bromide (78.4) and nitrate ions.

5-11. The Solvent Correction. The solvent correction is closely related to the presence of foreign salts in the solution being studied, and it imposes a practical lower limit on the concentration of a salt which can be determined conductometrically.

The solvent correction may be defined as the correction to the experimental data which is necessitated by the fact that the solvent itself makes a small contribution to the observed conductance. The equivalent conductance of dissociated water ($\Lambda_{H_2O} = \lambda_{H^+} + \lambda_{OH^-}$) is 546.4 mho-cm.[2] per equivalent; by Eq. (5-3), since c is 10^{-7} N, this corresponds to a specific conductance κ of 5.5×10^{-8} mho per cm. Water of this specific conductance has indeed been prepared[1] (by 43 successive distillations *in vacuo* in a quartz apparatus!), but ordinary laboratory distilled water has a very much higher specific conductance which is due to traces of dissolved ammonia, carbon dioxide, and other materials. Usually the distilled water available in the laboratory will have a specific conductance of about 2 to 3×10^{-6} mho per cm. This is just equal to the specific conductance of a 2×10^{-5} N solution of potassium chloride. It follows from this that the specific conductance of a very dilute solution of potassium chloride (or any other salt) will not appear to be proportional to its concentration, and that seriously erroneous results will be secured in analyses by direct conductometry unless some correction for the conductance of the water is made.

Provided that the ions of the salt involved do not react in any way with the ions (HCO_3^-, NH_4^+, etc.) present in the solvent, we may compute a "corrected" specific conductance which refers to the solute alone by means of the simple equation

$$\kappa_{corr.} = \kappa_{obs.} - \kappa_{solvent} \tag{5-19}$$

which follows from Eqs. (5-2) and (5-7). This equation can be used with fair confidence if the solute is a neutral salt. If, however, the specific conductance of the solvent is almost entirely due to the hydrogen and bicarbonate ions formed by the dissociation of dissolved CO_2, it is evident that the concentrations of the ions derived from this source would be greatly decreased by the addition of even quite small concentrations of hydrochloric acid, and so Eq. (5-19) would grossly overestimate the solvent correction which should be applied. An even more serious underestimation of the solvent correction would result if Eq. (5-19) were applied to sodium hydroxide solutions prepared with water containing an appreciable concentration of carbon dioxide. In such cases there is nothing to do but purify the solvent (*e.g.*, by a careful redistillation) to reduce its specific conductance to an entirely negligible value. No matter to what

[1] F. Kohlrausch and A. Heydweiller, *Ann. physik. Chem.*, **53**, 209 (1894).

lengths this purification is carried, however, it is clear that the finite conductance of even the purest water makes it impossible to derive any information from conductance measurements with extremely dilute solutions.

The significance of the solvent correction can be illustrated by some typical data. The resistance of a saturated silver chloride solution was found to be 220,000 ohms with a cell whose cell constant was 0.972, so that the specific resistance was 4.42×10^{-6} mho per cm. The equivalent conductance of silver chloride at infinite dilution is 138.2 mho-cm.^{2} per equivalent, so that if the conductance of the solution had been entirely due to dissolved silver chloride the concentration of this salt would have been 3.20×10^{-5} N, which corresponds to a solubility product of 1.03×10^{-9}, over five times the accepted value. However, the water used had a resistance of 372,000 ohms in the same cell. Consequently its specific conductance was 2.62×10^{-6} mho per cm., and the specific conductance due to the silver chloride was $(4.42 - 2.62) \times 10^{-6} = 1.80 \times 10^{-6}$ mho per cm. From this the concentration of dissolved silver chloride is calculated to be 1.30×10^{-5} N, which yields $K_{\text{AgCl}} = 1.69 \times 10^{-10}$, a value which is very close to the truth.

5-12. Electrical Apparatus for Conductance Measurements. Although it is possible in practice as well as in principle to measure the conductance of a solution by using a direct current, the electrode reactions which occur when direct current is used will change the compositions of the layers of solution near the electrode surfaces, and hence the measured resistance will tend to drift with time unless special and rather elaborate precautions are taken. For this reason it is customary to make the measurements with alternating rather than direct current. Most practical circuits are of the Wheatstone bridge type shown in Fig. 5-6.

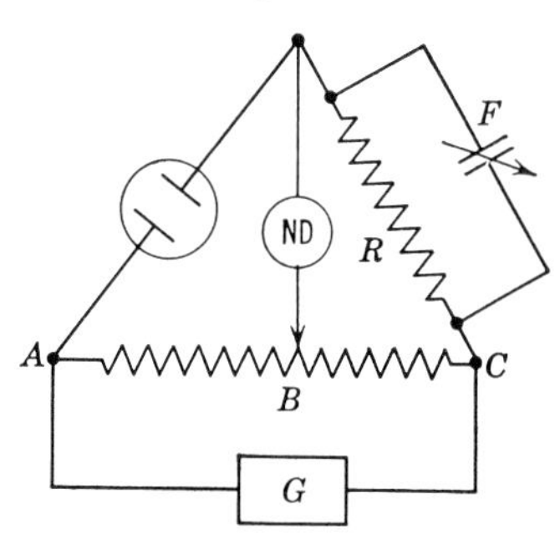

FIG. 5-6. Simple Wheatstone bridge circuit for conductance measurements.

In this circuit G is an a-c generator, ABC is an accurately linear voltage divider, and R is typically any one of a number of standard resistors whose resistances differ by successive factors of 10. The null detector, ND, which may be a pair of earphones (for measurements around 1000 cycles), an a-c galvanometer, or a 6E5 "magic eye" tube in an appropriate electronic circuit, serves to indicate the point at which no current flows between the points to which it is connected. This occurs when the voltage drop across AB is the same as the voltage drop through the cell:

$$i_{AB}R_{AB} = i_{\text{cell}}R_{\text{cell}}$$

At the same time

$$i_{BC}R_{BC} = i_R R_R$$

However, if no current flows through the null detector, we must have

$$i_{AB} = i_{BC}$$

and

$$i_{\text{cell}} = i_R$$

At balance, therefore

$$\frac{R_{AB}}{R_{BC}} = \frac{R_{\text{cell}}}{R_R}$$

or

$$R_{\text{cell}} = \frac{R_R R_{AB}}{R_{BC}} \tag{5-20}$$

which is the fundamental equation for the use of a Wheatstone bridge circuit.

It is obvious that the best precision will be secured if the ratio R_{AB}/R_{BC} is neither very large nor very small, and this is why provision is made for selecting different values of the standard resistance R. If at any setting of the latter the ratio R_{AB}/R_{BC} required to balance the bridge is less than about 0.2 or greater than about 5, the standard resistance is changed by a factor of 10 and the position of the slider B is varied until balance is again achieved.

When very high resistances are involved, trouble, in the form of an apparent lack of sensitivity in the null point detector, may be encountered because of capacitance effects. Without going deeply into the a-c theory involved, we may say that what is achieved in a-c measurements is a balancing of, not the *resistances*, but the *impedances* of the bridge. The impedance depends on the resistance, but it also depends on the capacitance. Since an unbalanced capacitance in one arm of the bridge (*i.e.*, in the cell, which can never be designed so as to have no capacitative reactance whatever) will introduce a phase shift into the a-c passing through it, this will make it impossible to reduce the current through the null detector to zero regardless of the value of R_{AB}/R_{BC}. It is to permit balancing the bridge when the resistance of the cell is high that the variable condenser F is employed. With F set at zero, the position of B is varied until an approximate balance is achieved; then F is slowly increased while B is moved back and forth through this point until a sharp minimum is indicated by the null detector. A further increase of F beyond this optimum value leads to a return of the apparent insensitivity of the detector.

Too many components can be used in setting up a Wheatstone bridge circuit to permit detailed consideration of all the possibilities. About the simplest that is worth considering is the slide wire of a Leeds and Northrup students' potentiometer. By making appropriate connections to the binding posts marked H, H', L, and L' the slide wire of this instrument can be isolated from the rest of the internal wiring and used as the bridge *ABC* in Fig. 5-6. A 1000-cycle audio oscillator, which should develop a reasonably pure sine-wave output, is used as the generator *G*. The assembly of standard resistors, *R*, may be a four- or five-dial resistance box, or more economically it may be a rotary switch on which are mounted a number of General Radio Co. (Cambridge, Mass.) Type 500 standard resistances, having values of 1, 10, 100, 1000, 10^4, and 10^5 ohms.

Several completely self-contained conductance bridges can be bought commercially. A typical one, the Type RC-16B available from Industrial Instruments, Inc. (Cedar Grove, N.J.), for about $150, has provision for making measurements at either 60 or 1000 cycles and a range from 0.2 to 2,500,000 ohms (which is considerably wider than is needed in most work); balance is indicated by a 6E5 tube. The slightly more expensive Serfass bridge sold by Arthur H. Thomas Co. (Philadelphia, Pa.) can be made to read directly in reciprocal ohms, which is a great convenience in practical work.

5-13. Conductance Cells. Several typical kinds of cells are shown in Fig. 5-7. Type *a* is the Ostwald cell.[1] Although much used at one

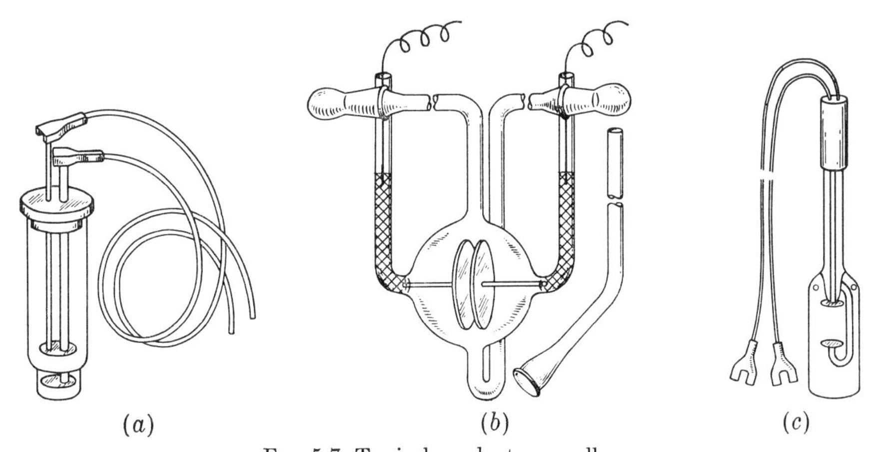

FIG. 5-7. Typical conductance cells.

time, it suffers from the disadvantage of having large capacity effects between the leads to the electrodes, and it has been replaced in accurate work by the Washburn cell,[2] *b*, in which these leads are so placed that the interelectrode capacitative effects are virtually eliminated. The "dip type" cell[3] shown in *c* is very convenient for student work; it is used by simply dipping it into a large beaker containing the solution. It is possi-

[1] E. H. Sargent and Co. (Chicago, Ill.), Catalog S-29855.

[2] Sargent Catalog S-29875.

[3] Sargent Catalog S-29975. These illustrations are reproduced through the courtesy of the manufacturer.

ble to buy these cells with different cell constants for use with solutions having different specific conductances, but this is quite unnecessary, since exactly the same purpose is served by the resistance R in Fig. 5-6.

The platinum electrodes used in conductance cells should be platinized. Directions for platinizing electrodes are given in Sec. 4-3. Of course no platinum wire anode is needed, for while one electrode of the cell is being platinized the other serves as the anode. The simple circuit shown in Fig. 5-8 is convenient for this purpose. Only a very light coat of platinum should be deposited on each electrode; the surface should appear light gray, not black. Too heavy a deposit of platinum leads to undesirable adsorption effects. After the platinum is deposited (which should consume only about 15 seconds of cathodic polarization of each electrode), the platinizing solution is replaced by dilute sulfuric acid and the electrodes are alternately made the cathode and anode for brief periods, exactly as described in Sec. 4-3. The cell is then stored in distilled water or dilute sulfuric acid whenever it is not in use. If the platinum black dries out or loses its activity for any other reason, a fresh coat can be deposited right on top of the old one.

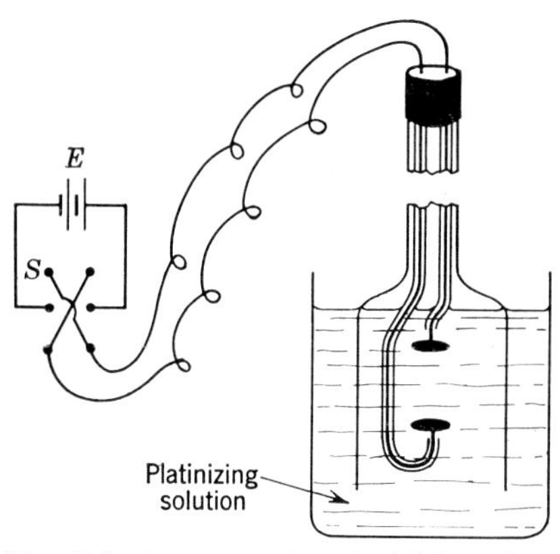

FIG. 5-8. Apparatus for platinizing electrodes of conductance cells.

5-14. Temperature Control in Conductance Measurements. As was mentioned above, the specific resistances of most solutions of strong electrolytes decrease about 2 per cent when the temperature is raised one degree. This change is a reflection of the increase in the mobilities of the ions with increasing temperature, and the variation of the conductance of a weak electrolyte with temperature may be considerably larger than this because of the additional variation of the degree of dissociation with changes of temperature. As it is a relatively simple matter to make conductance measurements with a precision of ± 0.5 per cent or so even with quite inexpensive apparatus, it is evident that the temperature should be controlled to within about ± 0.25°C. even for routine analytical purposes. This precision of temperature control is easily obtained by immersing the cell in a simple water thermostat. Occasional conductometric titrations can be carried out with the cell at room temperature, even though the heat evolved or absorbed in the titration reaction may cause a detectable distortion of the curve secured. For example, the quantity of heat liberated in the titration of 0.1 F hydrochloric acid with 1 F sodium hydroxide is great enough to raise the temperature of the solution more than a degree even if the two solutions are initially at the same temperature. Evidently even more serious variations of temperature might occur if one solution were appreciably warmer than the other. Usually the error which results from ordinary variations in room temperature is quite negligible in conductometric titrations having sharply defined end points. But in direct conductometry the error thus caused may be much larger than the error of the measurements themselves, and care must be taken to guard against this in practical work.

PROBLEMS

The answers to the starred problems will be found on page 403.

*1. The specific conductance of 0.02000 N potassium chloride is 2.768×10^{-3} reciprocal ohms per cm. at 25°C. What is the cell constant of a cell whose resistance is 407.9 ohms when filled with this solution?

2. The ohm is defined as the resistance at 0°C. of a column of mercury, having a uniform cross section and a length of 106.300 cm., and weighing 14.4521 g. The resistance of a certain conductance cell was 2.764 ohms when filled with mercury at 0°C., and 1017.6 ohms when filled with a strong sulfuric acid solution at the same temperature. What was the specific conductance of the sulfuric acid solution?

3. A typical conductance cell consists of two pieces of platinum foil which are mounted rigidly with respect to the two ends of a glass rod or tube. If the linear coefficients of thermal expansion of platinum and of the glass used in a certain cell are both equal to 9.0×10^{-6} per degree, what is the temperature coefficient of the cell constant for that cell?

*4. The cell of Problem 1 was filled with a solution of hydrochloric acid at 25°C. and was then found to have a resistance of 496.2 ohms. What was the specific conductance of the solution?

5. The equivalent conductance of a hydrochloric acid solution at concentrations up to about 0.03 N is given fairly accurately by the equation $\Lambda = 426.2 - 141\sqrt{C}$. What was the normality of the hydrochloric acid solution of Problem 4?

6. If the transference number of hydrogen ion in 0.01 N hydrochloric acid is found by the moving boundary method to be 0.8251, what is the equivalent conductance of chloride ion in this solution? Use the data given in Problem 5.

7. From the following data on the equivalent conductances and cation transference numbers of lithium chloride solutions of various concentrations, find the equivalent conductance of lithium ion at infinite dilution.

C, equiv./l.	Λ	t_+
0.02	104.65	0.3261
0.01	107.32	0.3289
0.005	109.40	0.3304
0.001	112.40	0.3337
0.0005	113.15	0.3345

8. The equivalent conductances of acetic acid solutions of various concentrations, according to MacInnes and Shedlovsky, are reproduced in the following table. Assuming that the equivalent conductances of hydrogen and acetate ions are equal to their values at infinite dilution, find the dissociation constant of the acid at each concentration, and by extrapolation find the thermodynamic dissociation constant of acetic acid.

[HOAc], *F*	Λ, *mho-cm.²/g.-equiv.*
0.0098421	16.371
0.0034407	27.199
0.0013634	42.227
0.00021844	96.493
0.00011135	127.75

***9.** Show that the conductance of an extremely dilute solution of any strong base in pure water must pass through a minimum as the concentration of base is increased, and find the concentration of lithium hydroxide at which this minimum occurs.

10. What information is required to permit the calculation of K_1 for a dibasic weak acid from data on the conductances of solutions of the acid?

***11.** The solubility product of thallous bromide is 3.6×10^{-6}. What would be the resistance of a saturated solution of thallous bromide, measured with a cell whose cell constant was 0.9331? (Neglect the solvent correction.)

12. The basic dissociation constant of pyridine, which is the equilibrium constant of the reaction

$$C_5H_5N + H_2O = C_5H_5NH^+ + OH^-$$

is 6.5×10^{-5}. Sketch the titration curves which would be obtained in conductometric titrations of (*a*) pyridine with hydrochloric acid, (*b*) pyridinium chloride with sodium hydroxide, and (*c*) pyridine with formic acid ($K_a = 1.8 \times 10^{-4}$). Which titration would be likely to yield the most precise results?

13. Exactly 100 ml. of a solution of hydrochloric acid is titrated with 0.1005 *F* sodium hydroxide, using a conductance cell for which l/A is 1.023. The end point is found to occur at 40.00 ml. of base. Calculate the resistance observed after the addition of 0, 20, 39, 40.01, and 50 ml. of base.

14. The equivalent conductance of an 0.0001952 *F* solution of anilinium chloride ($[C_6H_5NH_3]^+Cl^-$) is 138.4. In the presence of a large excess of aniline, the equivalent conductance of anilinium chloride at the same concentration is only 107.2. The equivalent conductance of 0.0001952 *F* hydrochloric acid is 419.2. Calculate the equilibrium constant of the reaction

$$C_6H_5NH_3^+ = C_6H_5NH_2 + H^+$$

***15.** The equivalent conductance of potassium ion at infinite dilution increases from 73.52 at 25°C. to 159 at 75°C. What is the mean temperature coefficient of the equivalent conductance of potassium ion over this range of temperatures?

16. The specific conductance of a certain sample of laboratory distilled water is found to be 2.1×10^{-6} mho per cm. If dissolved ammonia (for which K_b is 1.8×10^{-5}) is the only impurity present, what will be the pH of the water?

CHAPTER 6

POLAROGRAPHY AND AMPEROMETRIC TITRATIONS

6-1. Introduction. In the three preceding chapters we have discussed two special aspects of the relationships among the composition of a solution in an electrochemical cell, the voltage across the two electrodes of the cell, and the electric current which flows between the electrodes. Potentiometric measurements deal with the relationship between cell voltage and solution composition when the current flowing through the cell is zero. Conductometric measurements deal with the relationship between the composition of a solution and its electrolytic resistance, which in turn is related to the voltage across the cell and the current flowing through it.[1]

In this chapter we shall present a more general treatment of the relationships among these three variables. In particular, we shall discuss the entire course of the current–voltage curve, of which a potentiometric

[1] The relationship is given explicitly by an equation of the form of Eq. (7-1). In the measurement of the conductance of, say, a perchloric acid solution, the reactions taking place at the two electrodes are almost exclusively

$$2H^+ + 2e = H_2 \text{ (adsorbed)}$$

at the cathode and

$$H_2 \text{ (adsorbed)} = 2H^+ + 2e$$

at the anode; the alternation of the current is sufficiently rapid to prevent appreciable amounts of hydrogen from leaving the electrode surface during the cathodic half of the cycle. Because the overpotentials of these reactions at platinized electrodes are very small, E_d in Eq. (7-1) is virtually zero, and naturally i_r is also zero because no current will flow between two identical electrodes in the same solution when they are shorted together. So the cell obeys Ohm's law, and the perchlorate ions contribute to the flow of current between the electrodes without taking any appreciable part in the reactions by which electrons are transferred between the electrodes and the solution. No doubt the same reactions occur at the two electrodes during the measurement of the conductance of a solution of a salt like sodium sulfate, though in this case we might more properly write

$$2H_2O + 2e = H_2 \text{ (adsorbed)} + 2OH^-$$

With a salt such as cupric nitrate, on the other hand, some and perhaps even most of the electron transfer across the electrode–solution interfaces is attributable to the alternate reduction and oxidation of the copper.

measurement gives only one point. This more general study is known as *polarography* or *voltammetry*. *Voltammetry* is the generic term, and *polarography* refers to that portion of the general field which deals with the characteristics of current–voltage curves obtained by the use of a dropping mercury electrode. Because the area of such an electrode is very small, it follows for reasons which will appear shortly that the current flowing through the cell is also very small, generally only a few microamperes. Consequently the iR drop through the cell can easily be made small, and then the *voltage* impressed across the two electrodes will be virtually equal to the difference between their *potentials*. The other electrode of a polarographic cell is always a reference electrode whose potential is known and is unaffected by the flow of a small current.

As the potential of the small electrode or microelectrode is varied, the composition of the solution at its surface must change in accordance with the Nernst equation. Suppose, for example, that the electrode is immersed in a solution containing equal concentrations of chromic and chromous ions, and that the experimental conditions are such that the chromic–chromous equilibrium is rapid and thermodynamically reversible. If the microelectrode is simply shorted to an S.C.E., a current will flow through the circuit as the reaction

$$2Cr^{++} + Hg_2Cl_2 = 2Cr^{+++} + 2Hg + 2Cl^-$$

proceeds spontaneously. To prevent this reaction from occurring—*i.e.*, to decrease the current to zero—a potential which is equal but opposite to the potential of the cell must be applied across the electrodes from some external source. This, of course, is the fundamental principle of potentiometric measurements. The potential thus applied will in this case be simply equal to the formal potential of the chromic–chromous couple under the conditions of the experiment and with respect to whatever reference electrode is used.

Now, if the externally applied potential is changed slightly in either direction, the concentrations of chromic and chromous ions at the surface of the microelectrode must change to conform to the equation

$$E_{ind} = E^{0'}_{Cr^{+++},Cr^{++}} - 0.05915 \log \frac{[Cr^{++}]}{[Cr^{+++}]}$$

If the potential of the microelectrode is forced to become more negative than the formal potential of the chromic–chromous couple, the ratio $[Cr^{++}]/[Cr^{+++}]$ must increase, and this can only happen if chromic ion is reduced at the surface of the microelectrode. This will cause a current to flow through the cell, and if the reference electrode is an S.C.E. this reduction of chromic ion will be accompanied by an oxidation of mercury to mercurous chloride in the reference electrode.

The current which flows under these conditions will depend on two things. One is evidently the ratio of the chromous and chromic concentrations which must prevail at the surface of the microelectrode to satisfy the Nernst equation at the particular applied potential used. If this ratio is larger than 1, a cathodic current will be needed to reduce chromic ion; if the ratio is smaller than 1, an anodic current will flow to oxidize chromous ion. The result will be, in the first of these cases, that the concentration of chromic ion at the electrode surface will be smaller than in the bulk of the solution some distance away, and so chromic ions will diffuse toward the electrode surface in an attempt to equalize the chromic ion concentration. At the same time chromous ions will diffuse in the opposite direction because their concentration is higher at the electrode surface than in the body of the solution.

The other factor which will affect the current is the rate at which the substance being oxidized or reduced reaches the microelectrode surface by this diffusion process. If this rate is small, only a few electrons can be transferred to or from the electrode in a given length of time; but if it is large it will be necessary to oxidize or reduce a considerable number of ions during each second to maintain the equilibrium ratio of concentrations in the layer of solution surrounding the microelectrode. This rate in turn depends on several factors. It is proportional to the concentration of the oxidizable or reducible substance in the bulk of the solution, it is proportional to the area of the microelectrode, and it depends on the mobility of the ions or molecules being oxidized or reduced. In quantitative polarographic analysis it is the first of these which is of most immediate importance, for under the proper conditions the current measured in a polarographic experiment is directly proportional to the concentration of some one constituent of a solution. However, the other factors are also of vital importance to polarographic theory and practice.

We have seen qualitatively how the relationship between current and potential in a solution of constant bulk composition involves thermodynamic considerations similar to those which enter into potentiometric measurements, and we have also seen how the relationship between current and solution composition at constant potential can be employed in quantitative analysis. These applications of polarographic measurements will be discussed in detail in the following sections of this chapter.

6-2. The Dropping Mercury Electrode. The dropping mercury electrode (usually abbreviated d.m.e. or d.e.) has some peculiar characteristics which are reflected by the current–potential curves obtained with it, and consequently a brief description of the dropping electrode and its properties will be given here.

A dropping mercury electrode consists essentially of a length of very fine-bore (i.d. about 0.06 mm.) capillary tubing connected to a "stand tube" which contains a column of pure metallic mercury. A typical dropping electrode assembly is shown in Fig. 6-1. As mercury flows through the capillary, it forms a droplet on the capillary tip. This grows until it reaches a certain perfectly definite maximum size, which depends on the radius of the capillary tip and on the interfacial tension between mercury and the solution. Then the drop falls and another one begins to grow, and this process is repeated indefinitely. The measurement of the weight of a droplet of liquid formed under such conditions is, of course, the basis of a familiar method for the determination of surface and interfacial tensions.

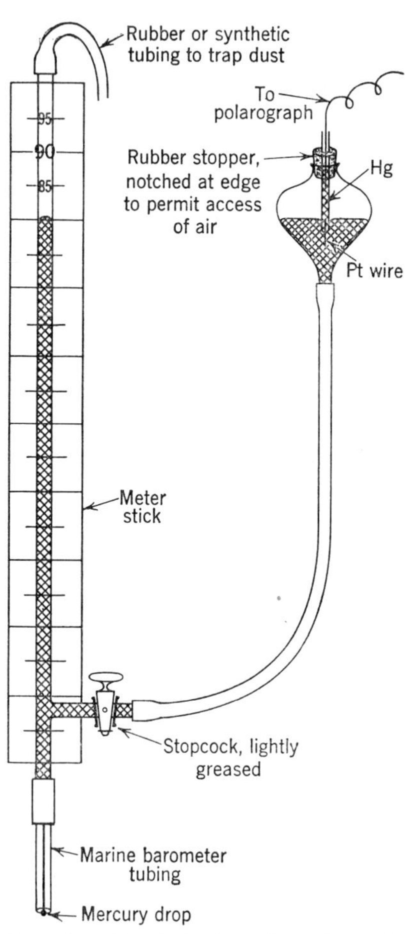

Fig. 6-1. Typical dropping electrode assembly.

Mercury has several advantages over a solid metal such as platinum as an indicator electrode for polarographic work. One of the most important of these is the very high overpotential (Sec. 7-2) for the evolution of hydrogen on mercury. This makes it possible to study the reductions of many substances, including even the alkali and alkaline earth metal ions, which could not possibly be deposited onto a platinum cathode without interference from the simultaneous reduction of water or hydrogen ion. Unlike a solid electrode, a mercury droplet has a perfectly smooth surface free from scratches or any other irregularities; this makes the accurate calculation of the electrode area a rather simple matter. Because the surface of the electrode is renewed every three or four seconds by the fall of one drop and the growth of another, adsorbed or deposited materials cannot accumulate on the electrode surface. The numerous metals which are soluble in mercury usually behave more reproducibly and reversibly when amalgamated than in the pure solid state, and this is why indicator electrodes made from such metals are frequently amalgamated for use in potentiometric work (Sec. 3-8). Finally, the dropping mercury electrode is much less sensitive to mechanical disturbances than a stationary solid microelectrode, because a shock which would cause the latter to vibrate for some time merely dislodges one drop prematurely from a dropping electrode without having much effect on the next. An even more serious defect of a stationary solid microelectrode is that the concentration gradient resulting from the flow of current extends farther and farther out into the solution as time goes on, so that ions must diffuse through a greater and greater distance to reach the electrode surface, resulting in a steady decrease of the current. With a dropping electrode it is found that the fall of each drop stirs the solution around the capillary tip, and therefore each new drop is born into a solution of almost exactly the same composition as was encountered by its predecessor.[1]

On the other hand, there are certain definite limitations associated with the dropping electrode. Metallic mercury is fairly readily oxidized, and so even in solutions containing only such noncomplexing anions as perchlorate and nitrate the oxidation of mercury begins at a potential near +0.4 v. *vs.* S.C.E. Since the activity of metallic mercury at the drop surface is naturally always equal to unity, the ordinary operation of the Nernst equation means that the concentration of dissolved mercurous or mercuric ion present around the drop must be quite large even at potentials only very little more positive than this. So a large negative (anodic) current flows, making it impossible to measure or even detect a small additional current due to the reduction or oxidation of some substance originally present in the solution. This limit of the range of potentials within which useful data can be gotten is even less positive in solutions containing ions like chloride, cyanide, and hydroxide, which combine with either mercurous or mercuric ion and thus facilitate the oxidation of metallic mercury. For work at potentials where mercury is oxidized, there is no way of avoiding the disadvantages of a solid microelectrode (*cf.* Sec. 6-18).

As was mentioned above, the current obtained in any given set of circumstances depends on, among other things, the area of the microelectrode. The area of a dropping electrode increases continuously during the life of each drop, then suddenly decreases almost to zero at the instant when the drop falls, and repeats this cycle again and again. This

[1] In this connection it may be mentioned that the ratio of electrode area to solution volume in a typical polarographic experiment (and with it the ratio of the current which flows to the concentration of material being reduced or oxidized) is so very small that the rate at which the composition of the bulk of the solution changes because of the flow of current is almost undetectable.

periodic variation of the electrode area produces a periodic variation of the current, so that the current-measuring instrument (which may be a microammeter, a galvanometer, or a potentiometer measuring the iR drop across a known resistance in series with the cell) will oscillate between a maximum and a minimum value. If the measuring instrument responds slowly enough,[1] the mean of its maximum and minimum deflections will be very closely equal to the average current during the drop life.

This regular oscillation usually makes the student feel rather badly treated when he first tries to measure the current in a polarographic experiment. With a little experience, however, he ceases to be troubled by this, and comes to regard the reproducibility of the successive oscillations as a valuable indication of the proper functioning of the capillary.

6-3. Form of the Current–Potential Curve. Let us consider the manner in which the current will vary with the potential of the dropping electrode in a solution of chromic ion in saturated calcium chloride.[2] In this medium the formal potential of the chromic–chromous couple is −0.51 v. *vs.* S.C.E. We shall temporarily neglect the small "residual currents" (Sec. 6-7) which would flow even in the absence of the chromic ion, and we shall assume that the solution contains no other substance which could be reduced or oxidized at a potential near −0.5 v. *vs.* S.C.E.

Suppose first that the dropping electrode and the S.C.E. are simply shorted together, so that the potential of the dropping electrode, $E_{d.e.}$, becomes 0 v. *vs.* S.C.E. At this potential the concentration of chromous ion around the drop surface must conform to the equation

$$0 = -0.51 - 0.05915 \log \frac{[Cr^{++}]}{[Cr^{+++}]}$$

From this we calculate that the fraction of the chromic ion which will have to be reduced to bring about this state of affairs is only 2.4×10^{-9}. Obviously only a very small current would have to flow through the cell to reduce this tiny fraction of the chromic ions present in the thin layer of solution right around the surface of the drop.

Now let us imagine that the cell is connected to an external source of potential such as that shown in Fig. 6-2. In this circuit E is a battery, V is a voltmeter which measures the voltage across the bridge R, and R itself is a voltage divider which can be used to apply a known fraction

[1] Its period should be at least equal to the drop life; to achieve this, it is usually necessary to provide "overdamping" in the form of a small resistance across the terminals of a galvanometer or a large condenser across the terminals of a microammeter.

[2] One of the functions of the calcium chloride is to provide an environment in which the chromic–chromous couple is rapid and reversible. Two other functions will be described in Sec. 6-6.

of the total voltage across it to the electrolysis cell.[1] If the dropping electrode is connected to the negative terminal of this circuit, and if the sliding contact B is set at such a position that the voltage across AB is 0.51 volt, then the value of $E_{d.e.}$ will be forced to become -0.51 v. *vs.* S.C.E. At this potential the ratio of the chromous to chromic concentrations at the surface of the drop must become 1 to satisfy the Nernst equation, and so exactly half of the chromic ions which reach the drop surface will be reduced.

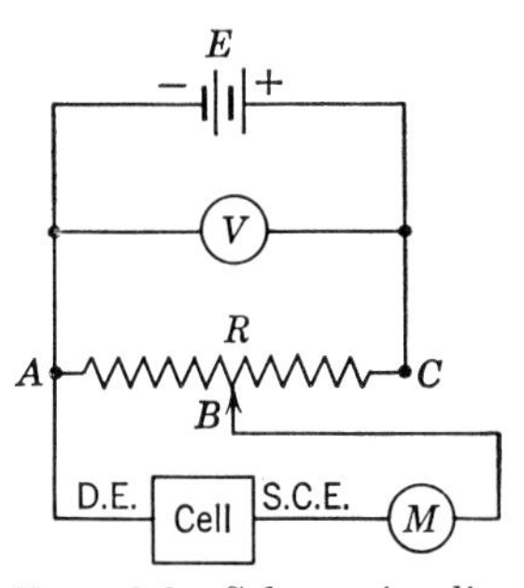

Fig. 6-2. Schematic diagram of a simple circuit for polarographic measurements.

Repeating these calculations at a number of other values of $E_{d.e.}$ gives the results shown in Table 6-1. In checking these calculations the student should note that the fraction of the chromium which is reduced is given by $[Cr^{++}]^0/([Cr^{++}]^0 + [Cr^{+++}]^0)$. The small superscript zeros here serve as a reminder that we are dealing with the concentrations of these ions *at the surface of the drop.*

Table 6-1. Fractions of Cr^{+++} Reduced at Various Values of $E_{d.e.}$*

$E_{d.e.}$, v., *vs.* S.C.E.	$\frac{[Cr^{++}]^0}{[Cr^{+++}]^0}$	Fraction reduced
±0.00	2.4×10^{-9}	2.4×10^{-9}
−0.10	1.2×10^{-7}	1.2×10^{-7}
−0.20	5.7×10^{-6}	5.7×10^{-6}
−0.30	2.8×10^{-4}	2.8×10^{-4}
−0.40	0.014	0.014
−0.43	0.044	0.043
−0.46	0.143	0.125
−0.49	0.459	0.315
−0.51	1.00	0.500
−0.53	2.18	0.685
−0.56	7.00	0.875
−0.59	22.5	0.957
−0.62	72.5	0.986
−0.65	233	0.996
−0.70	1630	0.9994
−0.80	8×10^4	0.99999

* See text for details and sample calculations.

[1] The student should note the similarity between this circuit and the ones shown in Figs. 3-4 and 3-5, and should realize that this similarity is the result of the fact that all of these circuits fulfill substantially the same purpose.

By Faraday's law, the average current during the life of each drop is proportional to the number of chromic ions which reach the surface of the drop and are reduced, and this number in turn depends on the difference between the concentrations of chromic ion in the solution and at the electrode surface. When this difference is small, very little diffusion of chromic ions toward the drop will occur; and as the difference increases, more and more chromic ions will diffuse from the solution into the layer surrounding the drop. Consequently

$$i = k_{Cr^{+++}}([Cr^{+++}] - [Cr^{+++}]^0) \qquad (6\text{-}1)$$

where i is the average current flowing through the cell and $[Cr^{+++}]$ is the concentration of chromic ion in the bulk of the solution. The factors which affect the proportionality constant $k_{Cr^{+++}}$ will be discussed in Sec. 6-8.

The maximum rate of diffusion of chromic ions up to the electrode surface, and consequently the greatest possible current resulting from the reduction of chromic ions, will be observed when the value of $[Cr^{+++}]^0$ is so low that virtually every chromic ion which reaches the electrode surface is reduced. This is the case at any potential more negative than about -0.65 v. under the conditions of Table 6-1. In this region the current which flows is limited by the number of chromic ions which can diffuse up to the electrode surface during the life of the drop. This *diffusion current*, which is given the symbol i_d, is proportional to the concentration of chromic ion in the bulk of the solution, because the value of $[Cr^{+++}]^0$ in Eq. (6-1) is virtually zero. Accordingly we have

$$i_d = k_{Cr^{+++}}[Cr^{+++}] \qquad (6\text{-}2)$$

Combining Eqs. (6-1) and (6-2) now gives an equation for the concentration of chromic ion at the surface of the microelectrode:

$$[Cr^{+++}]^0 = \frac{i_d - i}{k_{Cr^{+++}}} \qquad (6\text{-}3)$$

By exactly the same reasoning we see that the concentration of chromous ion at the electrode surface is proportional to the current i. For purposes of symmetry with Eq. (6-3) we shall write

$$[Cr^{++}]^0 = \frac{i}{k_{Cr^{++}}} \qquad (6\text{-}4)$$

where the constant $k_{Cr^{++}}$ is completely analogous to $k_{Cr^{+++}}$ in Eq. (6-3). Now, with the aid of the Nernst equation relating the concentrations of chromous and chromic ions at the surface of the electrode with the electrode potential, Eqs. (6-3) and (6-4) give

$$E_{\text{d.e.}} = E^{0\prime}_{Cr^{+++},Cr^{++}} - 0.05915 \log \frac{i}{i_d - i} - 0.05915 \log \frac{k_{Cr^{+++}}}{k_{Cr^{++}}} \qquad (6\text{-}5)$$

This can be still further simplified by introducing the concept of the *half-wave potential* $E_{1/2}$, which is the potential at which the current is just half of the diffusion current. According to Eq. (6-5), when $i = i_d/2$

$$E_{\text{d.e.}}(= E_{1/2}) = E^{0'}_{\text{Cr}^{+++},\text{Cr}^{++}} - 0.05915 \log \frac{k_{\text{Cr}^{+++}}}{k_{\text{Cr}^{++}}} \tag{6-6}$$

and when this definition of $E_{1/2}$ is substituted back into Eq. (6-5) we obtain the following equation for a reversible polarographic wave:

$$E_{\text{d.e.}} = E_{1/2} - \frac{0.05915}{n} \log \frac{i}{i_d - i} \tag{6-7}$$

It will be noted from Eq. (6-6) that $E_{1/2}$ depends on the ratio $k_{\text{Cr}^{+++}}/k_{\text{Cr}^{++}}$ as well as on the formal potential of the chromic–chromous couple. Anticipating the detailed discussion in Sec. 6-8 of the factors which affect these constants, we shall note that this ratio is equal to the ratio of the square roots of the diffusion coefficients of chromic and chromous ions; the other factors which influence $k_{\text{Cr}^{+++}}$ and $k_{\text{Cr}^{++}}$ affect them in the same way and so disappear from their ratio. Usually the diffusion coefficients of the oxidized and reduced forms of a couple are not very greatly different, and so, to a very good approximation, the half-wave potential of a polarographic wave is simply equal to the formal potential of the half-reaction which occurs at the dropping electrode.

A plot of the current, i, against the potential of the dropping electrode according to Eq. (6-7) is shown in Fig. 6-3*a*. This characteristic manner of variation of current with electrode potential is known as a polarographic *wave*.

6-4. Comparison of Polarography and Potentiometry. From the foregoing discussion it should be evident that the current–potential data secured in polarographic experiments merely serve to reflect the manner in which the potential of an electrode is related to the composition of the solution surrounding it. As the electrode potential is varied, the concentrations of the oxidized and reduced forms of a redox couple in a very thin layer of solution around the electrode must change accordingly. This can only occur if a current is allowed to flow through the cell, and the magnitude of this current indicates the extent to which the composition of the solution surrounding the electrode must be altered in order to achieve equilibrium at the particular applied potential used.

In potentiometric measurements, on the other hand, the composition of the solution is the primary variable. One must first prepare a solution of known bulk composition, either by weighing out known amounts of the oxidized and reduced forms or by partially oxidizing or reducing some known fraction of one of the two forms in a solution. The latter procedure was illustrated by Table 3-6. In either case, one then measures the potential of an electrode in equilibrium with this solution of known composition. This must be done in such a way that no current flows through the cell, for otherwise the concentrations of the oxidized and reduced forms at the surface of the electrode would differ from those in the bulk of the solution.

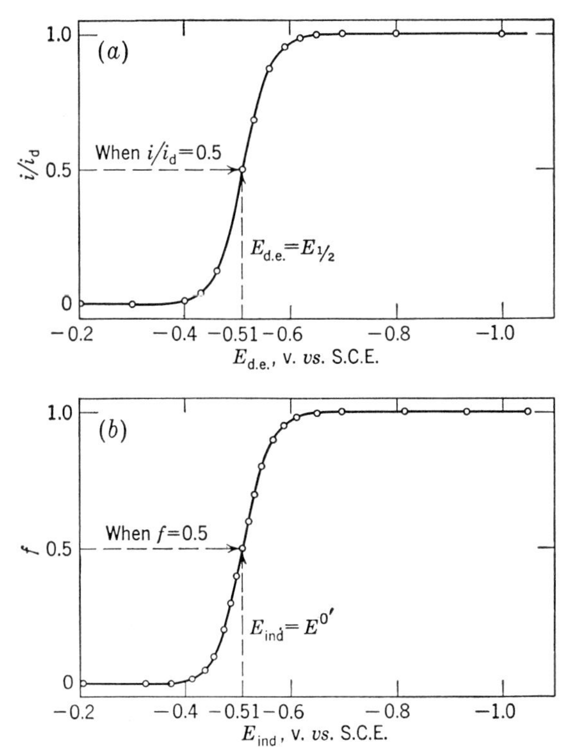

FIG. 6-3. (*a*) Polarogram of chromic ion in saturated calcium chloride; (*b*) titration curve for the potentiometric titration of chromic ion in saturated calcium chloride with a powerful chemical reducing agent, using an inert metal indicator electrode.

Figure 6-3*b* shows the data which would be obtained in the potentiometric titration of a solution of chromic ion in saturated calcium chloride with an extremely powerful reducing agent. The graph is plotted only up to $f = 1$, which is the equivalence point of the titration. Ordinarily, of course, one plots the potential of the indicator electrode against f, because the latter is the independent variable in potentiometric measurements. In this case, however, we have plotted f against the electrode potential to emphasize the similarity between this curve and that in Fig. 6-3*a*.

Although the techniques of measurement used in polarography and potentiometry may at first appear to be quite different, they are actually very intimately related. In fact, the very process of balancing a potentiometer is really nothing more than the location of the point at which the complete current–potential curve of the indicator electrode passes through zero current, and so a potentiometric measurement simply amounts to finding a single point on the current–potential curve.

In carrying out a quantitative analysis by means of a potentiometric titration, one measures the volume V of reagent required to reach the inflection point, where dE/dV (or, what is the same thing, dE/df) attains its maximum value. At this point—if, as is usually true, K for the titration reaction is large—virtually every ion of the starting material has reacted with the reagent. In quantitative polarographic analysis one measures the current on the flat portion of the wave, where dE/di is very large, and where virtually every ion or molecule of the starting material reaching the film of solution around the electrode surface is being consumed by the electrode reaction. We might regard the electric current in a polarographic experiment as a "reagent" which is used to alter the composition of a layer of solution surrounding the microelectrode.

A great deal of valuable thermodynamic information can be obtained by potentiometric measurements with reversible couples, and it is evident that similar information concerning such couples can be secured by polarographic measurements.[1] However, the fact that a couple is not reversible does not prohibit its use in potentiometric titrations; potassium dichromate is a valuable reagent for such titrations even though the dichromate–chromic ion couple is irreversible. Similarly, the fact that nickel ion is irreversibly reduced to metallic nickel does not affect the fact that nickel ion can be determined polarographically by measuring its diffusion current.

On the basis of Fig. 6-3 it is convenient to regard direct polarography, in which the concentration of some constituent of a solution is found from a measurement of its diffusion current, as the polarographic technique which corresponds to a potentiometric titration. The mathematical analysis of the rising portion of a polarographic wave (Sec. 6-12) corresponds to direct potentiometry. Finally, we can use polarographic measurements to carry out an "amperometric titration," in which the diffusion current of some substance involved in a titration is measured and used to find the end point of the titration.

[1] Because of the necessity of having a "supporting electrolyte" (Sec. 6-6) present in all polarographic measurements, extrapolation to zero ionic strength to find the standard potential is rather impractical with polarographic data, and so polarographic work is almost exclusively concerned with formal potentials, formal dissociation constants, and so on.

6-5. Polarographic Nomenclature. It is convenient here to introduce and define some of the terms peculiar to polarography which will be needed in the subsequent sections of this chapter.

Polarography itself is usually defined as that branch of electrochemistry which deals with the relationships among current, electrode potential, and solution composition in a cell of which one electrode is a dropping mercury electrode subject to *concentration polarization*. *Concentration polarization* denotes a situation in which the composition of the solution immediately adjacent to the surface of an electrode differs from the composition of the solution some distance away. In polarographic work concentration polarization is almost invariably simply the result of the flow of current through the cell and of the concentration gradients thus set up, though in some cases it may be due to the preferential adsorption of some constituent of the solution onto the electrode surface.[1]

A *polarogram* is a plot of the current flowing through a polarographic cell against the potential of the dropping electrode. A *polarograph* is an instrument used to obtain current–potential data. This word was coined by Heyrovsky and Shikata in 1927 to denote their newly invented instrument for the automatic recording of polarograms, and the name "Polarograph" is a registered trade-mark of E. H. Sargent and Co., who are the licensed American manufacturers of the Heyrovsky-Shikata instrument. Nevertheless, it is customary to use the terms *recording polarograph* and

[1] The generic term *polarization* refers to any situation in which the potential of an electrode differs from the value predicted by the Nernst equation on the assumption that the solution is homogeneous throughout. Concentration polarization results from the failure of this assumption, and evidently occurs whenever the composition of the layer of solution around the surface of an electrode is being altered by the flow of a current, for it could be prevented only by infinitely efficient stirring. *Ohmic polarization* results from the inclusion of an iR drop in the measured potential, and requires no further discussion here. *Activation polarization* results from the necessity of overcoming an activation energy during the course of an electrode reaction; it is this which is primarily responsible for the overpotential phenomena to be discussed in Sec. 7-2.

The student must also be prepared to find the term *polarization* used in the literature in quite a different sense to describe situations in which the value of dE/di is less than R, the cell resistance. It is in this sense that the chromic ion in Fig. 6-3 is referred to as a *depolarizer;* at potentials in the neighborhood of -0.5 v. *vs.* S.C.E. the discrepancy between dE/di and R is smaller in the presence of chromic ion than in its absence.

Because of these conflicting usages, electrochemists have come to regard the term *polarization* with much disfavor. A better usage is that in which one speaks of the *overpotential,* which is defined as the difference between the measured potential of an electrode and the potential which it would have according to the Nernst equation if the solution were homogeneous throughout. A major advantage of this terminology is that the concentration overpotential can be measured and expressed quantitatively, whereas the extent of concentration polarization cannot. Further details concerning concentration, ohmic, and activation overpotential can be found in any good modern treatise on electrochemistry.

manual polarograph, respectively, to designate any automatic or non-automatic instrument used to secure current–potential data.[1]

The virtually flat upper portion of a polarographic wave is called its *plateau*, and the total current which flows through the cell at a potential on the plateau is called the *limiting current* of the substance responsible for the wave, and is denoted by the symbol i_l. The difference between the limiting current and the *residual current*, i_r, which would flow through the cell at the same potential in the absence of the substance responsible for the wave is called the *wave height* of that substance.

A solution whose polarogram is to be recorded must always contain a *supporting electrolyte* in addition to the ion or molecule being studied. A supporting electrolyte is a mixture of wholly or partially ionized substances which are neither oxidized nor reduced in the range of potentials which is of interest. It may be a single acid, base, or salt, or it may be a mixture of two or more acids, bases, or salts. It may contain substances such as ammonia and pyridine which serve as complexing agents for metal ions, and it may contain substances such as acetic acid which serve to buffer the solution. The most important functions of the supporting electrolyte will be discussed in Sec. 6-6.

6-6. Theory of the Limiting Current. The current which flows through a cell at a potential on the plateau of a wave is the sum of three components. One is the residual current, which is due partly to the presence of small concentrations of impurities in the solution and partly to the fact that energy must be supplied continuously to maintain the "electrical double layer" (Sec. 6-7) around the constantly expanding drop. The residual current at any given potential may be measured directly by using a "blank" solution of the supporting electrolyte alone, or it may be corrected for by an extrapolation technique to be described in a later section.

Another component of the limiting current is the diffusion current, i_d, which is directly proportional to the concentration of the substance being reduced or oxidized. As its name implies, the diffusion current is due to the reduction or oxidation of ions or molecules which reach the surface of the dropping electrode by free diffusion caused by the existence of the concentration gradient around the electrode.

If no supporting electrolyte were present in the solution, there would be another force acting on the reducible[2] ions. This would result from

[1] In reading through the literature, the student will occasionally come across the word "polarograph" used as a verb: "Add 2 ml. of nitric acid and polarograph the resulting solution." This is not good usage, because one would never speak of potentiometering or conductance bridging a solution. What is meant is "Add 2 ml. of nitric acid and secure the polarogram of the resulting solution."

[2] For the sake of convenience, we shall henceforth usually refer to the substance responsible for a wave as the reducible constituent of the solution, and it must be understood that oxidizable substances will behave in exactly the same way as reducible ones.

the gradient of electrical potential between the two electrodes of the cell, which causes ions to move through the solution in order to carry the current through the cell. For example, imagine a dropping electrode at a potential of -1.0 v. *vs.* S.C.E. in a pure cadmium nitrate solution. At this potential cadmium ions are reduced at the dropping electrode (nearly completely, because this potential is considerably more negative than the half-wave potential of cadmium ion). So the concentration of cadmium ion at the surface of the dropping electrode decreases nearly to zero, and cadmium ions diffuse in from the body of the solution toward the electrode surface to replace the ions which have been reduced. This reduction of cadmium ions at the dropping electrode is accompanied by the oxidation of mercury in the S.C.E., and a current flows through the cell and around the external electrical circuit. There is only one way in which current can be carried through the solution, and this is by the migration of cadmium ions toward the negatively charged d.e. and of nitrate ions toward the positively charged S.C.E. Because of this migration, the number of cadmium ions reaching the surface of the dropping electrode during each second is greater than it would be under the influence of diffusion alone, and therefore the current due to the reduction of cadmium ion is greater than the diffusion current.

TABLE 6-2. EFFECT OF POTASSIUM NITRATE ON THE TRANSFERENCE NUMBER OF CADMIUM ION IN 0.001 F CADMIUM NITRATE

$[KNO_3]$, F	$t_{Cd^{++}}$
0	0.470
0.0001	0.446
0.0005	0.370
0.001	0.305
0.005	0.127
0.01	0.0736
0.05	0.0168
0.1	0.00857
0.5	0.00174
1	0.00087

The difference between the total reduction current of cadmium ion, which is equal to $i_l - i_r$, and the diffusion current of cadmium ion is called the *migration current*, i_m. The migration current depends on the transference number of cadmium ion in the solution; decreasing this to zero would cause the current to be carried through the solution by the movement of the other ions alone, so that only the diffusive forces would remain operative on the cadmium ions. This can be accomplished by the addition of a large excess of some other salt, as is illustrated by Table 6-2. The transference numbers of cadmium ion given in this table were calculated by means of Eq. (5-8) from the equivalent ionic conductances of the ions at infinite dilution. Although the values are only approximate, because the equivalent conductances would vary as the potassium nitrate concentration was changed, the table clearly demonstrates the fact that the transference number of an ion decreases practically to zero in the presence of even a hundredfold excess of a supporting electrolyte. In such solutions almost all of the current is carried through the cell by the potassium and nitrate ions. However, these cannot be reduced at -1.0 v. *vs.* S.C.E. in significant numbers, and so the reduction of cadmium ions remains the only process by which electrons can be transferred across the interface between the dropping electrode and the solution.

In this way the migration current is practically eliminated at the same time that the iR drop through the cell is reduced to a negligible value, and these are the two most important functions of the supporting electrolyte. It is evident that the wave height of cadmium ion in a cadmium nitrate solution containing little or no added potassium nitrate would depend not only on the concentration of cadmium ion, but also on the concentration of any other salt that happened to be present, and in addition it would depend to some extent on the particular foreign salt present. These phenomena would be seriously disadvantageous in practical analytical work.

Figure 6-4 shows the way in which the wave height of cadmium ion in a cadmium chloride solution decreases as the migration current is eliminated by successive additions of potassium chloride. Note the way in which the slope of the rising part of the wave increases as the cell resistance decreases with increasing electrolyte concentration. Each complete oscillation on these polarograms, which were secured with a recording polarograph, represents the growth and fall of one drop.

To eliminate the migration current and ensure that the resistance of the solution is so low that the iR drop through it can safely be neglected, it is desirable that the ionic strength of the supporting electrolyte be at least 50 or 100 times the normality of the ion responsible for a wave, but in any case not less than about 0.1 M.

6-7. The Residual Current. The residual current at any given potential arises from two sources. One is the reduction of traces of impurities in the supporting electrolyte, including oxygen as well as various heavy metal ions. The removal of oxygen from solutions will be discussed in Sec. 6-21. The heavy metal ions may be present in the distilled water used to prepare the solution, or may come from the chemicals used to prepare the supporting electrolyte. To illustrate the degree of purity required of the reagents, it may be mentioned that a 1 F sodium hydroxide solution prepared from reagent grade sodium hydroxide containing only 0.001 per cent zinc will contain 7×10^{-6} F zinc, which would give a wave large enough to be rather easily detected even with quite crude apparatus. The fraction of the residual current which is due to the reduction of such impurities is called the *faradaic* component of the residual current, because it obeys Faraday's laws of electrolysis, and it is given the symbol i_f.

Suppose that a dropping electrode is immersed in a pure 1 F potassium chloride solution in contact with an S.C.E., and that a fairly negative potential, say -1.0 v., is applied to the dropping electrode. The negatively charged d.e. will exert an electrostatic attraction for the potassium

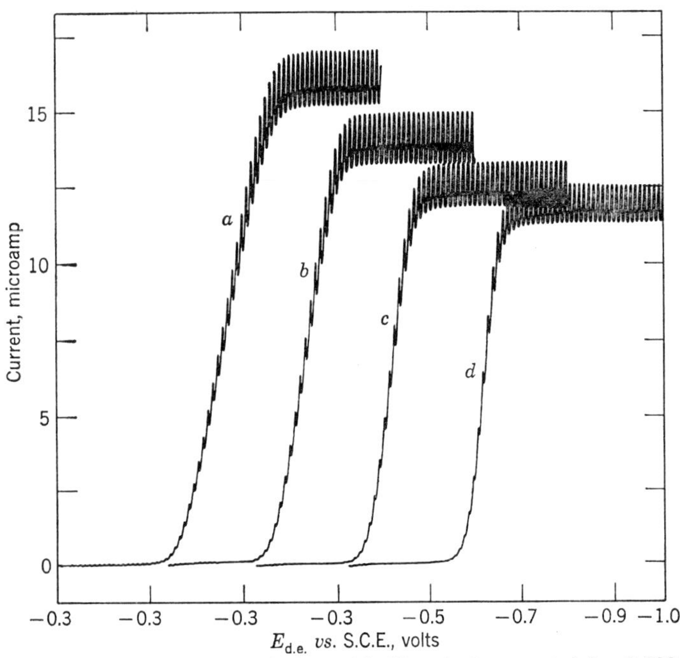

FIG. 6-4. Polarograms of 1.0 mF cadmium chloride solutions containing 0.002 per cent Triton X-100 (*cf.* Sec. 6-17) and (*a*) 0, (*b*) 0.67, (*c*) 4.0, and (*d*) 30 mF potassium chloride, illustrating the decrease of the limiting current of a cation as its migration current is suppressed. Each curve begins at -0.3 v. *vs.* S.C.E., and each interval along the abscissa corresponds to 0.2 volt. (*Reproduced by permission from L. Meites, "Polarographic Techniques," Interscience Publishers, Inc., New York,* 1955.)

ions in the solution. However, the value of $E_{d.e.}$ is not sufficiently negative to permit the reduction of any significant fraction of the potassium ions at its surface ($E_{1/2}$ for potassium ion is -2.14 v. *vs.* S.C.E.), and so the potassium ions will simply accumulate in a layer very close to the surface of the electrode. This positively charged shell will attract chloride ions from the solution, and will therefore be surrounded by a negatively charged layer of chloride ions a little farther away from the electrode. The "electrical double layer" thus created resembles an electrical condenser, and the current required to charge and maintain the electrical double layer is accordingly known as the *condenser* current, i_c.

The value of the condenser current depends on the potential of the dropping electrode. As this becomes less and less negative, the electrostatic attraction for potassium ion decreases, and so does the extent to which chloride ions are repelled from the electrode surface. Eventually a point will be reached at which potassium and chloride ions will be present at exactly equal concentrations at the surface of the dropping electrode. The potential at which this occurs is known as the *potential of the electrocapillary maximum* or *e.c. max. potential*, and it depends on both the identity and, to a lesser extent, the concentration of the salt present. It is -0.470 v. *vs.* S.C.E. in 0.1 F potassium chloride, and -0.592 v. *vs.* S.C.E. in 0.1 F potassium thiocyanate. The value is more negative in the thiocyanate solution because thiocyanate ions are more strongly adsorbed on a mercury surface than are chloride ions.

If $E_{\text{d.e.}}$ is made more positive than the potential of the electrocapillary maximum, the layer of solution nearest the surface of the dropping electrode will contain an excess of anions, and the outer layer will contain an excess of cations. A double layer is thus again set up, but it now has the opposite polarity from that which exists at potentials more negative than the e.c. max. potential, and hence the condenser current will have the opposite sign. At the e.c. max. potential itself the double layer does not exist, and so the condenser current is zero.

Figure 6-5 shows a typical residual current curve. The residual current is seen to be zero at a potential somewhat more positive than the e.c. max. potential, which is very nearly -0.47 v. *vs.* S.C.E. in this solution, and it is distinctly positive at the e.c. max. potential. This is due to the presence of traces of reducible impurities in the solution. The curve is approximately linear between -0.1 and -1.0 v. *vs.* S.C.E. The *initial current rise* reflects the cessation of the anodic oxidation of mercury to mercurous chloride; by convention anodic currents are given a negative sign and cathodic currents a positive sign. The *final current rise* beginning at about -1.0 v. reflects the onset of the reduction of hydrogen ion to hydrogen gas. The slope of the intermediate, nearly linear, portion of the curve depends on the capillary characteristics (Sec. 6-8), but it usually lies between 0.5 and 1 microamp. per v.

6-8. The Diffusion Current. The fundamental equation for the polarographic diffusion current was first derived by the Czech chemist Ilkovič in 1934, and is known as the Ilkovič[1] equation. It is

$$i_d = 607nD^{1/2}Cm^{2/3}t^{1/6} \tag{6-8}$$

where i_d is the average diffusion current (in microamperes) during the

[1] Pronounced as though it were Ilkovitch, with the accent on the first syllable.

life of a drop, n is the number of electrons consumed in the reduction of one ion or molecule of the substance responsible for the wave, D is the diffusion coefficient of that substance (cm.² per sec.) and C its concentration (millimoles per liter),[1] m is the rate of flow of mercury through the capillary (mg. per sec.), and t is the time (in seconds) which elapses between the fall of one drop and the fall of the next. The quantities m

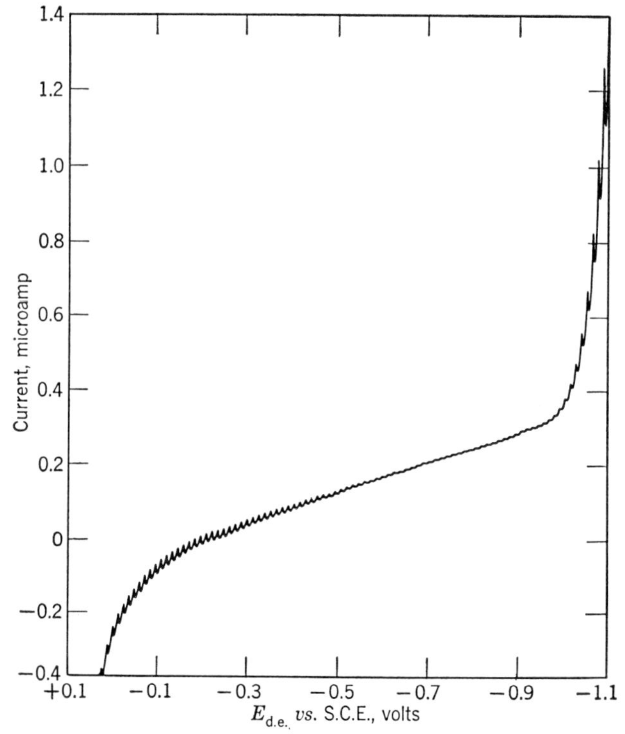

FIG. 6-5. Residual current curve for a solution containing 0.1 F potassium chloride and 0.1 F hydrochloric acid. (*Reproduced by permission from L. Meites, "Polarographic Techniques," Interscience Publishers, Inc., New York,* 1955.)

and t depend largely on the geometry of the capillary, and are therefore called the *capillary characteristics.*

For reasons which cannot be discussed here, Eq. (6-8) holds true *only* when the drop time is between about 2.5 and 7 seconds. In all practical work, therefore, care must be taken to ensure that t falls within this range.

For purposes of quantitative analysis, the most important single feature of the Ilkovič equation is its prediction that the diffusion current is pro-

[1] The concentrations ordinarily employed in polarographic work lie between 10^{-5} and 10^{-2} M (0.01 to 10 mM), so that the millimole per liter is a much more convenient unit than the mole per liter.

portional to the concentration of the substance being oxidized or reduced at the dropping electrode. After many experimental tests of the validity of this prediction, it is now believed to be accurate to within ± 0.1 per cent *if* the residual current (which must be subtracted from the limiting current actually measured to secure the value of i_d) is properly measured and *if* maxima (Sec. 6-17) are completely absent or properly dealt with.

Naturally the uncertainty of a routine polarographic analysis is considerably greater than ± 0.1 per cent, which is the precision obtained by skilled experimenters under optimum conditions; an average error of ± 1 to 2 per cent is much more nearly representative of ordinary practical work. Since polarographic analysis is generally used for the determination of quantities far too small to be handled conveniently by most other techniques, an error of this magnitude is by no means a deterrent to its use. No polarographer would have to think twice about determining the copper present in 10 ml. of an 0.1 mM solution, which contains just 63 micrograms of copper. Even the determination of 0.63 microgram of copper in 1 ml. of solution would probably involve no difficulty worth mentioning if a suitable cell were on hand. In this range of amounts and concentrations only spectrophotometry of all the instrumental techniques in common use today is capable of competing seriously with polarography. Probably this statement will no longer be true after a few years, when radioactivation analysis (Sec. 10-9) and coulometry at controlled current (Sec. 7-11) will have come into their rightful places as everyday tools of the analyst. At the present time, however, both of these techniques are still very near the frontiers of analytical chemistry.

Under constant experimental conditions (supporting electrolyte composition, $E_{\text{d.e.}}$, temperature, etc.), the quantities n and D in Eq. (6-8) are fixed for any one reducible ion or molecule. However, C, m, and t can all be varied at the will of the experimenter. It is therefore convenient to rewrite Eq. (6-8) in the form

$$\frac{i_d}{Cm^{2/3}t^{1/6}} = I = 607nD^{1/2} \tag{6-9}$$

The quantity I defined by this equation is called the "diffusion current constant," and it has a characteristic value for any one electrode process occurring under any defined set of experimental conditions. The values of i_d, m, and t are all easily measurable, and it is a simple matter to calculate the concentration of an unknown solution with the aid of these data and a value of I taken from the literature.[1] Suppose, for example, that the diffusion current of bismuth in an unknown solution containing

[1] A comprehensive critical tabulation of the diffusion current constants of inorganic substances is given in Appendix B of L. Meites, "Polarographic Techniques," Interscience Publishers, Inc., New York, 1955.

1 F nitric acid as the supporting electrolyte is found to be 3.50 microamperes, using a capillary for which m is 2.50 mg. per sec. and t is 3.00 sec. The diffusion current constant of bismuth in this supporting electrolyte is 4.59 (microamperes per millimole per liter per mg.$^{2/3}$ sec.$^{-1/2}$), and so the concentration of bismuth is

$$C = \frac{i_d}{Im^{2/3}t^{1/6}} = \frac{3.50}{(4.59)(2.50)^{2/3}(3.00)^{1/6}}$$
$$= 0.345 \text{ m}M$$

This is the so-called "absolute" method of polarographic analysis. The name is a misnomer, however, for the diffusion current constant cannot be calculated *a priori* from nonpolarographic data; it must be obtained experimentally by measuring the diffusion current obtained with a solution of known concentration and a capillary of known characteristics. This is partly because of an uncertainty in the numerical coefficient appearing in the Ilkovič equation, and partly because no good way of measuring the diffusion coefficient of an ion or molecule in the presence of a large excess of an indifferent salt has yet been developed. So the "absolute" method is simply a way of comparing diffusion current data secured with different capillaries, whereas the "comparative" methods to be discussed in a later section serve only for the comparison of data secured in different solutions with the same capillary.

6-9. Factors Affecting the Capillary Characteristics. The rate of flow of a liquid through a capillary is given by the Poiseuille equation, which for present purposes may be written

$$m = 4.64 \times 10^9 \frac{r_c^4}{l}\left(h_{\text{Hg}} - \frac{3.1}{m^{1/3}t^{1/3}}\right) \tag{6-10}$$

where r_c is the radius of the capillary and l its length (both in cm.), h_{Hg} is the height (in cm.) of the column of mercury between the tip of the capillary and the top of the mercury meniscus in the stand tube, and m and t have their ordinary polarographic meanings. The terms inside the parentheses express the pressure acting upon the mercury; the first term gives the hydrostatic pressure of the mercury column, and the second is due to a "back" pressure which reflects the interfacial tension between the mercury drop and the solution. Ordinarily h_{Hg} is about 50 cm., and the back pressure term is only about 1.5 cm. Equation (6-10) is often written in the form

$$m = k_1 h_{\text{corr}}$$

for any one capillary, since r_c and l are obviously constant. The quantity $h_{\text{corr.}}$, which is equal to the quantity enclosed in parentheses in Eq. (6-10), is called the "corrected" or "effective" pressure acting on the mercury drop. If the solution composition, dropping electrode potential, and temperature are all kept constant, mt, and with it the back pressure term, is constant and independent of the height of the mercury column. Varying the latter, therefore, changes t according to an equation of the form

$$t = \frac{k_2}{h_{\text{corr.}}}$$

Combining these two equations with Eq. (6-8) gives, when all other experimental variables are held constant,

$$i_d = km^{2/3}t^{1/6} = k(k_1 h_{\text{corr.}})^{2/3}\left(\frac{k_2}{h_{\text{corr.}}}\right)^{1/6} = Kh_{\text{corr.}}^{1/2}$$

This provides an easy method of testing the validity of the Ilkovič equation in any unknown case, which is often of importance in evaluating n for an electrode reaction (Sec. 6-11). Occasionally it is found that the wave height is not proportional to the square root of the corrected pressure, but instead is either independent of or proportional to the first power of $h_{\text{corr.}}$. It can be shown that a wave height which is independent of mercury pressure results when the species actually being reduced is formed by a slow chemical reaction in the diffusion layer. This is the case, for example, in the reduction of dextrose and other aldoses; these are present in solution almost entirely in the ring (pyranose) form, but only the open-chain (aldehydo) form is reducible. So the wave height is governed by the rate at which the ring form is transformed into the aldehydo form at the electrode surface. On the other hand, a wave height proportional to the first power of $h_{\text{corr.}}$ is observed when either the substance being reduced or its reduction product is adsorbed onto the surface of the drop to form an electrically insulating film. Under these conditions the current at any instant during the drop life is proportional to the rate at which a fresh mercury surface becomes exposed, *i.e.*, proportional to dA/dt, the rate of change of electrode area with respect to time. Since $(mt)^{2/3}$, which is proportional to the maximum area attained by the drop at the instant of its fall, does not depend on h, the number of coulombs which flows during the life of a drop also does not depend on h. Increasing h does, however, increase the number of drops which fall in a given interval, and therefore the height of such an "adsorption wave" is proportional to $h_{\text{corr.}}$.

Probably the most important of the factors which affect the interfacial tension between mercury and a solution is the potential applied to the mercury. As $E_{\text{d.e.}}$ is varied, the interfacial tension passes through a maximum at the e.c. max. potential, and so does the product mt, which is very nearly directly proportional to the interfacial tension. Because the second term inside the parentheses in Eq. (6-10) is much smaller than the first, this results in only a very small variation of m with applied potential. Consequently the variation of mt with potential is almost entirely due to changes in t. The effects of applied potential on m, t, mt, and $m^{2/3}t^{1/6}$ are shown for a typical case in Table 6-3. It will be

TABLE 6-3. EFFECTS OF APPLIED POTENTIAL ON CAPILLARY CHARACTERISTICS*

$E_{d.e.}$, v. *vs.* S.C.E.	m, mg./sec.	t, sec.	mt, mg.	$m^{2/3}t^{1/6}$, mg.$^{2/3}$ sec.$^{-1/2}$
±0.00	3.4042	2.700	9.191	2.6703
−0.20	3.3948	2.883	9.787	2.6947
−0.40	3.3831	2.925	9.895	2.6950
−0.60	3.3831	2.927	9.902	2.6953
−0.80	3.3872	2.882	9.762	2.6906
−1.00	3.3891	2.740	9.286	2.6690
−1.20	3.3915	2.551	8.652	2.6386
−1.40	3.4039	2.337	7.955	2.6068
−1.70	3.4218	1.933	6.614	2.5344
−2.00	3.4487	1.435	4.949	2.4243

* The data were secured with 0.1 F potassium chloride as the supporting electrolyte at 25°C.

observed that $m^{2/3}t^{1/6}$ decreases about 10 per cent between −0.5 v. (the e.c. max. potential in this solution is −0.47 v.) and −2.0 v. *vs.* S.C.E. Accordingly the diffusion current of a wave also decreases about 10 per cent between these potentials; this phenomenon is illustrated by Fig. 6-6. Because of this variation of $m^{2/3}t^{1/6}$ with applied potential, it is necessary to measure the value of this quantity at the same potential at which the wave height is measured when one is calculating or using a diffusion current constant.

The interfacial tension between mercury and an aqueous salt solution is not very greatly affected by the kind of salt present or by its concentration, and so the values of m secured with any one capillary at the same applied potential and the same corrected pressure of mercury are practically the same from one supporting electrolyte to another. However, the drop time is so much more sensitive to changes in the interfacial tension that it must be remeasured whenever any of the experimental conditions is altered. Moreover, appreciable changes even in m may result from the presence of a maximum suppressor (Sec. 6-17) or of substantial concentrations of nonaqueous solvents such as ethanol or dioxane, which are widely used in organic polarography because of the limited solubilities of many organic substances in water.

6-10. Polarographic Diffusion Coefficients. At infinite dilution, the equivalent conductance and diffusion coefficient of an ion are related by the following equation, first derived by Nernst:

$$D^0 = \frac{RT}{zF_y^2}\lambda^0 = 2.67 \times 10^{-7}\frac{\lambda^0}{z} \qquad (6\text{-}11)$$

where z is the electrical charge on the ion (*i.e.*, its valence). As this is strictly true only at infinite dilution, it cannot be used for the exact calculation of the diffusion coefficient of an ion under polarographic conditions, which is to say in the presence of a hundredfold or even greater

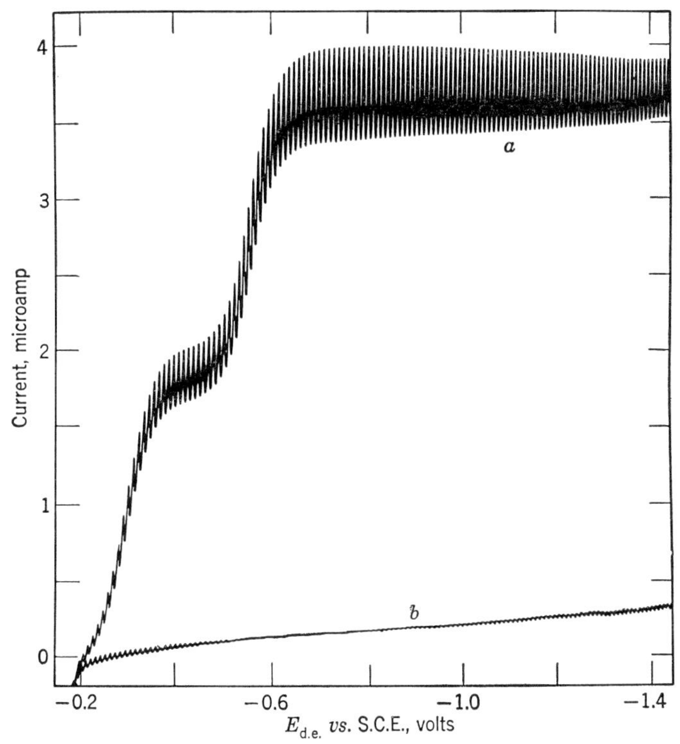

FIG. 6-6. (*a*) Polarogram of 0.45 mM cupric copper in 1 F ammonia–1 F ammonium chloride–0.002 per cent Triton X-100. (*b*) Residual current curve of the supporting electrolyte alone. (*Reproduced by permission from L. Meites, "Polarographic Techniques," Interscience Publishers, Inc., New York,* 1955.)

excess of an indifferent salt. Efforts have been made to measure these diffusion coefficients experimentally, but so far these efforts have not led to data which seem completely reliable.

Though we cannot use Eq. (6-11) to calculate the diffusion current constant of a reducible ion, it is useful as the basis of several important predictions. Inspection of a table of the equivalent conductances of various simple (*i.e.*, aquo-complex) ions reveals that λ^0/z for most such ions is not very far from 30; only the hydrogen and hydroxyl ions deviate greatly from this value. This corresponds to a diffusion coefficient of about 7×10^{-6} cm.^{2} per sec., and this in turn corresponds by virtue of Eq. (6-9) to a diffusion current constant of $1.6n$. Though this generalization is very rough, it is often very helpful in deciding between two possible products of an electrode reaction responsible for a wave. For instance, uranyl ion, UO_2^{++}, gives a single wave in a weakly acidic acetate buffer. When it was first discovered, this might easily have been presumed to represent reduction to uranium(IV), because this is the highest stable oxidation state of the element below the +6 state. However, the diffusion current constant of the wave was found to be only 1.7, which is much too small to correspond to a two-electron reduction. On the contrary, it strongly suggested that the reduction involves only one electron and proceeds to the unfamiliar +5 state, a suggestion that has been confirmed by much subsequent study of the redox chemistry of uranium in aqueous solutions.

Complexes of the metal ions with such common complexing agents as chloride, hydroxide, cyanide, thiocyanate, ammonia, and so on do not usually differ much in size from the corresponding aquo-complexes. Therefore the diffusion current constants of the metals in solutions containing these complexing agents are usually also close to $1.6n$. However, coordination with a large anion, such as tartrate, citrate, ethylenediaminetetraacetate, and the like, does result in a considerable increase of ionic radius, and the resulting large ion has a much smaller diffusion coefficient than the simple metal ion from which it is formed. Thus, the diffusion current constant of bismuth(III) in a slightly acidic tartrate buffer is only 3.1, and yet the reduction involves three electrons and gives bismuth amalgam as the product. The difference between this value and the one given in Sec. 6-8 for the three-electron reduction of bismuth from a nitric acid supporting electrolyte reflects the large size of the bismuth tartrate complex.

Equation (6-11) is also important in that it permits us to estimate the temperature coefficient of the diffusion coefficient. This is given by

$$\frac{1}{D^0}\frac{dD^0}{dT} = \frac{1}{\lambda^0}\frac{d\lambda^0}{dT} + \frac{1}{T}$$

The temperature coefficient of λ^0, which is the first term on the right-hand side of this equation, is about +2 per cent per degree for most ions, while at 25°C. $1/T$ is 0.0034. Consequently the temperature coefficient of D is approximately +2.3 per cent per degree. Since i_d is proportional to $D^{1/2}$, the variation of D alone with temperature would give rise to a temperature coefficient of about +1.2 per cent per degree for the diffusion current. All of the other terms in the Ilkovič equation, except of course

n, also vary with temperature, but their variations are very much smaller than the variation of D, and consequently the entire temperature coefficient of the diffusion current is generally about +1.3 per cent per degree.

In Sec. 6-8 it was mentioned that the ordinary accuracy of routine polarographic analyses is of the order of ±1 to 2 per cent. In order to allow for the uncertainties in measuring or controlling such other quantities as the height of the mercury column, the composition of the supporting electrolyte, and the wave height, which affect or enter into the final calculations, it is desirable to control the temperature to within about ±0.25°C.

The diffusion coefficient of a spherical particle which is very much larger than a solvent molecule is given by the Stokes-Einstein relation

$$D = \frac{RT}{6\pi\eta rK} \tag{6-12}$$

where η is the viscosity coefficient of the solution (in dyne-sec. per cm.^{2}), r is the radius of the diffusing particle, and K is Avogadro's number. Not very many of the inorganic ions or organic molecules encountered in the course of ordinary polarographic work have such large radii that they can be expected to obey this equation exactly, but the prediction that for any given diffusing species

$$D\eta = \text{constant}$$

is generally fairly closely satisfied. This amounts to saying that the product $i_d\eta^{1/2}$ remains constant *provided that* the change in viscosity is not accompanied by any change in the nature (*i.e.*, the extent of solvation or complexation) of the reducible ion or molecule. Naturally the viscosity of a supporting electrolyte solution depends on its concentration, but fortunately the relationship is not usually a very sensitive one. It is only necessary to keep the concentration of every constituent of the supporting electrolyte constant to within about ±10 per cent to ensure that the variation of i_d resulting from this effect will be negligibly small.

6-11. Evaluation of n for Polarographic Waves. One of the most widely used techniques of estimating the value of n for an electrode process responsible for a polarographic wave was described in Sec. 6-10. It depends on the fact that I/n for most aquo-metal ions, as well as for most complexes of metal ions with those complexing agents whose molecular or ionic radii are not much different from the radius of a water molecule, is generally close to 1.6.

Unfortunately, this criterion cannot be applied when the complexing agent is very large, because then the radius of the complex becomes much larger than that of the simple ion. In view of Eq. (6-12), this results in a decrease of D and thus of I/n. As there is no way of predicting the extent of this decrease with any reasonable assurance, it sometimes becomes difficult to decide between two possible values of n. For example, the diffusion current constant of the total double wave of +2 copper in an ammoniacal ammonium chloride solution (*cf.* Fig. 6-6) is 3.75, and the reduction is known to involve two electrons and give copper amalgam as the product. In weakly acidic 0.5 F sodium tartrate, however, the diffusion current constant of cupric copper is only 2.37. If we assumed that the diffusion coefficients of the cupric complexes in these solutions were the same, we could calculate n in the tartrate medium from the equation

$$\frac{I_1}{I_2} = \frac{n_1}{n_2} \tag{6-13}$$

which would give a value of 1.26. This might tempt us to conclude that n was actually 1, and that a cuprous tartrate complex was being formed by the electrode reaction, but this would be incorrect. We could take the much larger radius of the tartrate complex into account by comparing its diffusion current constant with that of a similar complex for which n is known, such as the lead tartrate complex. In weakly acidic 0.5 F sodium tartrate the diffusion current constant of lead is 2.37, and since the +1 oxidation state of lead is unknown in aqueous solutions this must represent a two-electron reduction. So, rewriting Eq. (6-13) in the form

$$n_{Cu} = n_{Pb} \frac{I_{Cu}}{I_{Pb}}$$

we would conclude correctly that the cupric tartrate complex undergoes a two-electron reduction to copper amalgam.

The polarograms obtained when a single substance is reduced stepwise to give a series of products are usually very easy to interpret. Figure 6-6 showed a simple example of this sort of polarogram. It consists of two waves, both of which are due to the reduction of cupric copper. The total height of the second wave, which we shall call $(i_d)_2$, and which could be measured at any potential more negative than about −0.7 v., is just twice the height of the first wave. Obviously the concentration and diffusion coefficient of the complex cupric ion are the same from the beginning to the end of the polarogram, and the value of $m^{2/3}t^{1/6}$ does not vary appreciably over the range of potentials involved. So

$$\frac{n_2}{n_1} = \frac{(i_d)_2}{(i_d)_1} = 2$$

Because we know enough about the chemistry of copper to be quite sure that the maximum possible value of n is 2, this shows that the first wave represents the addition of one electron and the formation of a cuprous ammonia complex, while on the plateau of the second wave each cupric complex ion picks up two electrons and is reduced to the metallic state.

The estimation of n for a wave is generally very simple when n is small. This is because n can have only integral values, and an uncertainty of even ± 10 per cent does not cause much trouble in choosing between two small integers, especially when some pertinent chemical information concerning the possible reaction products is available to guide the choice. Confusion may arise, however, when n is large (because then a small error in the explicitly or implicitly assumed value of D may result in a numerically large error in n), or when the electrode process involves such complications as adsorption or catalytic reactions.[1] In such cases the value of n is best determined by coulometry at controlled potential (Sec. 7-10).

6-12. Criteria of Reversibility. In Sec. 3-8 we saw that the reversibility of a couple could be tested potentiometrically by varying the concentrations of the oxidized and reduced forms of the couple and observing whether the resulting variations in the potential of an indicator electrode conform to the Nernst equation. This may be done by titrating a solution containing one of the forms of the couple with a suitable reagent, using the Nernst equation to calculate the formal potential of the couple from data obtained at a number of points preceding the equivalence point, and observing whether the resulting value of $E^{0\prime}$ is constant within experimental error.

Exactly the same thing is done in studying the reversibility of a couple polarographically. Instead of varying the concentrations of the oxidized and reduced forms of the couple by adding a chemical reagent, however, one simply takes advantage of the fact that their ratio is varied automatically by changing the potential of the dropping electrode. In general, the wave will be described by the equation

$$E_{\text{d.e.}} = E_{1/2} - \frac{0.05915}{n} \log \frac{i}{i_d - i} \tag{6-7}$$

As can be seen from its derivation (in Sec. 6-3), the validity of this equation rests on the implicit assumption that the Nernst equation is obeyed by the couple in question.[2] So the "reversibility" or "irreversibility"

[1] The theory of kinetic, catalytic, and adsorption waves cannot be discussed here; the student may be referred to any recent monograph on polarography. Suffice it to say that the foregoing methods for the estimation of n should be used *only* when the $i_d = kh_{\text{corr.}}^{1/2}$ prediction of the Ilkovič equation has been found to be obeyed by the wave in question.

[2] More accurately, it rests on the assumption that the Nernst equation is obeyed at every instant during the life of the drop. Here we are dealing with a truly dynamic equilibrium, because the current flows continuously throughout the life of each drop, in contrast to the much more static situation existing during potentiometric measurements. The indicator electrode in a potentiometric measurement may take a long

of a couple under polarographic conditions can be tested simply by observing whether or not Eq. (6-7) is obeyed by the current–potential data secured on the steeply rising portion of the wave. If the electrode reaction is reversible according to the definition in the preceding footnote, Eq. (6-7) must be obeyed, but on the other hand it is possible for Eq. (6-7) to be obeyed within experimental error even when the reaction is not truly reversible.[1]

A more rigorous treatment of the concept of polarographic reversibility is not within the scope of an introductory text. The interested student is advised to consult Delahay's monograph (Ref. 31, page 406).

The most straightforward way of applying this practical criterion of reversibility consists of plotting $E_{d.e.}$ against $\log i/(i_d - i)$. If this plot is a straight line whose slope is equal to $-0.05915/n$ volts, it is usually safe to conclude that the electrode reaction responsible for the wave is reversible. Little attention should be paid to any deviation from the straight line at potentials such that i is either smaller than about 5 per cent or larger than about 95 per cent of i_d, for in these regions the normal experimental error of measuring i leads to large relative errors in the ratio $i/(i_d - i)$.[2] A reasonable allowance for experimental error in the

---

time to reach a steady potential, but once a steady potential is reached it will persist indefinitely unless the solution undergoes a chemical decomposition. This is not so in polarography, where a steady-state situation at one instant is destroyed by the growth of the drop, the flow of the current, and the motions of the ions in the solution during the next instant. So, for example, the couple

$$VO^{++} + 2H^+ + e = V^{+++} + H_2O$$

is potentiometrically reversible, for a platinum indicator electrode immersed in a mixture of vanadyl (VO^{++}) and vanadic (V^{+++}) ions eventually attains a potential which obeys the Nernst equation. Polarographically, however, the couple is irreversible, because at any potential near the potentiometrically determined formal potential of the couple the reduction of vanadyl ion and the oxidation of vanadic ion proceed so slowly that neither ion gives rise to any significant current. Potentiometric reversibility is a necessary but not a sufficient condition for polarographic reversibility; it is also essential that equilibrium be reestablished virtually instantaneously after each successive infinitesimal displacement.

In view of these considerations, polarographic reversibility is defined in the following way. An electrode reaction is said to be polarographically reversible if:

1. The reducible species reaches, and the reduced species leaves, the electrode surface under the influence of diffusive forces alone (*i.e.*, there are no chemical side-reactions occurring in the layer of solution at the electrode surface).

2. The actual transfer of electrons between electrode and solution is very rapid compared to the diffusion processes, so that in effect the Nernst equation equilibrium is preserved at the electrode surface throughout the life of each drop.

[1] This is the case in the reduction of nickel(II) from an ammoniacal ammonium chloride buffer containing hydrazine, to cite only one example.

[2] For much the same reasons, data like those in Table 3-6 do not give reliable values of the formal potential at values of f which are very close to either 0 or 1.

slope of the line is about ± 10 per cent; any larger deviation from a value of $-0.05915/n$ v. is unambiguous proof of the existence of some slow step in the over-all electrode reaction. Cases are known[1] in which irreversible half-reactions give rise to log plots whose slopes are numerically smaller than the theoretical value, but these are very exceptional. Usually the slope of a log plot for an irreversible wave is numerically larger than the expected value. This means that irreversible waves are generally more drawn-out than reversible waves; the increment of potential required to produce a given change of current is larger for the irreversible wave.

In just the same way that the diffusion current must be obtained by subtracting the residual current from the limiting current at a potential on the plateau of the wave, so the current actually measured at each potential on the rising portion of the wave must be corrected for the residual current at the same potential in order to secure a value of i which accurately reflects the extent to which the couple being studied is reduced or oxidized at that potential.

It may be pointed out that $E_{1/2}$ is easily found from a log plot by interpolation to find the value of the potential when the log term is zero.

Another valuable criterion of reversibility involves securing a polarogram of the product of the half-reaction which is believed to occur at the dropping electrode. This is illustrated by Fig. 6-7. Curve a in this figure is a polarogram of ferric iron in a saturated oxalic acid medium, and curve c is a polarogram of ferrous iron in the same solution. If the ferric–ferrous couple is reversible in this medium, the composition of the solution at the surface of the electrode at any particular potential must naturally be independent of the composition of the bulk of the solution. For example, at the half-wave potential [which, by Eqs. (6-6) and (6-8), is almost exactly equal to the formal potential of the couple, because the diffusion current constants of the ferric and ferrous oxalate complexes are nearly the same], the solution at the surface of the electrode must contain equal concentrations of ferric and ferrous iron. If the bulk of the solution contains the iron exclusively in the ferric state, this equilibrium situation can be attained only by the reduction of half of the ferric complex ions which diffuse in to the electrode surface, and this means that electrons must flow from the reference electrode to the dropping electrode through the external circuit. On the other hand, if (as in curve c) the bulk of the solution contains only ferrous iron, the concentrations of ferric and ferrous iron at the drop surface can become equal only if half of the ferrous oxalate ions reaching the electrode surface are oxidized, and in order for this to happen electrons must flow from

[1] For example, the reduction of zinc(II) from a strongly alkaline supporting electrolyte containing pyridine.

the dropping electrode to the reference electrode through the external circuit. If the latter current is arbitrarily given a negative sign, the complete equation for the wave would have to be written

$$E_{\text{d.e.}} = E_{1/2} - \frac{0.05915}{n} \log \frac{i - (i_d)_a}{(i_d)_c - i} \tag{6-14}$$

where $(i_d)_a$ is the (negative) anodic diffusion current of the reduced form of the couple, and $(i_d)_c$ is the (positive) cathodic diffusion current of the

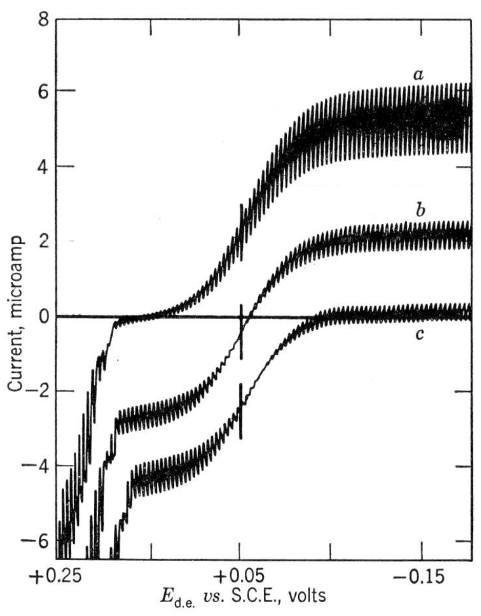

FIG. 6-7. Polarograms of saturated oxalic acid solutions containing 0.0002 per cent methyl red (as a maximum suppressor) and (*a*) 1.4 m*M* ferric iron, (*b*) 0.7 m*M* ferric iron and 0.7 m*M* ferrous iron, and (*c*) 1.4 m*M* ferrous iron. The short vertical lines indicate the half-wave potentials of the three waves. (*Reproduced by permission from L. Meites, "Polarographic Techniques," Interscience Publishers, Inc., New York,* 1955.)

oxidized form of the couple. Comparing this with Eq. (6-7) reveals that the latter corresponds to the special case in which the concentration of the reduced form in the bulk of the solution, and hence its anodic diffusion current, is zero.

When the solution initially contains both the oxidized and reduced forms of the couple, as in curve *b*, its polarogram will consist of a single "composite" anodic-cathodic wave if the couple behaves reversibly at the dropping electrode. If the concentrations of the two forms are equal (and if, as is usually the case, their diffusion coefficients are not very different), the polarogram will intersect the residual current curve at the formal potential of the couple. Even if the concentrations of these two forms in the body of the solution are not the same, the point of intersection of these two curves represents the point at which the layer of solution at the surface of the electrode has the same composition as the bulk of the solution. This is the one point on the current–potential curve which could be located by an ordinary potentiometric measurement.

From the foregoing discussion it is evident that for a reversible couple the half-wave potentials of the cathodic, anodic, and composite waves will all be the same, and this is actually the fundamental practical criterion of polarographic reversibility.

6-13. Polarographic Studies of Complex Ions. Once it has been established that the reductions of a simple metal ion and of a complex ion containing that metal in the same oxidation state are reversible and involve the same number of electrons, polarography becomes an extremely powerful technique for studying the chemistry of the complex ion.

The half-wave potential of simple cadmium ion in 0.1 F potassium nitrate is -0.578 v. *vs.* S.C.E. In this medium the cadmium is reduced to cadmium amalgam, so that $n = 2$. By applying Eq. (6-7) it is found that the reduction is reversible. When excess ammonia is added to the solution, the cadmium is converted to a complex ion which we may write as $Cd(NH_3)_p^{++}$, and this is also found to be reversibly reduced to cadmium amalgam. In a solution containing 0.1 F potassium nitrate and 1.00 F ammonia, the half-wave potential of cadmium is -0.784 v. *vs.* S.C.E.

On the basis of this information we can write one equation for the potential of a cadmium amalgam electrode in a solution containing simple cadmium ion

$$E = -0.578 - 0.02957 \log \frac{[Cd(Hg)_x]}{[Cd^{++}]}$$

where $[Cd(Hg)_x]$ is the concentration of metallic cadmium present in the amalgam. In a solution containing the ammino-cadmium complex ion, the potential of a cadmium amalgam electrode will be given by the equation

$$E = -0.784 - 0.02957 \log \frac{[Cd(Hg)_x][NH_3]^p}{[Cd(NH_3)_p^{++}]}$$

This may be rewritten

$$\begin{aligned} E &= -0.784 - 0.02957 \log \frac{[Cd(Hg)_x]}{[Cd^{++}]} \frac{[Cd^{++}][NH_3]^p}{[Cd(NH_3)_p^{++}]} \\ &= -0.784 + (E + 0.578) - 0.02957 \log K \end{aligned} \qquad (6\text{-}15)$$

where K is the dissociation constant of the ammino-cadmium complex.

This in turn gives

$$0.02957 \log K = -0.784 + 0.578 = -0.206$$

whence K is calculated to be 1×10^{-7}. This entire calculation amounts merely to combining the formal potentials of the half-reactions

$$Cd^{++} + x\text{Hg} + 2e = Cd(Hg)_x \qquad (a)$$
$$Cd(NH_3)_p^{++} + x\text{Hg} + 2e = Cd(Hg)_x + pNH_3 \qquad (b)$$

in such a way as to secure the equilibrium constant of the reaction

$$Cd(NH_3)_p^{++} = Cd^{++} + pNH_3 \qquad (b) - (a)$$

and the entire procedure is completely analogous to the calculations described in Sec. 3-3.

This has given us the dissociation constant of the complex; in order to find its formula (*i.e.*, the value of p) we need merely differentiate Eq. (6-15) with respect to the concentration of ammonia. We thus secure

$$\frac{\Delta E}{\Delta \log [NH_3]} = -\frac{0.05915}{n} p \qquad (6\text{-}16)$$

where E is the potential of a cadmium amalgam electrode in a solution of the ammino-cadmium complex. Implicit in the differentiation is the requirement that the ratio $[Cd(Hg)_x]/[Cd(NH_3)_p^{++}]$ be kept constant. One way of ensuring that this requirement is satisfied consists of taking E as the half-wave potential of the wave secured in the presence of any concentration of ammonia. At the half-wave potential the ratio $[Cd(Hg)_x]/[Cd(NH_3)_p^{++}]$ is constant and equal to unity regardless of the concentration of either ammonia or the ammino-cadmium complex ion in the bulk of the solution.

In the present case, experimental measurements show that $\Delta E_{1/2}/\Delta \log [NH_3]$ is -118 mv. for values of $[NH_3]$ between 0.1 and about 1.2 F. Since n is known to be 2, this proves that the formula of the complex ion in these solutions is $Cd(NH_3)_4^{++}$.

In all studies of this sort it is necessary to have a large excess of the complexing agent present in the solution. The reduction of the complex metal ion liberates the free complexing agent at the surface of the electrode, and so the concentration of the complexing agent at the electrode surface increases with increasing current at potentials along the rising part of the wave. This means that the concentration of free ammonia in the layer of solution right at the electrode surface varies as the polarogram is being recorded. However, when the concentration of the complexing agent already present in the bulk of the solution is much larger than the concentration of the complex metal ion, the relative increase thus produced becomes very small and can be neglected. For exactly the same reason it is essential to use a well-buffered supporting electrolyte whenever the half-reaction taking place at the dropping electrode either produces or consumes hydrogen ions. In such cases, of course, measuring the change of $E_{1/2}$ as the pH of the solution is varied permits one to calculate the number of hydrogen ions taking part in the electrode reaction. This is of particular importance in studying the polarographic behavior of an organic compound.

6-14. Qualitative Polarographic Analysis. The fact that the half-wave potential of a metal ion depends on the nature and concentration of any complexing agent present in the supporting electrolyte can be used to good advantage in carrying out qualitative analyses. Data are now available on the half-wave potentials of most of the common metal ions in about thirty different supporting electrolytes containing a wide variety of complexing agents, among them chloride, thiocyanate, acetate, hydroxide, tartrate, citrate, malonate, ammonia, pyridine, hydrazine, and a good many more.

In any one of these supporting electrolytes the waves due to any particular pair of metal ions may occur at so nearly the same potential that they overlap on a polarogram of a mixture. For example, the formal potentials of the copper(II)–copper amalgam and bismuth(III)–bismuth amalgam couples are almost exactly the same, and so the half-wave potentials of cupric and bismuth ions in a perchloric acid supporting electrolyte are so close together (+0.01 and +0.02 v. *vs.* S.C.E., respectively) that a perchloric acid solution containing these two ions gives a polarogram consisting of just one wave. (Naturally, the height of this wave is equal to the sum of the separate diffusion currents of the two ions.) From such a polarogram it would be impossible to tell whether both elements were present or not; if other evidence indicated that they were, the polarogram would not give any useful information concerning their relative concentrations.

Both copper and bismuth form tartrate complexes, but in an 0.5 F sodium tartrate solution at pH 4 the copper complex is much the weaker of the two. Therefore the half-wave potential of copper is shifted only to −0.06 v. *vs.* S.C.E. in this medium, whereas that of bismuth becomes −0.29 v. *vs.* S.C.E. A polarogram of a mixture of copper and bismuth in this supporting electrolyte will consist of two waves which are sufficiently far apart to permit the identification of each. Moreover, the concentration of each metal can easily be determined by measuring the height of its wave (except in the case discussed on pages 168 and 169).

In carrying out a qualitative polarographic analysis, one usually begins by securing a polarogram of an aliquot of the unknown in a comparatively simple supporting electrolyte, such as 1 F potassium chloride. Let us suppose that a certain unknown gives three waves in this medium, and that the half-wave potentials of these waves are −0.45, −0.65, and −1.00 v. *vs.* S.C.E. On consulting a table of half-wave potentials, such as Table D in the Appendix, we would see that the wave at −0.45 v. might be due to either thallous ion or lead ion or both. The wave at −0.65 v. might be due to chromic ion, but in that case we should also find a wave at −0.85 v., and the fact that no such wave appears on the polarogram indicates that chromic ion is actually absent, and that the wave at −0.65 v. can only be due to cadmium. The final wave, at −1.00 v., cannot be due to chromate ion, because the first chromate wave (at −0.3 v.) did not appear, and so this wave must be due to zinc alone.

Next we might obtain the polarogram of another portion of the sample in 1 *F* potassium cyanide, which forms complexes with all of these elements except thallous thallium. Let us suppose that this polarogram also shows three waves, at >0,† −0.71, and −1.20 v. The first wave might be due to ferric iron (present as ferricyanide), but ferric iron would also have given a wave starting from zero applied potential on the first polarogram. Since this did not appear, this wave must be due to thallous ion alone, and the wave at −0.45 v. in the chloride solution must have been at least partly due to thallous ion. The half-wave potential of the second wave is close to the half-wave potential of lead in 1 *F* cyanide (−0.72 v.), thus proving that the unknown contains lead as well as thallium. The wave at −1.20 v. is due to cadmium, confirming the interpretation of the wave at −0.65 v. in the chloride medium. No wave is obtained for zinc in a cyanide medium (because potassium ion is more easily reduced than the zinc cyanide complex), and the fact that no wave is found in the cyanide solution to correspond to the wave at −1.00 v. in the potassium chloride serves to confirm the presence of zinc.

Depending on the complexity of the unknown, it may sometimes be necessary to secure many more than two polarograms in various supporting electrolytes in order to identify every element present. However, the procedure just outlined is entirely typical of what is done even in a much more complicated case.

6-15. Quantitative Polarographic Analysis. The problems which must be solved during the development of a procedure for the polarographic

† This symbol indicates that the wave merges with the initial current rise; the current rises directly from large negative values (which reflect the anodic oxidation of mercury) to positive values (owing to the reduction of thallous ion) without passing through any inflection around zero current. Chemically this means that, in 1 *F* cyanide, thallous ion oxidizes metallic mercury spontaneously according to the equation

$$2Tl^+ + (2x + 1)Hg + 2CN^- = Hg(CN)_2 + 2Tl(Hg)_x$$

We describe such waves by saying that they "start from zero applied potential." The phrase is not very accurately descriptive, but nothing better is in common use.

analysis of a sample are very similar to those encountered in devising a volumetric or gravimetric method. The first and most obvious requirement is that the element in question must be capable of being accurately determined under the proposed conditions when it is present alone. This means that one must select a supporting electrolyte in which it gives a well-defined wave—a wave, that is, which has a fairly long flat plateau so that the diffusion current can be measured accurately—whose height is, if possible, directly proportional to the concentration of the substance being determined. Once in a while, because of complications such as adsorption, catalytic processes, and so on, one comes across a wave whose height is not proportional to the concentration of the substance responsible for it. Such waves can be used for analytical purposes if one constructs an empirical plot of wave height against concentration from data secured with known solutions, but it is desirable to avoid this whenever possible. In most methods for the interpretation of the wave height (Sec. 6-16), it is taken for granted that i_d is proportional to C, and so it is always advisable to make sure that this is true under the conditions finally selected for carrying out an analysis.

Let us suppose that a certain sample has to be analyzed for bismuth. On consulting a table of half-wave potentials, we would immediately reject such supporting electrolytes as alkaline citrate or tartrate (in which the bismuth waves are ill-defined), and potassium cyanide or sodium hydroxide (in which bismuth is only very slightly soluble); and we would look for media in which bismuth gives a good wave. These include nitric, hydrochloric, and sulfuric acids, acidic solutions of citrate, tartrate, malonate, or hydrazine, solutions containing triethanolamine or ethylenediaminetetraacetate, and many others.

If the sample is a pure bismuth salt or a solution containing no reducible substance except bismuth, this would be virtually the end of our problem. We might arbitrarily select 1 F nitric acid as a supporting electrolyte and, after making sure that the height of the bismuth wave is proportional to the concentration of bismuth in this solution, we would only have to secure a polarogram of a solution of the sample in 1 F nitric acid and calculate the concentration of bismuth from its measured diffusion current.

However, the problem may become considerably more complicated if the sample contains other reducible substances in addition to the bismuth. This naturally depends on the nature of these other substances. Hydrogen ion is reducible, but obviously it does not interfere with the measurement of the bismuth wave height (if it did, we could hardly use nitric acid as a supporting electrolyte in the analysis of pure bismuth solutions). This is because hydrogen ion is not reduced until $E_{d.e.}$ becomes over a volt more negative than the half-wave potential of bismuth. Such other metals as zinc, manganese, lead, and cadmium, which also give waves separated by some tenths of a volt from the bismuth wave, would certainly not interfere.[1] But when the solution contains one or more substances whose waves precede or coincide with the bismuth wave, it may be difficult or even impossible to measure the diffusion current of bismuth accurately. For example, in a nitric or perchloric acid supporting electrolyte the half-wave potentials of copper and bismuth are almost identical, and so even a relatively small amount of copper would cause an error in the determination of bismuth in either of these media. Difficulty would also be encountered if the solution contained a very large concentration of ferric iron, whose wave precedes that of bismuth in these supporting electrolytes. In principle we could measure the height of the ferric wave, and subtract this from the sum of the iron and bismuth diffusion currents, measured at a potential on the plateau of the bismuth wave. This works fairly well when the iron/bismuth ratio is smaller than about 10, but otherwise it leads to an intolerable relative error in the small diffusion current of the bismuth.

In either of these cases one would try to find a supporting electrolyte in which the desired wave precedes the wave of any other constituent of the solution, preferably by several tenths of a volt or more so that there will be plenty of room for the full development of the plateau of the wave one is interested in.

Tables of half-wave potentials show that a slightly acidic (to prevent hydrolysis and precipitation of the bismuth) solution of sodium fluoride should be well suited to the determination of bismuth in the presence of large concentrations of ferric iron. In 1 F fluoride the half-wave potentials of bismuth and ferric iron are -0.07 and -0.49 v. *vs.* S.C.E., respectively, so that the two waves are very well separated. This reversal of the relative positions of the waves merely reflects the fact that the ferric fluoride complex is far more stable than the bismuth fluoride complex.

Alternatively, one might replace the fluoride by a reducing agent, such as hydroxylamine, sufficiently powerful to reduce iron to the ferrous state without reacting with the bismuth.

The determination of bismuth in the presence of a relatively small concentration of copper presents no difficulty; we could use the acidic tartrate solution mentioned in Sec. 6-14. But if much more copper than bismuth were present, this would only lead us into the same problem

[1] This requires a minor qualification. A solution containing 1 F nitric acid and, say, 5 F zinc nitrate is much more viscous than pure 1 F nitric acid, and this would result in a considerable decrease in the diffusion coefficient of bismuth. However, this could be handled very easily, either by determining the diffusion current constant of bismuth in the nitric acid–zinc nitrate mixture or by using the "pilot ion" method to be described in Sec. 6-16.

caused by ferric iron in a mineral acid solution, for the copper wave precedes the bismuth wave. Unfortunately, no simple supporting electrolyte is known in which bismuth gives a wave preceding the copper wave, for bismuth complexes generally have smaller dissociation constants than the corresponding cupric complexes. This situation presents a real test of one's ingenuity, and an experienced chemist might rise to the occasion by preparing a supporting electrolyte containing sodium cyanide (which reacts with cupr*ous* ion to form a complex whose dissociation constant is so small that it gives no wave at all), hydrazine (to serve as a reducing agent and bring about the reaction

$$4Cu^{++} + 8CN^- + N_2H_4 + 4OH^- = 4Cu(CN)_2^- + N_2 + 4H_2O)$$

sodium hydroxide (to keep the solution alkaline and prevent liberation of cyanogen or hydrogen cyanide), and sodium ethylenediaminetetraacetate (to complex the bismuth and prevent precipitation of hydrous bismuth oxide). As the cyanide and hydrazine together would convert ferric iron to ferrocyanide, which is also not reducible, this supporting electrolyte could equally well be used for the determination of bismuth in samples containing large amounts of both ferric iron and copper.

In this way it is usually possible to find a supporting electrolyte in which any one metal (excepting, of course, the alkali, alkaline earth, and rare earth elements, which are not particularly amenable to differential complexation) can be determined polarographically in the presence of any other. However, this becomes more and more difficult as the number of elements present increases, and with complex samples it is often necessary to resort to some kind of chemical or electrolytic separation to remove the interfering constituents of the sample. On the basis of several examples similar to those we have described, some of its enthusiastic proponents during the early days of polarography suggested that the polarograph was about to render the analytical chemist obsolete. What has happened instead is that polarography has turned into one of the most fertile fields for the exercise of the analytical chemist's knowledge and imagination. Many analyses can be carried out polarographically which would be more difficult, more time-consuming, or more inaccurate if done by any other technique yet devised. But whether this is true of any one specific determination depends primarily on the resourcefulness of the chemist.

6-16. Methods of Interpreting the Wave Height. Assuming that one has found a supporting electrolyte which satisfies the criteria outlined in the preceding section, and has measured the diffusion current in that medium of the substance being determined, the only remaining problem in the execution of a quantitative polarographic analysis is the calculation of the concentration of that substance from its diffusion current.

One way of doing this, the so-called "absolute" method, was discussed in Sec. 6-8. It necessitates the measurement of m and t under exactly the same conditions employed for the measurement of the diffusion current. Then, using a diffusion current constant taken from the literature, the concentration of the solution is calculated from the relation

$$C = \frac{i_d}{Im^{2/3}t^{1/6}}$$

In student work, m is best measured by collecting a known number of mercury drops in a small glass cup or weighing bottle which is provided with a handle so that it can be easily manipulated under the surface of the solution. The mercury thus collected should be washed once or twice with water to remove dissolved salts, and then dried at room temperature after rinsing with either ethanol and ether or reagent grade acetone. This gives mt, from which m is easily calculated after t is measured. To do this, one merely determines the time which elapses between the fall of one drop and the fall of the fifth or tenth drop thereafter. It must be emphasized that m and t have to be measured with the applied potential, height of the mercury column, and all other experimental conditions unchanged from those which prevailed during the measurement of the wave height.

Very often a value of the diffusion current constant cannot be found in the literature. Then one can use the "standard solution" method instead. This involves the preparation of a number of known solutions of the ion being determined in the supporting electrolyte used. The concentrations of these solutions should cover a range which includes the expected concentration of the unknown. The diffusion currents of all of the solutions, including the unknown, are measured under identical experimental conditions. Generally the values of i_d/C for the known solutions will be constant, and then the calculation of the concentration of the unknown solution from its measured diffusion current is perfectly straightforward. Otherwise one can construct a plot of i_d against C for the known solutions, and use this to find the value of C for the unknown.

In both of these methods such factors as the temperature and viscosity of the solution and the value of $m^{2/3}t^{1/6}$ must be either known or kept constant. The so-called "pilot ion" method has the advantage of eliminating the necessity of controlling such variables. It involves the addition to the unknown solution of a known concentration of some other ion whose wave is well separated from that of any constituent of the sample. If a solution containing only cadmium is being analyzed in a supporting electrolyte containing 1 F ammonia and 1 F ammonium chloride, we might add zinc as a pilot ion, for the zinc ($E_{1/2} = -1.35$ v.)

and cadmium ($E_{1/2} = -0.81$ v.) waves are so well separated that both diffusion currents will be very easy to measure. In the resulting mixture the ratio of the cadmium and zinc wave heights will be

$$\frac{(i_d)_{\text{Cd}}}{(i_d)_{\text{Zn}}} = \frac{607 n_{\text{Cd}} D_{\text{Cd}}^{1/2} C_{\text{Cd}} (m^{2/3} t^{1/6})_{\text{Cd}}}{607 n_{\text{Zn}} D_{\text{Zn}}^{1/2} C_{\text{Zn}} (m^{2/3} t^{1/6})_{\text{Zn}}}$$

where $(m^{2/3}t^{1/6})_{\text{Cd}}$, for example, is the value of $m^{2/3}t^{1/6}$ at the potential at which the cadmium diffusion current is measured. The ratio of the $m^{2/3}t^{1/6}$ values at any two potentials will be constant for any given supporting electrolyte at a fixed temperature, and since the values of n and D are constant under any fixed set of experimental conditions, we have simply

$$\frac{(i_d)_{\text{Cd}}}{(i_d)_{\text{Zn}}} = k \frac{C_{\text{Cd}}}{C_{\text{Zn}}}$$

The value of k can be found by measuring the two diffusion currents in a solution containing known concentrations of both cadmium and zinc. Once k is known, the concentration of cadmium in any solution containing a known concentration of zinc can be calculated from the ratio of their diffusion currents.

The peculiar advantage of the pilot ion method is that a change in the experimental conditions which alters the diffusion current of one of the ions tends to affect the diffusion current of the other in exactly the same way, and so the effect disappears from the ratio of the diffusion currents. An increase in the temperature of the solution, for example, increases the diffusion coefficients of both ions, but it increases them to almost exactly the same extent. The same thing is true of variations in the pressure of mercury above the capillary, the viscosity of the solution, and so on. As a result, the pilot ion method is one of the most useful techniques of practical polarographic analysis.

6-17. Maxima and Maximum Suppressors. Not all polarographic waves are as well defined as those shown in the preceding illustrations in this chapter. In particular, some waves are distorted by "maxima" such as those shown on the first few curves of Fig. 6-8. These often cause great trouble in measuring the height of a wave, especially if a later wave begins before the maximum has completely disappeared. In addition, the presence of a maximum may lead to serious errors in the measurement of the half-wave potential as well as make it impossible to tell whether a wave is reversible or not. Consequently it is usually desirable to suppress a maximum when it is encountered, and this is done by adding a "maximum suppressor," which is a substance that is capable of being adsorbed on the surface of the drop.

One of the best maximum suppressors is Triton X-100, a nonionic detergent manufactured by the Rohm and Haas Company, Philadelphia, Pa. When about 0.002 to 0.004 per cent of this material is added to a solution—most conveniently in the form of a stock 0.1 per cent solution in water—the suppressive effect is illustrated by Fig. 6-8. Triton X-100 is preferable to most other maximum suppressors because its solutions are both very easy to prepare and very stable. Gelatin, which is perhaps even more widely used than the Triton, is considerably inferior in both respects.

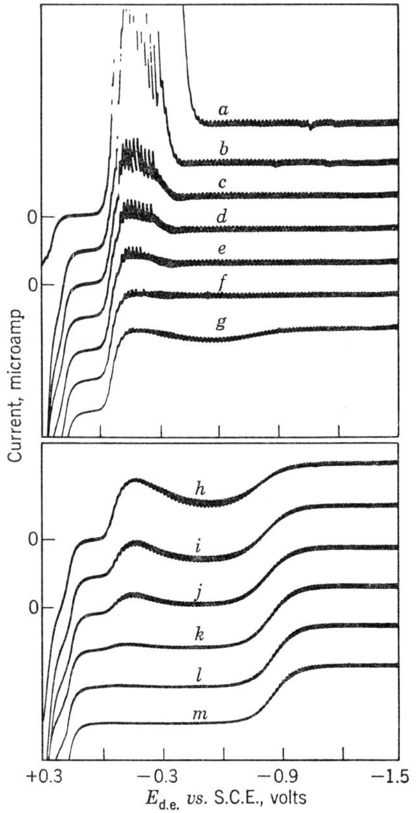

FIG. 6-8. Polarograms of 4 m*M* cupric copper in *ca.* 0.09 *F* (saturated) potassium sodium tartrate, pH 4.00, and (*a*) 0, (*b*) 0.300, (*c*) 0.595, (*d*) 0.884, (*e*) 1.16, (*f*) 1.72, (*g*) 2.24, (*h*) 3.26, (*i*) 4.72, (*j*) 6.37, (*k*) 11.1, (*l*) 24.8, and (*m*) 44.8 $\times 10^{-3}$ per cent Triton X-100. The zero current lines are indicated only for curves *a*, *c*, *h*, and *j*. [*L. Meites and T. Meites, J. Am. Chem. Soc.*, **73**, 177 (1951).]

The danger of adding too much Triton X-100 (or any other maximum suppressor) is demonstrated by curves *f* to *m* of Fig. 6-8.

It is well established that maxima reflect a streaming of the solution past the drop surface; this causes many more of the reducible ions to reach the electrode than could do so if the solution were completely quiet. Presumably the maximum suppressor serves to immobilize a very thin layer of solution around the electrode surface, but the mechanism of maximum formation and suppression cannot be discussed in detail here.

6-18. The Rotating Platinum Microelectrode. In Sec. 6-2 we discussed the reasons behind the fact that the dropping mercury electrode is by far the most widely used indicator electrode in voltammetric measurements. However, since mercury itself is fairly easily oxidized, it is impossible to obtain very much information with the dropping electrode about the many electrode processes which occur at potentials more positive than the initial current rise in any supporting electrolyte. Thus, ceric and ferric ions are reducible at widely different potentials in mineral acid media, and yet the fact that they both oxidize mercury makes it impossible to measure their half-wave potentials or even to distinguish between them when a dropping electrode is used.

For work with such strong oxidizing agents, or with organic compounds which cannot be reduced and can be oxidized only with difficulty, it is necessary to use a platinum electrode instead of the dropping electrode. This extends the range of available potentials on the positive side to values at which water itself begins to be oxidized to oxygen, usually around +1.1 v. *vs.* S.C.E., which is about 0.7 v. more positive than the most positive value that can be attained with a dropping electrode under any conditions.

The most widely used kind of platinum indicator electrode is the rotating platinum microelectrode, which may consist merely of a small piece of platinum wire sealed into a length of glass tubing in such a way that it projects through the side of the tubing perpendicular to its axis. Electrical connection to the platinum is made by filling the tube with mercury and dipping a nickel or tungsten wire into the mercury, and the entire electrode is rotated at a speed of about 600 r.p.m. during the measurements.

Unlike a stationary platinum microelectrode, this arrangement gives a current which does not depend perceptibly on the length of time for which the electrical circuit has been connected, for the local concentration gradients around the electrode are destroyed as the electrode rotates.

Unfortunately, the factors which govern the diffusion current at a rotating platinum microelectrode are not as yet completely understood. They certainly include the area and rate of rotation of the electrode and the concentration and diffusion coefficient of the substance being reduced or oxidized. Specifically, it is known that the diffusion current is directly proportional to the concentration of the diffusing ion or molecule, but we cannot write a complete equation which will account quantitatively for the effects of the remaining variables. In actual practice one simply uses the standard solution method described in Sec. 6-16 to construct an empirical wave height–concentration plot for the particular rotating electrode used in a given experiment, but of course it would be necessary to establish a new curve if the electrode had to be replaced for some reason.

The behavior of a platinum microelectrode is greatly influenced by the formation of thin films of platinum oxides or adsorbed hydrogen on its surface, which produce effects that are being very actively investigated. Meanwhile the primary use of the rotating platinum microelectrode is as an indicator electrode in amperometric (Sec. 6-19) and coulometric (Sec. 7-11) titrations involving powerful oxidizing agents such as bromine and ceric ion.

6-19. Amperometric Titrations. An amperometric titration is the polarographic analogue of a conductometric titration, and consists of using diffusion current measurements to locate the end point of a titration with a chemical reagent.

Let us consider the titration of lead ion with potassium chromate in an acetate buffer,[1] using a dropping mercury electrode at a potential of -0.75 v. *vs.* S.C.E. as the indicator electrode. At this potential both lead and chromate ions yield their diffusion currents. At the beginning of the titration the concentration of lead ion is $C^0_{Pb^{++}}$, and the measured diffusion current will be

$$i_d{}^0 = I_{Pb^{++}} C^0_{Pb^{++}} m^{2/3} t^{1/6}$$

If to V^0 ml. of this solution we add v ml. of $C_{CrO_4^=}$ mF potassium chromate, containing $vC_{CrO_4^=}$ micromoles of chromate, the concentration of lead ion will become

$$C_{Pb^{++}} = \frac{V^0 C^0_{Pb^{++}} - vC_{CrO_4^=}}{V^0 + v}$$

Then

$$i_d = I_{Pb^{++}} m^{2/3} t^{1/6} C_{Pb^{++}}$$

which becomes, on rearranging,

$$i_d \left(\frac{V^0 + v}{V^0} \right) = i_d{}^0 - \left(\frac{I_{Pb^{++}} m^{2/3} t^{1/6} C_{CrO_4^=}}{V^0} \right) v \qquad (6\text{-}17)$$

This is entirely analogous to Eq. (5-13), which expresses the change in conductance with the volume of reagent solution added before the end point of a conductometric titration. By entirely similar reasoning we could derive an equation for the diffusion current of the excess chromate ion in the solution after the equivalence point is passed; after correcting for dilution this will evidently be proportional to the volume of excess chromate added. In the idealized case where the product of the reaction is completely insoluble (or completely undissociated, etc.), therefore, the amperometric titration curve will consist simply of two straight lines intersecting at the equivalence point.

However, if the titration reaction is not entirely complete, the experimental points will deviate more and more from these extrapolated straight lines as the equivalence point is approached, and so there is often a good deal of curvature near the end point. This is handled in exactly the same way as in a conductometric titration, by ignoring the vicinity of the end point and only securing points sufficiently far before and after the end point (where the excess of one or the other of the reactants forces

[1] This serves two purposes: it suppresses the migration current of lead ion, which would tend to give a curved titration curve rather than a straight one, and it prevents the formation of a basic lead chromate, which would be accompanied by low results.

the reaction virtually to completion) to define the two straight lines which are then extrapolated to their point of intersection.

As with conductometric titrations, different kinds of amperometric titrations give rise to several different kinds of titration curves. For example, if in the titration of lead with chromate the indicator electrode potential had been ±0.0 v. *vs.* S.C.E. (a potential at which chromate ion is reduced in this medium but lead is not), the diffusion current would have remained practically zero up to the end point and would then have increased linearly with the volume of chromate in excess, giving a _/-shaped titration curve. Incidentally, one practical advantage of carrying out the titration at this potential rather than at −1.0 v., even though the shape of the titration curve secured at the latter potential is more conducive to accurate results, is that dissolved oxygen would not have to be removed from the solution because it is not reducible at 0 v. *vs.* S.C.E.

Most of the remarks made in Chap. 5 concerning the theory and execution of conductometric titrations apply to amperometric titrations as well. The most important difference between the two kinds of titrations results from the fact that the conductance of a solution depends on the concentration of every ion present, while the diffusion current at a given potential does not. We would not think of trying to determine a trace of lead in a concentrated zinc nitrate solution by a conductometric titration with sulfate, for the zinc nitrate would swamp out the small changes in conductance occurring as the lead was precipitated. However, lead ion is reducible at an applied potential of −0.60 v. *vs.* S.C.E., whereas zinc and nitrate ions are not. So the same titration could be carried out amperometrically without difficulty. This is the most important advantage of amperometric over conductometric titrations. In fact, the presence of an excess of some indifferent salt is actually desirable in an amperometric titration because it eliminates the migration current, which would distort the titration curve near the beginning. Moreover, this tolerance of indifferent salts renders amperometric redox titrations perfectly feasible even with very simple equipment (compare Sec. 5-9).

Of course this does not mean that amperometric titrations are completely free from interferences. Barium ion would certainly interfere in the amperometric titration of lead with sulfate. Furthermore, any other substance which contributed to the measured current at a potential on the plateau of the lead wave would decrease the accuracy and precision of the titration even if it did not react chemically with sulfate. For example, if the solution originally contained enough lead to give a diffusion current of 20 microamp., the measured current would decrease from 20 microamp. to virtually zero at the end point. But if the solution also contained enough ferric iron to give a diffusion current of 200 microamp., the total current on the plateau of the lead wave (after correction for dilution) would change only from 220 to 200 microamp. To be sure, this particular interference could be overcome by adding a reducing agent to convert the iron to the ferrous state before starting the titration.

It is unnecessary to know either $m^{2/3}t^{1/6}$, the temperature, or any of the other variables which affect the diffusion currents measured in an amperometric titration; it is necessary only to make sure that none of them varies appreciably during the titration. Partly because the number of important variables is greatly decreased, partly because the accurate measurement of a volume of reagent is much simpler than the equally accurate measurement of a fluctuating electrical current, and partly because the process of drawing the straight-line portions of a titration curve has the effect of averaging out the random errors in the individual current measurements, amperometric titrations tend to be much more accurate than ordinary polarographic analyses. An accuracy of ± 0.3 per cent or better, which is commonplace in amperometric titrations of solutions as concentrated as 1 mM, can be attained in direct polarography only by extremely meticulous work.

6-20. Electrical Apparatus for Polarographic Measurements. The electrical circuits used in polarographic work have just two main functions. One is the application of a known d-c potential across the cell, and the other is the measurement of the resulting current. The potentials required are never greater than about 2.7 v., and so they may easily be obtained from a pair of dry cells in series. The currents may be as large as 100 microamp., though this is a little unusual, and in practical polarographic analysis it is desirable to be able to secure an accuracy of the order of ± 0.005 microamp. when measuring very small diffusion currents.

Figure 6-2 shows what is probably the simplest possible circuit for making polarographic measurements. A linear voltage divider is used to apply a known fraction of the voltage indicated by the voltmeter V across the cell, and a microammeter is used to measure the current at the resulting potential. Most microammeters have extremely short periods, and as a result the average of their maximum and minimum deflections during the life of a drop is considerably smaller than the true average current. It is therefore advisable to damp the meter by connecting a high-quality electrolytic capacitor of about 250 microfarads across its terminals. The measurements are made by adjusting the voltage divider R to secure some known value of the applied potential, then reading the average deflection of the microammeter needle, and repeating this at as many applied potentials as may be necessary to secure the requisite data.

It is impossible to cover a very wide range of currents with a microammeter, and so the circuit shown in Fig. 6-9 is much more useful in practical measurements. Here R_s is a standard 10000-ohm resistor in series with the cell. The iR drop across this resistor is measured with a students' potentiometer, which permits the easy calculation of the current flowing through the cell. The galvanometer used in the potentiometer circuit must be considerably overdamped by connecting a small resistance across its terminals, so that the periodic variation of the iR drop being measured as the drops grow and fall results in a to-and-fro oscillation no greater than 1 or 2 cm. The setting of the potentiometer at the point of balance, where the galvanometer deflects equally far away from zero in each direction, corresponds to the average current flowing through the cell.

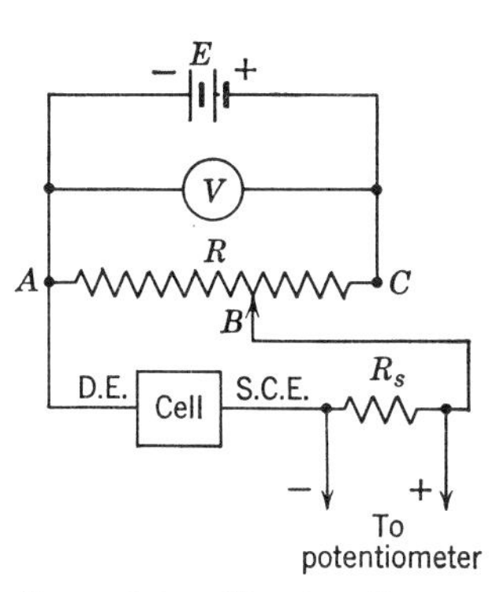

FIG. 6-9. Circuit diagram of a polarograph employing a linear voltage divider and a "resistance-potentiometer" arrangement for current measurements.

In this circuit there is, of course, a difference (which is equal to the iR drop across the resistor R_s) between the *voltage* provided by the bridge and the *potential* actually applied across the cell. By applying the considerations outlined in Sec. 3-10, it is easily shown that the value of $E_{\text{d.e.}}$ is more *positive* than the output of the voltage divider R whenever a positive current (corresponding to reduction at the dropping electrode) flows through the cell circuit. On the other hand, when i is negative (oxidation at the dropping electrode), $E_{\text{d.e.}}$ is more *negative* than the voltage provided by R. Suppose, for example, that the voltmeter V reads 2.90 v., that R is set to apply exactly four-tenths of this voltage to the cell circuit, and that the dropping electrode is made the negative electrode of the cell. Then the applied voltage is

$$-(0.400)(2.90) = -1.160 \text{ v.}$$

and if the measured iR drop across R_s is +0.0325 v., the cell potential is

$$E_{\text{d.e.}} = -1.160 + 0.0325 = -1.127_5 \text{ v.}$$

while the current is

$$i = \frac{0.0325}{10000} \times 10^6 = 3.25 \text{ microamp.}$$

Both $E_{\text{d.e.}}$ and i naturally vary during the life of a drop, and this procedure yields an average value of each quantity.

If a suitable linear voltage divider is not available, the circuit of Fig. 6-10 may be used. Here the potentiometer can be connected across either the cell itself or the standard resistor R_s by appropriate manipulation of the double-pole–double-throw switch. When the potentiometer is connected across R_s, one measures the average iR drop across this resistor in exactly the same way as with the circuit of Fig. 6-9. Throwing the switch in the other direction permits a direct measure of the potential across the cell. This circuit is suitable for the most refined polarographic measurements; its sole disadvantage is the fact that the measurements proceed rather slowly, because the potentiometer must be balanced twice at each point on the current–potential curve. With the addition of more switches, which are included to facilitate making the connections required in measuring negative currents and positive values of $E_{d.e.}$, this is the circuit recommended in Expt. VI.

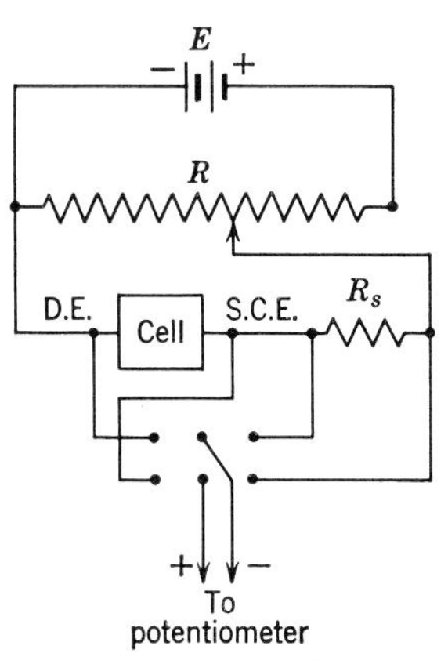

FIG. 6-10. Circuit diagram of a "resistance-potentiometer" polarograph employing a nonlinear voltage divider.

Several automatically recording polarographs are now available commercially. The most recent ones employ what is essentially the circuit of Fig. 6-9. However, the voltage divider R is motor-driven at a constant rate so that the value of $E_{d.e.}$ increases linearly with time, and a recording potentiometer connected across the standard resistor in series with the cell is used to draw a continuous plot of the iR drop against time (*i.e.*, applied potential). A number of standard resistors are included to permit varying the sensitivity over a wide range.

A galvanometer can also be used as the current-measuring element of a polarographic circuit; the cell current is caused to flow directly through the galvanometer coil, producing a deflection which is directly proportional to the current. In our opinion the requisite circuitry is too difficult for beginners to construct to warrant detailed description here.

Students often have trouble in deciding how many measurements must be made in securing a polarogram. As a general rule, data should be secured at rather small intervals, preferably about 0.02 v. in $E_{d.e.}$, on the rising portion of a wave, where the current is changing rapidly, but points 0.1 or even 0.2 v. apart in $E_{d.e.}$ will suffice on the plateau or before the start of a wave, where the current is almost independent of the applied potential. Even this depends on the purpose for which the data will be used. If all that is wanted is an approximate value of $E_{1/2}$ for a qualitative analysis, points 0.05 v. apart on the rising part of the wave will suffice. But if one wants to use Eq. (6-7) to determine whether a wave is reversible or not, it is desirable to have data at intervals as small as 0.01 v. Finally, in certain quantitative methods (*e.g.*, the one used in Expt. III-12) all that is needed is a single measurement of the current at one potential on the plateau of a wave. When securing a complete polarogram, it is convenient to plot the data roughly as soon as they are secured, because then one can easily tell where additional points are needed to define the curve.

6-21. Dropping Electrodes and Cells. A typical dropping electrode assembly was shown in Fig. 6-1, and its construction is so simple as to need no description beyond a warning that, for obvious reasons, all of the Tygon-glass connections must be securely wired.[1]

The dropping electrode itself is most conveniently made from the "marine barometer tubing" sold by the Corning Glass Works. This has a reasonably uniform i.d. of 0.06 to 0.08 mm. and an o.d. of about 6 mm. A 10-cm. length will usually give drop times in the desired range of 2.5 to 7 sec. at convenient heights of the mercury column. The capillary should be freshly cut from the stock of tubing, with care to ensure that the tip is cut off perfectly square, and must be perfectly vertical when in use. Erratic results will be secured if the tip of the capillary is inclined even a few degrees away from a horizontal plane.

The radius of the capillary is so small that it is easily stopped up by dust, dried salts, or any other foreign matter. An experienced polarographer can use a single capillary for many years, but a capillary can, and in the hands of beginners often does, become unusable in an hour when carelessly treated.

When the capillary is first connected to the stand tube and the assem-

[1] Despite its innocuous appearance, and despite the contempt bred by familiarity, metallic mercury is a cumulative and extraordinarily unpleasant poison. It is the responsibility of the laboratory supervisor to ensure that any laboratory where extensive polarographic work is done is sufficiently well ventilated to reduce the concentration of mercury vapor to a wholly negligible value. In addition, every student must exercise unremitting care to make sure that no mercury is spilled onto the laboratory bench or floor. If any mercury is spilled, it must be promptly and completely picked up.

In carrying out polarographic work it should become a habit to keep the tip of the dropping electrode constantly over a beaker or stainless steel tray except when it is actually immersed in the cell, and all such vessels used to collect mercury must be thoroughly cleaned at the end of each laboratory period. This, as well as the removal of mercury from cells, etc., is best accomplished by using a 125-ml. suction flask, connected to a good aspirator, as a trap. The mercury which collects in the flask can then be collected periodically for repurification and reuse. Mercury must never be poured into the laboratory drains or sinks. Quantities of mercury in excess of a few milliliters should never be collected in glass vessels; heavy polyethylene or stainless steel should be used instead.

bly is filled with mercury, the leveling bulb should be raised to a height sufficient to cause mercury to flow through the capillary, then slowly lowered again until droplets barely cease forming on the dry capillary tip. A mark should be placed on the stand tube at this point. Thereafter the capillary must never be lowered into a solution unless the mercury is above that mark. At the end of an experiment the mercury should be allowed to remain above the mark while the capillary is washed with no less than 100 ml. of water from a wash bottle. Then the capillary may be tapped gently to dislodge the last drop of water, and finally the leveling bulb may be lowered until the mercury is just at the mark. A pinchcock may be applied to the tubing connecting the leveling bulb and the stand tube to prevent excessive loss of mercury on standing overnight or longer. When a capillary is left in this way, mercury will continue to flow very slowly through it as the last trace of water evaporates from its tip,[1] and when the tip becomes completely dry the flow of mercury will cease, leaving the capillary filled with mercury and protected against the ingress of dust particles.

A dirty capillary gives erratic drop times and currents, and should generally be replaced without further ado, taking care to avoid spilling the mercury in the stand tube.

A polarographic cell must perform a number of functions. It must, of course, contain the solution being studied as well as the reference electrode being used, and it must be furnished with a salt bridge which keeps the mercurous ions in the S.C.E. out of the sample solution. In order to keep the iR drop through the cell down to a negligible value, this salt bridge must have a resistance of less than 1000 ohms. Otherwise there would be a substantial difference between the applied voltage and the actual value of $E_{\text{d.e.}}$. The increasing slopes of the curves in Fig. 6-4 as the supporting electrolyte concentration is increased reflect the decreasing resistance of the cell and demonstrate the necessity of eliminating the iR drop if meaningful half-wave potentials are to be secured.

In addition, a polarographic cell almost always has to include some provision for removing dissolved air from the sample solution and for preventing re-solution of air during the measurements. This is because oxygen is reducible in the range of potentials employed in polarographic work, and gives two waves which are due to the successive half-reactions

$$O_2 + 2H^+ + 2e = H_2O_2$$
$$H_2O_2 + 2H^+ + 2e = 2H_2O$$

The half-wave potentials of these waves are nearly independent of the composition of the supporting electrolyte, and are approximately -0.05

[1] The interfacial tension between mercury and water is smaller than that between mercury and air.

and −0.9 v. *vs.* S.C.E., respectively. The second wave is extremely irreversible, extending over nearly half a volt. Consequently these waves would seriously interfere in most polarographic measurements.

Dissolved oxygen is most easily removed from a solution by bubbling a stream of pure nitrogen through it for a suitable length of time, usually between 5 and 20 min., depending on the volume of solution, the size of the nitrogen bubbles, and the rate of flow of nitrogen. Since bubbling gas through a solution while measurements are being made leads to high and erratic currents because of the stirring produced, the gas stream

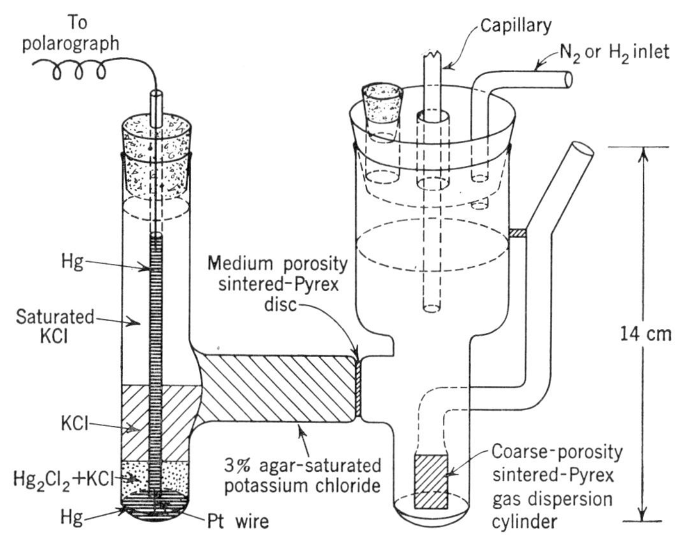

FIG. 6-11. Polarographic H-cell. A very similar cell is available from E. H. Sargent & Co., Chicago (Catalog S-29438). (*Reproduced by permission from L. Meites, "Polarographic Techniques," Interscience Publishers, Inc., New York,* 1955.)

should then be deflected over the surface of the solution to produce a blanket of nitrogen which serves to exclude air.

A typical cell which performs all of these functions is shown in Fig. 6-11. The horizontal cross-member of this "H cell" is filled with a 4 per cent agar gel saturated with potassium chloride; the sintered-glass disc serves merely to hold the molten gel in place while the cell is being filled. Detailed directions for this are given in Expt. VI. The gas-dispersion cylinder breaks up the entering gas stream into a large number of tiny bubbles, thus permitting the solution to be completely deaerated within a minute or two. After the deaeration is complete, the capillary is inserted through one hole in the rubber stopper of the right-hand compartment of the cell, and the gas stream is diverted from the gas-dispersion cylinder to a small tube projecting through another hole in the stopper. A third hole in the stopper allows the experimenter to add reagents during an experiment (*e.g.*, in an amperometric titration) without disturbing the dropping electrode.

The proper care of a cell demands that it be cleaned immediately and thoroughly with distilled water at the end of every experiment. Metallic mercury should never be allowed to stand in the cell exposed to the air, for it is easily air-oxidized to basic mercurous salts which will adhere to the cell walls and interfere in later work. Once the solution compartment of the cell has been cleaned, it may be filled with water if the cell will be used again within 24 hours; otherwise saturated potassium chloride should be used. If this is not done, the gel will dry out and shrink away from the walls of the cross-member, making it necessary to replace both the reference electrode and the agar bridge.

PROBLEMS

The answers to the starred problems will be found on page 403.

***1.** The formal potential of the titanium(IV)–titanium(III) couple in 0.2 F citric acid is -0.12 v. *vs.* N.H.E. What is the equilibrium composition of an 0.2 F citric acid solution containing 0.002 F titanium in contact with an electrode whose potential is -0.30 v. *vs.* S.C.E.?

2. If the formal potential of the cadmium ion–cadmium amalgam couple in 1 F hydrochloric acid is -0.648 v. *vs.* S.C.E., what potential must be applied to a mercury electrode in such a solution to give a cadmium amalgam in which the concentration of dissolved cadmium metal is 1000 times the concentration of cadmium ion remaining in the solution?

3. The formal potential of the mercurous ion–metallic mercury couple in 1 F perchloric acid is $+0.776$ v. *vs.* N.H.E. Calculate the potential, in volts *vs.* S.C.E., at which a mercury electrode in 1 F perchloric acid will be in equilibrium with 0.01 M mercurous ion.

4. The diffusion current of cadmium ion in 1 F potassium nitrate under certain experimental conditions is 4.00 microamp. What would be the height of the cadmium wave under the same conditions but in the absence of any added potassium nitrate? (Assume that the potassium nitrate does not change the viscosity of the solution.)

***5.** The diffusion current of a 2 mM solution of lead ion in 0.1 F potassium chloride, measured with a capillary for which $m^{2/3}t^{1/6}$ is 2.50 mg.$^{2/3}$ sec.$^{-1/2}$, is 20.0 microamp. If the lead ion is reduced to the metallic state, calculate the diffusion coefficient of lead ion in this medium.

6. The equivalent conductance of thallous ion at infinite dilution is 74.7 mho-cm.^{2} per g.-equiv. What would be the diffusion current constant of thallous ion at infinite dilution, assuming that the migration current were completely suppressed?

7. The diffusion current constant of zinc in 1 F sodium hydroxide is 3.14. An unknown solution in 1 F sodium hydroxide gives a zinc diffusion current of 7.00 microamp., measured at $E_{d.e.} = -1.70$ v. *vs.* S.C.E. The values of m and t at this potential are 2.83 mg. per sec. and 3.02 sec., respectively. What is the concentration of zinc in the solution?

8. The diffusion coefficient of oxygen in dilute aqueous solutions is 2.6×10^{-5} cm.^{2} per sec. An 0.25 mM solution of oxygen in 0.01 F potassium nitrate gives a diffusion current of 5.8 microamp. at $E_{d.e.} = -1.50$ v. *vs.* S.C.E., where m and t are 1.85 mg. per sec. and 4.09 sec., respectively. To what is the oxygen reduced under these conditions?

***9.** A solution gave a lead wave whose diffusion current was 6.70 microamp. when m was 2.50 mg. per sec. and t was 3.40 sec. The height of the column of mercury above the capillary was then changed so that t was 4.00 sec. What was the diffusion current of the lead wave under these new conditions?

10. The diffusion current constant of ferricyanide ion in 0.1 F potassium chloride is 1.789 at 25°C. The diffusion current of ferricyanide in an unknown solution in 0.1 F potassium chloride, measured with a capillary for which $m^{2/3}t^{1/6}$ under the conditions of the measurement was 2.141, was 11.10 microamp. when the temperature of the solution was 28.5°C. What was the concentration of ferricyanide in the solution?

11. A solution containing lead and zinc in 0.1 F potassium chloride was analyzed with the capillary used to secure the data of Table 6-3. The total current was 4.30 microamp. at $E_{d.e.} = -0.80$ v. *vs.* S.C.E., which is on the plateau of the lead wave but before the beginning of the zinc wave, and 4.50 microamp. at $E_{d.e.} = -1.70$ v. *vs.* S.C.E., which is on the plateau of the zinc wave. The residual currents at these two potentials were measured separately and found to be 0.20 and 0.50 microamp., respectively. The diffusion current constant of lead in this medium is 3.85, and that of zinc is 3.42. Calculate the concentrations of lead and zinc present.

12. The diffusion current constant of lead ion in an aqueous 0.1 F potassium chloride solution, whose viscosity coefficient is 0.90 centipoise, is 3.85. The viscosity coefficient of 0.1 F potassium chloride containing 40 per cent by weight of sucrose is 5.19 centipoises, and the diffusion current constant of lead ion in this medium is 1.62. Interpret these observations.

***13.** The diffusion current constant of osmium tetroxide, OsO_4, in 0.1 F hydrochloric acid is 8.5 ± 0.2. To what oxidation state does this suggest that osmium is reduced?

14. The diffusion current constant of tellurate ion, $TeO_4^{=}$, in 1 F ammonia–1 F ammonium chloride is 17.5. What is the most probable reduction product?

15. Arsenic(III) gives a double wave in 1 F hydrochloric acid; the diffusion current constants are 6.04 for the first wave and 12.00 on the plateau of the second. What are the electrode reactions responsible for the two waves, assuming that arsenic(III) exists as $HAsO_2$ in 1 F hydrochloric acid solutions?

16. Derive an equation for the difference between $E_{3/4}$, which is the potential where (after correction for the residual current) the current on the rising part of a wave is three-fourths of the diffusion current of the wave, and $E_{1/4}$, where $i = i_d/4$, assuming that the wave is reversible.

***17.** Chromate ion gives a single wave in 1 F sodium hydroxide. When the polarogram of a 2.00 mM solution of chromate ion in 1 F sodium hydroxide was obtained with a capillary for which $m^{2/3}t^{1/6}$ at $E_{d.e.} = -1.10$ v. *vs.* S.C.E. was 2.00, the diffusion current at that potential was found to be 23.2 microamp. The current due to the reduction of chromate was 4.45 microamp. at -0.80 v. *vs.* S.C.E., and was 22.0 microamp. at -0.95 v. *vs.* S.C.E.

a. What electrode reaction is responsible for the wave?

b. What is the half-wave potential of the wave?

c. Is the electrode reaction polarographically reversible?

18. The vanadic (V^{+++})–vanadous (V^{++}) couple is found to be reversible in 1 F potassium oxalate at pH 4.5. The diffusion current constants of the vanadic and vanadous complex ions formed in this medium are 1.95 and -1.43, respectively; the latter value is negative because the wave is anodic and i_d is therefore negative. Calculate the formal potential of the couple in this medium if $E_{1/2}$ is -1.136 v. *vs.* S.C.E.

19. The half-wave potentials of lead in 0.1 and 1 F sodium hydroxide are -0.681 and -0.764 v. *vs.* S.C.E., respectively. The waves in these media are reversible. What is the formula of the lead–hydroxide complex in these solutions?

20. The half-wave potential of zinc in 0.1 *F* sodium perchlorate is -0.998 v. *vs.* S.C.E.; after adding 0.04 *M* ethylenediamine ($H_2NCH_2CH_2NH_2$), $E_{1/2}$ is found to be -1.319 v.; and in 0.1 *F* sodium perchlorate containing 1.96 *M* ethylenediamine, $E_{1/2}$ for zinc is -1.450 v. Find the formula and the dissociation constant of the zinc–ethylenediamine complex.

***21.** An unknown lead solution is made 1 m*M* in cadmium ion and 0.1 *F* in potassium chloride. From a polarogram of the resulting solution the diffusion current of lead is found to be 0.79 microamp., while that of cadmium is 8.61 microamp. If the diffusion current constants of lead and cadmium in 0.1 *F* potassium chloride are 3.85 and 3.53, respectively, what was the concentration of lead in the unknown?

22. The diffusion current of nickel in an unknown solution is 3.75 microamp. Exactly 1.00 ml. of a standard 10.0 m*M* nickel solution is added to 20.0 ml. of the unknown, and the height of the nickel wave in the mixture is found to be 6.80 microamp. Find the concentration of nickel in the unknown solution.

23. In an amperometric titration of lead with chromate in an acetate buffer at $E_{d.e.} = -1.0$ v. *vs.* S.C.E., the current (after correction for the residual current) was found to be 8.5 microamp. at the end point. The diffusion current constants of lead and chromate ions under these conditions are 2.9 and 6.0, respectively, and $m^{2/3}t^{1/6}$ was 2.40. Calculate the molar solubility of lead chromate in the acetate buffer.

24. In 0.1 *F* potassium thiocyanate cupric copper gives two reversible waves of equal heights whose half-wave potentials are -0.14 and -0.42 v. *vs.* S.C.E. Chromous ion gives a single anodic wave for which $E_{1/2}$ is -0.73 v. *vs.* S.C.E., and chromic ion gives a single cathodic wave at $E_{1/2} = -0.91$ v. *vs.* S.C.E. Sketch the titration curves which would be obtained at -0.28, -0.6, and -1.2 v. *vs.* S.C.E. in the amperometric titration of cupric copper in 0.1 *F* thiocyanate with chromous ion.

CHAPTER 7

ELECTROLYTIC METHODS

7-1. Introduction. In the preceding chapters we have discussed three electroanalytical techniques—potentiometry, conductometry, and polarography—which involve various aspects of the relationships among the potential of an electrode, the composition of a solution, and the electric current which flows through a cell. In none of these techniques does any appreciable change in the composition of the solution result from carrying out the experimental measurements. The apparatus used for potentiometry is so designed that virtually no current at all flows between the electrodes when the point of balance is reached. In conductometry, the reactions that occur at each electrode during one half of the a-c cycle are reversed during the next half. Of these three techniques, only polarography involves any permanent change in the composition of the solution. But the currents which flow during polarographic measurements are so small that the rate of change of concentration throughout the whole of the solution is ordinarily almost too small to measure. The composition of a layer of solution right around the electrode surface is indeed changed, but the electrode is so small and the layer so thin that this affects only a very tiny fraction of the entire solution.

We shall now turn to a consideration of the theory, practice, and application of several electrolytic techniques. Unlike potentiometry, conductometry, and polarography, these techniques involve the passage of an electric current through a solution in order to carry some electrochemical reaction virtually to completion.

The first such technique was devised by Wolcott Gibbs and by M. C. Luckow in 1864. It consists of electrolyzing a solution under such conditions that the substance being determined (or, less frequently, some compound of that substance) is quantitatively deposited on one of the electrodes, then weighing the material deposited. For example, an acidic solution of a cupric salt may be electrolyzed between two platinum electrodes; copper is quantitatively deposited on the cathode, and the weight of copper in the original solution can be determined by subtracting the initial weight of the platinum cathode from the total weight of the cathode at the end of the experiment. This is a popular experiment in beginning courses in quantitative analysis. Halides may be determined by electrolyzing their solutions between a platinum cathode and a silver anode; with chloride ion the reaction occurring at the anode is

$$Ag + Cl^- = AgCl + e$$

and the gain in weight of the anode is equal to the weight of chloride in the original solution.

By such electrogravimetric methods it is a rather simple matter to determine the amount of almost any metal, or of any of a considerable number of nonmetals, present in a practically pure solution, and to do so with an accuracy and precision comparable to those of the best volumetric and gravimetric procedures. This is by no means valueless, as anyone who has had to standardize a solution of a cobalt salt will attest. In such cases the electrogravimetric method has two very important advantages over the classical procedures: the deposition can be made to proceed without supervision from the chemist (which greatly decreases the amount of time he must spend on the analysis), and the much greater simplicity of the manipulations greatly decreases the danger of mechanical loss.

However, no analytical method would be of much use if it could be applied only to pure solutions. If we had to determine copper in a copper-zinc alloy, we would shun an electrogravimetric method—no matter how accurate the results it gave when applied to a pure copper sulfate solution—if there were no way to keep zinc from depositing along with the copper on the cathode.

There have now been developed a number of kinds of electrolytic methods, of which we shall discuss only a few. Much of the research which has dealt with the theoretical foundations and practical applications of these methods has been done in this country during the last dozen or so years, and the story of this work would constitute one of the brightest chapters in the history of American analytical chemistry.

These methods are classified according to the electrical quantity to which primary attention is paid during the course of the electrolysis. This may be the voltage applied across the two electrodes of the electrolysis cell, the current which flows through the cell, the potential of the electrode at which the electrochemical reaction of interest is taking place, or the total quantity of electricity which is consumed during the course of that reaction. These methods are very closely interrelated, but the variety of the separations possible by the use of each of them differs greatly from one to another. We shall discuss them in turn in the following sections, beginning with the "constant current" method. Although the theory of this method is actually somewhat more complex than that of some of the others, it is the oldest and, because it requires the least intricate and expensive equipment, still the most widely used of the methods mentioned.

7-2. Electrolysis at Constant Current. Let us suppose that a solution containing about 0.1 N (0.05 F) cupric sulfate and about 1 N (0.5 F) sulfuric acid is electrolyzed between two large platinum gauze electrodes, and that the solution is stirred so efficiently that no concentration polarization (Sec. 6-3) exists at either electrode.

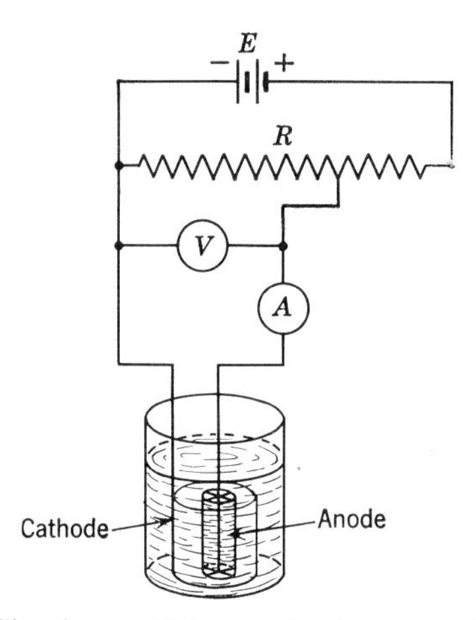

FIG. 7-1. Simple circuit for constant-current electrolysis.

If these electrodes are connected to an external circuit such as is shown in Fig. 7-1, and if the voltage applied across the electrodes is varied by means of the rheostat R, a plot of the current (indicated by the ammeter A) against the applied voltage (indicated by the voltmeter V) would resemble Fig. 7-2*a*.

When the applied voltage is zero (*i.e.*, when the two electrodes are shorted together), the current will also be zero. The two identical electrodes are, of course, at the same potential, and any flow of electricity between them under these conditions is clearly impossible. However, when energy is supplied to the cell by applying a finite voltage across it from the external circuit, the flow of current through the cell becomes possible. At low values of the applied voltage, the current will be very small. It will be due to the presence of small concentrations of impurities, such as ferric iron, which can be reduced at the cathode ($Fe^{+++} + e = Fe^{++}$) to give ferrous iron which would then be reoxidized at the anode ($Fe^{++} = Fe^{+++} + e$), or oxygen, which could be reduced at the cathode ($O_2 + 2H^+ + 2e = H_2O_2$) to give hydrogen peroxide which would be reoxidized at the anode. The current which flows through the cell in this range of applied voltages is more or less analogous to the residual current on polarograms secured with the dropping mercury electrode (Sec. 6-7), but of course the periodic change in area of the latter,

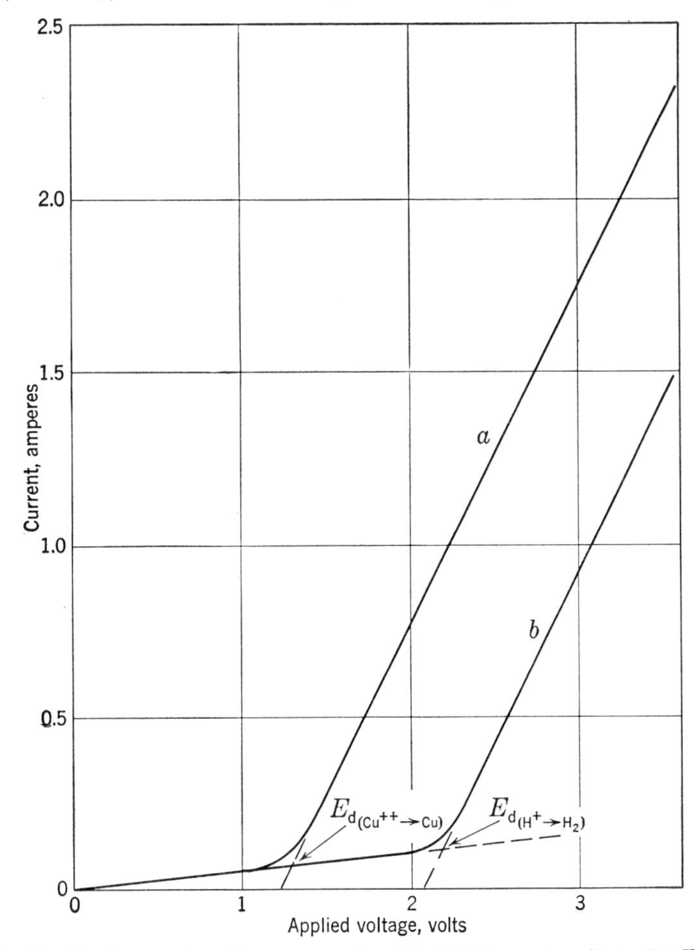

FIG. 7-2. Idealized current–voltage curves for (*a*) 0.05 *F* cupric sulfate–0.5 *F* sulfuric acid, and (*b*) 0.5 *F* sulfuric acid alone, electrolyzed between two large platinum electrodes in the absence of any concentration polarization. The cell resistance was assumed to be 1 ohm.

which is responsible for the condenser-current component of the residual current, has no parallel in current–voltage curves secured with electrodes of constant area.

As the applied voltage is increased, a point is eventually reached at which copper begins to deposit on the cathode.[1] The applied voltage at

[1] This, when closely examined, turns out to be considerably less susceptible to exact definition and location than one might perhaps think. Instead of being constant and

which this occurs is usually called the "decomposition potential," E_d, of the solution.[1] Strictly speaking, it ought to be called the "decomposition potential with respect to copper deposition," $E_{d_{(Cu^{++} \rightarrow Cu)}}$, because this solution has another decomposition potential as well: the potential required to initiate the cathodic evolution of hydrogen.

The decomposition potential is best defined in the following way. Suppose that the dependence of the residual current (the current which would be observed in the absence of the copper sulfate but under otherwise identical conditions) on the applied voltage is known, either from direct measurement or by extrapolation as indicated by the dashed line in Fig. 7-2. If the residual current i_r at any applied voltage E in the range in which copper is deposited on the cathode is subtracted from the total current i at that voltage, the current due to the reduction of copper will be given by

$$i - i_r = \frac{E_{d_{(Cu^{++} \rightarrow Cu)}} - E}{R} \tag{7-1}$$

where R is the resistance of the solution. This resistance naturally depends on the areas of the electrodes, the distance between them, and the composition of the solution, in accordance with Eq. (5-7).

In order to calculate the decomposition potential with respect to copper deposition, assume that a layer of copper several atoms thick has already been deposited onto the cathode. (This corresponds to so small an amount of copper that the concentration of cupric ion in the solution will not have been appreciably altered from its original value, 0.05 M.)

---

equal to 1, the activity of the solid copper deposited on the cathode varies with the amount of copper deposited. This activity is very small for the first few atoms of metallic copper produced, and it increases steadily as a larger and larger fraction of the electrode surface becomes covered with copper. Finally, when the entire surface of the electrode is covered with copper to a depth of several atoms, the activity of the deposited copper does become equal to 1. The effect of this variation in activity is to "smear out" what might be expected to be a sharp discontinuity into a much more gradual change, and so a close examination of an actual current–voltage curve reveals a distinct curvature, rather than a sudden break, around the decomposition potential. An entirely similar phenomenon, observed in the potentiometric titration of mixtures of halides with silver nitrate, and likewise due to the variable activity of a solid present in very small amounts, was encountered in Sec. 3-13.

[1] Equating the decomposition potential to an applied voltage in this fashion is tantamount to assuming that the iR drop corresponding to the residual current at this point is negligibly small. The error thus introduced is rarely troublesome, but it would clearly be more accurate to define the decomposition potential in terms of the potential across the cell. The definition given in the text is, however, the one universally encountered; this is partly because of tradition and partly because the applied voltage is much easier and cheaper to measure than the cell potential.

Meanwhile oxygen will have been evolved at the anode by the reaction $2H_2O = 4H^+ + O_2 + 4e$. If the solution is exposed to the air, the partial pressure of oxygen will be very nearly 0.20 atmosphere. In this way we shall have set up the cell

$$Cu|Cu^{++}(0.05\ M),\ H^+(0.5\ M)|O_2(0.20\ \text{atm.}),\ Pt$$

The reaction which will occur in this cell is

$$2Cu + O_2 + 4H^+ = 2Cu^{++} + 2H_2O$$

which is just the reverse of the reaction we are trying to bring about by electrolysis. With the following values of the standard potentials involved

$$Cu^{++} + 2e = Cu \qquad E^0 = +0.345\ \text{v.}$$
$$O_2 + 4H^+ + 4e = 2H_2O \qquad E^0 = +1.229\ \text{v.}$$

it is a simple matter to calculate the potentials of the two electrodes of this cell. These are

$$E_{Cu} = 0.345 - 0.02957 \log \frac{1}{0.05} = +0.307\ \text{v.}$$

$$E_{O_2} = 1.229 - 0.01478 \log \frac{1}{(1)^4(0.2)} = +1.219\ \text{v.}$$

so that the potential of the cell (often referred to as the "back e.m.f.") is $E_{O_2} - E_{Cu} = 0.912$ v. Since this is positive, the cell reaction proceeds spontaneously as written. This, of course, means that energy must be supplied from the external circuit to reverse the direction of this reaction and cause the deposition of copper and the evolution of oxygen.

It is immediately obvious that the voltage which must be supplied to the cell must be at least equal to the opposing cell potential. Otherwise we should merely decrease the rate at which the copper was redissolved. If we apply exactly 0.912 v. across the electrodes (connecting the copper-plated electrode to the negative terminal of the external circuit), we shall succeed in decreasing this rate to zero, and then no current will flow through the cell in either direction. (The student will recognize this as the principle of potentiometric measurements.) However, an applied voltage greater than this must be used to cause the reaction to proceed in the opposite, nonspontaneous direction. These considerations would lead one to expect that a voltage even very slightly in excess of 0.912 v. would suffice to initiate the desired reaction, but experimentally it is found that this is not the case. Actually a considerably larger applied voltage will be required.

This difference between the back e.m.f. and the measured decomposition potential is known as the "overpotential."[1] Overpotential is a phenomenon associated with a slow step in one or both of the electrode reactions,[2] and it is affected by many factors. Among these are the nature of the electrode (platinum, gold, copper, mercury, etc.) and the condition of the electrode surface (smooth, rough, or platinized[3]), the current density (which is the ratio of electrolysis current to electrode area[3]), the temperature, and numerous others. Unfortunately a complete discussion of these factors cannot be undertaken here, and the student must be referred to textbooks of electrochemistry for further information.[4]

However, a word of caution must be added with reference to the frequently repeated statement that overpotential is a phenomenon primarily associated with the evolution of a gas (hydrogen and oxygen are the two of greatest importance), and that the overpotential involved in the deposition of a solid metal is negligibly small. This is true in the case we are considering, which is merely another way of saying that the cupric ion–copper electrode behaves reversibly. In general, it is also true for the deposition of any other metal which forms a reversible electrode of the first order with its ions (or which gives a polarographically reversible wave); these include such metals as silver, lead, cadmium, zinc, and many others. But it is very far from true of such hard metals as cobalt, nickel, iron, chromium, molybdenum, and tungsten, which do not form reversible electrodes of the first order and do not take part in reversible reactions at a dropping mercury electrode. Indeed, the overpotentials associated with the deposition of molybdenum and tungsten are so great that these metals have never been deposited in pure form from aqueous solutions in

[1] The terms "overpotential" and "overvoltage" have often been used more or less interchangeably. However, "overpotential" is the more precise term, for the quantity referred to is really a difference between two potentials: the actual potential of the electrode in an electrolytic cell, and the potential calculated from the Nernst equation on the assumption that the solution is homogeneous throughout. *Cf.* the footnote on page 144.

[2] We are assuming that the solution is so efficiently stirred that there is no concentration overpotential at either electrode. In this case the cathode overpotential will be very small, and the overpotential of the cell will be almost entirely the activation overpotential associated with the evolution of oxygen at the anode.

[3] Roughening or platinizing the surface of an electrode increases its effective area, as distinct from its geometrical or projected area, and hence affects the current density which corresponds to any given value of the electrolysis current.

[4] An especially valuable discussion of overpotential in electrolytic cells is given in J. J. Lingane, "Electroanalytical Chemistry," Interscience Publishers, Inc., New York, 1953, pp. 174–178.

spite of the relatively low values of the back e.m.f. which would have to be overcome.[1]

For present purposes, we shall take the anodic overpotential of oxygen on smooth platinum in acid solutions as being approximately 0.40 v. This value must be added to the back e.m.f. of the cell, 0.91 v., to secure the actual decomposition potential of the solution with respect to copper deposition. Consequently copper will not begin to deposit until the applied voltage becomes at least 1.31 v.

7-3. Completeness of Constant-current Depositions. The equation for the potential of the copper electrode in the example just discussed indicates that the back e.m.f. of the copper-oxygen cell will increase 0.02957 v. for each tenfold decrease in the concentration of cupric ion in the solution. As the electrolysis proceeds, oxygen will be produced at the anode, but the concentration of dissolved oxygen will remain constant by virtue of the equilibrium with the atmosphere. At the same time hydrogen ions will be produced. However, the equation

$$2Cu^{++} + 2H_2O = 2Cu + 4H^+ + O_2$$

shows that the hydrogen ion concentration, initially 1 M, will increase only to 1.1 M as the concentration of cupric ion is decreased from 0.05 M to zero. The effect of this increase in the concentration of hydrogen ion on the back e.m.f. of the cell would be only 2.4 mv. (= 59.15 log 1.1 − 59.15 log 1.0), which is small enough to ignore. Of course it would have to be taken into account if we had started with a cupric sulfate solution containing little or no sulfuric acid, because then the production of 0.1 mole of hydrogen ion per liter would result in a relatively large change in its concentration.

We may conclude from this that the decomposition potential of the solution with respect to copper deposition will increase very nearly 30 mv. for each tenfold decrease in the concentration of the cupric ion remaining undeposited. When 90 per cent of the copper has been deposited, $E_{d_{(Cu^{++} \to Cu)}}$ will be $1.31 + 0.03 = 1.34$ v.; when 99 per cent has been deposited, $E_{d_{(Cu^{++} \to Cu)}}$ will be $1.31 + 2\ (0.03) = 1.37$ v., and so on. In order to secure quantitative deposition of the copper, we must know the relationship between the applied voltage and the concentration of cupric ion which will remain undeposited when the flow of current ceases.

A little thought will reveal that equilibrium will be reached, so that the current due to copper deposition will fall to zero and the deposition of copper will stop, when the decomposition potential of the solution with

[1] An interesting discussion of the problems involved in the electrodeposition of these so-called "reluctant" metals is given by A. Brenner, *Record Chem. Progr.*, **16,** 241–269 (1955).

respect to copper deposition becomes just equal to the voltage actually being applied across the electrodes. For example, if the applied voltage is 1.34 v., copper will stop being deposited when only 90 per cent of the cupric ions initially present have been reduced. Or, more generally,

$$E_a - E_d^0 = \frac{0.05915}{n} \log \frac{[Cu^{++}]^0}{[Cu^{++}]} \tag{7-2}$$

where E_a is the applied voltage, E_d^0 the decomposition potential of the original solution with respect to copper deposition, and $[Cu^{++}]^0$ and $[Cu^{++}]$ the initial and final concentrations of cupric ion in the solution.

From Eq. 7-2, we can calculate that the concentration of cupric ion in our solution would be reduced to 0.01 per cent of its initial value ($[Cu^{++}]^0/[Cu^{++}] = 10^4$), which is a reasonable definition of quantitative deposition, by an electrolysis at an applied voltage of 1.43 v. This, it will be noted, is only 0.12 v. greater than the decomposition potential of the original solution. On the other hand, the error due to undeposited copper will increase tenfold if the actual applied voltage differs from this theoretical figure by even as little as 0.03 v. So small an error might result from any of a number of causes: an error in the measuring voltmeter, an error in our rather arbitrary value for the oxygen overpotential, a variation of either of the standard potentials with temperature, a difference between activities and concentrations (which alone would cause a substantial error in 0.5 F sulfuric acid with an ionic strength of 1.5), and many others. In view of these many uncertainties, the analyst would probably be well advised to allow about half a volt as a margin of error. In fact, few if any texts on quantitative analysis recommend an applied voltage less than 2 v. for the electrogravimetric determination of copper.

7-4. Separations in Constant-current Electrolyses. I. If electrolyses of this kind are to be carried out at applied voltages so much larger than those which should be theoretically satisfactory, careful attention must be paid to the possibility that the applied voltage will exceed the decomposition potential of the solution with respect to some other substance.

The way in which one can predict whether another substance known or thought to be present in the solution will deposit together with the desired one may be illustrated by considering the conditions under which hydrogen will be deposited along with copper. If hydrogen were evolved at the cathode (and oxygen at the anode), the back e.m.f. of our cell

$$Cu, H_2(g)|H^+(1\ M)|O_2(0.2\ \text{atm.}), Pt$$

would be approximately 1.219 v. (and, be it noted, independent of the hydrogen ion concentration). This figure is only approximate because the partial pressure of hydrogen gas is not 1 atm., as was assumed in calculating it, but is actually much lower. However, its exact value is unknown, and this assumption furnishes at least a rough guide to what may be expected.

Now, in order to deposit hydrogen on the cathode while evolving oxygen on the anode, we would have to apply to the cell a voltage at least equal to the sum of this back e.m.f. (1.22 v.), the overpotential associated with the evolution of oxygen on our smooth platinum anode (roughly 0.40 v.), and a second overpotential associated with the evolution of hydrogen on a *copper* cathode (roughly 0.60 v.). So the decomposition potential of this solution with respect to hydrogen evolution will be approximately 2.2 v. The course of the current–voltage curve at applied voltages higher than this is indicated in Fig. 7-2.

If, therefore, our cupric sulfate–sulfuric acid solution is electrolyzed at an applied voltage of 2.2 v. or greater, hydrogen and copper will deposit together on the cathode. Although the simultaneous evolution of hydrogen gas will certainly not affect the weight of copper secured—as attested by the countless successful electrogravimetric determinations of copper which have been carried out under just such conditions—it does have an adverse effect on the character of the deposited copper. The continual evolution of bubbles of hydrogen on the cathode disturbs the orderly growth of the crystal structure of the metallic copper, and so one gets a deposit of copper which is much more spongy and porous, and which adheres to the platinum much less tightly, than would have been obtained if the deposition of hydrogen had been prevented. This increase of the surface area of the deposited copper greatly increases its susceptibility to air-oxidation while it is being washed and dried, and of course this would lead to results which are somewhat high. On the other hand, the deposit may in an extreme case be so poorly adherent that particles of copper fall off while the electrode is being carried from the laboratory bench to the balance room.

It is clear that errors very much larger than these will result if the co-deposited substance is another metal. Suppose that one wishes to deposit copper from a sulfuric acid solution containing equal concentrations of copper and bismuth. The standard potential of the bismuthyl ion–bismuth couple

$$BiO^+ + 2H^+ + 3e = Bi + H_2O \qquad E^0 = +0.32 \text{ v.}$$

is only 0.025 v. less positive than that of the cupric ion–copper couple. Consequently one would predict that only about two-thirds of the copper could be deposited from this solution without initiating the deposition of bismuth. Any attempt to deposit the copper quantitatively would of course result in the quantitative deposition of the bismuth as well.

7-5. Current–Potential–Time Curves in Constant-current Electrolyses. A constant-current electrolysis may now be defined as an electrolysis in which the applied voltage exceeds the decomposition potential of the solution with respect to the deposition of some ion whose concentration is much larger than that of the ion in which we are interested.[1] For example, the electrolysis of the 0.05 *F* cupric sulfate–0.5 *F* sulfuric acid solution discussed above, if it were carried out at an applied voltage of 2.5 v., would be a constant-current electrolysis because this voltage is larger than the decomposition potential of the solution with respect to hydrogen evolution. (The student who wonders what name would be applied to an electrolysis of this solution at an applied voltage of, say, 2.0 v. will find out in Sec. 7-8.)

Under these conditions the total current flowing through the cell will be the sum of a current due to the reduction of cupric ions and of a current due to the reduction of hydrogen ions. Because the concentration of hydrogen ion in the solution is very much larger than that of cupric ion—it is ten times as large at the beginning of the electrolysis, and becomes still larger as the electrolysis proceeds and more and more of the copper is deposited—many more hydrogen ions than cupric ions will reach the surface of the cathode in any given small interval of time. Consequently the current due to the reduction of hydrogen ion even at the beginning of the electrolysis is much larger than that due to the reduction of cupric ion. Even though the latter drops continuously as the electrolysis proceeds, and finally approaches zero when the copper is nearly all deposited, this could result in only a very small change in the total current flowing through the cell. As a matter of fact, the decrease of the current due to the decreasing concentration of cupric ion is just compensated by an increase in the current due to the reduction of hydrogen ion as the hydrogen ion concentration increases from 1.0 to 1.1 *M* during the electrolysis.

Although it may seem at first glance that the reduction of hydrogen ions at the cathode would decrease the concentration of hydrogen ion in the solution, this is not true. Each hydrogen ion removed from the solution at the cathode by the reaction $H^+ + e = \frac{1}{2}H_2$ is simultaneously replaced at the anode by the reaction $\frac{1}{2}H_2O = H^+ + \frac{1}{4}O_2 + e$. It may help the student to understand this point if he writes the electroneutrality rule expressions for the composition of the solution at the beginning and at the end of the electrolysis, remembering as he does so that sulfate ion does not take part in any reaction at either electrode.

[1] In actual fact, the current may be far from constant during the practical execution of a constant-current electrolysis, due to the variations in back e.m.f., overpotentials, and cell resistance which result from the changing composition and temperature of the solution.

Let us now see whether we can make use of this information to calculate the way in which the concentration of copper remaining in the solution will change as the electrolysis proceeds. To do this, we shall have to employ Faraday's law, which states that one faraday of electricity (96493 coulombs, a coulomb being the quantity of electricity which corresponds to a current of 1 ampere flowing for 1 second) is required to reduce or oxidize one equivalent of any substance. From the equation for the half-reaction representing the reduction of copper, $Cu^{++} + 2e = Cu$, we see that two faradays of electricity will be needed to deposit one mole (63.54 g.) of copper.

Suppose that the volume of solution being electrolyzed is 100 ml., and that the current which flows through the cell is 1 ampere. One hundred milliliters of 0.05 F cupric sulfate contains 0.005 mole or 5 millimoles of cupric ion, which will consume 10 millifaradays of electricity or 965 coulombs. If the entire current were consumed in reducing cupric ion at the cathode, the complete deposition of copper would require only 965 sec. or about 16 min. Of course this is unrealistic, because we have neglected the fact that most of the current is actually consumed by the reduction of hydrogen ion.

We can obtain a better estimate of the situation by making the reasonable assumptions that the number of cupric ions reaching the surface of the cathode (and being reduced) during any brief interval of time is proportional to the concentration of cupric ion remaining undeposited, and that the current due to the reduction of hydrogen ion at any instant is likewise proportional to the concentration of hydrogen ion at that instant.[1] Together with the fact that the total current remains constant at 1 ampere, these assumptions give

$$k_{H^+}[H^+] + k_{Cu^{++}}[Cu^{++}] = 1$$

where the concentrations are expressed in equivalents per liter for convenience in applying Faraday's law. At the end of the electrolysis, when $[Cu^{++}]$ is virtually zero and $[H^+] = 1.1\ N$, this gives $k_{H^+} = 0.909$. At the start of the electrolysis $[Cu^{++}] = 0.1\ N$ and $[H^+] = 1.0\ N$; from this we can calculate that $k_{Cu^{++}}$ is also equal to 0.909. We therefore have

$$i_{Cu^{++}} = 0.909[Cu^{++}] \qquad (7\text{-}3)$$

where $i_{Cu^{++}}$ is the current due to the reduction of cupric ion.

[1] The sophisticated electrochemist, or the student after he has read the portion of this section which deals with the variation of the cathode potential with time, may object to these assumptions on the entirely valid ground that the current efficiency for the reduction of cupric ion is always somewhat greater than these assumptions predict. Although this is true, the equation for the current–time relationship which results from these assumptions is of the correct form, and any more rigorous derivation would be much too complicated to present here.

Now, if $dQ_{Cu^{++}}$ faradays of electricity is used to reduce cupric ion during any given interval of time, the number of equivalents of cupric ion "added" to the solution, $-dN_{Cu^{++}}$, will be equal to $dQ_{Cu^{++}}$. This in turn is given by the equation

$$dQ_{Cu^{++}} = i_{Cu^{++}} \frac{dt}{F_y} \tag{7-4}$$

where dt is the length of the interval of time (in seconds) and $i_{Cu^{++}}$ is the (average) current due to the reduction of cupric ion during that interval. Meanwhile $dN_{Cu^{++}}$ is given by

$$dN_{Cu^{++}} = V(d[Cu^{++}]) \tag{7-5}$$

where V is the volume of the solution in liters. Combining Eqs. (7-3), (7-4), and (7-5) gives

$$\frac{d[Cu^{++}]}{[Cu^{++}]} = -\frac{0.909dt}{VF_y} \tag{7-6}$$

which on integration becomes

$$\ln [Cu^{++}] = -\frac{0.909t}{VF_y} + C$$

where C is a constant of integration. In view of Eq. (7-3), we may write

$$\ln \frac{i_{Cu^{++}}}{0.909} = -\frac{0.909t}{VF_y} + C$$

where C is now obviously equal to $\ln (i^0_{Cu^{++}}/0.909)$, $i^0_{Cu^{++}}$ being the current which is due to the reduction of copper at $t = 0$. Making this substitution and replacing the natural logarithms by common logarithms gives

$$\log \frac{i_{Cu^{++}}}{i^0_{Cu^{++}}} = -\frac{0.395t}{VF_y} \tag{7-7}$$

From Eq. (7-7) we can calculate the time required to decrease the current due to the reduction of copper, and hence the concentration of copper remaining undeposited, to any desired fraction of its initial value. To deposit 99.9 per cent of the copper, if our initial assumptions were correct, should require a little over 20 hours. Approximately 6¾ hours will be required to deposit the first 90 per cent of the copper, another 6¾ hours for the deposition of an additional 9 per cent, a third 6¾ hours for the deposition of an additional 0.9 per cent, and so on. In practice, of course, this time may be greatly decreased by such expedients as increasing the areas of the electrodes and placing them closer together, warming the solution, etc.; all of these serve to decrease the resistance of the cell and thus to increase the values of $i_{Cu^{++}}$ in accordance with Eq. (7-1). (Would the deposition of copper be speeded up if the resistance of the cell were decreased by adding more sulfuric acid?)

The change of the reduction current of cupric ion with time is shown graphically in Fig. 7-3*a*. The other curves in the figure show the effects of time on the current due to the reduction of hydrogen ion and on the total electrolysis current.

At the beginning of the electrolysis, or as soon thereafter as the layer of deposited copper has become several atoms thick, the cathode may be regarded simply as a copper electrode. Although the current flowing

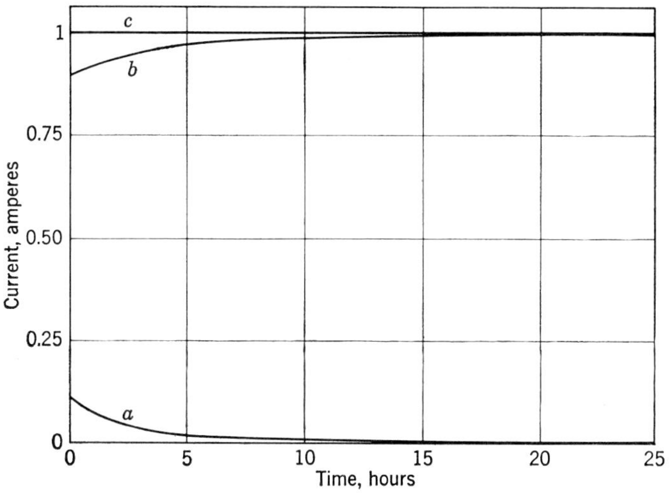

FIG. 7-3. Variation with time of (*a*) the current due to the reduction of cupric ion, (*b*) the current due to the reduction of hydrogen ion, and (*c*) the total current during a constant-current electrolysis of a cupric sulfate–sulfuric acid solution under the conditions described in the text.

through the cell, as well as the deposition of hydrogen occurring at the same time, will cause the potential of the cathode to differ considerably from the expected zero-current potential of a reversible copper electrode, at least a qualitative picture of events may be easily deduced. Early in the electrolysis the cathode will behave, to some extent at least, as a copper electrode. As the concentration of cupric ion decreases, the electrode potential will slowly become more negative, until eventually the cupric ion concentration becomes so small that the electrode functions simply as a hydrogen electrode (irreversible because of the overpotential involved).

In the foregoing we have assumed infinitely efficient stirring, which is to say the total absence of any concentration gradients around the electrode. This is not particularly easy to secure, and in most electrolyses there is a certain amount of concentration polarization. This arises in a very thin but more or less quiet layer of solution surrounding the electrode, and it leads to many situations in which a large electrode in a stirred solution and a microelectrode in a quiet solution behave in exactly similar fashions. For example, consider the curves shown in Fig. 7-4. They were obtained with an electrode which consisted of a brass cylinder about 20 cm. long and 2.5 cm. in diameter. The area of this electrode was 142 cm.^{2}, whereas the maximum electrode area of a typical dropping electrode is only about 0.03 cm.^{2}, and the solution around the brass electrode in these experiments was stirred at the full speed of a heavy-duty

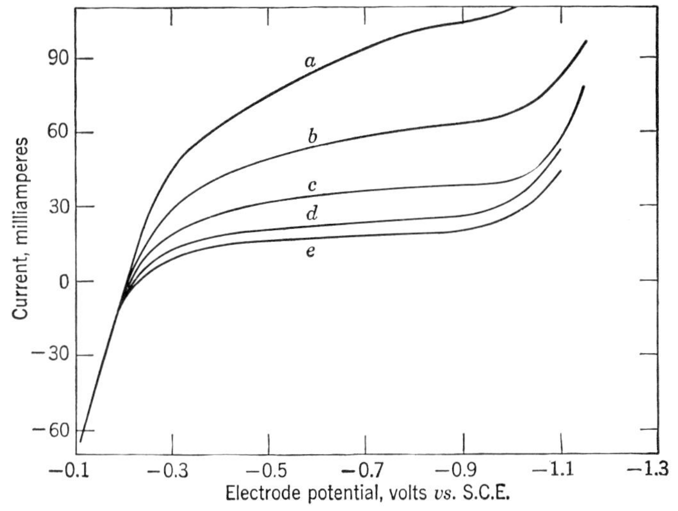

FIG. 7-4. Current–potential curves secured in the electrolysis of a rapidly stirred 0.2 *F* sodium chloride solution containing (*a*) 0.20, (*b*) 0.125, (*c*) 0.072, (*d*) 0.044, and (*e*) 0.03 m*M* dissolved oxygen, using a cylindrical brass indicator electrode having an area of 142 cm.^{2}

stirring motor. In spite of its very much larger area, and in spite of the very rapid stirring used, the large electrode gave a set of current–potential curves having the characteristic features—a limiting current proportional to the concentration of oxygen, a half-wave potential independent of the oxygen concentration, etc.—of polarograms secured with concentration-polarized microelectrodes.

In reality, therefore, the current which is due to the reduction of any one ion at a large electrode during any instant depends, to an extent which in turn depends on the efficiency of the stirring, on the number of those ions which can penetrate the quiet layer of solution around the surface of the electrode. This number is reflected by the magnitude of the limiting current for that ion. If the current actually flowing through the cell at any instant is smaller than the limiting current of the most easily reducible substance in the solution, then the entire current will be consumed in reducing that one substance. On the other hand, if the current exceeds the limiting current of that substance, then a second substance will also have to be reduced at the same time. In discussing the manner in which the changing composition of a solution during electrolysis affects the working electrode potential, and with it the occurrence and extent of any possible competing reaction, it will be helpful to refer to the curves of Fig. 7-4.

Let us first suppose that a constant current of 60 milliamperes (60 ma. = 0.060 amp.) had been forced through the original solution (curve *a*). The potential of the cathode would then have been approximately −0.39 v. *vs.* S.C.E., and, because this is much less negative than the potential (*ca.* −0.95 v. *vs.* S.C.E.) at which hydrogen evolution begins,[1] all of the current would have been consumed in reducing oxygen.

Now, if the solution were isolated from the air, and if the anode (where oxygen is produced by the reaction $2H_2O = 4H^+ + O_2 + 4e$) were separated from the solution by a salt bridge or porous diaphragm, the reduction of oxygen at the cathode would naturally cause the concentration of oxygen in the solution to decrease. When the oxygen concentration has decreased to 0.125 m*M*, curve *b* shows that the potential of the brass cathode will have had to become −0.77 v. to maintain the constant current of 60 ma. This is still less negative than the potential at which hydrogen evolution begins, and so oxygen reduction remains the only possible cathode process. However, the current is now only very slightly smaller than the limiting current of oxygen, and because the current–potential curve in this region is very nearly flat, a slight further decrease in the concentration of oxygen will cause the cathode potential to undergo a rapid shift toward more negative values. From curve *c* we find that the cathode potential will be approximately −1.12 v. *vs.* S.C.E. when the oxygen concentration has been decreased to 0.072 m*M*, about 35 per cent of its initial value. At this point only a little more than half the current is consumed by the reduction of oxygen, because only enough oxygen molecules reach the cathode surface to use up 33 ma. out of the total constant current of 60 ma., and so the remainder of the current will be consumed by the reduction of hydrogen ions. If the electrolysis is continued until the oxygen concentration has dropped to 0.03 m*M* (curve *e*), the cathode potential will change only to −1.15 v., but now only about a quarter of the total current will be consumed by the reduction of oxygen, and three-quarters of it will be consumed by the reduction of hydrogen ion.

[1] This is less negative than the potential at which hydrogen evolution would begin on a mercury cathode in the same solution, because the overpotential of hydrogen on brass is considerably smaller than the overpotential of hydrogen on mercury.

Exactly the same phenomena would have been observed if we had used a constant current of 15 ma., except, of course, that the electrolysis would have lasted four times as long. The initial cathode potential would have been −0.23 v., and this would change only to −0.35 v. as the oxygen concentration decreased from 0.2 m*M* at the start of the electrolysis to 0.044 m*M* (curve *d*). Then it would begin to change more rapidly, and would be −0.48 v. when the concentration of oxygen had decreased to 0.03 m*M*. Since at this concentration the limiting current of oxygen is only about 17 ma., a further decrease of the oxygen

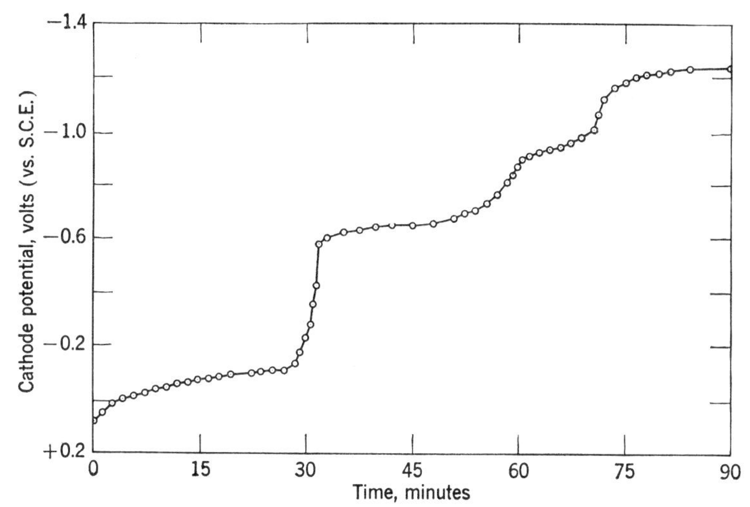

FIG. 7-5. Variation with time of the cathode potential required to maintain a constant current of 0.260 ampere during the electrolysis at a large mercury electrode of 100 ml. of 1 *F* sulfuric acid containing approximately 2.5 m*M* each of ferric, cupric, cadmium, and nickel ions. The cathode potential after 180 min. was only 0.045 volt more negative than after 90 min. (*Reproduced by permission from L. Meites, "Polarographic Techniques," Interscience Publishers, Inc., New York,* 1955.)

concentration to 0.025 m*M* [which would give a limiting current of 17 (0.025/0.03) = 14.2 ma.] would initiate the reduction of hydrogen ion and change the cathode potential to about −1.0 v. Another example of this behavior is shown in Fig. 7-5.

On the other hand, if, in the example of Fig. 7-4, the constant current forced through the cell had been, say, 110 ma., the limiting current of oxygen would have been exceeded right from the start of the electrolysis. The reductions of oxygen and of hydrogen ion would then have proceeded simultaneously throughout the experiment. The cathode potential would have been about −1.02 v. at the beginning of the electrolysis, and would have increased only very slowly and gradually, without any sudden change, until the oxygen had been completely reduced. Exactly the same phenomena would have been observed in the electrolysis of Fig. 7-5 if the constant current used had been larger than the combined limiting currents of ferric, cupric, cadmium, and nickel ions in the original solution.[1] In any case, the drift of the cathode potential to more negative values is eventually virtually halted by the fact that the concentration of hydrogen ion[2] is so high that it remains practically unchanged over long periods of time. Indeed, if the anode is not isolated from the solution, the depletion of hydrogen ion at the cathode is exactly counterbalanced by its production at the anode, and so no further change of the hydrogen ion concentration will result if the electrolysis is prolonged indefinitely[3] after the removal of the easily deposited constituents of the solution.

7-6. Current Efficiency. At this point it is convenient to pause briefly to introduce the concept of current efficiency. The *instantaneous current efficiency* for any reaction taking place in an electrochemical cell is defined as the ratio of the current being consumed by that reaction to the total current flowing through the cell at any given instant during the electrolysis. For example, suppose that the solution of Fig. 7-4*a* is electrolyzed with the brass cathode described above at a constant current of 60 ma. The instantaneous current efficiency for the reduction of oxygen will be 100 per cent until about 40 per cent of the oxygen has been removed. Then hydrogen ion will begin to be reduced along with the oxygen, and when the oxygen concentration has been decreased to 0.072 mM the instantaneous current efficiency for the reduction of oxygen will have decreased to $(100)(33/60) = 55$ per cent. At the same time, the instantaneous current efficiency for the reduction of hydrogen ion, which is the only other process that can occur at the cathode, will be $100 - 55 = 45$ per cent. Note that the sum of the instantaneous current efficiencies for all of the electrochemical reactions taking place at either of the electrodes must always be 100 per cent.

When two or more competing processes are occurring at one electrode, the instantaneous current efficiency is not easy to determine experimentally. What is usually referred to as "the current efficiency" is therefore the *average* efficiency over the whole duration of the electrolysis. Let us suppose that a current of 60 ma. flows through 100 ml. of the

[1] In fact, no break is observed near the end of the ferric ion reduction in Fig. 7-5 simply because the limiting current of ferric ion in the solution was smaller than 0.260 amp., so that the reductions of ferric and cupric ions occurred simultaneously from the very start of the electrolysis.

[2] Or of the acidic component of a buffer system, *e.g.*, $2CO_2 + 2H_2O + 2e = 2HCO_3^- + H_2$.

[3] Within limits, of course, because the net reaction, $2H_2O = 2H_2 + O_2$, while leaving the amount of hydrogen ion in the solution unchanged, will eventually carry the solution to dryness if the electrolysis is sufficiently prolonged.

solution of Fig. 7-4 for 24 hours, and that at the end of this time the concentration of oxygen is found to have decreased from 0.2 mM to virtually zero. The total quantity of electricity consumed is

$$0.06 \times 60 \times 60 \times 24 = 5184 \text{ coulombs}$$

or 0.0537 faraday. The number of moles of oxygen in the original solution was $0.1 \times 2 \times 10^{-4} = 2 \times 10^{-5}$; if the oxygen is reduced to water, four faradays will be required per mole of oxygen, or a total of 8×10^{-5} faraday. Consequently $(8 \times 10^{-5}/5.4 \times 10^{-2})(100) = 0.15$ per cent of the total quantity of electricity will have been consumed in reducing oxygen, and so the average current efficiency for the reduction of oxygen will have been 0.15 per cent. Then the average current efficiency for the reduction of hydrogen ion will have been 99.85 per cent. If 48 hours had elapsed from the beginning of the electrolysis until the time at which it was found that practically all of the oxygen had been reduced, the average current efficiency for the reduction of oxygen would have dropped to 0.075 per cent, because the total quantity of electricity used would have doubled without any measurable change in the amount of oxygen reduced.

7-7. Separations in Constant-current Electrolyses. II. We have seen that the cathode potential in a constant-current electrolysis is eventually limited by the evolution of hydrogen.[1] This, in effect, serves to divide the various metals into two groups: those which are deposited on a cathode at potentials less negative than the potential required for the evolution of hydrogen, and those which could be deposited only at potentials more negative than this.

This division is not a perfectly sharp one in any case, because metals which can just be deposited at potentials sufficiently negative to cause hydrogen evolution obviously fall into neither category, and moreover the classification of any particular metal is greatly dependent on the experimental conditions (including especially the hydrogen overpotential and the composition of the "supporting electrolyte"). For example, manganese is fairly readily deposited from a neutral ammonium sulfate solution, but in alkaline tartrate media a complex ion is formed which cannot be reduced under the conditions of commercial electroplating. However, as a very rough guide it does serve to indicate the possibility of effecting such separations as that of copper from manganese in strongly acidic solutions, of iron from aluminum in nearly neutral or weakly acidic oxalate media, and so on.

The most widespread analytical use of separations by constant-current

[1] In exactly the same way, the anode potential during, say, the electrodeposition of lead dioxide onto a platinum anode is limited by the anodic evolution of oxygen (or of chlorine if the solution contains chloride ion at high concentration).

electrolysis is the employment of the mercury cathode for the separation of such elements as iron, copper, cobalt, nickel, and (under the proper conditions) manganese, which can all be deposited into a mercury cathode from a dilute sulfuric acid solution, from such others as tungsten, titanium, vanadium, aluminum, and molybdenum which are not deposited

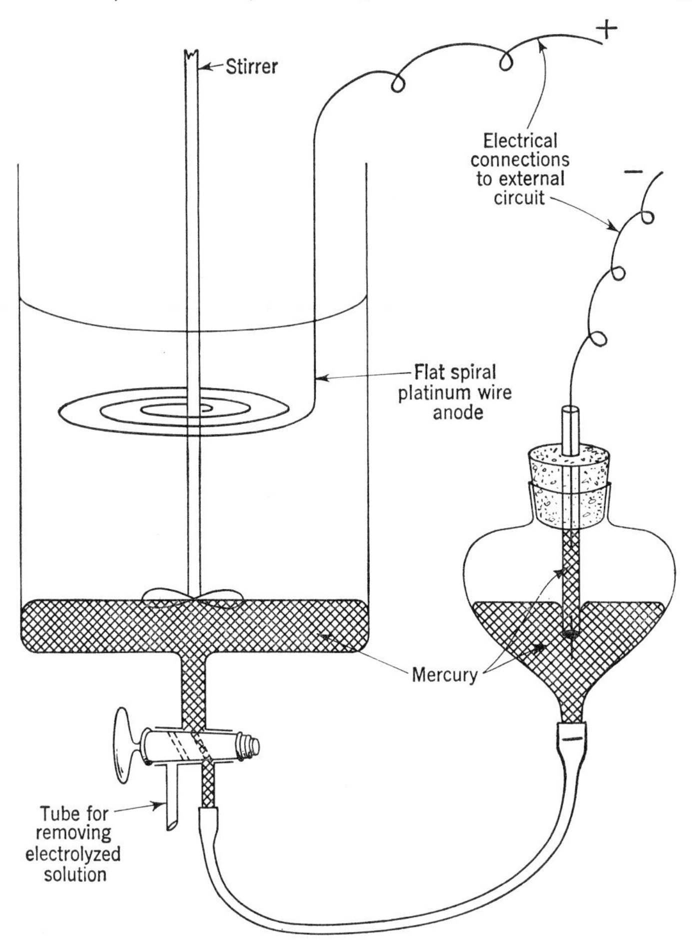

Fig. 7-6. Melaven cell for constant-current electrolysis with a mercury cathode.

under these conditions. This provides a very useful method of preparing a solution of a steel or other ferro-alloy for the determination of small percentages of these important alloying elements. The cathode in such electrolyses is a pool of mercury with an exposed area of about 50 $cm.^2$, and the anode is generally a coil of stout platinum wire; a typical cell is shown in Fig. 7-6. It is particularly important to be able to remove the cathode from the solution without interrupting the electrolysis current;

otherwise some of the deposited elements might be reoxidized by a reaction like $Fe(Hg) + 2H^+ + \frac{1}{2}O_2 = Fe^{++} + H_2O$, occurring at the interface between the amalgam and the air-saturated solution. In work with the mercury cathode one ordinarily uses applied voltages of the order of 6 to 12 v.; since both the cell constant and (because of the relatively high concentrations of acid and metal salts present) the specific resistance of the solution (Sec. 5-2) are very low, currents of several amperes are caused to flow through the cell. Ordinarily this at least equals the total limiting current of all the depositable metals present, so that these are all reduced simultaneously. By using very efficient stirring and taking advantage of the heating effect of so large a current, these limiting currents may be raised to such a point that the sum of the current efficiencies of the processes in which the various metals are deposited becomes quite large during most of the electrolysis. Thus it is possible to separate as much as a gram of iron, etc., from a solution within as little as 15 or 20 min.

It should be obvious that two metals cannot be separated by constant-current electrolysis if both can be deposited at potentials lower than the stable final value which is reached when hydrogen evolution begins to proceed with 100 per cent current efficiency. This is the case, for example, with copper and cadmium. Although copper could be completely deposited from a sulfuric acid solution at a cathode potential which is much too low to initiate deposition of cadmium, the cathode potential cannot be prevented from drifting to a value which is sufficiently negative to deposit cadmium. This problem may be solved by the use of constant-voltage electrolysis (Sec. 7-8) or, more elegantly and conveniently, by electrolysis at controlled potential (Sec. 7-9). However, it is worth pointing out that, in principle, it can also be solved by the addition of a suitable "depolarizer" to the solution being electrolyzed.

A depolarizer may be defined for present purposes as a substance which prevents the potential of an electrode in an electrolysis cell from attaining as large a value as it would reach in the absence of the depolarizer. Thus, in the electrolysis of a solution of cupric sulfate in sulfuric acid, the cupric ion serves as a cathodic depolarizer because the potential of the cathode corresponding to any given electrolysis current is less negative than would be the case if no cupric ion were present, so that the entire current had to be consumed by the reduction of hydrogen ion. Or we might add hydrazine to the solution to prevent the evolution of oxygen at the anode; the oxidation of hydrazine, which takes place in accordance with the half-reaction $N_2H_5^+ = N_2 + 5H^+ + 4e$, proceeds in preference to the oxidation of water to oxygen, and so the anode potential would be less positive in the presence of hydrazine than in its absence.

Returning to the copper–cadmium separation, we might add a relatively large quantity of vanadic ion, V^{+++}, to the solution. Vanadic ion is reduced to vanadous ion, V^{++}, and the formal potential for this reduction is approximately -0.5 v. *vs.* S.C.E. This value is negative enough for the cupric ion to be practically completely reduced before the cathode potential becomes sufficiently negative to initiate the reduction of vanadic ion, and yet is positive enough to prevent cadmium from depositing until nearly all of the vanadic ion has been reduced. On prolonged electrolysis, however, the solution will reach a steady state in which vanadic ion is re-formed at the anode ($V^{++} = V^{+++} + e$) just as rapidly as it is consumed at the cathode ($V^{+++} + e = V^{++}$). Consequently, the cathode potential could never become negative enough for cadmium to deposit, provided only that the amount of vanadic ion added to the solution is great enough so that the limiting current of vanadic ion always exceeds the electrolysis current.[1]

7-8. Electrolysis at Constant Voltage. Let us now return to a consideration of the electrolysis of the 0.05 F cupric sulfate–0.5 F sulfuric acid solution discussed in several of the preceding sections of this chapter. We saw in Sec. 7-3 that the deposition of copper would proceed practically to completion when a voltage of 1.43 v. was applied across a pair of platinum electrodes in the solution, and in Sec. 7-4 we found that hydrogen would not be evolved at any applied voltage smaller than about 2.2 v. If a voltage of, say, 2.0 v. is applied across the two electrodes, it is obvious that copper will be completely deposited, but that no hydrogen could be evolved. Under these conditions hydrogen ion will not contribute to the observed current, so that the value of k_{H^+} in the equations leading to Eq. (7-7) will be zero. The current will decrease exponentially with time as the electrolysis proceeds and will eventually become equal to the very small residual current at the applied voltage used, and the current efficiency for the deposition of copper (neglecting the small constant residual current) will be 100 per cent.

Or suppose that we wish to determine copper electrogravimetrically in a solution containing 0.05 F cupric nitrate, 1 F nitric acid, and 1 F lead nitrate. The decomposition potential of the solution with respect to the deposition of lead may be calculated by the method shown in Sec. 7-2. It is equal to the sum of the back e.m.f. of the cell

$$\mathrm{Pb}|\mathrm{Pb}^{++}(1\ M),\ \mathrm{H}^{+}(1\ M)|\mathrm{O}_2(0.2\ \mathrm{atm.}),\ \mathrm{Pt}$$

[1] One would naturally also have to add some *anodic* depolarizer to the solution to limit the anode potential, during the early part of the electrolysis when cupric ion was being reduced at the cathode, to a value insufficiently positive to bring about the anodic oxidation of the vanadic ion according to the half-reaction $V^{+++} + H_2O = VO^{++} + 2H^+ + e$. Otherwise considerable complication might result from the presence of the vanadyl ions thus produced.

which is

$$E_{\text{back}} = E_{O_2} - E_{\text{Pb}} = E^0_{O_2} - \frac{0.05915}{4} \log \frac{1}{(1)^4(0.2)} - E^0_{\text{Pb}}$$
$$= 1.229 - 0.010 - (-0.126) = 1.345 \text{ v.}$$

and of the overpotential, approximately 0.40 v., for the deposition of oxygen on a smooth platinum anode. So lead will begin to deposit if the applied voltage exceeds 1.74 v. Since copper is practically completely deposited even at an applied voltage of 1.43 v., it follows that the quantitative separation of copper from lead can be accomplished by an electrolysis at any applied voltage between these two values.

Electrolysis at constant applied voltage is described (but not always named) in nearly every text on elementary quantitative analysis, but it is very doubtful whether it is actually widely enough used to merit so much attention. To see why this is so, we must write an equation for the total voltage E applied across an electrolysis cell. This is

$$E = E_a - E_c + iR \tag{7-8}$$

where E_a and E_c are the potentials (including any overpotentials) of the anode and the cathode, respectively, expressed relative to the same reference electrode, and where i is the electrolysis current in amperes and R the cell resistance in ohms.

Suppose that 100 ml. of the above cupric nitrate–lead nitrate–nitric acid solution is electrolyzed at a constant voltage of 1.60 v., and that the resistance of the cell is 5 ohms, which is a fairly typical value. At the start of the electrolysis E_a will be $+1.219 + 0.40 = 1.62$ v. At the same time, as was calculated in Sec. 7-2, E_c will be $+0.307$ v. Substituting these values into Eq. (7-8) gives for the electrolysis current

$$i = \tfrac{1}{5}(1.60 - 1.62 + 0.31) = 0.058 \text{ amp.}$$

This is not a very large current, and as a result the complete deposition of copper will consume much more time than is ordinarily desirable. In particular, when the limiting current of cupric ion under the experimental conditions employed is larger than the initial current calculated from these considerations—and this is the case in the great majority of practical experiments—placing this arbitrary restriction on the initial current will greatly increase the duration of the electrolysis. A further difficulty arises from the fact that the value of E_a varies considerably during the electrolysis because of the change of the anodic overpotential of oxygen with changing current density. The necessity of taking this variation of E_a into account may often introduce an intolerable uncertainty into the calculated value of the applied voltage to be used. Electrolyses at constant applied voltage are indeed used on rare occasions when a difficult analysis or separation must be carried out with a bare minimum of equipment, but their most frequent use is by the authors of textbooks for illustrating the theoretical principles of electrolytic separations.

7-9. Electrolysis at Controlled Potential. The primary factor which determines the reaction which will take place at one electrode of an electrolytic cell is neither the current flowing through the cell nor the total voltage applied across the cell, but the potential of the electrode under consideration. Once the electrode potential is known, whether or not a given reaction will occur at that electrode can be foretold with complete certainty; then the rate at which it will take place is determined by such factors as the composition and temperature of the solution, the rate of stirring, the resistance of the cell, and so on. On the other hand, we have seen how the occurrence of a reaction (*e.g.*, the deposition of cadmium from a solution containing copper, cadmium, and vanadic sulfates) can be prevented—regardless of the current consumed in reducing and reoxidizing vanadic ion, and regardless of the voltage applied across the cell and consumed by iR drop in accordance with Eq. (7-8)—by limiting the cathode potential to a value which is too low to initiate the deposition of cadmium.

The practical difference between electrolyses at constant voltage and at controlled potential may be illustrated by a single example. Consider again the separation of copper from lead by the electrolysis of a nitric acid solution which was discussed in Sec. 7-8. We have previously seen that the deposition of copper onto a platinum electrode from an 0.05 M solution of cupric ion will begin at a cathode potential of $+0.307$ v. (*vs.* N.H.E.), and that the deposition of lead from a 1 M solution of lead ion will begin when the cathode potential becomes as negative as -0.126 v. (*vs.* N.H.E.). Suppose that under a certain set of experimental conditions (cathode area, rate of stirring, temperature, etc.) the limiting current of cupric ion in the solution is 1 amp. If the cell resistance is 5 ohms, as assumed previously, the total voltage which can be applied across the cell

$$\mathrm{Pt|Cu^{++}}(0.05\ M),\ \mathrm{Pb^{++}}(1\ M),\ \mathrm{H^+}(1\ M)|\mathrm{O_2}(0.2\ \mathrm{atm.}),\ \mathrm{Pt}$$

to give a cathode potential of, say, ± 0.00 v. *vs.* N.H.E. will be, referring to Eq. (7-8),

$$E = (1.219 + 0.40) - 0.00 + (1)(5) = 6.62 \text{ v.}$$

This is considerably higher than the figure of 1.6 v. selected for an electrolysis at constant applied voltage in Sec. 7-8, and it results in an electrolysis current some seventeen times as great. This in turn would cause the electrolysis to proceed practically to completion in a small fraction of the time that would be required in the electrolysis at constant voltage.

Of course it is obvious that the electrolysis cannot be allowed to proceed for very long with a voltage of 6.62 v. applied across the cell. When 10 per cent of the copper has deposited, its limiting current will be only 0.9 amp. If the applied voltage is still 6.62 v., the cathode potential will be

$$E_c = -(E - E_a - iR) = -[6.62 - 1.62 - (0.9)(5)]$$
$$= -0.50 \text{ v. } (vs. \text{ N.H.E.})$$

This is nearly 0.4 v. more negative than the potential at which lead will begin to deposit along with the copper. To prevent this, it is necessary

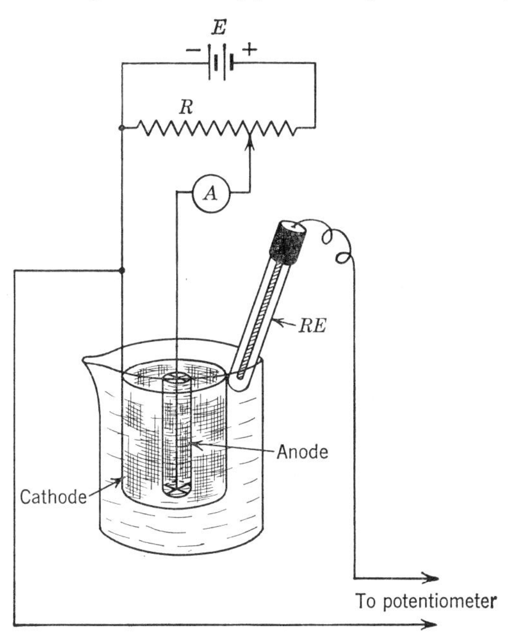

FIG. 7-7. Apparatus for controlled potential electrolysis at a working platinum cathode. A stirrer is required but is not shown.

to counteract the 0.5-v. decrease of the iR drop during this portion of the electrolysis by decreasing the applied voltage from 6.62 v. at the start of the electrolysis to 6.12 v. by the time that 10 per cent of the copper has been deposited. In this way the applied voltage must be continuously readjusted during the electrolysis; eventually, when all of the copper has deposited and the current has fallen to zero, the applied voltage will be decreased to 1.62 v.

A simple circuit suitable for carrying out such an electrolysis is shown in Fig. 7-7. The electrolysis circuit proper is identical with that of Fig. 7-1, but to it has been added the equipment necessary for observing the potential of the cathode at various points during the electrolysis. This consists of a reference electrode, *RE*, which is usually a saturated calomel electrode, and a pair of wires leading to a potentiometer circuit such as that shown in Fig. 3-8.

To conduct an electrolysis at controlled potential with the apparatus of Fig. 7-7, one starts with the solution in the cell, the stirring motor turned on, and the electrodes in position. The rheostat *R* is then set to deliver a voltage of approximately 2 v. across the cell (this should be greater than zero to avoid burning out the first few turns of the rheostat, but considerably smaller than the anticipated voltage at the start of the electrolysis, to avoid initiating an undesired reaction). The potentiometer is set to indicate the desired potential, which is chosen in accordance with the considerations outlined above. With the tapping key in the potentiometer circuit locked down, the setting of *R* is then varied slowly until the galvanometer in the potentiometer circuit indicates approximate balance. No attention is paid to the voltage required to achieve this balance. As the electrolysis proceeds, the cathode potential will tend to drift toward more negative values, and *R* is continuously readjusted to maintain this potential at very nearly the desired value. As this is somewhat difficult for the beginner, it is fortunate that a temporary deviation of as much as 0.1 v. from the desired value of the control potential is generally quite unimportant.

In view of the logarithmic relationship between the electrolysis current and the time [Eq. (7-7)], it is clear that the applied voltage will need to be adjusted most frequently at the very beginning of the electrolysis, when the rate of change of the current is greatest. As the electrolysis proceeds and the current decreases, however, adjustments will need to be made less and less frequently, and near the end of the electrolysis only quite occasional attention will be required.

Incidentally, the observation of the cathode potential is greatly facilitated if a direct-reading vacuum-tube voltmeter is substituted for the potentiometer. This has the advantage of giving a continuous indication of the actual value of the cathode potential. It must be emphasized that an ordinary moving-coil d-c voltmeter cannot be used, because it will draw so large a current that its reading would be considerably in error. The vacuum-tube voltmeter, on the other hand, draws only a very small current from the circuit, and consequently the error [which is practically equal to the iR drop through the meter, in accordance with Eq. (7-8)] will be negligibly small.

However, there are some separations whose success depends on the operator's ability to keep the cathode potential constant to ± 0.05 v. or better, and this is rather difficult to do with this simple apparatus. Even when the potential need not be controlled particularly closely, the constant attention and frequent readjustment demanded by the procedure just described tend to tax the operator's patience and alertness. For this reason, as well as in the interest of freeing the operator for other duties as the electrolysis proceeds, much effort has been devoted to the design of instruments known as "potentiostats," which are devices which automatically maintain the potential of an electrode constant at any predetermined value.

In a typical potentiostat, the potential difference between the electrode whose potential is to be controlled and the reference electrode is connected in series-opposition with an accurately known d-c voltage obtained from a voltmeter-bridge arrangement like that in Fig. 6-9. "Series-opposition" means that the two voltages are connected together in such a way that they tend to counteract each other [(− +) —— (+ −)] rather than reinforce each other as is the case when two dry cells are connected "in series" [(− +) —— (− +)]. The latter arrangement gives the sum, whereas the former gives the difference, of the two voltages.

The difference between the desired potential and the actual value at any instant is then converted from direct to alternating current, and the resulting a-c signal is electronically amplified to a level at which it is large enough to actuate a small motor. This is attached to a heavy-duty voltage divider whose output is applied across the cathode and anode of the electrolysis cell. The polarities and electrical phase relationships are such that, if the actual potential of the cathode becomes slightly more negative than the desired value at any instant, the motor turns in the direction that will drive the slider of the heavy-duty voltage divider downscale, thus decreasing the voltage applied across the cell. On the other hand, if the actual cathode potential momentarily becomes slightly too positive, the motor is turned in the other direction and the applied voltage is increased until the cathode potential becomes equal to the desired value.

Unlike ordinary constant-current or constant-voltage electrolyses, electrolyses at controlled potential are always carried out in cells provided with three electrodes. One of these is the electrode whose potential is being controlled and at whose surface the reaction of greatest interest is taking place. In the foregoing discussion we have tacitly assumed that this is the cathode, but of course there is no reason why one could not control the potential of the anode in, say, the deposition of lead dioxide onto the anode from a nitric acid solution. Consequently the electrode whose potential is being controlled is usually called the "working electrode" of the cell. The second electrode merely serves to conduct the electrolysis current; this would be the anode in an electrodeposition of copper, or the cathode in an electrodeposition of lead dioxide. It is called the "auxiliary electrode" of the cell. The third electrode is a reference electrode, which does not conduct any of the electrolysis current and which serves merely to permit observation of the potential of the working electrode during the electrolysis.

A typical cell for controlled-potential electrolyses with a mercury working electrode is shown in Fig. 7-8.

The conical compartment on the left contains the working and reference electrodes, as well as a motor-driven stirrer, whose blades should trail in the mercury to provide the most efficient possible stirring of the interface between the mercury and the solution, and a sintered-glass gas-dispersion cylinder through which nitrogen or hydrogen can be bubbled to remove air from the solution being electrolyzed. This is often desirable for the same reason as in ordinary polarographic work. A stopcock is

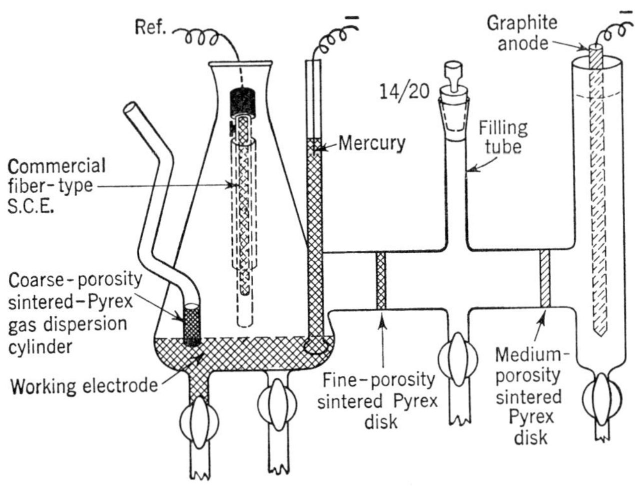

FIG. 7-8. Double-diaphragm cell for controlled potential electrolysis and coulometry at controlled potential with a mercury electrode in air-free solutions. [*L. Meites*, *Anal. Chem.*, **27**, 1116 (1955).]

provided for draining off the solution at the end of an electrolysis without interrupting the electrolysis circuit, and another stopcock at the bottom permits the removal of the mercury and the draining and washing of the cell. The compartment at the right contains the auxiliary electrode, which may be either a carbon rod or a platinum wire helix; if platinum is used when solutions containing chloride are being electrolyzed, a little hydrazine dihydrochloride should be added to this compartment to act as a depolarizer and prevent the liberation of chlorine and the anodic solution of the platinum. The stoppered central compartment, which is separated from each of the other compartments by a sintered-glass disc, merely serves to prevent oxygen, chlorine, or other products of the reaction at the auxiliary electrode from reaching (and taking part in a reaction at) the working electrode. This is important if 100 per cent current efficiency is to be attained in any single process occurring at the working electrode.

Detailed directions for the practical use of this cell are given in Expt. VII.

Many simpler cells for controlled-potential electrolysis have been described by various authors, and the interested student is advised to consult the literature for details. In particular, it may be mentioned that a very much simpler cell, such as that shown in Fig. 7-7, serves very well for controlled-potential electrogravimetric analyses in which platinum gauze working and auxiliary electrodes are used.

Much work has been devoted to the development of procedures for controlled-potential electrogravimetric analysis, and with the aid of these procedures it is possible to carry out highly accurate analyses of many mixtures whose separation by older methods would be difficult or even impossible. One example of such a mixture is the pair of elements copper and bismuth; these had never been separated by electrogravimetric methods until the brilliant work of Lingane and Jones,[1] which the student should examine as an outstanding example of the combination of a meticulous attention to detail and an imaginative synthesis of information drawn from many fields which is characteristic of the best of modern analytical chemistry.

However, it is quite probable that even more effort has been devoted to the use of mercury working electrodes in effecting various electro-separations at controlled potential. This is due in part to the same considerations that lead one to prefer the dropping mercury electrode to the platinum microelectrode for most voltammetric work. These have been summarized in Sec. 6-2. But it also reflects the fact that our information concerning the electrochemical behaviors of various substances at mercury electrodes is incomparably more extensive than our similar knowledge concerning platinum or any other solid electrodes.

Suppose, for example, that we wished to determine the percentage of cobalt present in a sample of reagent-grade nickel sulfate. By consulting a table of polarographic half-wave potentials we could find that the half-wave potentials of nickel and cobalt are separated by nearly 0.3 v. in a supporting electrolyte consisting of 0.1 F pyridine and 0.1 F pyridinium chloride. Unfortunately, the nickel wave ($E_{1/2} = -0.78$ v.) precedes the cobalt wave ($E_{1/2} = -1.06$ v.), so that a polarogram of a solution of the sample would consist of an enormous nickel wave followed by a tiny cobalt wave. As the diffusion current of cobalt would have to be found from the very small difference between two very large numbers,

[1] J. J. Lingane and S. L. Jones, *Anal. Chem.*, **23**, 1798 (1951); an excellent summary is given by J. J. Lingane, "Electroanalytical Chemistry," Interscience Publishers, Inc., New York, 1953, pp. 297–308.

we could hardly expect to be able to secure an accurate determination of the cobalt. In this situation we might very profitably resort to a separation of the nickel by controlled-potential electrolysis of a solution of the sample at a mercury working electrode in the pyridine–pyridinium chloride medium. A suitable potential for carrying out this electrolysis would be the mean of the two half-wave potentials, or -0.92 v. *vs.* S.C.E. This is well up on the plateau of the nickel wave, so that practically every nickel ion reaching the surface of the electrode would be reduced and removed from the solution, and yet it is well before the start of the cobalt wave, so that none of the cobalt would be reduced to the metallic state. Before attempting this separation at a platinum working electrode we would have to measure the decomposition potentials of cobalt and nickel in this medium, for these are unknown and their calculation from the standard potentials of cobalt and nickel would be very dangerous, both because of the formation of pyridine complexes of unknown stabilities and because of the known irreversibilities of the cobalt and nickel electrodes.

Many similar examples of controlled-potential electroseparations at mercury working electrodes prior to polarographic analysis of the residual solution have been described in the literature. These include the separation of large amounts of cadmium from traces (down to 0.0001 per cent) of zinc, of copper from traces (down to 0.00001 per cent) of nickel and zinc, of nitro compounds from phthalate esters, and many others. Although there is no reason why the residual solution could not be analyzed by techniques other than polarography, the close relationship between polarography and controlled-potential electrolysis means that a solution which is well adapted to an electroseparation at controlled potential is equally well adapted to a polarographic analysis after the separation is complete.

It is apparent that these separations can be used for preparative as well as for analytical purposes. For example, controlled-potential electrolysis at a mercury cathode serves very well for the preparation of cobalt salts entirely free from any detectable trace of nickel, of zinc salts practically free from cadmium, and of thallium salts practically free from lead (by the anodic re-solution of the metallic thallium deposited in the controlled-potential electrolysis of a sodium hydroxide solution containing thallous and lead ions). These are all very difficult to accomplish by older methods. Moreover, the technique is extremely useful in the preparation of solutions containing such relatively unfamiliar substances as +3 osmium, +3 vanadium, and +1 manganese. Solutions thus prepared are entirely free from any excess chemical reducing agent (or from the oxidized form of any such reducing agent), and so they are ideally suited to the study of the chemistries of these uncommon ions.

Electrolysis at controlled potential is inherently much more selective than any other method of bringing about a reduction or an oxidation. The best example of this was studied by Lingane, Swain, and Fields.[1] These authors wished to reduce 9-(*o*-iodophenyl)-acridine (I) to the dihydro compound (II). Polarograms of the starting material revealed that it gave two waves under appropriate conditions. The first of these, $E_{1/2} = -1.32$ v. *vs.* S.C.E., corresponded to the desired reaction:

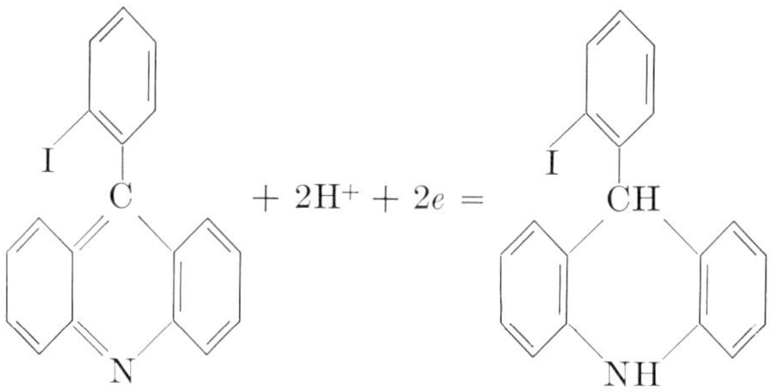

I C N (I) $+ 2H^+ + 2e =$ I CH NH (II)

The second, $E_{1/2} = -1.62$ v. *vs.* S.C.E., represented the complete reduction of the molecule, with the elimination of the iodine:

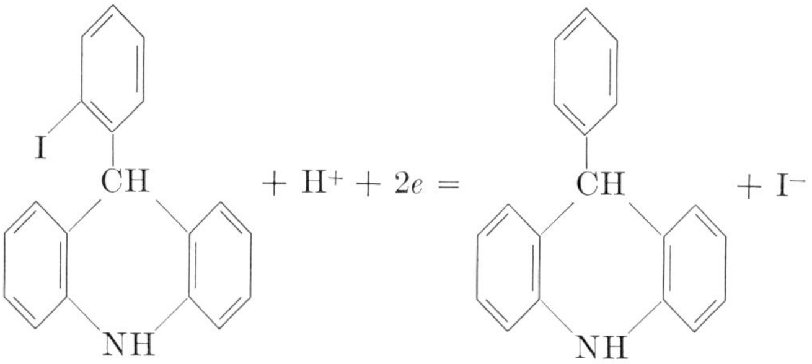

I CH NH (II) $+ H^+ + 2e =$ CH NH (III) $+ I^-$

The desired reaction was very easily brought about, and the product was isolated in 90 per cent yield, by a controlled-potential electrolysis at a mercury cathode at -1.39 v. *vs.* S.C.E. Chemical reducing agents, on the other hand, either failed to react at all or else removed the iodine atom in addition to reducing the double bond.

7-10. Coulometry at Controlled Potential. Coulometry at controlled potential is a very recently developed technique which dates back only to Hickling's pioneering experiments in 1942. It depends on the ease with which a single electrode reaction can be made to take place with 100 per cent current efficiency during a controlled-potential electrolysis.

[1] J. J. Lingane, C. G. Swain, and M. Fields, *J. Am. Chem. Soc.*, **65**, 1348 (1943).

The fundamental theoretical basis of coulometry at controlled potential involves no more than a restatement of Faraday's law: when a single electrode reaction occurs with 100 per cent current efficiency, the total quantity of electricity consumed (measured in faradays) is equal to the number of equivalents of the substance involved in the reaction. That is,

$$\int_0^\infty i\,dt = Q = nF_yN^0 \qquad (7\text{-}9)$$

where i is the current (in amperes) t seconds after the electrolysis has begun. The integral of the current–time curve gives Q, the number of coulombs consumed in the reduction (or oxidation) of N^0 moles of material. As usual, n is the number of electrons appearing in the half-reaction which involves one ion or molecule of the starting material; it is simply the number of equivalents per mole for the reaction at the working electrode.

Two things must be done to secure accurate analyses by coulometry at controlled potential. One is to carry out the electrolysis in such a way that only one reaction can take place at the working electrode (and, of course, for a long enough time to ensure that the reaction goes to completion). This ordinarily involves very little more than the judicious selection of the supporting electrolyte and working electrode potential on the basis of polarographic data. The other is to perform an accurate integration of the current–time curve to find the number of coulombs.

The easiest way to perform this integration is to measure the electrolysis current at a number of exactly known times during the electrolysis. This can be done quite simply with the aid of a precision ammeter and a stopwatch. In view of Eq. (7-7), these data will serve to define a straight line of the form

$$\log i = \log i^0 - kt \qquad (7\text{-}10)$$

which upon integration gives

$$Q = \int_0^\infty i\,dt = \frac{i^0}{2.303k}$$

Here i^0 is simply the initial current, secured by extrapolating the current–time curve, plotted in accordance with Eq. (7-10), back to zero time. A typical plot is shown in Fig. 7-9.

This simple technique is attractive in principle, but its accuracy even under optimum conditions seems to be limited to about ± 2 per cent. When the electrolysis current is so high that the temperature of the solution changes appreciably during the experiment, or when the current–time curve does not have the theoretical form (which is quite likely to be the case if the working electrode potential is controlled manually rather than by a potentiostat), even more serious errors may result.

Better results may be secured by the use of a coulometer. This is a device through which the electrolysis current is passed, and in which an electrochemical reaction occurs whose extent is related by Faraday's law to the quantity of electricity used. The best-known and most venerable coulometers are the silver and iodine coulometers. The former consists essentially of a platinum cathode and a silver anode in a silver nitrate or, better, silver perchlorate solution; when a current is passed through this

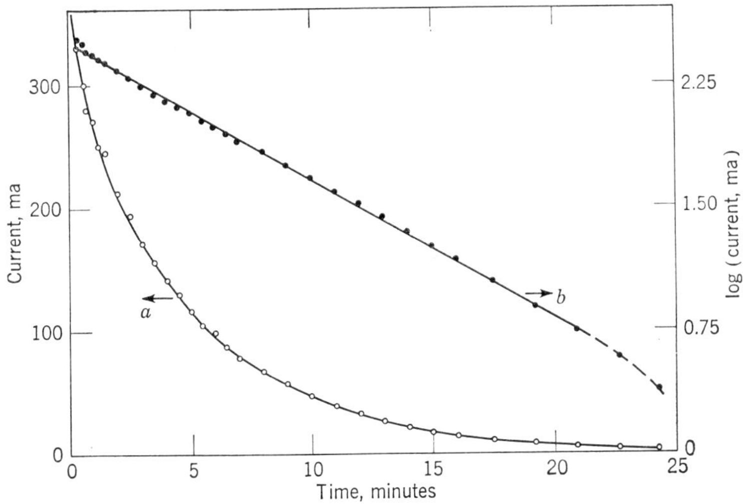

FIG. 7-9. Variation of current with time during the reduction of 1.004 millimole of picric acid in 60 ml. of 0.7 F hydrochloric acid–0.7 F potassium chloride at a mercury cathode whose potential was kept constant at −0.40 v. *vs.* S.C.E.: (*a*) current (on left-hand ordinate scale) *vs.* time, and (*b*) log current (on right-hand ordinate scale) *vs.* time.

cell, each faraday of electricity causes 107.880 g. of silver to deposit on the cathode. The iodine coulometer consists of two platinum electrodes in separate compartments of a cell containing an iodine–potassium iodide solution; the amount of iodine produced at the anode or consumed at the cathode of this coulometer is determined by titration with standard arsenite or thiosulfate.

A coulometer which is much more convenient, and which has been much more widely used, than either of these is the hydrogen-oxygen coulometer shown in Fig. 7-10. This consists essentially of a buret which is filled with an electrolyte such as 0.5 F potassium sulfate, and which contains two platinum electrodes. When a current is passed through these electrodes, hydrogen is evolved at one and oxygen at the other:

$$2H_2O = 4H^+ + O_2 + 4e$$
$$4H^+ + 4e = 2H_2$$

and the mixture of hydrogen and oxygen is collected above the solution in the buret. At standard temperature and pressure 0.1733 ml. of gas is evolved per coulomb of electricity. A leveling bulb is provided so that the pressure of the confined gas can be made equal to the atmospheric pressure, which must be measured at the time of the experiment. (During the electrolysis the height of this leveling bulb must be readjusted occasionally to ensure that the pressure exerted on the gas mixture is never very different from the atmospheric pressure, for solubility equilibrium is attained only very slowly after a large change of pressure. Of course care must be taken to presaturate the potassium sulfate solution with the gas mixture by passing a current through the coulometer for some time, with the stopcock open to the atmosphere, before beginning an experiment.) A water jacket is used to permit the measurement, and ensure the constancy, of the temperature of the gas collected. A detailed description of the assembly and use of this coulometer has been published by Lingane,[1] and the student should consult this paper before beginning any experiment in which a hydrogen-oxygen coulometer is used.

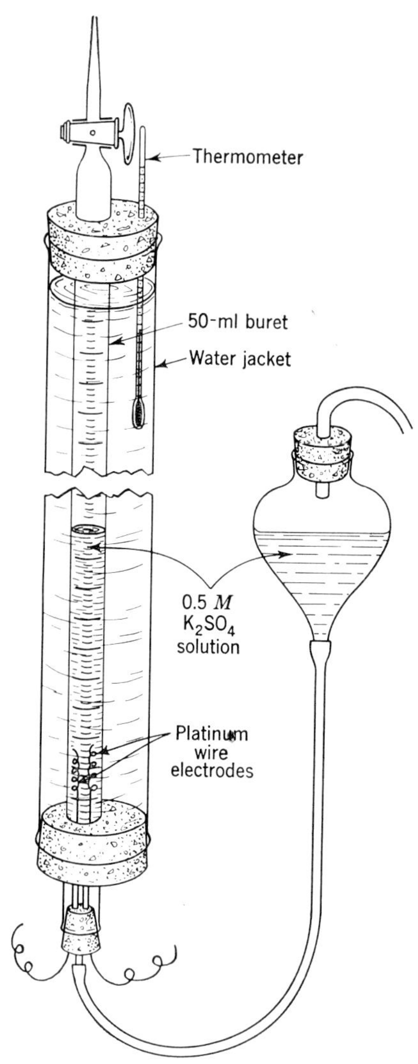

FIG. 7-10. Hydrogen-oxygen coulometer.

In addition to the previously mentioned advantages of electrolytic methods in general—the saving of operator time which results from the possibility of letting the experiment proceed unattended, and the greatly reduced chance of loss of material which results from the frequent elimination of separations by precipitation, extraction, and the like—and in addition to the enormous selectivity and specificity of controlled-potential

[1] J. J. Lingane, *J. Am. Chem. Soc.*, **67**, 1916 (1945).

electrogravimetric analysis, coulometry at controlled potential has the very great advantage of being applicable to electrolytic reactions in which no solid product is formed. This makes it possible to determine such elements as tungsten, vanadium, molybdenum, titanium, chromium, and many others by merely measuring the quantity of electricity consumed in a quantitative reduction or oxidation from one valence state to another which remains in solution. Furthermore, and in the long run probably even more important, coulometry at controlled potential is easily applicable to quantitative organic analysis. For example, such compounds as picric acid and trichloroacetic acid can be assayed with an ease and accuracy unrivaled by any other method merely by taking advantage of the fact that the reactions

$$C_6H_2(OH)(NO_2)_3 + 18H^+ + 18e = C_6H_2(OH)(NH_2)_3 + 6H_2O$$

(2,4,6-trinitrophenol → 2,4,6-triaminophenol)

and $$Cl_3CCOO^- + H^+ + 2e = Cl_2HCCOO^- + Cl^-$$

can be made to proceed quantitatively and with 100 per cent current efficiency by controlled-potential electrolysis with a mercury cathode.

Coulometry at controlled potential is also a valuable technique for studying reactions which take place under polarographic conditions. For example, manganese(II) in a potassium cyanide supporting electrolyte gives a polarographic wave which its discoverer believed to represent a two-electron reduction to manganese amalgam. However, the half-wave potential of this wave is considerably less negative than that for the wave of simple manganous ion in a sodium perchlorate medium, and since the latter wave is known to involve only a relatively small activation overpotential it is difficult to credit this explanation.[1] When a cyanide solution of +2 manganese is electrolyzed at a large stirred mercury cathode at a potential on the plateau of the wave, coulometric measurements reveal that only one faraday of electricity is consumed in the reduction of each mole of manganese. This is conclusive proof that manganese is reduced only to the +1 state rather than to the metal.

[1] It may be pointed out, however, that in some cases it is possible for the half-wave potential of a cathodic wave to become less negative as a result of complex formation, contrary to what one might perhaps expect from the discussion of Sec. 6-13. For example, the half-wave potential for the reduction of the simple nickel ion from a perchlorate supporting electrolyte is about −1.1 v. *vs.* S.C.E., but in a thiocyanate medium it is only about −0.7 v. This is because the activation overpotential involved in the reduction of the nickel–thiocyanate complex is so much smaller than that involved in the reduction of the simple nickel ion as to outweigh the expected shift to more negative values resulting from complex formation.

7-11. Coulometry at Controlled Current. Like coulometry at controlled potential, coulometry at controlled current is a technique in which one measures the quantity of electricity consumed in the course of a quantitative reaction involving the substance being determined. In each case the data are interpreted by applying Faraday's law, Eq. (7-9), and in each case the flow of current through the solution must cause some known reaction to take place with 100 per cent current efficiency.

The experimental difference between the two methods lies in the way in which the current integral is secured. In coulometry at controlled potential, the current decreases continuously during the electrolysis, and so it must be integrated by a chemical coulometer, an electronic current integrator, or in some other way which is able to cope with a varying current. In coulometry at controlled current, the current is kept constant throughout the electrolysis, and the quantity of electricity consumed is found from the product it, where i is the constant current in amperes and t is the number of seconds required to reach the equivalence point of the reaction which takes place. This naturally implies, and correctly so, that in the constant-current procedure we must have a way of locating the equivalence point of the reaction,[1] whereas no such problem exists in coulometry at controlled potential. The various methods which have been used for the location of the equivalence point are mentioned briefly in a later paragraph; for the present let us merely note that they are all quite simple experimentally.

In coulometry at controlled potential, then, we need a potentiostat and a coulometer or current integrator; in coulometry at controlled current we need a source of constant current, an electric stop clock so arranged that it measures the time during which the current flows through the solution, and a device for locating the equivalence point. Simple but quite accurate circuits for producing constant currents can be very easily and cheaply constructed from components readily available in any laboratory, and the accurate measurement of time presents no difficulties whatever, so that the quantity of electricity consumed in an analysis by coulometry at controlled current is not at all difficult to measure to ± 0.1 per cent or even better. However, the design and construction of a reasonably simple and equally accurate coulometer or current integrator proved for many years to be a different matter altogether.

[1] Because this problem of locating the equivalence point of a reaction is, in classical analytical methods, peculiar to the titration procedures of volumetric analysis, coulometry at controlled current has often been termed "coulometric titration." This is a short and convenient name, but it is not particularly apt, because the word "titration" is generally understood to imply the addition of a measured volume of a standard solution, as in "potentiometric titration," and so it is slowly being replaced by the longer but more exact and descriptive nomenclature.

Consequently, for almost a decade the development of coulometry at controlled potential as a practical analytical technique lagged far behind that of coulometry at controlled current. Nevertheless, there is little doubt that coulometry at controlled potential is inherently the more versatile of these two techniques, and the attention devoted to it will quite certainly increase enormously in the next few years.

In Sec. 7-5 it was stressed that it is impossible to bring about a single reaction *at an electrode* in a constant-current electrolysis. No matter how the solution composition and the magnitude of the current are varied, there comes a point at which the constant current being forced through the solution exceeds the limiting current of the substance being determined, and at this point some other reaction must begin *at the surface of the electrode* if the current is to remain constant. The student may profit from re-reading the discussion of Fig. 7-4 (page 198). How, then, is it possible to bring about a single reaction quantitatively and with 100 per cent current efficiency in an analysis by coulometry at controlled current?

The answer to this apparent dilemma is indicated by the words italicized in the preceding paragraph. If, that is, the product of the second reaction *at the surface of the electrode* immediately reacts chemically *in the body of the solution* with the substance being determined, and if the latter is thus converted into the same product which would have been formed if it had reacted directly at the electrode surface, no error will result from the initiation of the second reaction.

To illustrate this principle, let us consider Fig. 7-11. Curve a in this figure is a somewhat idealized current–potential curve for a platinum electrode in a solution containing 0.001 M ferrous ion, 0.001 M cerous ion, and 1 F sulfuric acid. It is assumed that the ionic mobilities of Fe^{++} and Ce^{+++} are very nearly equal, and that the area of the anode and the rate of stirring are such that each ion gives a limiting current of 50 ma. (milliamperes). Suppose that a current of -25 ma. is forced through the solution, as indicated by the horizontal dashed line (remember from Chap. 6 that an anodic or oxidation current is given a negative sign). This is smaller than the limiting current of ferrous ion under these conditions, and so the entire current is consumed in oxidizing ferrous ion at the surface of the anode.

This will continue to be true until just half of the ferrous ion has been oxidized and the limiting current of the remaining ferrous ion has decreased to 25 ma. At this point the current–potential curve for the platinum electrode will be given by curve b in the figure. Note the appearance of a cathodic portion of the curve which reflects the near-reversibility of the ferric–ferrous half-reaction. The potential of the electrode will now drift rapidly toward more positive values in order to allow the current to remain constant at 25 ma.; this corresponds exactly to the behavior depicted in Fig. 7-5. Now, since ferrous ions can no longer reach the surface of the electrode in sufficient numbers to consume the entire 25 ma. of current, cerous ion will begin to be oxidized. At the point when a total of three-quarters of the ferrous ion initially present has been oxidized (curve *c*), the limiting current of ferrous ion will have

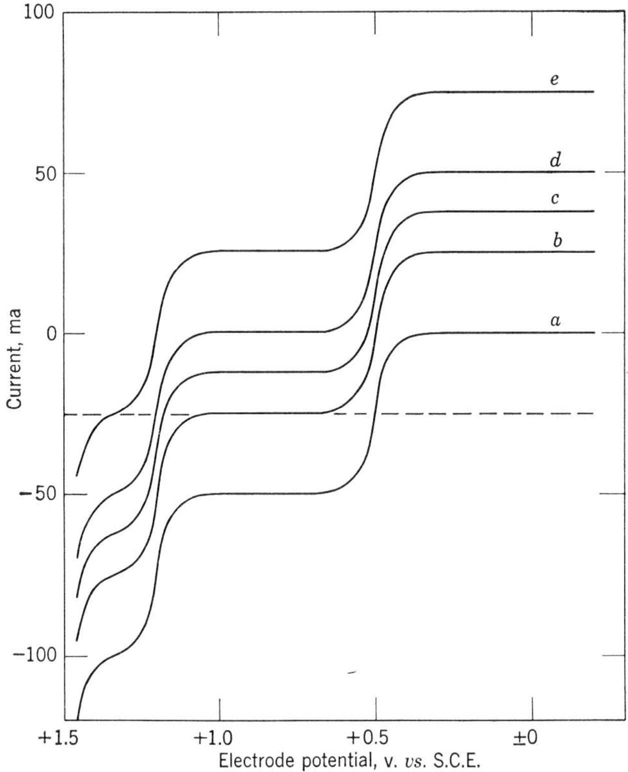

Fig. 7-11. Idealized current-potential curves for a platinum electrode in a solution originally containing 0.001 *M* ferrous ion, 0.001 *M* cerous ion, and 1 *F* sulfuric acid, after the passage of a current of 25 milliamp. through the solution for a length of time corresponding to the oxidation of (*a*) 0, (*b*) 50, (*c*) 75, (*d*) 100, and (*e*) 150 per cent of the amount of ferrous ion initially present.

decreased to 12.5 ma. So 12.5 ma. of the current will be consumed in oxidizing ferrous ion, and the remaining 12.5 ma. will be consumed in oxidizing cerous ion; the instantaneous current efficiency for the oxidation of ferrous ion *at the electrode surface* at that instant will be 50 per cent, and it will decrease more and more nearly to zero as still more of the ferrous ion is oxidized.

Actually, however, the cerium(IV) produced when cerous ions are oxidized at the electrode surface is capable of reacting, and does react, with ferrous ions in the bulk of the solution. The oxidation of a cerous ion at the surface of the electrode proceeds according to the equation

$$Ce^{+++} + 3SO_4^{=} = Ce(SO_4)_3^{=} + e$$

and the ceric complex thus produced reacts with ferrous ion in the solution according to the equation

$$Ce(SO_4)_3^{=} + Fe^{++} = Ce^{+++} + 3SO_4^{=} + Fe^{+++}$$

The sum of these equations, which is the equation for the net reaction which occurs when one electron is transferred from the solution to the anode, is

$$Fe^{++} = Fe^{+++} + e$$

and this is of course identical with the equation which represents the direct oxidation of ferrous ion at the anode. Regardless of the actual path of the reaction, one mole of ferrous ion will be oxidized by each faraday of electricity which flows through the cell.

After a length of time which depends on the electrolysis current and on the volume and concentration of the solution, the equivalence point of the oxidation of ferrous ion will have been reached. The current–potential curve of the solution will then be given by curve *d*, and the student should satisfy himself that the composition of the solution will be given by the equations

$$[Ce^{+++}] = [Fe^{+++}] = 0.001$$
$$[Ce(SO_4)_3^{=}] = [Fe^{++}]$$

This corresponds to the equivalence point of a titration of 0.002 *M* ferrous ion with 0.002 *F* ceric sulfate, and so this titration might be described as a "coulometric titration of ferrous ion with electrolytically generated ceric sulfate."[1] The ceric complex is referred to as the "intermediate" in the over-all process.

If the electrolysis is prolonged beyond this point, ceric ion will be generated with 100 per cent current efficiency until the limiting current of the remaining cerous ion just becomes equal to the electrolysis current. During this interval the amount, and hence the concentration, of ceric ion will increase linearly with the electrolysis time, and at the end of this period the current–potential curve of the solution will be given by curve *e* of the figure. The potential of the anode will now again drift more rapidly toward a more positive value, and in order that the current may continue to remain at 25 ma. oxygen will begin to be evolved in accordance with the equation

$$2H_2O = 4H^+ + O_2 + 4e$$

[1] But see the footnote on p. 220.

Because not all of the current is consumed in oxidizing cerous ion, the rate of increase of the ceric ion concentration will begin to fall off, and eventually this concentration will become constant when all of the cerous ion has been oxidized.

Exactly similar phenomena would have been observed if the electrolysis current had been 10 ma. or 40 ma., but a little thought will reveal that oxygen would have generated right from the start of the electrolysis if the constant current used had been, say, 120 ma. Because most of the oxygen produced would have simply escaped from the solution without oxidizing either ferrous or cerous ion, this would have resulted in a serious error in the quantity of electricity used. Therefore it is customary to use a concentration of the "reagent precursor," which in this case is the cerous ion, which is very much larger than the concentration of the material being determined. As the latter may vary over quite a wide range from one analysis to the next, it is not uncommon to use a concentration of the reagent precursor which is 10^4 or even 10^5 times that of the substance being determined.

The end point of a "coulometric titration" can be located by either potentiometric, amperometric, or spectrophotometric measurements, and as far as is known there is, in general, little if any reason for preferring any one of these to the others. In any specific case, of course, a clear basis for choice may exist. Thus, one would certainly hesitate to apply amperometric procedures to the coulometric determination of an acid or base. Such decisions are straightforwardly and reliably guided by an inquiry into the utility of each titration technique in the execution of the same titration with a standard solution of the reagent. Conductometric methods seem not to have been used at all in "coulometric titrations"; although one can envision a few cases in which they might be of service, the usual need for a relatively very large concentration of an ionic reagent precursor is antipathetic to the fundamental requirement of a successful conductometric titration.

When the equivalence point is to be located potentiometrically, it is necessary to place two electrodes in the solution in addition to the cathode and anode of the electrolysis cell, which are usually referred to as the "generating" electrodes. These additional electrodes are usually a platinum indicator electrode (or, for acid-base "titrations," a glass electrode) and a saturated calomel reference electrode, and the potentiometric measurements are made with an ordinary potentiometric apparatus. The solution must be well stirred throughout the experiment, and this is generally done by means of a magnetic stirrer because using a stirring motor in addition to the four electrodes would unduly complicate the superstructure above the cell. The "titration" is carried out in a manner quite analogous to that used in an ordinary potentiometric titration: after measuring the potential of the indicator electrode at the start of the experiment, the generating current is turned on for 30 sec. or so and then turned off again. After allowing a few seconds for chemical and electrical equilibrium to be attained, the potential is measured again. Both the potential and the time indicated by a stop clock synchronized with the generating current are recorded (Fig. 7-12 shows one method of accomplishing this synchronization), and the generating current is turned on for another 30 sec. It is convenient to secure data at closely and equally spaced times in the vicinity of the equivalence point. The time which corresponds to the equivalence point is found either from a plot of the potential of the indicator electrode against time, or analytically by a simple modification of the procedure described in Sec. 3-12.

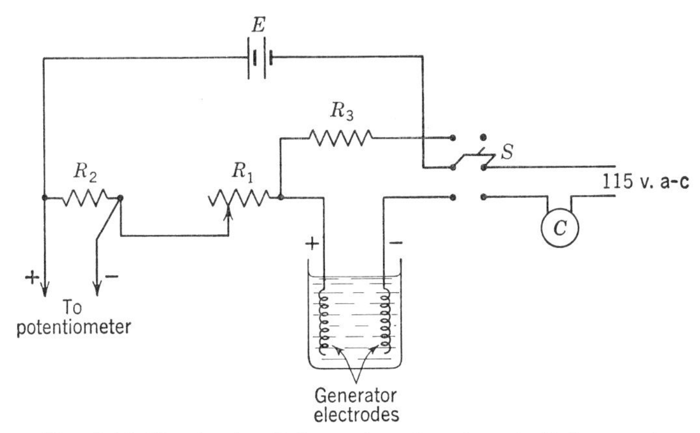

FIG. 7-12. Simple circuit for coulometry at controlled current.

The way in which the equivalence point potential is affected by the presence of the large excess of the reagent precursor was discussed in Sec. 3-16.

When the equivalence point of the reaction is to be located amperometrically, the indicator electrode is usually a rotating platinum microelectrode. This and a saturated calomel electrode are connected to the output terminals of a polarograph, and a potential equal to some predetermined value—generally on the plateau of the wave due to the product formed as the reagent precursor is oxidized or reduced at the surface of the working electrode—is impressed across them. An indicator electrode potential of +0.8 v. *vs.* S.C.E. might be employed in the titration of Fig. 7-11. Up to the equivalence point, the concentration of ceric ion would be negligibly small, and the current measured would be simply the anodic diffusion current of the ferrous ion remaining unoxidized. This will decrease with time linearly toward zero, and after the equivalence point is passed the cathodic diffusion current of the excess ceric ion will increase linearly with time. So a plot of the current against the time of the electrolysis will allow the equivalence point to be found as the point at which the current becomes equal to the residual current of the supporting electrolyte alone.

Figure 7-12 illustrates the simplest way of obtaining a small constant current. Two or more 45-volt B batteries, E, are connected in series with the generating electrodes of the cell, with the variable dropping resistor R_1 (which may typically have a maximum value of about 20000 ohms), and with the precision resistor R_2 (100 ohms $\pm$ 0.05 per cent). From Ohm's law and Eq. (7-8), the current flowing through this circuit is given by

$$i = \frac{V - (E_a - E_c)}{R_1 + R_2 + R_{\text{cell}}}$$

where V is the voltage provided by the batteries. Since $R_1 + R_2$ is of the order of 10^4 ohms, whereas R_{cell} is unlikely to be greater than about 20 ohms, even a relatively large variation of the cell resistance during the electrolysis will have hardly any effect on the current. Much more important is the variation of $E_a - E_c$; in the titration illustrated by Fig. 7-11 this might vary by as much as 0.7 v. Therefore it is necessary to use a large value of V to minimize changes in the current as the electrolysis proceeds, and R_1 is made correspondingly large to provide a current which is of the desired order of magnitude. It is clearly advantageous to use an electrolysis current which exceeds the limiting current of the substance being determined even at the start of the titration (*cf.* the last paragraph of Sec. 7-5). The precision resistor R_2 is used to permit the accurate measurement of the current; when the voltage drop, E_2, across this resistor is measured with a precision potentiometer, the electrolysis current is easily calculated from the equation

$$i = \frac{E_2}{R_2}$$

The resistor R_3, which is so connected that the electrolysis current flows through it whenever the cell is disconnected from the circuit, is used to keep a nearly constant current flowing through R_1 at all times. This keeps R_1 at a constant temperature, and so minimizes the variations in its resistance which would occur if the current through R_1 were interrupted periodically. Before beginning an experiment, R_1 should be allowed to warm up for at least ten minutes by throwing switch S on and connecting R_3 into the circuit. The resistance of R_3 should not differ greatly from that of the cell; a value of about 20 ohms is usually satisfactory.

It is usually desirable to isolate the auxiliary generating electrode from the bulk of the solution to avoid undesirable reactions such as the reduction of ceric and ferric ions in the experiment described above. This may be done fairly satisfactorily without unduly increasing the cell resistance by placing the auxiliary electrode inside the tube of a sintered-glass gas-dispersion cylinder which is immersed in the solution. When the highest precision is essential, this simple arrangement may lead to some error because of diffusion through the sintered cylinder, thus necessitating some rather more complicated expedient.

Coulometry at controlled current is especially well adapted to the determination of very small amounts of material, because it is only necessary to employ a correspondingly small generating current. By the exercise of appropriate precautions, and with the use of specially sensitive methods of locating the equivalence point, it has actually been possible to determine as little as 0.005 microgram (5×10^{-9} g.) of manganese (as permanganate, using ferrous iron as the intermediate) with an accuracy and precision of ± 10 per cent. In determining such minuscule amounts of any substance, it is of course generally necessary to resort to rather heroic measures to ensure the purity of the acids, salts, and other reagents added in large excess. This, however, is not a problem peculiar to coulometry at controlled current, for it arises in trace analysis by any method whatever.

In general, any titration which can be performed volumetrically can be used as the basis of a method employing coulometry at controlled current. Acid-base titrations are carried out by generating hydroxyl ion at the cathode ($2H_2O + 2e = H_2 + 2OH^-$), or hydrogen ion at the anode ($2H_2O = 4H^+ + O_2 + 4e$), in the electrolysis of a neutral salt solution which is flowed past the working generator electrode into the solution being "titrated." Bromine can be generated at a working anode in the electrolysis of an acidified bromide solution, and used for the oxidation of weak reducing agents such as arsenite or for the bromination of such organic substances as phenol, aniline, and 8-hydroxyquinoline. Ferrous ion can be generated at a platinum cathode in a solution of a ferric salt, and used in the determination of oxidizing agents such as ceric, permanganate, and dichromate ions. Ceric ion has been employed in the determination of many reducing agents, including several phenols and hydroquinones. Silver ion, produced at a silver anode, has been used in the determination of chloride, bromide, iodide, and organic mercaptans. Many additional examples might be cited.

Coulometry at controlled current is inherently more accurate and convenient than the older volumetric techniques, because it entirely eliminates all of the problems associated with the preparation, standardization, and storage of standard solutions. This is especially important when very small amounts of material have to be determined, for extremely dilute solutions of most reagents frequently undergo rapid changes of normality. In addition, "coulometric titrations" are obviously just as readily adaptable to automation as are potentiometric titrations (Sec. 3-17). On the other hand, coulometry at controlled current is inherently a less versatile and selective procedure than coulometry at controlled potential, because it is not possible to bring about a reaction by coulometry at controlled current unless a reagent can be found which will cause the same reaction to occur quantitatively. To cite two examples which were mentioned previously, it is most improbable that either the reaction

$$Cl_3CCOO^- + H^+ + 2e = Cl_2HCCOO^- + Cl^-$$

or the reduction of 9-(*o*-iodophenyl)-acridine to the dihydro compound (page 215, II) could be employed in an analysis by coulometry at controlled current, because no chemical reagent is known which will effect either of these reductions quantitatively.

PROBLEMS

The answers to the starred problems will be found on page 403.

***1.** What percentage of the lead ion originally present will be deposited if 50 ml. of a 1 *F* hydrochloric acid solution of lead ion is electrolyzed for an hour at a potential on the plateau of the lead wave, using a dropping mercury electrode for which $m^{2/3}t^{1/6}$ is 3.00 mg.$^{2/3}$ sec.$^{-1/2}$? The diffusion current constant of lead ion in this medium is 3.86.

2. The internal resistance of a certain electrolysis cell is 0.50 ohm. What voltage must be applied to the cell to cause a current of 0.75 amp. to flow in the electrolysis of a cupric sulfate solution under such conditions that the decomposition potential with respect to copper deposition is 1.350 v.?

3. An 0.1 *F* solution of cupric sulfate in 1 *F* sulfuric acid is electrolyzed between two platinum electrodes. The overpotentials for the deposition of oxygen on copper and on platinum are 0.85 and 0.40 v., respectively, and the overpotential for the deposition of hydrogen on copper is 0.60 v.

a. At what applied voltage will copper begin to deposit on the cathode?

b. What concentration of cupric ion will remain undeposited if the electrolysis is carried to completion at a voltage which is just equal to the decomposition potential of the final solution with respect to hydrogen evolution?

***4.** For the electrolysis of a 3 *F* solution of sulfuric acid containing 0.001 *F* cupric sulfate, calculate the voltage which must be applied to the cell to cause 99.99 per cent complete deposition of the copper on the cathode. Use the data given in Problem 3.

5. An analyst wishes to determine the concentration of thallous ion in a solution containing 2 *F* perchloric acid, 0.1 *F* cadmium perchlorate, and 0.001 *F* thallous perchlorate. The pertinent standard potentials are

$$O_2 + 4H^+ + 4e = 2H_2O \qquad E^0 = +1.229 \text{ v.}$$
$$Tl^+ + e = Tl \qquad E^0 = -0.337 \text{ v.}$$
$$Cd^{++} + 2e = Cd \qquad E^0 = -0.403 \text{ v.}$$

The overpotentials for the depositions of oxygen on platinum and of hydrogen on thallium are 0.40 and 0.80 v., respectively.

a. What voltage will have to be applied across two platinum electrodes immersed in the solution to initiate the deposition of thallium? (Assume that no activation overpotential is involved at the cathode.)

b. What will be the highest voltage that can be applied to such a cell without resulting in the deposition of any other substance at the cathode?

c. What is the maximum percentage of the thallium present that can be recovered under the conditions of part *b*?

***6.** A solution of lead nitrate in dilute nitric acid was electrolyzed under such conditions that both lead dioxide and oxygen were produced at the anode. After a current of 1 amp. had flowed through the cell for 50 min. it was found that 2.000 g. of lead dioxide had been deposited. What volume of oxygen, measured at standard conditions, was evolved during this period?

7. Exactly 100 ml. of a solution containing 0.100 M cupric ion and 1 F sulfuric acid is electrolyzed at a voltage which is smaller than the decomposition voltage with respect to hydrogen evolution. Fifteen minutes after the start of the electrolysis it is found that 0.1000 g. of copper has been deposited on the cathode. For how much longer must the electrolysis be carried out to ensure that no more than 0.1 mg. of copper remains undeposited?

***8.** It is proposed to separate lead from cadmium by the electrolysis of a mixture of these two ions at a mercury working electrode. In the supporting electrolyte employed, the half-wave potentials of lead and cadmium are -0.43 and -0.64 v. *vs.* S.C.E., respectively. Between what limits must the potential of the working electrode be maintained if at least 99.9 per cent of the lead, but no more than 0.1 per cent of the cadmium, is to be deposited? Assume that the volume of the working electrode is equal to that of the solution.

9. Exactly 50 ml. of a mixture of lead and cadmium ions was electrolyzed under the conditions of Problem 8, using a cell whose resistance was 10 ohms. A platinum electrode in 1 F sulfuric acid was employed as the anode. If the overpotential for the deposition of oxygen on platinum is 0.40 v., and if the initial current was 0.50 amp., what voltage had to be applied across the cell at the start of the electrolysis?

10. An unknown solution of picric acid (2,4,6-trinitrophenol) in 1 F hydrochloric acid was electrolyzed at a mercury cathode under such conditions that the 18-electron reduction to the triaminophenol took place with 100 per cent current efficiency. Exactly 100 sec. after the start of the electrolysis the current flowing through the cell was 0.222 amp.; 500 sec. later it was 0.032 amp. What weight of picric acid was present in the unknown solution?

***11.** A solution of cobalt(II) in 1 F ammonia–1 F ammonium chloride was electrolyzed with a mercury cathode whose potential was kept constant at -1.50 v. *vs.* S.C.E. When the cobalt had been completely reduced to the metallic state and the current had fallen to zero, a hydrogen-oxygen coulometer in series with the cell showed that 27.00 ml. of gas had been evolved. The barometric pressure was 745.5 mm. of mercury, and the temperature of the gas in the coulometer was 24.1°C. What weight of cobalt was present in the original solution?

12. The half-wave potential for the reduction of vanadic ion, V^{+++}, to vanadous ion, V^{++}, in 1 F perchloric acid is -0.508 v. *vs.* S.C.E., and the wave is thermodynamically reversible. When 50.0 ml. of a solution of vanadic ion in 1 F perchloric acid was electrolyzed with a mercury cathode whose potential was kept constant at -0.60 v. *vs.* S.C.E., the current fell to zero after 250.0 coulombs of electricity had been consumed. What was the concentration of vanadic ion in the solution?

13. In the determination of phenol by coulometry at controlled current, the bromine which is produced electrolytically at a platinum anode in an acidified sodium bromide solution reacts with the phenol according to the equation

$$C_6H_5OH + 3Br_2 = C_6H_2Br_3OH + 3H^+ + 3Br^-$$

What weight of phenol corresponds to the flow of a current of 1 ma. for 1 sec.?

CHAPTER 8

THE ABSORPTION OF VISIBLE AND ULTRAVIOLET LIGHT

8-1. Introduction. In the preceding chapters we have discussed a number of instrumental techniques which depend on the measurement of the electrochemical properties of a solution. In this and the following chapter, we shall describe some ways in which the composition of a sample can be studied by measuring its ability to absorb light. Methods which depend on the absorption of light[1] are called *absorptimetric;* for historical reasons which are no longer entirely valid they are divided into two classes. *Colorimetric* methods are those in which one measures the extent to which radiation comprised within a single band of wavelengths is absorbed in passing through a sample. This band may be so broad as to include the entire visible portion of the spectrum (*i.e.*, white light), and the intensity of the transmitted beam may be estimated either visually (*visual colorimetry*) or by a photoelectric device. On the other hand, some "colorimeters" employ filters to permit the observation of the extent to which the sample absorbs radiation within a comparatively narrow band of wavelengths; we shall refer to such instruments as "filter photometers." A *spectrophotometer* (or *spectrometer*) is an instrument which includes provisions for *continuously* varying the central wavelength of the band being employed, and is therefore well adapted for studying the way in which the absorption by the sample varies with wavelength. However, many practical analyses are nowadays made by using a spectrophotometer to carry out a measurement at a single predetermined wavelength setting; on the other hand, by the successive use of a number of filters it is possible to secure absorption–wavelength curves with a filter photometer, so that the distinction between colorimetry and spectrophotometry is no longer of much value.

The regions of the electromagnetic spectrum most commonly employed in absorptimetric measurements are the ultraviolet, visible, and infrared.[2] The physical laws which govern the absorption of radiant energy in these

[1] More properly, electromagnetic radiation; it will very shortly become apparent that these methods are by no means confined to the visible portion of the spectrum.

[2] However, both X-ray and microwave absorption techniques are rapidly increasing in popularity.

regions, as well as the instruments and experimental procedures used to measure this absorption, are very closely related. However, they do differ in many details, and so it will be convenient to consider ultraviolet and visible absorptimetry together in this chapter and to discuss infrared measurements separately in Chap. 9.

8-2. The Nature of Light and Its Interaction with Matter. Light is a form of radiant energy, that is, energy which is propagated as an electromagnetic wave. Such waves may be described in terms of either their wavelength (which is the distance between two successive corresponding points on the wave train) or their frequency (the number of complete waves which pass a fixed point in unit time). The wavelength λ (in cm.) and the frequency ν (in sec.$^{-1}$) are related to the speed of propagation c (in cm. per sec.) of the wave in any given medium by the equation

$$\lambda\nu = c \tag{8-1}$$

In a vacuum c is 2.99776×10^{10} cm. per sec., but in any material medium it is smaller than this and is given by

$$nc = 2.99776 \times 10^{10}$$

where n is the index of refraction of the medium. Light is the electromagnetic radiation which falls within a certain more or less arbitrary range of wavelengths or frequencies. Its relation to the other kinds of electromagnetic radiation is illustrated by the "electromagnetic spectrum" represented in Fig. 8-1.

Chemists most frequently express wavelengths in microns ($1\ \mu = 10^{-6}$ m. = 0.001 mm.) when referring to the infrared portion of the spectrum, and in millimicrons ($1\ \text{m}\mu = 10^{-9}$ m. $= 0.001\ \mu$) when referring to the visible and ultraviolet. The angstrom unit ($1\ \text{Å} = 10^{-10}$ m. $= 0.1\ \text{m}\mu$) is rarely used in analytical absorptimetry. Frequencies are always expressed in cycles per second. Especially in the interpretation of infrared spectra, it is often convenient to speak of the *wavenumber*, which is the number of waves per centimeter in the wave train. This is equal to ν/c, where c is the speed of light *in vacuo*. Wavenumbers are expressed in reciprocal centimeters (cm.$^{-1}$), called kaysers.

The portion of the electromagnetic spectrum used in absorptimetric measurements is customarily and somewhat arbitrarily divided into the ultraviolet, which extends from about 100 to about 400 mμ; the visible, which extends from about 400 to about 800 mμ; and the infrared, which extends from about 800 mμ to 400 μ. Wavelengths in the "vacuum ultraviolet" portion of the spectrum (so called because air itself absorbs in this region) below about 200 mμ are only beginning to be used in analytical work, partly because they have not been accessible with commercial spectrophotometers heretofore available. The same is true of the far infrared, beyond 25 μ. We shall therefore usually speak of the ultraviolet and the infrared as meaning the portions of the spectrum between 200 and 400 mμ and between 0.8 and 25 μ, respectively.

There are two properties of light which are of interest in absorptimetry. One is its quality or kind, which is described by its wavelength or frequency; the other is its quantity or amount, which is usually referred to as its intensity. It is the first of these which governs the extent to which a beam of light is capable of interacting with any particular kind of matter; by measuring the intensity of the light transmitted by the

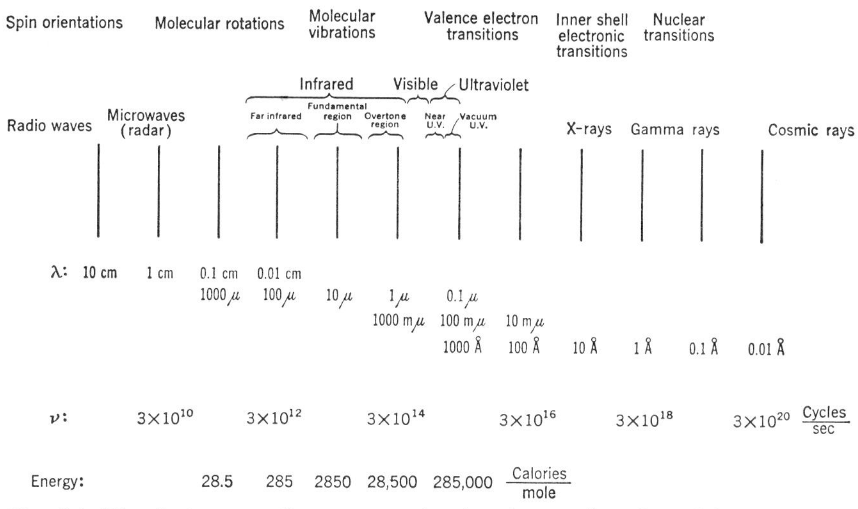

FIG. 8-1. The electromagnetic spectrum, showing the wavelengths and frequencies of various kinds of radiation, the energies represented by radiation of various frequencies, and the physical processes responsible for absorption in different regions.

sample and comparing this with the intensity of the light with which the sample was illuminated, we can secure a quantitative measure of the extent of the interaction between the sample and the energy contained in the light beam.

In considering the ways in which light and matter can interact, it is convenient to regard the energy contained in a beam of light as being divided into discrete amounts contained in units called photons or quanta. The energy content of a photon depends on the frequency of the radiation, and is given by the equation

$$E = h\nu \tag{8-2}$$

where E is the energy content of a photon in ergs, ν is the frequency in cycles per second, and h is Planck's constant, 6.624×10^{-27} erg-sec. It follows that radiation of any one wavelength or frequency[1] is composed of photons having exactly equal energy contents. The intensity, or radiant power, of a beam of radiation is proportional to the number of photons which in unit time pass through a plane of unit area perpendicular to the direction of the beam.

According to the quantum theory, radiation is absorbed by matter only when the energy content of the photon corresponds to some energy requirement of the substance with which it comes in contact. These requirements depend on the electronic structures of the molecules being irradiated. The total energy of a molecule may be considered to be the sum of its translational, electronic, rotational, and vibrational energies.

The first of these, the translational or kinetic energy, is of very little interest in absorptimetric theory, because the selection rules of quantum mechanics forbid the direct interchange of energy between an oscillating electromagnetic field and the translational energy component of the total energy of a molecule. This is to say that the absorption of electromagnetic radiation can only accompany a change in the molecule which involves a change of electrical moment, such as a change of the electronic distribution (*i.e.*, the electronic energy), the separation between two nuclei (the vibrational energy), or the rotation of a dipole (the rotational energy). Of course, to take one of these as an example, a vibration once initiated does not continue forever; the energy originally gained is eventually lost by radiation or by collision with another molecule, and thus it is finally transformed into kinetic or thermal energy. So the end result of the absorption of light by a molecule is the liberation of an equivalent amount of heat. But (except in fluorimetry, which is beyond

[1] From Eqs. (8-1) and (8-2) it is evident that the energy of a photon is defined unambiguously by stating the frequency of the radiation, but not by stating its wavelength; in the latter case we must also know the speed of light in the medium involved. In other words, any particular energy content always corresponds to radiation of the same frequency, but the corresponding wavelength varies from one medium (*e.g.*, water, *n*-hexane, carbon tetrachloride) to another. Since it is the energy of the photon which determines the extent of interaction between matter and a beam of radiation, it is apparent that frequency is really far more fundamental than wavelength. Workers in the field of infrared absorptimetry have had considerable success in their attempts to discover the specific atomic and molecular processes responsible for absorption at various particular frequencies, and in order to correlate such data they find it both necessary and desirable to refer to the frequency at which absorption occurs. The corresponding attempts to interpret visible and ultraviolet spectra have as yet been much less successful, and this explains, though it may not excuse, the habit of speaking of, for example, a wavelength of maximum absorption in either of these regions. To avoid misleading the student with regard to current usage, we shall usually speak in this chapter of the wavelength at which a phenomenon occurs, but in Chap. 9 we shall speak of its frequency instead.

the scope of this text) we are not concerned with the final fate of the energy which is absorbed; all that interests us is the process responsible for the original absorption.

Each of the remaining components of the total energy of a molecule can have only certain definite values called energy levels. A molecule whose electronic, rotational, and vibrational energies are all at their lowest values is said to be in the ground state. In this state it is capable of absorbing energy, but only in certain definite amounts. When a molecule is irradiated by photons whose energies just correspond to the difference in energy between the ground state and some higher or excited state of the molecule, absorption occurs and the molecule is raised to the higher energy level. On the other hand, if the energy of a photon impinging on a molecule differs appreciably from the energy separation between the ground state and any excited state of the molecule, no absorption can occur. These ideas are summarized briefly by saying that the electronic, rotational, and vibrational energies of a molecule are *quantized;* they can assume only certain definite discrete values.

The space available here prohibits a more extensive treatment of the very important but rather intricate subject of permissible energy levels. For further information on this score the advanced student will do well to consult a treatise devoted to quantum theory.

It follows from these considerations that a sample exposed to heterochromatic radiation should absorb only at those sharply defined frequencies (or wavelengths) which correspond to the permissible energy transitions within the molecule. But in reality the molecules whose electronic energy levels are being raised by the incident radiation are not all in the same rotational and vibrational energy states. To effect a given electronic transition in a molecule which is in, say, an excited vibrational state requires a slightly but definitely different amount of energy from that which would be required if the molecule were in its vibrational ground state. This phenomenon is responsible for the fact that the absorption spectra even of gases at very low pressures consist of extremely narrow bands, rather than of lines as one might perhaps expect. Moreover, these very narrow bands are greatly increased in width by increasing the pressure, which is to say the magnitude of the forces existing between the molecules. This is known as "pressure broadening," and it is also, of course, evident in the spectra of substances in the condensed (liquid or solid) states. The intermolecular forces existing in solutions are not confined to the molecules of solute alone, for there are also intermolecular forces between the molecules of the solute and those of the solvent. Indeed, the latter are much larger except in fairly concentrated solutions. Consequently the absorption spectra of solutions, as well as of pure solids and liquids, always consist of more or less broad bands instead of sharp lines. These bands represent the absorption of frequencies lying outside the very narrow frequency range which corresponds to the amount of energy required to bring about the same energy transition if intermolecular forces were completely absent. We shall refer to these considerations again in discussing the infrared spectrophotometry of gaseous samples (Sec. 9-6).

The energy levels which correspond to electronic transitions in a molecule are much farther apart than those which represent rotational or vibrational transitions. By virtue of Eqs. (8-1) and (8-2), therefore, electronic transitions can be brought about only by radiation in the visible and ultraviolet portions of the spectrum, whereas infrared radiation suffices only to bring about transitions from one rotational or vibrational energy level to another. Whether or not any electronic transition in a molecule can actually be brought about by the amount of energy corresponding to a photon in the visible or near ultraviolet[1] depends, of course, on the electronic structure of the molecule in its ground state. For example, the electronic structure of a CH_3 group is such that its lowest permissible excited energy level could be attained only by the absorption of much more energy than is represented by a photon at even 200 mμ, and as a result of this fact all saturated hydrocarbons are optically transparent throughout the visible and near ultraviolet. The same thing is true of such other organic compounds as alcohols and ethers, and also of such inorganic ions as the halides, alkalies, and alkaline earths.

8-3. Constitution and Selective Absorption. In general, the absorption of visible or ultraviolet radiation by an organic compound can occur only when there is some locus of unsaturation in the molecule. For example, ethylene and acetylene give absorption bands at about 190 and 170 mμ, respectively. (Each also gives other bands at still shorter wavelengths.) A compound which contains two ethylenic or acetylenic bonds well separated from each other, as does 1,5-hexadiene, $CH_2{=}CH{-}CH_2{-}CH_2{-}CH{=}CH_2$, behaves chemically just as though the two unsaturated portions of the molecule were completely independent of each other. Treatment with bromine, for instance, results in 1,2 addition across each of the double bonds; in this respect 1,5-hexadiene will behave just as does propene, except that of course each molecule of the diene is capable of adding twice as much bromine as will react with a molecule of propene. In the same way, the optical properties of the diene will be very closely similar to those of propene: the two compounds will give absorption bands at almost exactly the same wavelength,

[1] The term "near ultraviolet" is used here to emphasize the fact that the portion of the ultraviolet between 100 and 200 mμ is being specifically excluded from the discussion.

although at equal concentrations the diene will show the stronger absorption[1] because it contains two double bonds instead of only one.

On the other hand, quite different behavior is exhibited by a compound containing two or more *conjugated* double bonds. Thus, 1,3-butadiene, $CH_2{=}CH{-}CH{=}CH_2$, undergoes 1,4 rather than 1,2 addition on treatment with bromine, and in general its chemical and physical properties are distinctly different from those which could be predicted by assuming that the two double bonds behaved quite independently. In short, the existence of a conjugated system of alternate double and single bonds in a molecule has a profound effect on its electronic structure. It is no surprise, therefore, to find that conjugation also has a considerable effect on the optical properties of a molecule. Although ethylene has an absorption band at 193 mμ, 1,3-butadiene gives one about twice as strong at 217 mμ, and we should expect 1,3,5-hexatriene to give a still stronger absorption band at a still longer wavelength. The effect of the conjugation, in other words, is to decrease the energy difference between the ground state and the lowest permissible excited state, and this difference is progressively lessened by an increase in the length of the conjugated chain. An interesting example of the cumulative effect on the absorption spectrum of a molecule resulting from an increase in the length of a conjugated system is provided by the data in Table 8-1. The absorptivity values quoted in this table are probably not very exact, but their general trend is clearly evident.

TABLE 8-1. PRINCIPAL ULTRAVIOLET ABSORPTION BANDS OF THE DIPHENYLPOLYENES*

n	$\lambda_{max.}$, mμ	Relative absorptivity
1	306	1.0
2	334	1.6
3	358	3.1
4	384	3.5
5	403	3.9
6	420	4.7
7	435	5.5

* The first column of the table gives the value of n in the formula $C_6H_5(CH{=}CH)_nC_6H_5$; the second gives the wavelength at the maximum of the principal absorption band; and the third gives the ratio of the absorptivity at that wavelength to the absorptivity of the parent compound, 1,2-diphenylethylene ($n = 1$), at its wavelength of maximum absorption. The data refer to solutions in benzene, and were secured by Hausser, Kuhn, and Smakula, *Z. physik. Chem.*, **B29**, 384 (1935).

[1] Anticipating the quantitative terminology to be introduced in Sec. 8-4, we may say that the molar absorptivity of the diene will be approximately twice that of propene.

Of course there are many other unsaturated groups which, when introduced into a nonabsorbing molecule (such as a saturated hydrocarbon, an alcohol, or an ether), will produce a compound which does absorb visible or ultraviolet radiation. Such a group is known as a *chromophore* or *chromophoric group.* The common chromophoric groups include, in addition to carbon–carbon double and triple bonds, the carbonyl, carboxyl, amido, azo, nitrile, nitroso, nitro, and thiocarbonyl groups. In general, any molecule containing one such group will show an absorption band somewhere in the visible or ultraviolet portion of the spectrum, and both the position and the height of the band will be characteristic of the chromophoric group which is present. This means that, for example, the absorption bands of a series of compounds like acetic, propanoic, and butanoic acids will be very nearly identical. Because of this fact, ultraviolet and visible spectrophotometry are of great use in differentiating among various types of compounds—between benzene and anthracene derivatives, for example, or between nitro and nitroso compounds.

A molecule which contains two or more chromophoric groups, provided that these are well isolated from each other, will show the absorption band characteristic of each group. Thus, ω-nitrohexanoic acid would be expected to have an absorption spectrum which is merely the sum of the spectra of nitromethane and acetic acid, for the nitro and carboxyl groups are so far removed from each other that neither will have any significant effect on the electronic structure of the other.

However, when a new chromophoric group is introduced into a molecule in such a position that it is conjugated with another such group already present, the electronic structures of both groups may be considerably altered from their original states. In that case the absorption of the resulting compound will differ greatly from a simple summation of the absorptions of the two isolated chromophores.

For example, benzene gives one major absorption band at 198 mμ followed by a highly characteristic closely spaced series of peaks centered around a wavelength of about 255 mμ. These peaks reflect the vibrational motions of the nuclei in the benzene ring, and they appear not only in the spectrum of benzene itself but also in the spectra of its derivatives. However, the spectrum of a benzene derivative in which the substituting group is chromophoric is not at all a mere summation of the absorption curves of benzene and of the isolated chromophoric substituent. The characteristic absorption band of the nitro group occurs at 271 mμ in a compound such as nitromethane. But the spectrum of nitrobenzene shows absorption at 330 mμ rather than at 271 mμ because of the mutual interaction between the electronic structure of the nitro group and that of the aromatic ring which is conjugated with it. At the same time the absorption band which occurs at 198 mμ in benzene is replaced by one at 252 mμ, and the fine-structure band appears at about 280 mμ rather than at 255 mμ.

It should be apparent from this discussion that many organic compounds have rather characteristic absorption spectra in the visible and ultraviolet. However, since the absorption in these regions is due to the electronic structures of certain parts of the molecule, any change which does not affect the chromophoric group(s) will have little or no effect on the visible and ultraviolet absorption. For example, the spectra of toluene and ethylbenzene are virtually identical, and they are very similar to the spectra of benzene and the xylenes. Visible and ultraviolet absorptimetry have been, and unquestionably will continue to be, of enormous use not only in the identification of organic compounds by comparing their spectra with those of similar compounds of known constitution, but also in the solution of many problems of molecular structure. For a further discussion of these fascinating and important applications of spectrophotometric techniques, the student is referred to the excellent monograph of Gillam and Stern.[1]

In the inorganic field, it is found that the ions which absorb in the visible or ultraviolet are those which have electronic structures such that a transition from one energy level to another can be brought about by the absorption of only the relatively small amount of energy represented by a photon in the visible or ultraviolet. These include ions like chromate, nitrate, and permanganate and also the ions of the rare earths (the lanthanides and actinides) and the transition elements. The fact that an ion does not absorb in the visible portion of the spectrum, and is therefore colorless to the eye, naturally does not prove that it does not absorb in the ultraviolet. Lead(II), bismuth(III), and thallium(I) in hydrochloric acid solutions are all transparent in the visible, but each has an absorption band in the ultraviolet which can be turned to good analytical use.

It goes without saying that absorptimetric techniques can be used directly for the determination of any substance which is responsible for the absorption of either visible or ultraviolet radiation. In the latter case, of course, visual colorimetry is useless, and an ultraviolet spectrophotometer becomes a necessity in spite of its rather substantial cost. On the other hand, it must not be inferred that absorptimetry is impossible to apply to the determination of a substance which does not absorb in either the visible or the ultraviolet. For dealing with such materials we actually have two very powerful absorptimetric techniques. One involves the formation of an absorbing compound between the substance

[1] A. E. Gillam and E. S. Stern, "An Introduction to Electronic Absorption Spectroscopy in Organic Chemistry," Edward Arnold & Co., London, 1954; especially chaps. 11, 14, and 15.

to be determined and some reagent (usually organic); several examples of this will be found in the experiments at the end of this text. The other involves a titration with an absorbing reagent, and will be discussed further in Sec. 8-13. In a few cases it is possible to determine a nonabsorbing substance by taking advantage of a chemical reaction which occurs when it is added to an absorbing solution; fluoride is often determined by measuring the extent to which a yellow solution of the peroxytitanium(IV) complex is bleached (because of the formation of a colorless fluotitanate complex) by addition of the unknown.

8-4. The Laws of Absorption. We must now turn to a consideration of the physical laws which govern the extent to which the intensity of a beam of radiation is decreased on passing through a sample. It is probably fair to say that the effects of the thickness and concentration of a solution upon the intensity of a beam of light passing through it are, qualitatively at least, intuitively obvious even to the person who puts these effects to use only in estimating the strength of his morning cup of coffee. It is therefore no surprise to find that the quantitative expressions of these relationships were among the earliest formulated scientific laws, and that the use of these laws for the purposes of quantitative analysis antedated the development of almost every other instrumental technique of analytical chemistry.

The first of these laws expresses the relationship between the thickness of an absorbing substance and the fraction of the light incident upon it which will be absorbed. This law was first formulated by Bouguer in 1729, but it is often attributed to Lambert, who restated it in 1768. If P_0 is the intensity (or, more properly, the *radiant power*) of a beam of light incident upon a sample b cm. thick, the Bouguer-Lambert law states that the radiant power P of the beam transmitted by the sample will be given by the equation

$$\log \frac{P}{P_0} = -k_1 b$$

which may also be written in the form

$$P = P_0 10^{-k_1 b} \tag{8-3}$$

where k_1 is a numerical constant which depends on such experimental conditions as the concentration of the solution, the identity of the absorbing material, the frequency of the light, and so on.

The second fundamental law of absorptimetry was formulated independently by Beer and Bernard in 1852, and is now usually attributed to Beer. This law relates the intensities of the light beams incident upon and transmitted by a solution of fixed thickness to the concentration of the solution:

$$\log \frac{P}{P_0} = -k_2c$$

or

$$P = P_0 10^{-k_2c} \tag{8-4}$$

Equations (8-3) and (8-4) may be combined to give one expression of what is variously known as the Beer-Bouguer-Lambert law, the Beer-Lambert law, or, most simply, Beer's law:

$$\log \frac{P}{P_0} = -abc$$

or

$$P = P_0 10^{-abc} \tag{8-5}$$

which is the fundamental equation of quantitative absorptimetry. The terms employed in these algebraic statements of Beer's law, and certain combinations of these terms which are employed for the sake of convenience, have in the past been given many different names and expressed in many different units. In this chapter and the next we shall follow the recommendations made in 1952 by the Joint Committee on Nomenclature in Applied Spectroscopy Established by the Society for Applied Spectroscopy and the American Society for Testing Materials.[1]

According to these recommendations, P_0 is called the *radiant power* of the beam striking the sample, and P is the radiant power of the beam transmitted by the sample. These quantities may be expressed in any convenient arbitrary units, for all practical absorptimetric measurements depend on the ratio of P and P_0 rather than on the absolute value of either. The symbol b represents the thickness of the absorbing sample, which is always expressed in centimeters, while c represents the concentration of the absorbing constituent of the sample. The quantity a, whose value depends on the identity of the absorbing species, the frequency of the light, the nature of the solvent, the temperature, etc., is called the *absorptivity*. When c is expressed in moles per liter, the corresponding value of a is called the *molar absorptivity* and is given the symbol ϵ. Occasionally it may be convenient to express c in different units (such as micrograms per milliliter or grams per liter), and this naturally necessitates the use of an appropriately different numerical value of a.

The ratio of P to P_0, which is the fraction of the incident radiant power that is transmitted by the sample, is called the *transmittance* and given the symbol T:

$$T = \frac{P}{P_0} \tag{8-6}$$

[1] H. K. Hughes (chairman) *et al.*, *Anal. Chem.*, **24**, 1349 (1952).

Another way of expressing the relative intensities or radiant powers of the incident and transmitted beams is provided by the equation

$$A = -\log T = \log \frac{P_o}{P} = abc \tag{8-7}$$

where A is known as the *absorbance.* It may be noted that, unlike the transmittance, the absorbance is directly proportional to the concentration of the absorbing species.

Beer's law may be regarded as a limiting law in the sense that it describes an idealized behavior toward strictly monochromatic radiation, *i.e.*, radiation which consists of but a single frequency or wavelength.[1] In reality, however, perfectly monochromatic light is impossible to secure. It is therefore found that experimental data obey Beer's law more and more closely as the width of the band of frequencies with which the sample is illuminated is decreased. This phenomenon will be examined more closely in Sec. 8-7.

One serious problem in quantitative absorptimetric measurements has to do with the definition and measurement of P_0. We have defined this as the radiant power of the beam striking the sample itself, but the quantity thus defined is impossible to measure in work with liquids or solutions. Consider a beam of radiation which is incident on one wall of a glass cell containing a solution. Some loss of radiant power results from reflection at the air–glass interface, and a further loss results from reflection at the glass–solution interface. Then the radiation passes through the sample, but before it can emerge from the cell and impinge upon a detector it must undergo two additional reflection losses at the rear wall of the cell. In an attempt to compensate for these losses, it is customary to compare the intensity of a beam which has passed through a sample with that of a beam which has passed through either the pure solvent or a suitable "blank" solution contained in a cell as nearly similar as possible to the cell containing the sample. For this reason, such quantities as the transmittance and absorbance which involve P_0 explicitly, as well as such other quantities as the absorptivity which involve P_0 implicitly, are understood to be referred to a value of P_0 which is determined by measuring the intensity of the beam after passage through a reference cell.

The total absorbance of a sample which contains two or more absorbing species is given by an equation of the form

$$\begin{aligned} A_t &= A_1 + A_2 + \cdots + A_n \\ &= b(a_1c_1 + a_2c_2 + \cdots + a_nc_n) \end{aligned}$$

[1] Moreover, we have neglected a correction term which involves the index of refraction of the sample, but this term is so little dependent on concentration that it may be ignored except in extremely rare instances.

where A_t is the total absorbance of the sample at some wavelength and A_1, A_2, etc., are the respective contributions to the absorbance by each substance present. As we shall see in Sec. 8-10, this is of considerable importance in the quantitative spectrophotometric analysis of mixtures.

8-5. Theory of Visual Colorimetry. Visual colorimetry, which is the simplest kind of absorptimetric procedure, consists of using the eye to compare the intensities of two beams of light. There are several different experimental techniques of visual colorimetry, but all are based on the same principle. Two solutions—one the unknown, the other a "reference" solution of the same colored substance at a known concentration—are illuminated with light beams of equal intensity. The beams emerging from the two solutions are compared visually, and the conditions of the "measurement" are so adjusted that no difference can be perceived between the intensities of these two beams.

Under these conditions, if we assume that Beer's law is obeyed under the conditions of the experiment, the intensity or radiant power of the beam transmitted by the unknown solution will be given by the equation

$$P_u = P_0 10^{-ab_uc_u}$$

and the radiant power of the beam transmitted by the reference solution will be

$$P_r = P_0 10^{-ab_rc_r}$$

Since P_0 is the same in these two equations, because the two solutions are equally intensely illuminated, it follows immediately that when P_u and P_r are equal

$$b_uc_u = b_rc_r \tag{8-8}$$

which is the fundamental equation of visual colorimetry. In the simplest techniques for intensity matching—those using Nessler tubes or slide comparators—b_u and b_r are made equal to each other, and the "analysis" consists merely of preparing a solution whose concentration is known and equal to that of the sample. A more complicated but more precise technique involves the use of a Duboscq colorimeter, which is an instrument that permits the variation and measurement of b_u and b_r. Here the value of c_u is found by measuring the ratio b_r/b_u required to achieve intensity matching with a reference solution whose concentration differs from that of the unknown. These techniques will be discussed in more detail in Sec. 8-6.

It may be remarked that this sort of comparison procedure is necessitated by the facts that the eye is incapable of measuring light intensity directly, and that it is also incapable of any accurate estimate of the magnitude of a difference in intensities. (These are the reasons why it is advantageous to use a comparison solution in a titration like that of sodium carbonate to a bicarbonate end point with phenolphthalein as the indicator.) Useful results can be secured only if the eye is used merely to detect a difference between the intensities of two light beams, no attention being paid to the absolute value of the intensity of either beam or to the absolute value of any difference between their intensities.

The ultimate precision which can be attained in analyses by visual colorimetry can be deduced from the fact that, under optimum conditions, the average observer can just detect a difference of about 6 per cent between the radiant powers of two light beams, so that the values of P_u and P_r may be considered to be equal when they actually differ by as much as 6 per cent. If we take this as being the *maximum* difference that will ever be encountered under these optimum conditions, we may reasonably expect that the *average* difference between the actual values of P for a large number of pairs of solutions which appear to the eye to be identical will not be very far from ± 3 per cent. Because of the logarithmic relationship between P and c, the average error in c_u resulting from this cause will be about ± 1.3 per cent. Accordingly this figure represents the best precision attainable in analyses by visual colorimetry.

However, there are many factors which combine to raise the mean error of such analyses from this figure to the ± 3 to 5 per cent which is generally achieved in practice. Among these are the omnipresent chemical and instrumental errors, the fact that the sensitivity of the eye is dependent on the color of the transmitted light, and the extremely insidious effect of visual and mental fatigue.

Moreover, the precision with which two intensities can be matched is greatly dependent on their absolute value, as can be seen from a single rather crude example. It is very easy to see the change produced by adding a second drop of 0.02 M permanganate to a beaker full of water containing one drop of the permanganate. But if two gallon jugs, one filled with 0.02 M and the other with 0.04 M permanganate, are placed side by side, many people will find it very difficult to tell which of the two solutions is more concentrated.

The problem encountered in dealing with very strongly absorbing solutions is not often troublesome, for one can always dilute them or take smaller samples. More important, because it imposes an irreducible lower limit on the range of applicability of visual colorimetry, is the loss of precision resulting from attempts to work with very weakly colored solutions. It is convenient to define a quantity S as the weight of a colored substance which must be present in a column of solution of unit area to produce a color which can barely be detected by comparison with a blank solution. If the amount Q of that substance which is actually present in such a column is greater than about 15 or 20 times S, the precision of intensity matching generally becomes comparable with the ± 6 per cent figure mentioned above. But at lower values of Q the error of matching increases rapidly, and it becomes ± 100 per cent for values of Q smaller than S. Some typical values of S for a number of familiar colored substances are given in Table 8-2. (Both S and Q are commonly

TABLE 8-2. SENSITIVITIES OF DETECTION OF SOME COMMON COLORED SUBSTANCES

Metal and oxidation state	Medium	Color of transmitted light	S, micrograms/cm.^{2}
Ce(IV)	H_2SO_4	Orange	15
Cu(II)	NH_3	Blue	10
Co(II)	KSCN in acetone	Blue	5
$CrO_4^=$	NaOH	Yellow	1
Cu(II)	9 F HCl	Yellow	0.5
Fe(III)	0.3 F KSCN	Red	0.1
MnO_4^-	H_2O	Purple	0.1*

* In terms that may be more readily appreciated, this figure means that a typical observer could just detect the presence of permanganate ion in a 30-cm. thick layer of a 3×10^{-8} M solution.

expressed in micrograms/cm.^{2}; they are the products of a path length in centimeters by a concentration in micrograms/cm.^{3})

In an earlier paragraph we mentioned some experimental techniques of visual colorimetry in which both b_u and b_r and c_u and c_r are made equal in the intensity-matching process. A little thought will reveal that the validity of these techniques does not depend on whether Beer's law is followed or not. On the contrary, all that is necessary is that the absorbance be a monotonic function of concentration, which means simply that any two solutions having different concentrations must also have different absorbances. However, this is not true of the techniques (such as that in which a Duboscq colorimeter is used) in which b_u and b_r are permitted to vary independently. In some published applications of these latter techniques one sees the validity of Beer's law under the particular experimental conditions used merely taken for granted. However, as will become apparent in subsequent sections of this chapter, this is an extremely risky assumption. It is always desirable to demonstrate experimentally the constancy of the product bc under any conditions being contemplated for use in an analytical procedure.

8-6. Techniques of Visual Colorimetry. The simplest technique of intensity matching is that in which a series of known solutions of the colored substance in question is prepared in addition to the unknown solution. The solutions are then transferred to (or they may be prepared in) clear glass tubes, care being taken to ensure that the heights of the columns of solution (*i.e.*, the values of b) are all identical. It is convenient to use a set of Nessler tubes (Fig. 8-2*a*). These are made from optically clear glass of uniform bore, have flat bottoms to eliminate distracting reflections, and are calibrated at one or two points so that solutions can be made up to definite volumes by dilution. Tubes of 50- and 100-ml. capacities are in common use. They are placed in a rack with holes in the bottom through which beams of light, reflected from an evenly illuminated surface, pass upward through the tubes. The tubes are viewed from above. Generally it will be found that the depth of color of the unknown lies between those of two neighboring known solutions instead of matching any of the knowns exactly. Then one may either estimate the unknown concentration by interpolation if the concentrations of the known solutions are sufficiently close together to make the interpolation error satisfactorily small, or prepare more standard solutions until one is found which appears to match the unknown exactly. Obviously the second technique is likely to give better results, but it does so at the expense of a relatively large additional amount of time and effort.

In using such a "standard series" method it is essential that the concentration of the unknown fall within the range covered by the standard solutions, because the eye is quite useless for extrapolation. The permissible difference between the concentrations of successive standard solutions depends, of course, on the accuracy which is desired, but it is customary to have each of these solutions between 25 and 50 per cent more concentrated than the one before it. This means that a relatively large number of standard solutions is needed unless all of the unknowns are about equally concentrated; if the standard solutions are not quite stable over long periods of time, the labor involved in preparing fresh standards for the analysis of each new group of unknowns may become by far the most time-consuming step in the analysis.

For this reason much effort has been expended on the development of artificial standards which will give colors indistinguishable from those of various substances commonly determined, but which are much more permanent than the actual samples. For example, mixtures of potassium chloroplatinate and cobaltous chloride in suitable proportions can be used as standards in the determination of ferric iron in thiocyanate media. This is a problem which is of particular importance in colorimetric pH measurements because many indicator solutions are somewhat unstable on long standing.

The colorimetric determination of pH (Sec. 4-2) is most easily accomplished by the use of a "slide comparator." This consists of a row of

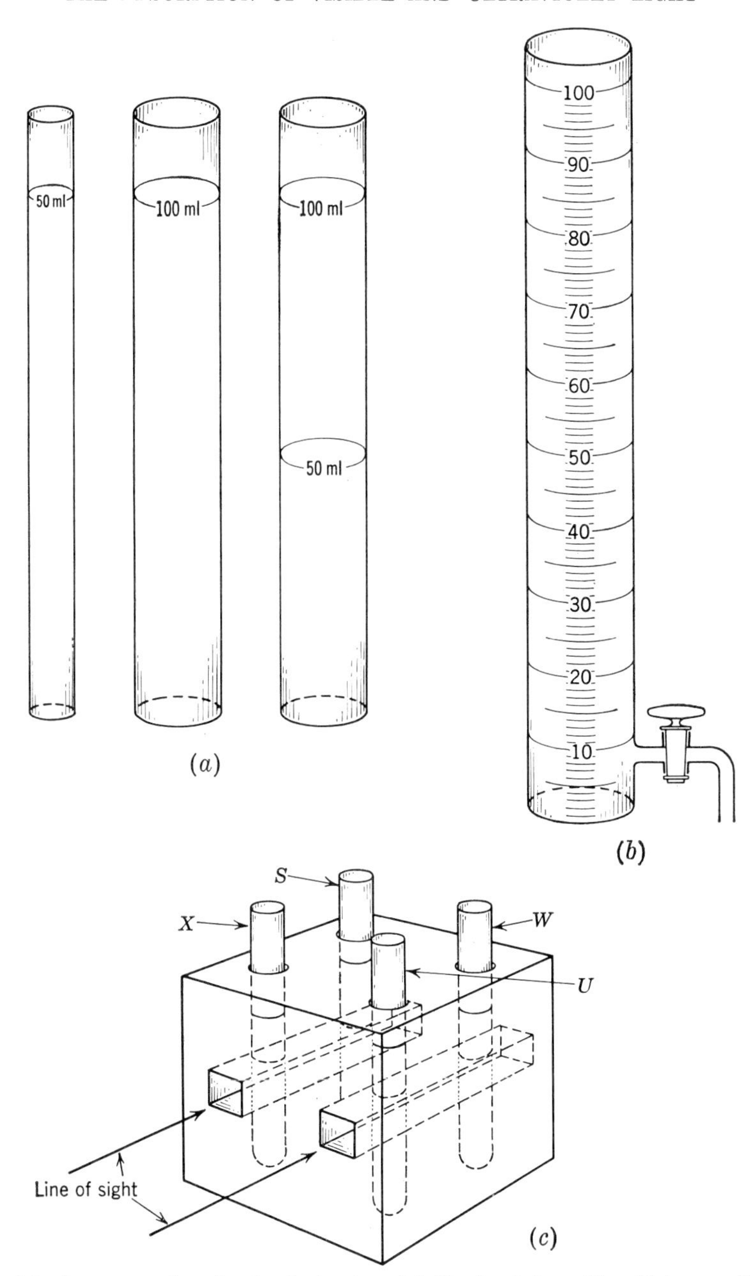

Fig. 8-2. Apparatus for visual colorimetry: (*a*) Nessler tubes, (*b*) a Hehner cylinder, and (*c*) a Walpole or block comparator.

sealed glass tubes, each of which contains a solution whose color is identical with that produced by a definite concentration of an indicator in a solution of known pH, mounted in a rack in the order of the pH values to which they correspond. When the specified concentration of the indicator is added to a colorless unknown and a portion of this mixture is transferred to a similar tube, the pH of the unknown is easily found with fair accuracy by comparison with (and, if necessary, interpolation between) the two standards which it most nearly resembles. A suitable series of standard comparison solutions can be secured commercially for nearly every one of the common acid-base indicators. In some comparators, for the sake of permanence, the standard solutions are replaced by pieces of suitably colored glass.

The need for a large number of standards can be avoided by the use of Hehner cylinders (Fig. 8-2*b*). These are graduated cylinders with uniform bores, optically clear flat bottoms, and stopcocks at the side near the bottom. They are placed in a rack similar to that used in work with Nessler tubes. One is filled with the unknown solution and another is filled with a standard solution of the same colored substance at only very roughly the same concentration. Solution is then withdrawn from one cylinder or the other until the colors appear the same when viewed from above. Then the heights of the two columns are measured (these are obviously proportional to the volumes of solution remaining in the respective cylinders), and the concentration of the unknown is computed from Eq. (8-8).

It is often difficult to decide when two colors compared in this way are exactly matched, largely because one must remember how the first color looked while viewing the second. This source of error can be eliminated by viewing the two solutions simultaneously, which is made possible by the Duboscq colorimeter, whose construction is illustrated by Fig. 8-3. In this instrument two identical cells, whose sides are made of dark-colored glass to reduce stray light, are illuminated equally from below by some suitable source, usually either reflected daylight or a source of diffuse incandescent light. One of the cells is filled with the unknown solution; the other contains a standard solution of roughly the same concentration, say within ± 50 per cent. A carefully ground and polished cylindrical glass plunger is partially immersed in each cell. Each cell can be moved up and down by means of a rack-and-pinion mechanism so that the experimenter can vary the heights of the columns of solutions through which the light beams must pass before entering the plungers. These heights can be measured by a vernier device, and a mechanism is provided so that each height can be adjusted to zero (*i.e.*, so that each plunger just touches the bottom of its cell) when the corresponding vernier is set at zero.

So far these details of construction amount merely to refinements over the Hehner cylinder. However, in the Duboscq colorimeter the two light beams which have passed through the solutions and the corresponding plungers are recombined by a set of mirrors or prisms, and then appear in a split-field eyepiece. This is so arranged that the light which has passed through the right-hand cell illuminates the left-hand half of the circular field of view, and *vice versa*. This produces a sharp line of demarcation between the two halves, which can then easily be brought to equal intensity by adjusting the heights of the columns of solution between the bottoms of the two cells and the corresponding plungers. Once this has been done, Eq. (8-8) can be used to calculate the concentration of the unknown solution. It is safest to repeat the comparison several times, approaching the balance point alternately from above and below by moving one cell while the other remains stationary. This tends to eliminate the effect of any backlash in the gear mechanism, and also to average out the random errors involved in the intensity-matching process. The mean value of b_r/b_u thus obtained is used in the final calculations.

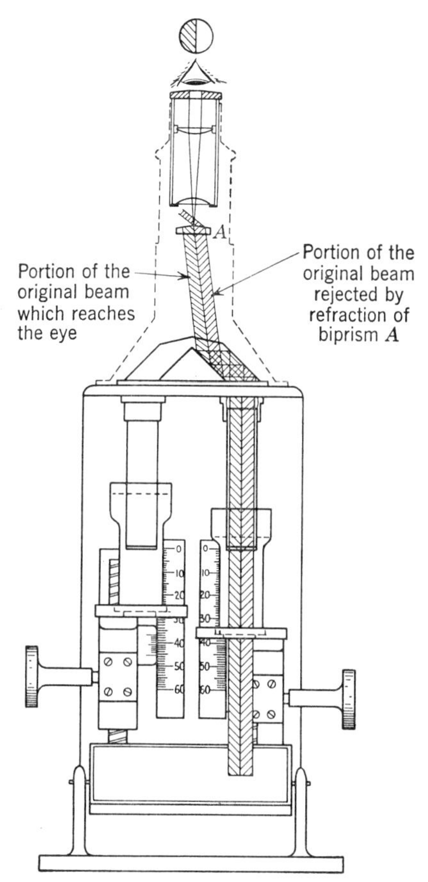

FIG. 8-3. The Duboscq colorimeter. (*Reproduced by permission from H. Diehl and G. F. Smith, "Quantitative Analysis," John Wiley & Sons, Inc., New York, 1952.*)

A still better procedure, designed to reduce errors due to possible small imperfections in the construction of the colorimeter, is recommended in the experiments at the end of this text and merits brief description here. After the light source, mirror, and cells have been adjusted so that the field of view is absolutely uniform and both scales read zero with the plungers touching the bottoms of the empty cells, both cells are filled with the same standard solution. One cell is then set at some arbitrary point, b_r mm. below the bottom of its plunger, where it remains throughout the subsequent manipulations. About ten readings of the position b_s of the other cell at the point of balance are then taken by the procedure described in the preceding paragraph. (Usually it will be found that the mean value of b_s is slightly different from b_r.) The second cell is emptied, rinsed thoroughly, and filled with the unknown solution, and another set of ten readings is made by the same procedure. Calling their mean b_u and the concentration of the standard solution c_r, we have

$$c_u = \frac{b_s}{b_u} c_r$$

Although the limiting accuracy obtainable with a Duboscq colorimeter probably differs very little, if at all, from that which can be secured by the standard series method under optimum conditions, the latter technique is by far the more tedious in practice. This is especially conducive to increased operator fatigue, which, as we have already mentioned, is one of the most dangerous and important sources of error in analyses by visual colorimetry.

One of the most serious disadvantages of visual colorimetry, as compared with some other absorptimetric techniques to be described below, is the difficulty of dealing with samples containing colored substances other than the one being determined. Suppose, for example, that one wished to determine the pH of a weakly acidic cupric sulfate solution, using methyl orange as an indicator. Methyl orange is red in solutions of pH 3.1 or below, yellow at pH 4.4 or above, and orange at pH 4.0. In the presence of a moderate concentration of cupric ion, however, the color change would appear to be from purple at low pH values to green at higher ones, and the intermediate colors would be more or less grayish.[1] These colors would certainly be extremely difficult to match with the colors of comparison buffers containing methyl orange alone.

In such cases, provided that the colors of the impurity and of the substance being determined are completely independent of each other,[2] one may resort to the use of a Walpole or block comparator, such as is shown in Fig. 8-2*c*. The unknown cupric solution alone is placed in cell X, and a comparison buffer of known pH containing some known concentration of methyl orange (but no other colored substance) is placed in cell S. The same concentration of methyl orange is added to another portion of the unknown cupric solution, and the resulting mixture is

[1] This is the same sequence of color changes observed with the familiar mixed indicators prepared by mixing methyl orange with an inert blue dye such as xylene cyanole FF or indigo carmine.

[2] This means that the colored impurity must not appear in any conservation equation which also includes the colored substance being determined.

placed in cell *U*. Cell *W* is filled with distilled water. If the color viewed through *X* and *S* is the same as that viewed through *U* and *W*, the pH of the comparison buffer is the same as that of the sample. Cell *W* serves merely to compensate for reflection losses at the walls of cell *S*.

This technique would serve admirably for the measurement of the pH of a relatively dilute cupric sulfate solution. If, however, the color of the cupric ion were very much more intense than that due to the methyl orange, neither this nor any other visual technique would give results of much value. This is a fairly serious limitation, and because of it visual colorimetry is hardly ever used for the analysis of solutions containing more than one colored substance. Indeed, the advances in photometric instrumentation during the last 20 years or so have so far displaced visual colorimetry from the preeminent position it once held that it is now used for almost none but the simplest of routine analyses.

8-7. Photoelectric and Filter Photometers. Many of the difficulties involved in color matching with the eye can be surmounted by the use of a photoelectric cell for intensity matching. A photoelectric or barrier-layer cell consists essentially of a sheet of copper or iron covered by a semiconducting film of cuprous oxide or elemental selenium, and this in turn is covered by a layer of a second metal, such as lead, platinum, or copper, which is so thin as to be virtually transparent. The transparent or "collector" electrode is electrically connected to the base metal by an external circuit which sometimes consists of nothing more than a microammeter. When the collector electrode is illuminated, a current flows through the external circuit and is measured by the microammeter. If the resistance of the external circuit is sufficiently low, this current will be very nearly proportional to the intensity of the light incident on the cell.

In the simplest kind of photoelectric photometer the beam of light obtained from a suitable source, such as an automobile headlight bulb, passes directly through an absorption cell containing a solution and is focused onto the collector electrode of a barrier-layer photocell. The photocurrent indicated by the microammeter when the absorption cell is filled with pure solvent (or with a blank solution) is proportional to P_0, and the reading obtained when the absorption cell is filled with the solution being studied is proportional to P. If neither the intensity of the source nor the sensitivity of the photocell has changed between the two measurements, the constants of proportionality will be the same, and the ratio of the two meter readings will give the transmittance P/P_0 directly.

Descriptions of these as well as of more elaborate instruments may be found in monographs on absorptimetry. For our purposes it will suffice to note that all photoelectric photometers suffer from one fairly serious disadvantage which may be illustrated by considering the data shown in

Fig. 8-4. This is a plot of transmittance *vs.* wavelength for a 1-cm. thick layer of a 4.8×10^{-4} *F* solution of *m*-nitroaniline-N,N-diacetic acid:

CH_2COOH

—N

CH_2COOH

NO_2

The principal absorption band of this compound in the visible and very near ultraviolet occurs between 340 and 600 mμ; at longer wavelengths

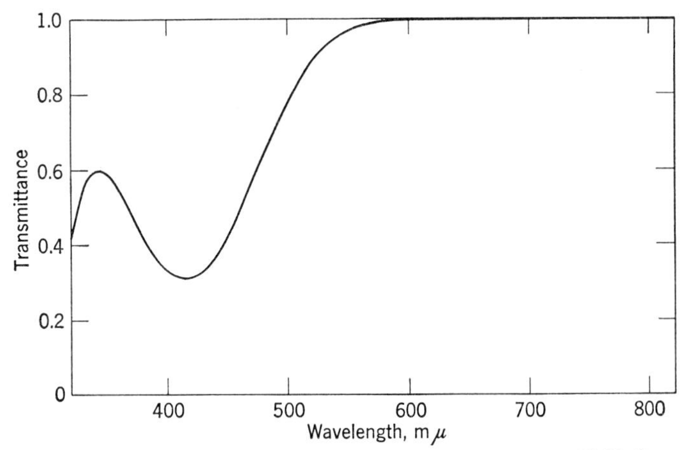

FIG. 8-4. Absorption spectrum of 4.8×10^{-4} *F* *m*-nitroaniline-N,N-diacetic acid, secured with a 1-cm. cell.

the solution is completely transparent. The wavelength of maximum absorption is 415 mμ, and the transmittance at that wavelength is 0.307. This means that the solution transmits only 30.7 per cent of the radiant power of a beam of light consisting of a very narrow band of wavelengths centered around 415 mμ. What is measured by a photoelectric photometer, however, is much more nearly the *average* transmittance over the entire range of wavelengths provided by the source and responded to by the photocell. If the source output and photocell sensitivity are assumed to be limited to the visible portion of the spectrum between 400 and 800 mμ and to be constant throughout that range, the transmittance indicated by a photoelectric photometer will be about 0.8 or 0.9. In other words, this particular solution absorbs only about 10 or 20 per cent of a beam of uniformly distributed white light, compared to its 70 per cent absorption of wavelengths very close to 415 mμ. So, in effect, we have paid for the convenience of using white light by sacrificing four-fifths of the sensitivity we might secure by using nearly monochromatic light. Worse still, if the photocell happened to be insensitive to radiation of wavelengths below about 600 mμ, the indicated transmittance would be unity regardless of the concentration of *m*-nitroaniline-N,N-diacetic acid in the sample. To be sure, this last contingency is not very plausible, but it does illustrate the fact that the loss in sensitivity is greatly aggravated if the substance being determined absorbs in a portion of the spectrum where either the source output or the photocell sensitivity is lower than in some other portion of the spectrum where the transmittance of the sample is high.

Considerations of this sort prompted the development of instruments known as filter photometers or abridged spectrophotometers. These are instruments in which the beam of white light secured from the source is passed through a filter either before or after traversing the absorption cell containing the solution. The filter serves to remove the unwanted wavelengths at which the transmittance of the sample is high, and to transmit only a more or less narrow band of wavelengths including the wavelength of maximum absorption by the sample. To analyze a solution of *m*-nitroaniline-N,N-diacetic acid, for example, we would employ a filter which transmitted light of wavelengths near 415 mμ but absorbed the remainder of the visible portion of the spectrum. Since light having a wavelength of 415 mμ appears blue to the eye, we would be using a blue filter. In the same way, if we wanted to analyze a permanganate solution, which appears purple to the eye because it absorbs light of wavelengths around 525 mμ, we would use a filter which transmitted light in that portion of the spectrum and absorbed elsewhere: such a filter would appear green to the eye.

Except for the introduction of a filter into the light path, the filter photometer is completely similar to the photoelectric photometer.

There are three kinds of filters, of which the most widely used are cutoff filters and bandpass filters. Their respective effects on a beam of white light are illustrated by Fig. 8-5. From this figure it is evident that a single bandpass filter suffices to isolate a band of transmitted wavelengths, whereas two cutoff filters must be used to achieve the same result. Often a pair of cutoff filters can be used together with a bandpass filter, or a pair of bandpass filters may be used, to provide a further reduction in the width of the band of wavelengths which is transmitted. The third type of filter, the "interference" filter, consists essentially of two parallel thin uniform films of metal supported on a glass plate so thin that interference effects result from reflections at the two metal films. Such filters are capable of providing very narrow bands of transmitted radiation.

It is now possible to secure filters or combinations of filters which will serve to isolate almost any desired wavelength with an effective band width (Sec. 8-8) as small as 5 mμ. However, the number of such filters needed to cover the entire visible portion of the spectrum is so large that such a set would be quite expensive. Furthermore, the radiant power which is contained in so narrow a band is so small that the photocells and electrical circuits employed in many photometers would be too insensitive to yield useful results.

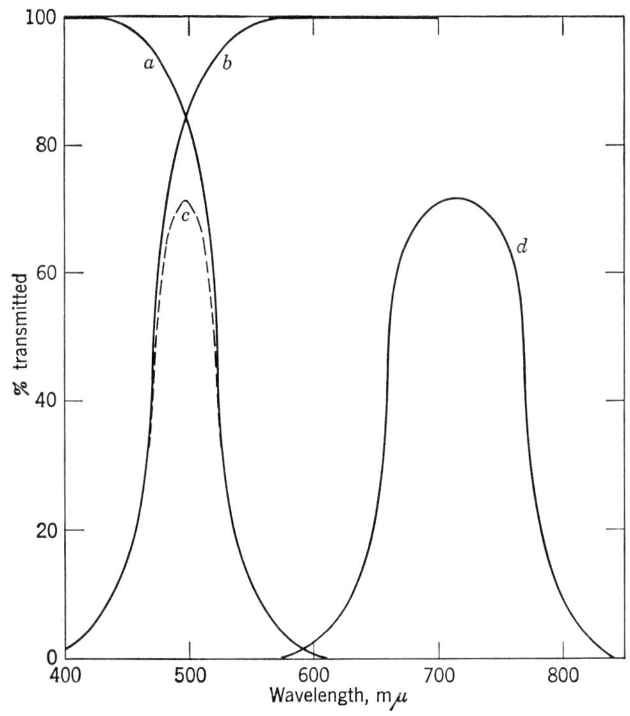

FIG. 8-5. Schematic transmission curves for two cut-off filters (*a*) and (*b*); (*c*) the combination of filters (*a*) and (*b*); (*d*) a bandpass filter.

Therefore the band width ordinarily employed in work with filter photometers is much larger than this; a typical value is as much as 20 or even 35 mμ. The effects of using such wide bands are illustrated by Fig. 8-6, and consist of both a distortion of the transmittance–wavelength curve and an apparent failure of transmittance–concentration data to obey Beer's law. The first of these is merely the result of the "averaging" of the true transmittance values over the range of wavelengths passed by the filter, but the reason for the nonlinear relationship between absorbance and concentration is more subtle. It may be most readily understood by considering the absorbing material to be a succession of thin layers. In passing through the first layer, the more strongly absorbed wavelengths in the band passed by the filter are decreased in intensity more than the other wavelengths. So the radiation striking the second layer will be richer in the less strongly absorbed wavelengths, and therefore the second layer will absorb a smaller fraction of the light incident upon it than did the first. This effect is sufficiently important to deserve a more detailed mathematical analysis.

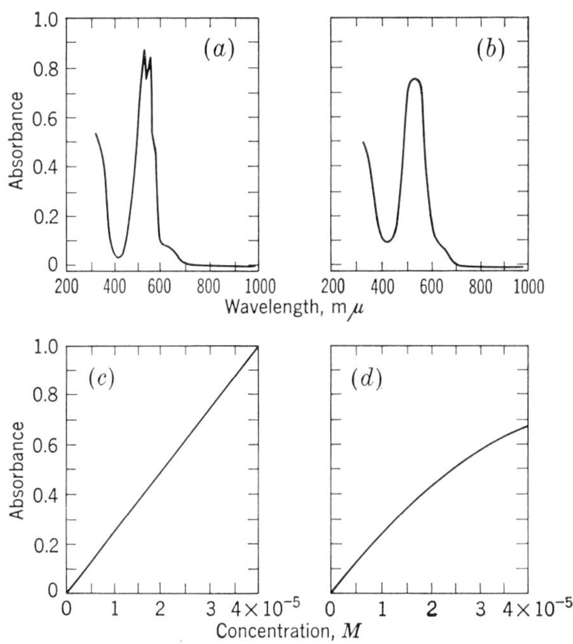

FIG. 8-6. Effects of band width on absorbance–wavelength and absorbance–concentration curves for aqueous potassium permanganate solutions: (*a*) and (*c*) curves obtained with effective band widths of 2 to 3 mμ, (*b*) and (*d*) curves obtained with effective band widths of 30 to 35 mμ.

Let us suppose for the sake of simplicity that the beam of radiation passed by the filter consists of just two wavelengths, λ_1 and λ_2. If f_1 is the fraction of the total radiant power of this beam represented by the wavelength λ_1, then the radiant power at that wavelength will be related to the total radiant power of the light incident upon the sample, P_0, by the equation

$$P_{0,\lambda_1} = f_1 P_0$$

and similarly

$$P_{0,\lambda_2} = f_2 P_0$$

The total radiant power of the beam transmitted by the sample will be

$$P = P_{\lambda_1} + P_{\lambda_2}$$

Assuming that Beer's law is obeyed at each of these wavelengths separately, we have

$$P = P_{0,\lambda_1}10^{-a_1bc} + P_{0,\lambda_2}10^{-a_2bc}$$

where a_1 and a_2 are the absorptivities corresponding to the two wavelengths λ_1 and λ_2. Combining these equations, the absorbance is found to be

$$A = -\log\frac{P}{P_0} = -\log(f_1 10^{-a_1bc} + f_2 10^{-a_2bc})$$

Differentiating this with respect to the concentration c gives

$$\frac{dA}{dc} = \frac{a_1f_1b10^{-a_1bc} + a_2f_2b10^{-a_2bc}}{f_1 10^{-a_1bc} + f_2 10^{-a_2bc}} \tag{8-9}$$

Now, if the absorptivity happens to be the same throughout the range of wavelengths transmitted by the filter, $a_1 = a_2$ and we have simply

$$\frac{dA}{dc} = ab$$

Under these conditions a plot of absorbance *vs.* concentration will be a straight line whose slope is equal to ab.

However, if the absorptivity varies over the range of wavelengths in question, so that a_1 and a_2 are not equal, differentiating Eq. (8-9) again gives

$$\frac{d^2A}{dc^2} = -\frac{2.303f_1f_2b^2(a_1 - a_2)^2 10^{-(a_1+a_2)bc}}{(f_1 10^{-a_1bc} + f_2 10^{-a_2bc})^2} \tag{8-10}$$

Regardless of the relative values of a_1 and a_2, the right-hand side of this equation must always be negative, and so a plot of absorbance *vs.* concentration will be concave toward the concentration axis, as in Fig. 8-6*d*. Evidently the degree of curvature will increase as the difference between a_1 and a_2 increases, because of the $(a_1 - a_2)^2$ term in the numerator. This means that the curvature of an absorbance–concentration plot increases as the width of the band of wavelengths transmitted by the filter increases. Moreover, the shape of such a plot will depend on how nearly the band of wavelengths being used is centered around the wavelength of maximum absorbance for the solution being studied. Suppose, for example, that a filter photometer is used to measure the absorbances of a number of solutions of *m*-nitroaniline-N,N-diacetic acid (Fig. 8-4). If the filter employed transmits only those wavelengths between 400 and 430 mμ, the curvature of the resulting absorbance–concentration plot will be slight, for the absorptivity of this compound does not vary greatly over this range of wavelengths. On the other hand, if the filter transmitted only the wavelengths between, say, 470 and 500 mμ, the much larger variation of the absorptivity over this range would result in a pronounced curvature of the absorbance–concentration plot obtained.

From the same assumptions which were made in the derivation of Eq. (8-10) it is a simple matter to secure an equation for d^2A/db^2. This turns out to be identical with Eq. (8-10) except for a transposition of the b's and c's. Consequently the use of heterochromatic radiation causes the measured absorbance values to deviate from the ideal absorbance–path length relation in exactly the same way that they deviate from the ideal absorbance–concentration relation. It is worth noting that this is the only known cause having any practical importance of a systematic deviation from Eq. (8-3). In Sec. 8-9 we shall discuss a number of other reasons why absorbance–concentration data may seem to fail to obey Beer's law, but in none of these cases is there an accompanying deviation of a plot of absorbance *vs.* path length from strict linearity.

This quite general failure of data secured with filter photometers to obey Beer's law as exactly or over as wide a range of experimental conditions as would be the case if a much smaller band width were used has several consequences in practical work. First of all, such data cannot be used for the direct calculation of the concentration of an unknown solution from its measured absorbance together with the concentration and absorbance of a single known, as could be done from the equation

$$c_u = \frac{A_u b_s c_s}{A_s b_u}$$

if Beer's law were obeyed. This makes it necessary, when using a filter photometer, to construct an empirical plot of absorbance *vs.* concentration from data obtained *with the same instrument* on a series of known solutions covering the entire concentration range of interest. Moreover, in all but the very crudest work it is necessary to construct a new calibration curve for each substance being determined whenever any component of the instrument (*i.e.*, the light source, the filter, or the photocell) is changed, and to check this calibration curve at one or two points occasionally to guard against any unexpected change in the instrumental characteristics. Finally, the fact that the slope of the absorbance–concentration curve decreases with increasing concentration means that the precision which can be attained is considerably poorer at high than at low concentrations.

Primarily because of the necessity for a high degree of electronic amplification of the very small photocurrents which result from the use of very narrow band widths, the spectrophotometer, like the glass electrode pH meter, has come of age only relatively recently. The first modern spectrophotometer was designed by Hardy in 1935, and the immensely popular Beckman Model DU spectrophotometer was designed by Cary and Beckman in 1941. Largely because of the considerations mentioned in the preceding paragraph, the spectrophotometer has since largely displaced the filter photometer, and filter photometers are being used less and less frequently as time goes on.

8-8. Spectrophotometers. The distinguishing feature of a true spectrophotometer is its ability to provide radiant energy of continuously variable wavelength. Along with this, in most instruments, goes the ability to provide a very narrow band of wavelengths in any portion of the spectrum. There are many commercial spectrophotometers available, but all of them have certain essential elements in common. These are:

1. A source of continuous radiant energy
2.[1] An entrance slit, which transmits a sharply defined beam of heterochromatic radiation to
3.[1] A prism or grating, which disperses this radiation to produce a continuous spectrum
4.[1] A means of rotating the prism or grating in order to vary the wavelength corresponding to the center of the beam which strikes
5.[1] An exit slit, which transmits that wavelength, together with a band of wavelengths on either side of it
6. The sample container
7. A receptor or detector, which develops an electrical current proportional to the radiant power of the beam transmitted by the sample
8. An electronic amplifier
9. A means of measuring the amplified photocurrent

We shall first discuss the functions of these components, in the order in which they have been named; then we shall briefly describe the two most common ways in which they are combined in present-day instruments.

The source used for measurements in the visible portion of the spectrum is always an incandescent tungsten filament lamp. However, this has very little emission in the ultraviolet, and so it is replaced by a hydrogen discharge tube for work below about 375 mμ. The latter consists of a small but intense arc between two electrodes in a transparent tube filled with hydrogen gas at low pressure, and it provides continuous radiation from below 200 to about 400 mμ.

Although these two sources together will provide any wavelength between 200 and 1000 mμ (the tungsten lamp can actually be used out to 3 μ in the near infrared), the radiant power which will be secured is not the same at all wavelengths. Since the sensitivity of the detector is also dependent on wavelength, some method must be provided to compensate

[1] These components, together with the lenses and mirrors which collimate and focus the beam, comprise the *monochromator* of the instrument.

for these variations so that the instrument can be set to read 100 per cent transmittance with a nonabsorbing solution in the light path regardless of wavelength. This is usually done by adjusting the widths of the slits in order to alter the width of the band of wavelengths emerging from the exit slit; in a portion of the spectrum where the source output and detector sensitivity are low, the slit widths would be increased. This would naturally result in an increase in the total radiant energy striking the detector, which would counteract the decreasing response of the instrument and tend to return the photocurrent to its original value.

It is usually desirable to ensure that the source output remains constant throughout any single experiment or between successive experiments whose results are to be compared. This is most easily done by regulating the alternating voltage supplied to the lamp or hydrogen discharge tube,[1] or by using a heavy-duty lead storage battery to power the lamp.

It is not our purpose here to discuss in detail the many different optical systems which are used in spectrophotometers; information concerning any specific commercial instrument can be secured from its manufacturer's literature. However, it is necessary to mention certain aspects of these systems which influence the ultimately measured transmittance data.

A beam of monochromatic radiation striking one face of a prism is bent in passing from air to the interior of the prism. The angle between its original direction and the direction in which it passes through the prism depends on the index of refraction of the material from which the prism is made. The index of refraction of any material medium varies with wavelength, and so when a beam of heterochromatic radiation enters the interior of a prism its constituent wavelengths are bent through different angles. The same process is repeated as the beam leaves the prism, and therefore the various wavelengths which entered the prism in the same direction travel in different directions on leaving it. There is thus produced a spectrum in which the shorter wavelengths have been deviated more from their original direction than the longer ones. The extent of this difference in deviations—that is, the angular difference between the positions of any two specified wavelengths in the emergent beam—depends on the *dispersion* of the material from which the prism is made. The dispersion in turn depends on the rate of change of refractive index with respect to frequency or wavelength, $dn/d\nu$ or $dn/d\lambda$.

Different materials have different dispersions. Quartz has a consider-

[1] It will be apparent from the discussion at the end of this section that variations of source output are automatically compensated in the operation of a double-beam instrument, and consequently that no special precautions need be taken to stabilize the source output of such an instrument.

ably smaller dispersion than glass in the visible portion of the spectrum, but quartz (or fused silica) prisms are nevertheless essential for work in the ultraviolet because glass is much less transparent in much of that portion of the spectrum. On the other hand, spectrophotometers (such as the Beckman Model B) which are designed for use only in the visible portion of the spectrum usually employ glass prisms to take advantage of the higher dispersion of glass.

A diffraction grating consists of either a transparent or a reflecting plate whose surface is ruled with a large number of closely and equally spaced fine parallel straight lines. Most gratings used in commercial spectrophotometers are actually replica gratings made by coating an original grating with a thin film of some suitable plastic material, allowing this to solidify, then stripping it off, mounting it on a glass or metal backing, and aluminizing its surface to make it a good reflector. Gratings disperse heterochromatic radiation into its constituent wavelengths by virtue of interference effects.

Although a detailed treatment of the optics of prisms and gratings is not necessary for our purposes, it is essential to compare some of their characteristics. A suitably mounted grating will yield a "normal" spectrum, that is, a spectrum in which the angular difference between the directions of any two wavelengths which differ by a fixed increment of wavelength is constant and independent of the portion of the spectrum in which the two wavelengths lie. Consequently a grating used with fixed slit widths will isolate a band whose width in mμ is the same regardless of the wavelength around which the band is centered. On a frequency scale, however, the band emerging from a grating monochromator will be much narrower at low frequencies (long wavelengths) than at high frequencies, unless this effect is counteracted by varying the slit widths. The reflectance (or transmittance, depending on the type of grating used) of a grating varies relatively little with wavelength, so that the distribution of intensities secured from a grating monochromator is essentially the same as the distribution provided by the source.

Prisms, on the other hand, have markedly nonlinear dispersions, and so the wavelengths in a spectrum produced by a prism will be much more crowded together in the red and near infrared than they are in the violet and ultraviolet. In a prism instrument, therefore, the slit widths which serve to isolate a band only 1 mμ wide at 300 mμ may isolate a band nearly 35 mμ wide at 1000 mμ. Thus the band width obtained when the slit widths are kept constant increases with increasing wavelength, and this is true whether the band width is expressed on a wavelength scale or a frequency scale. For this reason it is essential that the slit widths be adjustable in an instrument employing a prism monochromator, although some inexpensive spectrophotometers using grating monochromators have fixed slit widths. Finally, a prism absorbs a significant fraction of the radiation incident upon it at wavelengths where its dispersion is changing rapidly, and this may lead to an appreciable loss of sensitivity in those portions of the spectrum.

In practice one can obtain from the manufacturer of any spectrophotometer a graph showing how the "spectral slit width" of the instrument varies with wavelength and mechanical slit width. From this it is a simple matter to estimate the width of the band of wavelengths which will be incident on the sample at any combination of instrument settings. The spectral slit width is usually estimated by assuming that the radiation passing through the exit slit has a triangular distribution of intensity with wavelength or frequency, as illustrated by Fig. 8-7;

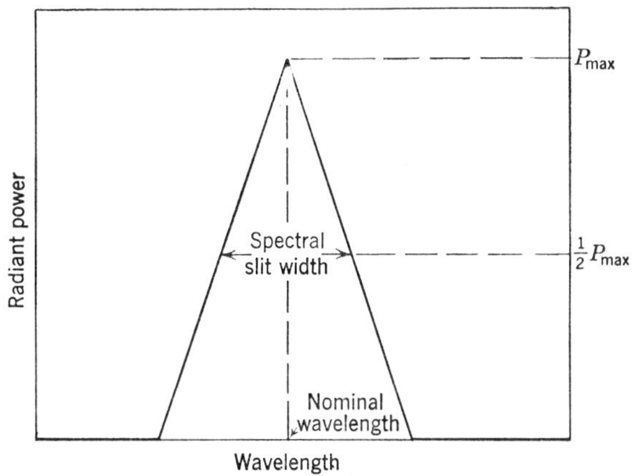

FIG. 8-7. Idealized schematic representation of the band of wavelengths emerging from the exit slit of a monochromator.

it is the width of this distribution curve where the intensity is one-half that at the nominal or central wavelength. This width is also sometimes referred to as the half-intensity or effective band width.

It is important to distinguish among the related terms which refer to the ability of a monochromator to separate adjacent portions of the spectrum. A detailed discussion of these is impossible here, but we shall briefly indicate the significance of each and the broad distinctions among them. A beam of radiation passed through a prism is deviated from its original direction; on being reflected by a grating it is deviated from the path of specular reflection. The extent of either of these deviations is called the *angular deviation.* The difference between the angular deviations of different wavelengths or frequencies is called the *angular dispersion,* and may be expressed as either $d\theta/d\nu$ or $d\theta/d\lambda$. The angular dispersion of a prism depends on the material from which the prism is made, the size of the prism, and the angle of incidence of the beam on the prism; the angular dispersion of a grating depends on the width of the grating lines and the distance between them. If the dispersed spectrum is focused onto a plane surface, such as the plane of the exit slit, the linear distance between the positions of any two wavelengths may be expressed in mm./mμ and referred to as the *linear dispersion*. This will depend not only on the angular dispersion of the prism but also on the distance between the prism and the exit slit and on the focusing employed.

The dispersion of a prism or grating is often confused with the *resolving power* or *resolution* of a monochromator. The resolution is defined as the difference between two frequencies or wavelengths which can just be completely separated by the monochromator. Let us suppose that the light source of a spectrophotometer is removed and replaced by a yellow sodium flame. The radiation produced by such a flame includes two sharply defined wavelengths at 589.0 and 589.6 mμ. Suppose that, after passing through the optics of the instrument, the images of these wavelengths appear as lines whose centers are 0.005 mm. apart at the plane of the exit slit, but that the width of each line is 0.01 mm. We would not even be able to distinguish the lines, let alone measure their intensities. If we increase the linear dispersion fourfold by placing the exit slit four times as far away from the prism or grating, the centers of the lines will be 0.02 mm. apart but each line will now be 0.04 mm. wide; the increase of dispersion will have resulted in no gain of resolution. From this simplified picture it is easy to see that dispersion and resolution are not the same. The resolution depends on the dispersion of the prism or grating, the widths of the slits, and the length of the light path between the prism or grating and the exit slit, but it is also affected by diffraction phenomena. As the slit widths are decreased, the resolution will be improved up to a certain point, but at that point the diffraction pattern produced by the slits becomes so broad that no further improvement of resolution can be produced by a further decrease of slit width. This is the so-called "Rayleigh diffraction limit" of the resolving power of a monochromator. In absorption measurements, because of the limited intensity of the source and sensitivity of the detector, the slit width is always considerably in excess of the Rayleigh diffraction limit. The resolving power is the value of either $\nu/\Delta\nu$ or $\lambda/\Delta\lambda$ (where, for example, ν is the central frequency of the band passing through the exit slit, and $\Delta\nu$ is the resolution at that frequency) which corresponds to the Rayleigh diffraction limit, and is often calculated for the prism or grating in an ideal system. The angular dispersion, resolution, and resolving power of a prism all vary with frequency; but for a grating the angular dispersion and resolving power are constant while the resolution varies with frequency.

Unfortunately, the term "resolution" is often given a different meaning in the absorptimetric literature, such as when it is stated that measurements were made "at the maximum resolution" of a spectrophotometer. This highly empirical phrase means merely that the absorbances were measured at slit widths which were so narrow that no detectable change resulted from decreasing them still further, so that, for example, the "resolution" or separation of two closely adjacent absorption bands could not be bettered by altering the settings of the particular instrument employed.

The absorption cells used in spectrophotometric work are of many different kinds; two popular types are shown in Fig. 8-8.[1] The ideal absorption cell has optical faces which are perfectly flat, transparent, and parallel; they should be completely free from striations, scratches, or other blemishes; and the length of the light path should be accurately known to permit ready interchangeability with other cells in use in the same laboratory. Cells may be made of glass or even plastic (which, however, is all too easily scratched) for work in the visible portion of the spectrum, but Vycor, quartz, or fused silica cells must be used in the ultraviolet. Sometimes it is convenient for rough measurements to use a test tube, a small beaker, or some other similar vessel as an absorption cell. It then becomes necessary to exercise great care in positioning the cell in the instrument, for small variations in the relative positions of the axis of the tube and the center of the light beam, or even a small rotation of a tube made from glass which is not quite optically perfect, may result in substantial changes in the measured absorbance.

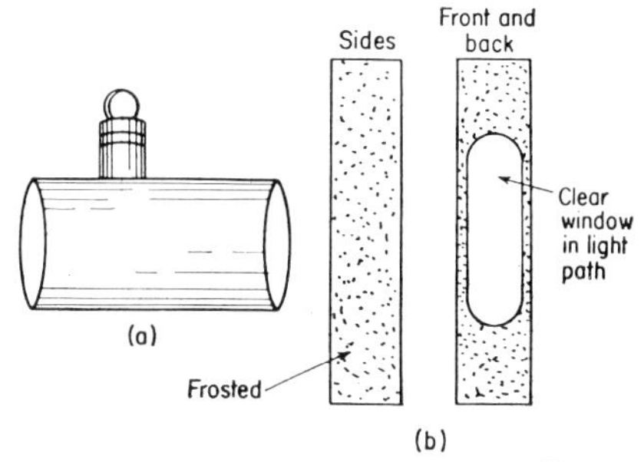

FIG. 8-8. Typical absorption cells.

Instead of employing a barrier-layer photocell (Sec. 8-7) as the detector, as is done in photoelectric and filter photometers, a phototube is almost always used in a spectrophotometer. There are three reasons for this. One is the fact that the low internal resistance of a photocell makes its output current somewhat difficult to amplify. Because the much narrower band widths used in spectrophotometers cause the detector to receive much less radiant power than in a photometer, the output current of a photocell would be too small for convenient measurement. However, this reason is now much less important than it once was,

[1] These illustrations are reproduced through the courtesy of the Beckman Division, Beckman Instruments, Inc., Fullerton, Calif.

because feedback amplifiers have been designed which can easily be adapted to use with photocells. More important is the fact that the output current of a photocell subjected to constant illumination decreases slowly with time. This can be ignored in many routine photometric analyses, but even there it is often a nuisance, and it would certainly be intolerable in really precise work. Finally, the spectral response of the photocells available is virtually limited to the visible portion of the spectrum; they are nearly useless for measurements in the ultraviolet.

A phototube consists of an evacuated glass tube containing two electrodes. One of these, the cathode, is a sheet of metal which is sufficiently large to intercept the entire beam of radiation which has passed through the sample, and which is coated with a layer of a second metal which emits electrons when illuminated. The nature of the coating determines the range of wavelengths over which the tube can be used. A layer of metallic sodium has photoemissive properties which are confined to the region between about 300 and 500 mμ, whereas a thin layer of potassium on a base of massive silver is photoemissive from 200 to 700 mμ. The electrons emitted by the cathode upon illumination are attracted to the anode, which consists of a wire maintained at a moderately positive potential with respect to the cathode. In this way a current, which is proportional to the radiant power of the beam striking the cathode, flows through the tube. It is amplified by a circuit basically very similar to that employed in a glass electrode pH meter, and may eventually be presented for measurement to either a direct-reading meter or a potentiometric circuit.

Unlike barrier-layer photocells, phototubes are virtually free from fatigue and retain nearly constant spectral response throughout their useful lives. It is common practice to use two phototubes to secure high sensitivity over the whole spectral range, and so one "blue-sensitive" tube may be used between 200 and 650 mμ while a "red-sensitive" tube is used from 600 to 1000 mμ.

In some applications, such as "precision colorimetry" (Sec. 8-12), where it is necessary to deal with samples having very high absorbances, even a phototube may be too insensitive to cope with the extremely low levels of radiant power involved. It is then advantageous to use a photomultiplier tube instead. This is a kind of phototube in which the original anode, when bombarded with the electrons emitted by the cathode, emits a larger number of electrons which in turn are attracted to the surface of a third electrode maintained at a still more positive potential. When a number of such electrodes ("dynodes") at successively more positive potentials are properly positioned in an evacuated tube, the photocurrent obtained from the last dynode may be as much as a million times the tiny current secured from a conventional phototube. Photomultiplier tubes are being used more and more frequently even in ordinary spectrophotometric work, where the phototube long reigned supreme, for their much greater sensitivity often makes it possible to use much narrower slit widths than could be used with phototubes and thus to decrease greatly the errors resulting from this cause (Sec. 8-7).

There are two kinds of spectrophotometers: single-beam and double-beam. In a single-beam instrument the beam of radiation emerging from the monochromator passes through a single absorption cell and then strikes the detector. This arrangement makes it necessary to carry out measurements in the following way. First the instrument is set to read zero transmittance (infinite absorbance) with the detector in darkness; this serves to compensate for the small current which flows even when no radiation at all is incident on the phototube cathode, and which is due to the emission of thermal electrons. Next a reference cell containing a nonabsorbing solution is placed in the beam, and the instrument is adjusted to read unit transmittance (zero absorbance) at the wavelength of interest. Finally the reference cell is removed from the beam and replaced by the cell containing the solution whose absorbance is to be measured.

This procedure allows one to measure two photocurrents—one proportional to the radiant power of the beam transmitted by the reference solution, the other proportional to the radiant power of the beam transmitted by the sample. In order for their ratio to be equal to the transmittance of the sample, both the output of the source and the sensitivity of the detector must remain constant between the instant at which the reference transmittance is set at 1 and the instant at which the transmittance of the sample is read off the scale. Of these two quantities, the phototube sensitivity is unlikely to vary appreciably over short periods of time, but the source output is dependent on the voltage supplied to the lamp, and in many localities the voltage of the a-c line is anything but constant. Varying the voltage supplied to an incandescent bulb has two effects. It affects the total radiant power emitted by the bulb, and it also changes the distribution of this radiant power with respect to wavelength. Accordingly a change of the voltage which has only a relatively small effect on the total emission may have a far larger influence on the fraction of that total which is represented by any particular narrow band of wavelengths. Therefore it is customary to stabilize the voltage applied to the lamp, either by using a voltage regulator in the a-c supply line or by operating the lamp from a carefully maintained heavy-duty storage battery.

A double-beam spectrophotometer overcomes this problem in the following way. The beam of radiation emerging from the monochromator is split into two beams, which naturally have identical intensities and spectral distributions. One of these passes through a reference cell containing a nonabsorbing solution; the other passes through a similar cell containing the sample. The ratio of the radiant powers of the beams emerging from the two cells is, of course, wholly unaffected by fluctuations of the source output. Two methods are available for measuring this ratio.

In a double-beam–double-detector instrument the beams emerging from the two cells are directed onto the cathodes of two phototubes or photomultipliers. The outputs of these detectors are connected in series-opposition, and the resulting difference between the two photocurrents is amplified and recorded.[1] One adjustment must be provided to bring the recorder pen to zero when both phototubes are illuminated by beams of the same radiant power, and another to cause a full-scale recorder deflection when the reference phototube is exposed to a beam which has passed through a nonabsorbing solution while the indicating phototube is kept in darkness. Since it is very unlikely that the two phototubes will be equally sensitive at all wavelengths, some provision must also be made for compensating for the differences between their characteristics.

In a double-beam–single-detector instrument the beams of radiation emerging from the two cells are caused to fall onto the same portion of the cathode of a single phototube. Now an opaque disc is rotated rapidly so that it cuts off one beam, then the other. In this way the detector is exposed to radiation whose intensity alternates between P and P_0, and so its output varies with the same frequency as the rate of rotation of the disc. This produces an alternating voltage whose amplitude will be proportional to the difference between P and P_0 provided that the phototube response is linear. In some instruments this alternating voltage is amplified and measured directly. In others the intensity of the reference beam is reduced by a neutral (gray) wedge, a polarizing prism, or some other device until the alternating voltage is just reduced to zero; then the fraction by which the intensity of this beam had to be decreased is determined from the prior calibration of the attenuator. From this, of course, it is a simple matter to compute the transmittance or absorbance of the sample. The computation is usually performed electrically by the instrument so that the recorder will draw either transmittance– or absorbance–wavelength curves directly.

8-9. "Deviations" from Beer's Law. It is sometimes found that the absorbances of a series of standard samples are not proportional to their concentrations, which is to say that the absorptivity varies (or appears to vary) with concentration. For the reasons described in Sec. 8-7, this behavior is always observed when the absorbances are measured with a

[1] Recorded rather than measured; the greatest single advantage of the double-beam design is its much readier adaptability to automatic recording.

filter photometer (unless, of course, the absorption band is so nearly flat over the range of wavelengths transmitted by the filter, or unless the range of concentrations examined is so small, that the variation of the absorptivity is smaller than the probable error of the data), but it is also often encountered in data secured with the most elaborate spectrophotometers. It has a number of possible causes, both instrumental and chemical.

Although the mathematical statement of Beer's law is founded on the assumption that the beam of radiation striking the sample is truly monochromatic, such a beam could never be secured in practice (*cf.* Sec. 8-8). One might perhaps suppose that a more general law of absorption could be derived for heterochromatic radiation. Unfortunately, to be truly rigorous such a law would have to include not only the variation of the absorptivity with wavelength, but also the accompanying changes of the output of the source and the sensitivity of the detector. These quantities would differ from one absorbing substance to another and also from one spectrophotometer to another, and consequently the difficulty of using an expression which would take these factors into account would quite certainly be prohibitively great.

As there appears to be no practical way in which theory can be of much help in this situation, we can only make every effort to secure as nearly monochromatic radiation as the optics of our apparatus permit, and then suffer under any consequences that the remaining heterochromaticity may cause. Instrument manufacturers are working toward the mitigation of these consequences by designing spectrophotometers whose monochromators have increased resolving powers, whose slits can be used at smaller and smaller widths, and whose detectors respond reliably to ever lower radiant powers; all of these permit a nearer and nearer approach to the unattainable ideal.

The practical consequences of using heterochromatic rather than monochromatic radiation were discussed at some length in Sec. 8-7 and were illustrated by Fig. 8-6. They consist of a broadening of the absorption band, a lowering of the absorbance measured at the wavelength of maximum absorption, and a decrease of the absorbance/concentration ratio with increasing concentration. From this it should be evident that in practical measurements the slit widths must be maintained as nearly constant as possible if reproducible absorbance data are to be secured. This is most important when the absorption band of the material being studied displays a sharp maximum at the wavelength of maximum absorption, for then the variation of absorptivity with wavelength in the neighborhood of the maximum is much more pronounced than it is, for example, on the relatively broad band shown in Fig. 8-4. Fortunately, most absorption bands in the visible and ultraviolet are so broad compared to the spectral slit widths obtainable with good modern spectrophotometers that the deviations from Beer's law due to this effect can usually be made smaller than the random errors of the absorbance measurements over relatively wide ranges of concentration.

It might be mentioned in passing that the integrated area under an absorbance–wavelength curve seems to be nearly independent of slit width; the maximum absorbance is decreased by increasing the slit widths, but this is compensated by an increase in the apparent width of the absorption band. Further investigation of this suggestion may eventually provide us with a technique which will yield reliable absorbance data even for very sharp absorption bands. This idea is being rather actively pursued by workers in the field of infrared spectrophotometry, where extremely sharp absorption bands are far more common than they are in the visible and ultraviolet, and where in consequence reliable values of the absorbance at the peak of an absorption band are much more difficult to secure.

In addition to this slit-width effect, there are a number of other instrumental sources of error in absorbance measurements. The errors due to variations in the positions of the absorption cells in the light beam, or to differences among absorption cells which are assumed to be identical and are used interchangeably, are relatively simple to overcome by the exercise of reasonable care. Errors due to nonlinearity of the amplifier or to drifts in the amplifier characteristics are largely beyond the control of the operator, but at least the first of these can be estimated and corrected for if necessary by constructing an empirical calibration curve for the instrument. Errors resulting from irreproducible wavelength settings or from backlash or excessive play in the mechanical assembly of the prism or grating are often very serious, especially when one is working with a very narrow absorption band or when it is necessary to make measurements on the steep side of an absorption band. The errors in the manufacturer's calibration of the wavelength dial are frequently fairly large; they are immaterial in a series of measurements made with a single instrument, but may become important if an absorbance–wavelength curve secured with one spectrophotometer is used to select a wavelength and measure an absorptivity for routine absorbance measurements with another instrument. For this reason it is often desirable to calibrate the wavelength dial, which may be done by substituting a mercury arc for the light source of the spectrophotometer and comparing the readings of the wavelength dial at the successive settings which yield maximum "transmittance" readings with the known wavelengths emitted by the mercury arc. Another problem, resulting from stray light within the instrument, is discussed in Sec. 9-6.

In constructing absorbance–concentration plots from data secured with known solutions, one usually finds that the absorptivity appears to decrease when the concentration exceeds a certain value, which varies from one absorbing substance to another. The graph is a straight line up to this concentration but then it becomes increasingly concave toward the concentration axis. This phenomenon is quite easily distinguished from the continuous curvature which results from the slit-width error previously discussed, and it is observed even when extremely narrow slit widths are employed. In many cases the effect may be due to interactions between the absorbing species and other molecules, either of solvent or solute, present in the solution. In addition, there is a marked change of the refractive index of a solution at an absorption band, and at high concentrations this may be large enough to cause significant deviations from Beer's law. Although other factors may also be involved, the instrumental problems involved in the accurate measurement of high absorbance values have only quite recently been largely overcome, and the question has not yet received the detailed attention it deserves.

There are a number of chemical phenomena which may cause apparent deviations from Beer's law. Naturally any chemical equilibrium which involves the concentration of the absorbing substance being studied will affect the absorbance finally measured. Consequently such variables as pH, reagent concentrations, temperature, and ionic strength frequently become of paramount importance.

The way in which the absorbance of a solution of picric acid in water varies with concentration constitutes a simple but striking example of the necessity for adequate control of these variables. Undissociated picric acid is colorless, and the yellow color of a picric acid solution is due to the picrate ion. Let us consider the absorbance of a c F solution of picric acid in water. Even if this absorbance is proportional to the concentration of picrate ion, which we may call Pi^-, it will not be proportional to c because the degree of dissociation of the acid (*i.e.*, the ratio $[Pi^-]/c$) decreases as c increases. Hence the absorbance of a $2c$ F solution of picric acid will be less than twice that of a c F solution. To be sure, data on the absorbances of solutions of picric acid in pure water would be very useful for the calculation of the dissociation constant of picric acid. But to analyze a picric acid solution it would be much better to buffer it at a pH of, say, 5 or 6, which is so much larger than pK_a ($= 0.8$) for picric acid that to all intents and purposes the acid is completely dissociated and $[Pi^-] = c$.

Entirely similar behavior is exhibited by weakly acidic solutions of chromium(VI), whose colors change from orange to yellow on dilution with water. The equilibrium involved here is

$$Cr_2O_7^{=} + H_2O = 2HCrO_4^- = 2CrO_4^{=} + 2H^+$$

Because this shifts to the right as the solution is diluted, it is obvious that the concentration of dichromate ion would not be simply halved by adding an equal volume of water to a dichromate solution. Therefore the absorbance of an aqueous solution of potassium dichromate is not proportional to the total concentration of chromium(VI) in the solution; the absorbance of a $2c$ F dichromate solution, measured at the wavelength of maximum absorption for chromate ion, will be less than twice that of a c F solution. This is referred to as a "negative deviation" from Beer's law.

We might suppose that this problem could be overcome merely by buffering the solution, but in fact this remedy may prove to be little better than the disease unless it is applied with care. Let us for the sake of simplicity assume that the concentration of $HCrO_4^-$ is always negligibly small, so that we can write

$$K = \frac{[CrO_4^{=}]^2[H^+]^2}{[Cr_2O_7^{=}]}$$

and

$$[CrO_4^{=}] + 2[Cr_2O_7^{=}] = c$$

where c is the total or formal concentration of chromium(VI) in the solution. Now, if the solution happens to be buffered at, for example, such a pH that $[H^+]^2 = K$, it is easy to show that an 0.1 F chromium(VI) solution will actually have an equilibrium concentration of dichromate ion equal to 0.0073 M, whereas an 0.2 F chromium(VI) solution will contain 0.0234 M dichromate at equilibrium. In this case, doubling the total concentration of chromium(VI) has more than tripled the equilibrium concentration of dichromate ion, and so the absorbance at the wavelength of maximum absorption for dichromate ion will increase more rapidly than the total analytical concentration of dichromate. This is known as a "positive deviation" from Beer's law. In order to secure absorbance values which are proportional to the formal concentration of chromium(VI) we would have to adopt one of two expedients: either we could make the solution strongly alkaline [so that virtually all of the chromium(VI) is present as chromate ion] and measure the absorbance at the wavelength of maximum absorption for chromate ion, or we could make the solution so strongly acidic that virtually all of the chromium(VI) is converted to dichromate ion and then measure the absorbance at the wavelength of maximum absorption for dichromate ion.

Phenomena of the same nature are encountered fairly frequently in work with complex metal ions or metal chelates. For example, Tiron (1,2-dihydroxybenzene-3,5-disulfonic acid) forms three distinct chelates with ferric iron in different pH ranges. A blue complex, containing one mole of Tiron per mole of ferric iron and having an absorbance maximum at 620 mμ, is formed at pH values below 5.6. A purple complex containing two moles of Tiron per mole of iron and having its absorbance maximum at 560 mμ is formed when the pH lies between 5.7 and 6.9. When the pH is above 7, a red complex is formed which contains three moles of Tiron per mole of iron and which has an absorbance maximum at 480 mμ. Evidently the absorbance at any one of these wavelengths will be anything but proportional to the total concentration of ferric iron if the pH is allowed to vary randomly. Because each of the three chelates is stable over a fairly wide range of pH values, it is certainly unnecessary to know the pH with great precision. But in routine analyses it would equally certainly be desirable to make sure that all of the samples fell within the same range of pH values so that all of their absorbances could be measured at the same wavelength with reasonable assurance that Beer's law would be at least approximately obeyed by the total concentration of iron present.

When the metal ion forms a series of successive moderately dissociated complexes with a reagent—which is the case with most monodentate ligands such as ammonia, cyanide, thiocyanate, and the like—it is usually found that both the position and the height of the absorption band undergo a more or less continuous shift as the concentration of the ligand is increased while that of the metal ion remains constant. Generally the band shifts toward longer wavelengths and the maximum absorbance increases. These changes correspond to the changes in the equilibrium composition of the solution as successively higher complexes are formed. In such a situation Beer's law will apply to the total concentration of metal present only if the ligand concentration is so high that virtually all of the metal ions are converted to the highest complex.[1] The literature contains many descriptions of analytical methods based on systems of this sort in which it is assumed that Beer's law is obeyed at a lower ligand concentration if such factors as the ligand concentration, ionic strength, pH, and so on are sufficiently closely controlled. This may be true within experimental error when the metal concentration varies relatively little among samples, but otherwise trouble is sure to result, for the positions of the various equilibria depend on the metal/ligand ratio as well as on these other factors. If Beer's law is to be obeyed by a system of this sort over a substantial range of metal concentrations, there is no alternative to making the ligand concentration so high that the dissociation of the highest complex is practically completely repressed.

A little thought will reveal that none of the examples which have been cited involves a real deviation from Beer's law. In every case the phe-

[1] The ligand concentration required to bring this about is the concentration above which a further increase causes no further change of either the wavelength of maximum absorption or the absorbance at that wavelength. This concentration is occasionally as much as 10^3 or even 10^4 times that of the metal ion involved.

nomena we have described would be observed even if Beer's law were obeyed with perfect rigor by the species actually responsible for the absorption, and the apparent failure of the law results from the simple fact that the concentration of that species is not in turn proportional to the concentration one is trying to determine. The solutions to these problems follow perfectly directly from the elementary principles of ionic equilibria. To make sure that none of these principles has been overlooked in the development of a spectrophotometric method of analysis, however, it is certainly advisable to test the validity of Beer's law experimentally for the particular system employed instead of merely assuming that it will be obeyed.

Of course Beer's law will also appear to fail if some of the samples contain an impurity which happens to absorb radiation of the wavelength used in the absorbance measurements. More elusive effects may be produced by impurities which do not absorb but which react chemically with the substance being determined. For example, ascorbic acid can be determined spectrophotometrically by measuring its absorbance at 265 mμ. However, it is slowly oxidized by dissolved oxygen, and so the absorbance decreases on standing in contact with air. This behavior is usually best handled by making a number of successive measurements of the absorbance of the slowly decomposing solution, plotting these values against the age of the solution, and extrapolating to find the absorbance at the instant the solution was prepared. Another common technique involves only a single measurement of the absorbance at some rigidly standardized time after the solution was made up, but this requires much closer control over the experimental conditions which govern the rate of the decomposition.

Some solutions containing both oxidizing and reducing agents, such as oxalate solutions containing either ferric or uranyl (UO_2^{++}) ion, appear to be perfectly stable as long as they are kept in the dark, but undergo oxidation and reduction when exposed to light. Consequently such solutions may be employed to measure the integral of radiant power with respect to time in photochemical experiments; the extent of the redox reaction is proportional to the total number of quanta of radiant energy falling upon the solution. Fortunately this is rarely important at the low levels of radiant power to which solutions are exposed during spectrophotometric measurements.

Certain other effects can only be classified as extraneous. These include the presence of colloidal or suspended solid materials (including dust!), which scatter light and thus decrease the transmitted radiant power. Gas bubbles, dirt, and fog on the cell walls have the same effect. The adsorption of an absorbing substance from one solution onto the cell walls may cause errors in the absorbances measured for solutions examined later. The evaporation of a volatile solvent leads to a spurious increase in concentration. Errors may arise because of fluorescence induced by the incident radiation.

Apart from the obvious effects of temperature on equilibrium constants and concentrations, a change of the temperature results in a change of the position of the absorption band. A change in temperature alters the energy state of the absorbing species in accordance with the Boltzmann distribution law. This will have no detectable effect on the distribution between the electronic states, for these are far too widely separated, but it may change the energy distribution among the vibrational modes corresponding to the lower electronic state and thus affect both the position and the intensity of the electronic transition involved. The direction of the effect is such that the wavelength of maximum absorption increases with increasing temperature. Consequently, the absorbance of a solution, measured at the wavelength of maximum absorption at one temperature T, will decrease if the temperature becomes either higher or lower than T.† The magnitude of this effect is often inappreciable, but this cannot be taken for granted in any specific case. It may be mentioned that thermostatting the absorption cell housing is coming to be common practice in careful work.

Changes of absorbance with time may also lead to apparent deviations from Beer's law. At the very low concentrations used in much spectrophotometric work, long periods of time are often required for equilibria to be reached in the reactions which produce the species whose absorbance is finally measured. To secure reproducible and accurate data it may be necessary to wait an hour or more between the addition of the reagents and the measurement of the absorbance to ensure that the substance being determined has been nearly completely converted to the absorbing form. If the absorbing species is chemically stable under the conditions employed, the absorbance can be measured at any convenient time after it has attained a constant value. If, however, the absorbing species is formed slowly and then decomposes slowly, one must proceed as described on page 272, delaying the first absorbance measurement until it is quite certain that the reactions responsible for its formation have gone to completion. Under such circumstances one could scarcely hope to secure very accurate results; only in desperation would an analytical method be erected on such a shaky foundation.

8-10. Quantitative Spectrophotometric Analysis. The first step in the development of a method of quantitative spectrophotometric analysis is

† On the other hand, the absorbance measured at a slightly longer wavelength λ', which is the wavelength of maximum absorption at some higher temperature T', will *increase* as the temperature is raised from T to T', then *decrease* as the temperature is raised above T'.

the selection of the absorbing species whose absorbance will eventually be measured. This involves two considerations. One is the convenience of the analyst—which includes the length of time he must spend on the chemical manipulations preceding the absorbance measurement—and the other is the objective suitability of the absorbing species for quantitative absorptimetry.

What this means in practice is that one first imagines a number of absorbing species into which the substance being determined could be converted, and then compares the accuracies with which it could be determined in each of these forms with the difficulties accompanying the preparation of each. Suppose, for instance, that the original sample contains manganese(III) in a weakly acidic pyrophosphate medium, and that what is wanted is a determination of the manganese concentration. The manganese(III) pyrophosphate complex in such media is purple, and so one could simply measure the absorption by this complex. On the other hand, one might consider measuring the absorbance of the permanganate resulting from a quantitative oxidation of the manganese with a suitable reagent, such as periodate. Permanganate ion is much more suitable for the purpose than the manganic pyrophosphate complex, because its molar absorptivity is far greater. Hence a method based on the measurement of the permanganate absorbance would be much more sensitive than one based on the measurement of the absorbance of the original complex. Unless this additional sensitivity were urgently needed, however, it would be so much simpler to measure the absorbance of the original solution than to go through the manipulations required to oxidize the manganese quantitatively that the former alternative would always be the one selected. Similarly, if one had to determine the concentration of nitrobenzene in a mixture with benzene, one might envision a technique for converting the nitrobenzene to a much more strongly absorbing azo dye, but instead of doing this one would probably merely measure the absorbance of the nitrobenzene itself at a wavelength, such as 330 mμ, where benzene shows no measurable absorption.

Very often, however, there is no way of avoiding some sort of chemical pretreatment. This may involve nothing more complicated than the addition of an indicator; or it may involve oxidation, reduction, the addition of a complexing or chelating agent, or the use of a catalytic or induced reaction. Sometimes several successive steps may be required. For example, calcium might be determined by precipitating it as the oxalate, dissolving this in acid and oxidizing the oxalate with a known amount of ceric sulfate, then adding excess iodide and measuring the absorbance of the triiodide formed by the reduction of the excess ceric ion. The possibilities along these lines are too many and varied to permit an extended discussion here; the student may profitably consult monographs devoted to the subject.[1]

Because spectrophotometric procedures are widely used for the determination of very small concentrations of materials, much attention must be paid to methods of separating the desired constituent from other constituents of the sample, for these often interfere by absorbing at the wavelength used for the absorbance measurements, reacting with the reagents added, and so on. An excellent discussion of this subject is given in chapter 2 of Sandell's monograph.

The ideal absorbing species would have a well-defined absorption band with a broad maximum (to minimize the slit-width errors previously discussed) and a very high absorptivity at the maximum (so that the sensitivity of the analysis would be correspondingly high).[2] The absorption band should occur in a portion of the spectrum where neither the solvent nor any of the reagents absorbs appreciably; otherwise the correction for their absorption may become a source of considerable error. The absorbing species itself should form rapidly and remain stable over long periods of time.

The ultimate sensitivity of a spectrophotometric analysis is largely governed by the absorptivity of the substance whose absorbance is being measured. The highest value of ϵ, the molar absorptivity, for any substance yet encountered is very roughly 1×10^5. If the spectrophotometer can conveniently accommodate a cell with a 5-cm. optical path length, and if it is realistically though more or less arbitrarily assumed that the smallest definitely detectable value of the absorbance—under optimum conditions and with the most scrupulous possible care to guard against dust, fog on the cell windows, and other equally extraneous phenomena—is of the order of 0.005, it follows that the smallest concentration of any substance that can be detected with certainty by spectrophotometric measurements is

$$c = \frac{A}{\epsilon b} = \frac{0.005}{(1 \times 10^5)(5)} = 1 \times 10^{-8}\ M$$

This is considerably smaller than the corresponding figures for potentio-

[1] M. L. Moss, Chemistry: Preparation of Systems for Absorptimetric Measurement, in M. G. Mellon (ed.), "Analytical Absorption Spectroscopy," chap. 7, John Wiley & Sons, Inc., New York, 1950. See also E. B. Sandell, "Colorimetric Determination of Traces of Metals," 2d ed., Interscience Publishers, Inc., New York, 1950.

[2] However, it may be noted that if one wants to analyze a moderately concentrated solution of the unknown substance, it may actually be advantageous to select a rather weakly absorbing colored form in order to avoid the necessity for extensive dilution prior to the absorbance measurement.

metric, conductometric, or polarographic measurements,[1] and it is this fact which is largely responsible for the present predominance of spectrophotometric procedures in our repertoire of practical methods of trace analysis. It is, however, a limiting figure which cannot by any means be attained with all absorbing substances, and in addition it appears unlikely to be decreased significantly in the foreseeable future. Accordingly there are very many cases in which the sensitivity of a spectrophotometric procedure is actually much inferior to the sensitivity attainable in the same determination by the use of some other instrumental technique.

Once a suitable absorbing substance has been selected, its absorbance–wavelength curve must be secured under the conditions contemplated for use in the analyses. This is necessary in order to permit the wavelength of maximum absorption to be located. Analytical absorbance measurements are, almost without exception, carried out at the wavelength of maximum absorption. There are several reasons for this which merit discussion in some detail.

First of all, it is obvious that not only the sensitivity but also the precision and accuracy which can be attained in trace analysis will be comparatively poor if the absorbance is measured at a wavelength where the species of interest is much more nearly transparent than at the wavelength of maximum absorption. Moreover, the fact that the absorbance–wavelength curve is nearly flat around the maximum but not at any other wavelength means that, in view of the discussion of Sec. 8-7, better conformance to Beer's law will be secured by measuring the absorbance at the wavelength of maximum absorption than by measuring it at any other point on the absorption band. Finally, a small error in setting the wavelength dial of the instrument or a small amount of backlash in the mechanical linkages in the prism or grating drive will naturally cause a much larger error when the absorbance is sensitively dependent on wavelength, as it is on the side of a band, than when the absorbance is nearly independent of wavelength, as is the case at the peak of a broad band.

Nevertheless, there are two situations in which it may be advantageous to measure the absorbance on the side of an absorption band rather than at its peak. One of these occurs when the sample contains some impurity which absorbs at the wavelength of maximum absorption by the substance being determined. Then one may prefer to use a wavelength at which the substance of interest absorbs less strongly but at which the offending impurity is completely transparent. This certainly results in some loss of accuracy, precision, and sensitivity, but so would an attempt to correct for the absorption by the impurity at the wavelength of maxi-

[1] On the other hand, there are already available some fairly simple modifications of conventional polarographic techniques which in some instances permit the detection and determination of substances present at concentrations as low as 10^{-9} M.

mum absorption by the substance being determined. This is merely a matter of choosing the lesser of two evils.

A situation in which it is definitely advantageous to measure the absorbance at a wavelength other than that at the maximum arises when the substance being determined is present in two forms which are in equilibrium with each other. This is fairly frequently encountered in connection with the determination of the total concentration of a substance which behaves as an acid-base or redox indicator.

Suppose, for example, that we wished to determine the total concentration of a dye such as bromthymol blue in a solution of unknown pH. The acidic form of bromthymol blue has an absorption maximum at about 440 mμ and a minimum at 600 mμ, while the maximum and minimum for the alkaline form occur at 625 and 450 mμ, respectively. We might, of course, treat this unknown as a two-component system (page 280) by measuring its absorbances at 440 and 625 mμ, using these data to calculate the concentration of each form of the indicator, and adding these values to find the total concentration present. However, there is a much simpler alternative which is based on a well-known mathematical theorem of continuity. This is Bolzano's theorem, which states essentially that if the difference, $a_A - a_B$, between the absorptivities of the acidic and basic forms is positive at 440 mμ and negative at 625 mμ, and is a continuous function of wavelength between these points (which is true because the individual spectra have no points of discontinuity), then $a_A - a_B$ must be zero at some intermediate wavelength. In this case the absorptivities of the acidic and basic forms are found by experiment to be equal at 501 mμ, and so the absorbance at 501 mμ will be constant for any given total concentration of the indicator regardless of the way in which this is divided between the acidic and basic forms. Therefore it would be possible to secure the desired information merely by measuring the absorbance of our unknown bromthymol blue solution at 501 mμ.

It follows from what was said in the preceding paragraph that the spectra of a number of solutions having equal formal concentrations of bromthymol blue but different pH values will all intersect at 501 mμ. Such a point is known as an *isosbestic point*, and the appearance of such a point on a family of spectra is a necessary (but not quite sufficient) criterion of the presence of two and only two forms of the absorbing substance in equilibrium with each other. This is of frequent importance in studies of complex metal ions.

In the example we have selected it would also have been possible to determine the formal concentration of the indicator from a single absorbance measurement if a large excess of either acid or base had first been added to the solution to convert the indicator nearly completely into one or the other of the two forms. However, when the chemical properties of the sample make this impossible, much time and effort may be saved by taking advantage of the existence of an isosbestic point.

Having located the wavelength best suited to the measurement of the absorbance of the substance being determined, it is advisable to investigate the way in which the absorbance of a blank solution varies with wavelength. So far as the solvent itself is concerned, its transmittance in a solution of the absorbing substance will be virtually identical with the transmittance of the pure solvent—and will therefore be accurately corrected for by setting the spectrophotometer to read 100 per cent transmittance with a cell filled with pure solvent in the light path—if the solution is sufficiently dilute. In visible and ultraviolet spectrophotometry the concentration of the substance being determined is most often between about 10^{-5} and 10^{-3} M. Even after a reasonable allowance is made for the presence of an excess of the reagents in the final solution, the effect on the concentration of the solvent will certainly be inappreciable except in very rare cases. So the correction for any absorption by the solvent, using the procedure just outlined, will involve no error worth mentioning, especially since one would never deliberately choose a solvent which was not perfectly transparent in the desired portion of the spectrum. The same principle applies whenever the sample contains any other absorbing impurities (including the reagents) if their concentrations in the sample and in the reference solution are the same, although a truly accurate compensation is often much more difficult in this case.

At this point it becomes necessary to determine whether Beer's law is obeyed and, if so, over what range of concentrations. This may be accomplished by plotting the absorbances of a number of known solutions against their concentrations; evidently these points will fall on a straight line if Beer's law is obeyed. A much more accurate technique[1] consists of computing A/c for each of the standard solutions and observing the range of concentrations over which the quotient is constant within the estimated experimental error of the absorbance measurements. If either of these tests shows that Beer's law is not obeyed over any fairly substantial range of concentrations, the data obtained may be used empirically to find the concentrations of subsequent unknowns. This is inherently a rather dangerous situation, however, for there is no guarantee that the curvature of the Beer's law plot is not due to a variation of some chemical

[1] The student will do well to look back to the results he secured in Expt. I-6, II-6, III-6, or IV-6 to note the way in which plotting Λc *vs.* c has the effect of concealing the substantial variation of Λ which is clearly shown by a plot of Λ *vs.* c (or $c^{1/2}$). In exactly the same way, a plot of $A\,(= abc)$ *vs.* c can often lead to an erroneous conclusion that Beer's law is obeyed when a plot of A/c *vs.* c would reveal a relatively large variation of a.

factor whose effect has been overlooked; if this factor happens to assume entirely different values during later analyses, the results obtained may be seriously in error. It is always wise to consider the possible causes of deviations from Beer's law (Sec. 8-9) to see whether the chemistry of the system can be modified in such a way as to yield data which do obey Beer's law.

If Beer's law is found to be obeyed, the analysis of an unknown becomes a very simple matter. If A_u is the absorbance of the unknown and c_u its concentration, we have

$$c_u = \frac{A_u}{A/c}$$

where A/c is the constant ratio of absorbance to concentration found with known solutions under the same experimental conditions. For reasons which will be discussed in Sec. 8-11, the best precision is generally secured when the measured absorbances of both the unknown and the known solutions fall between about 0.2 and 0.85. If a wide range of concentrations must be covered, it will be necessary to adjust either the size of the original sample, the path length of the cell, or both to bring all of the absorbances between these limits.

Absolute methods of analysis, in which concentrations are calculated from Beer's law with the aid of a literature value of the absorptivity, are almost never used in practical absorptimetric analysis. This is because different spectrophotometers (and even different instruments of the same make and model) give sensibly different values for the absorptivity of the same standard solution. Consequently the absorbances of both the known and unknown solutions must be measured with the same spectrophotometer if the best accuracy is desired. In an extended series of analyses it is also advisable to repeat the measurements of the absorbances of the standard solutions at occasional intervals to guard against any unexpected change of the characteristics of the spectrophotometer.

Provided that all of the chemical and instrumental factors influencing the absorbance are recognized and adequately controlled, quantitative spectrophotometric analyses are possible even if a plot of absorbance *vs.* concentration is a smooth curve rather than a straight line. Having measured the absorbance of an unknown, one need only read its concentration off the absorbance–concentration graph secured from measurements with known solutions under identical conditions.

Spectrophotometric methods can be used to determine the concentrations of two or more absorbing substances in a mixture if the spectrum of the mixture is simply the sum of the spectra of the individual components—if, that is, the absorbing substances do not interact in any way when they are mixed. It is highly desirable that each component obey

Beer's law at all wavelengths. There is a trivial case in which each substance present in the mixture gives an absorption band which is entirely separate and distinct from that of any other component. This is treated in the same way as a one-component analysis, repeated as many times as there are absorbing substances to be determined in the sample.

More frequently the absorption bands will overlap to some extent, so that the total absorbance measured at the wavelength of maximum adsorption of one component will include some absorption by another component. This is illustrated for a two-component mixture in Fig. 8-9.

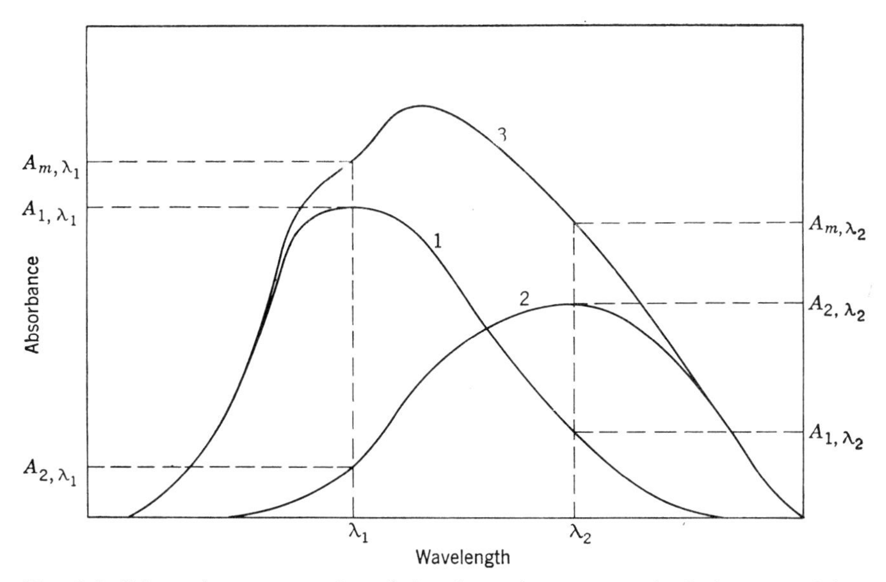

FIG. 8-9. Schematic representation of the absorption spectra of solutions containing (1) c_1 moles per liter of substance 1, (2) c_2 moles per liter of substance 2, and (3) c_1 moles per liter of substance 1 and c_2 moles per liter of substance 2.

If there is no interaction between the two components, the absorbance of the mixture at any wavelength, such as λ_1, will be simply the sum of the absorbances which they would yield at that wavelength if they were present separately:

$$A_{m,\lambda_1} = A_{1,\lambda_1} + A_{2,\lambda_1}$$

From the known absorbance–wavelength curves of the individual components in separate solutions, two wavelengths are selected. On purely mathematical grounds it can be shown that the optimum wavelengths are those at which the ratio of absorptivities a_1/a_2 attains its maximum and minimum values, but in practice one almost invariably selects the wavelengths of maximum absorption on the spectra of the separate components, for except in very unusual cases these will give about equally accurate results.[1] Then the absorptivities of both substances at each of these two wavelengths are evaluated from absorbance measurements with known solutions. When the absorbance of an unknown mixture is finally measured at each of these wavelengths, the concentrations of both components can be found by solving the simultaneous equations

$$A_{m,\lambda_1} = (a_{1,\lambda_1}c_1 + a_{2,\lambda_1}c_2)b$$
$$A_{m,\lambda_2} = (a_{1,\lambda_2}c_1 + a_{2,\lambda_2}c_2)b$$

Like any other "indirect" method of analysis, this should not be used without some understanding of its limitations. It never gives results quite as good as are obtained from an ordinary one-component spectrophotometric analysis, except, of course, in the trivial case previously mentioned where neither component absorbs at the wavelength of maximum absorption of the other (*i.e.*, where both a_{1,λ_2} and a_{2,λ_1} are zero). It is most accurate when component 1 is responsible for most of the absorption at λ_1 while component 2 is responsible for most of the absorption at λ_2, and it fails completely if the concentrations and/or absorptivities are so disparate that a_1c_1 is either very large or very small compared with a_2c_2 at both wavelengths.

The method may be extended to the analysis of mixtures of three or even more components by measuring the absorbance of the unknown at as many wavelengths as there are concentrations to be determined. How practical this actually is in any specific case depends on the relative values of the *ac* products at each of the wavelengths selected, and usually has to be judged on the basis of experiments with the particular mixtures in question.

8-11. Photometric Errors. The photometric error of an analysis is the error in the concentration which results from the instrumental error of the absorbance measurement. If does not include the chemical or instrumental sources of error discussed in Sec. 8-9. On the contrary, its theoretical treatment is founded on the assumption that the system adheres rigidly to Beer's law.

Let us express the error in the concentration ultimately computed from an absorbance measurement in terms of the deviation Δc from the true concentration c. The relative error in c will, of course, be $\Delta c/c$. This error will be due to such factors as nonlinear phototube and amplifier

[1] Measuring the absorbances of the mixture at its own wavelengths of maximum absorbance naturally minimizes the errors resulting from small uncertainties in the wavelength setting. However, these wavelengths may not be the most favorable for measuring the absorptivities of the separate components, so that what is gained on the one hand may easily be lost on the other. Moreover, as is shown by Fig. 8-9, the wavelengths of maximum absorbance for a mixture will vary with the ratio of concentrations of the components unless there is very little overlapping of the absorption bands. This would be very troublesome in practical work.

response (either of which causes a deviation from exact proportionality between the radiant power of the beam transmitted by the sample and the voltage or current ultimately measured), the electrical and mechanical effects which limit the sensitivity and accuracy of the circuit used to measure the output of the amplifier, and the errors in scale reading. Most nonrecording spectrophotometers are so designed that their readings would be proportional to transmittance if these errors could be completely eliminated. In other words, the quantity which is finally measured is intended to be directly proportional to the radiant power transmitted by the sample, and the meter or potentiometer scale is linear in transmittance. Such spectrophotometers are generally also provided with logarithmic scales from which the absorbance can be read directly, but this is merely for the operator's convenience and naturally does not affect the errors involved. It is therefore convenient to relate the error in c to the error, ΔT, in the transmittance indicated by the instrument, and to assume that this error will be nearly constant over the entire range of transmittance values from 0 to 1.

We may now write Beer's law in the form

$$c = -\frac{\log T}{ab} \tag{8-11}$$

Differentiating this with respect to T gives

$$\frac{\Delta c}{\Delta T} = -\frac{0.4343}{abT} \tag{8-12}$$

By combining Eq. (8-11) and (8-12) we secure the following expression for the relative error in c:

$$\frac{\Delta c}{c} = \frac{0.4343(\Delta T)/T}{\log T} \tag{8-13}$$

In this equation ΔT represents the absolute error in the transmittance, which is usually called the *absolute photometric error*. In practice ΔT is best taken as twice the mean deviation of a number of replicate transmittance measurements made with the same solution under identical experimental conditions. It will usually be found to lie between 0.002 and 0.01; a value of 0.01 corresponds to an absolute photometric error of 1 per cent in the transmittance of the sample.

It is now a simple matter to determine the way in which the relative error in c resulting from any given absolute photometric error depends on the transmittance of a solution. To do this we need only rewrite Eq. (8-13) in the form

$$\frac{(\Delta c)/c}{\Delta T} = \frac{0.4343}{T \log T} \tag{8-14}$$

and evaluate the right-hand side of this expression for various values of T. The results of such calculations are given in Table 8-3, from which it is apparent that the relative error in c has a minimum value of about 2.7 times the absolute photometric error, that this minimum occurs at a transmittance of about 35 per cent (more exactly, 36.8 per cent), and that the relative error in c is very little larger than this minimum value if the transmittance is between about 0.15 and 0.65 but increases rapidly if the transmittance is outside this optimum range.

TABLE 8-3. THE RATIO OF THE RELATIVE ERROR IN CONCENTRATION TO THE ABSOLUTE PHOTOMETRIC ERROR AT VARIOUS VALUES OF THE TRANSMITTANCE

T	$-\frac{(\Delta c)/c}{\Delta T}$†	T	$-\frac{(\Delta c)/c}{\Delta T}$†
0.95	20.5	0.40	2.73
0.90	10.6	0.35	2.72
0.85	7.2	0.30	2.77
0.80	5.6	0.25	2.89
0.75	4.64	0.20	3.11
0.70	4.01	0.15	3.51
0.65	3.57	0.10	4.34
0.60	3.26	0.05	6.7
0.55	3.04	0.02	12.8
0.50	2.88	0.01	43.4
0.45	2.78		

† The minus sign here reflects the fact, which is implicit in Eq. (8–14), that an erroneously high value of T will cause the calculated value of c to be too low.

The data in Table 8-3 permit the direct calculation of the best attainable relative error in c once an estimate of the absolute photometric error has been obtained for any one chemist working with any one spectrophotometer. If the absolute photometric error is 0.01, for example, the corresponding relative error in c is immediately seen to be about 0.03 (*i.e.*, 3 per cent) for transmittances between 0.15 and 0.65. On the other hand, to secure values of c which would be in error by no more than 0.6 per cent one would have to keep the absolute photometric error down to 0.002, and this is about the best reproducibility attainable with most present-day instruments.

The practical conclusion to be drawn from this is that one should always attempt to maintain the transmittances of unknown solutions between 0.15 and 0.65 (corresponding to absorbance values between 0.85 and 0.2). A very few experiments with known solutions will suffice to show the range of concentrations to which this corresponds for any particular absorbing substance. Then, in the analyses which follow, such variables as the optical path length, sample size, and volume of the final solution should be so adjusted as to yield solutions whose absorbances and concentrations fall within these limits. To be sure, this is usually inconvenient, and sometimes it may even be impossible, but it is clear from Table 8-3 that the analyst who ignores this injunction will often pay a heavy toll in the form of a seriously decreased precision.

Some spectrophotometers, such as the Cary automatically recording instruments, are so designed that the ultimate measurement is really one of absorbance rather than of transmittance. This is generally accomplished by transforming the output of the phototube (or, more often in such instruments, the photomultiplier) into a logarithmic function of the photocurrent before amplification. When this is done, the photometric error becomes much more difficult to analyze, because the portion of the over-all error which is due to the detector is proportional to the radiant power of the transmitted beam, whereas the amplifier and recorder errors are proportional to the logarithm of the transmitted radiant power. If the linearity of the detector is much better than that of either the amplifier or the recorder, which appears to be the usual case, the over-all error of the instrument will be a nearly constant fraction of an absorbance unit. So the relative error in c will decrease continuously with increasing absorbance, and the best precision is attained in work with relatively strongly absorbing solutions. Some such instruments are capable of yielding absorbance data which are precise to about 0.005 absorbance unit; when one is dealing with a sample whose absorbance is, say, 2.5 this corresponds to a relative error of only 0.2 per cent in c. It is only fair to mention that this threefold improvement over the relative error referred to in the preceding paragraph is not within the financial reach of every laboratory.

8-12. "Precision Colorimetry." As was mentioned in the preceding section, the precision attainable in a spectrophotometric analysis depends on the linearity of the phototube, amplifier, and measuring circuit of the instrument employed. Let us suppose for the sake of simplicity that the measuring circuit of a certain instrument consists simply of a microammeter connected to the output stage of the amplifier,[1] and let us assume that the errors due to the phototube and amplifier are much smaller than those attributable to the meter. The exact nature of this assumption will be dealt with at greater length in a later paragraph. For the moment, however, it will suffice to note that the unavoidable electrical and mechanical imperfections of the meter—including, for example, inhomogeneities in the magnetic field surrounding the moving coil, slight defects in the bearings and pivots which support the moving parts, and so on—will result in an error whose absolute value will be very nearly constant over the entire width of the scale. Thus, a meter having a 0 to 200 microamp. scale is said to have a linearity of ± 1 per cent if its reading

[1] This happens to be the measuring element of a Beckman Model B spectrophotometer, but the present argument is by no means restricted to that instrument alone.

may be in error by 2 microamp. at any point on its scale. (It is evident that the linearity and the relative error are two quite different things; this is true not only for meters but also for variable resistors and capacitors, amplifiers, etc., whose linearities are all expressed as the ratio of the error at any point to the full-scale value of the variable quantity.)

In reality the meter is likely to be calibrated in transmittance units rather than microamperes, but this is merely a disguise. Erasing the microampere scale and replacing it with another linear scale labeled "transmittance" clearly has no effect on the performance of the meter. So if in using the instrument we set the meter needle on zero transmittance (*i.e.*, zero microamperes) with the phototube in darkness, and on unit transmittance (*i.e.*, 200 microamp.) with the phototube exposed to a beam which has passed through a nonabsorbing reference solution, its absolute error of 2 microamp. will correspond to an absolute error of 0.01 transmittance unit. It is on this situation that the discussion of Sec. 8-11 was based.

However, this is a very unfavorable state of affairs for the analyst who must determine the concentration of a solution which transmits, say, only 5 per cent as much radiation as the reference solution, for the relative error in the measured transmittance of such a solution will be ± 20 per cent under our assumed conditions. (Actually, of course, it is quite likely to be even larger, because of the errors involved in making the two preliminary adjustments.)

One way around this difficulty consists of changing the number of transmittance units represented by the constant error of 2 microamp. We might, for example, replace the nonabsorbing reference solution with another solution whose transmittance at the wavelength in question is, say, 0.1. Now, by increasing the slit widths or the gain of the amplifier, we can arrange matters so that a current of 200 microamp. is again presented to the meter. Under these new conditions a full-scale meter deflection will represent only 0.1 transmittance unit, and so the constant error of 2 microamp. will correspond to an absolute photometric error of only 0.001. We cannot now measure any transmittance which is larger than 0.1, for the meter will go off scale, but we have achieved a tenfold reduction of the photometric error in dealing with solutions having transmittances smaller than that of our new reference solution.

This is the principle of the technique known as "precision colorimetry" or "differential spectrophotometry." It depends on a measurement of the ratio of the radiant powers of the beams transmitted by the sample and by a slightly less concentrated solution of the same absorbing substance. If it is assumed that Beer's law is obeyed under the conditions employed, it can be shown that

$$c_u - c_s = -\frac{\log (P_u/P_s)}{ab}$$

where the subscripts u and s refer to the unknown and standard solutions, respectively. As long as c_s is known exactly, the absolute value of P_s is quite immaterial, for the instrument is so adjusted as to "define" P_s arbitrarily as unity.

This idea can be carried one step further by using another known solution to define the other end of the meter scale. Instead of setting the meter to read zero transmittance (*i.e.*, 0 microamp.) with the phototube in darkness, one may expose the phototube to a beam which has passed through a solution of the same absorbing substance at a concentration slightly *higher* than that of the solution to be analyzed. After adjusting the instrument so that the meter reads zero under these conditions, it is adjusted to read unit transmittance with a solution slightly more dilute than the unknown in the light path. So, for example, if we had to analyze a solution having a transmittance of 0.09, we might prepare two known reference solutions of such concentrations that the two ends of the meter scale correspond to transmittances of 0.08 and 0.10. (Again, the exact transmittances of the reference solutions are of no concern; only their concentrations need be known.) In this way the photometric error corresponding to the constant electrical error of 2 microamp. would be reduced to only 0.0002 transmittance unit. This is called the "maximum precision" technique.

The gain in precision resulting from the use of these techniques is such as to entitle them to a place among the most precise analytical methods. By "precision colorimetry" it has been possible to achieve a precision of about ± 0.05 per cent in the analysis of aqueous solutions of 2,4-dinitrophenol, and even more spectacular results could undoubtedly have been obtained by the maximum precision technique.

We must now return to a more careful examination of the assumption on which this discussion has been based. What we have assumed is essentially that the error in the current finally measured is a constant fraction of the full-scale range of the meter. Let us suppose, however, that the maximum useful output of the phototube is about 10^{-4} microamp. and that its linearity is ± 0.5 per cent. This corresponds to a maximum error of 5×10^{-7} microamp. in the photocurrent. If we work with solutions whose transmittances are as small as, say, 0.01, the photocurrent will be only 10^{-6} microamp. and the relative error becomes enormous. Moreover, it is not possible to amplify any indefinitely small photocurrent up to 200 microamp. (or any other arbitrary level), for the Johnson "noise" in the phototube and amplifier, which results from the random thermal emission of electrons, will swamp out a very tiny photocurrent. This means that it is necessary to cause the phototube to develop as large a current when exposed to the beam transmitted by the strongly absorbing reference solution as it does in ordinary spectrophotometry when it is exposed to a beam which has traversed a nonabsorbing solution. In principle this might be done by using a more intense source, but this would seriously aggravate the problem of stray radiation. The only alternative, then, is to increase the slit widths, but this immediately returns us to the familiar subject of the deviations from Beer's law caused by the use of heterochromatic radiation. Although this is what is always done in practice, it is clear that it must be done with discrimination.

A more comprehensive treatment of these techniques would be beyond the scope of the present text. The student is urged to consult the original paper by Hiskey[1] and the more recent one by Reilley and Crawford[2] for further information.

8-13. Spectrophotometric Titrations. A spectrophotometric titration is a titration whose end point is found from data on the absorbance of the solution. It is a general feature of all titration methods, and one which we have stressed in several preceding chapters, that they are unaffected by many changes in experimental conditions which would seriously disturb analyses by the corresponding direct methods. For example, in analyses by direct conductometry we must either know the cell constant of the cell being used or be quite certain that it is the same as it was during a series of measurements with known solutions; but in conductometric titrations the value of the cell constant is a matter of total unconcern, provided only that it does not change during the course of a titration. Moreover, we have repeatedly said that titration methods are inherently much more accurate than the corresponding direct methods.

These things are also true of spectrophotometric titrations as compared to what, by analogy with the electrical methods, we might call direct spectrophotometry. Whereas the usual error of analyses by direct spectrophotometry is about ± 1 to 3 per cent, an accuracy and precision of the order of a few tenths of a per cent are attainable with comparative ease by spectrophotometric titrations. In direct spectrophotometry it is essential to know something about the path length of the absorption cell used; in a spectrophotometric titration it is only necessary that the path length should not change during the course of the titration,[3] and the same

[1] C. F. Hiskey, *Anal. Chem.*, **21**, 1440 (1949).

[2] C. N. Reilley and C. M. Crawford, *Anal. Chem.*, **27**, 716 (1955).

[3] This is not quite as trivial as it may seem at first glance. For reasons of convenience, spectrophotometric titrations are often carried out in beakers rather than in the much smaller absorption cells used in direct spectrophotometry. Of course the optical path length through a round vessel is considerably affected by even a relatively slight sideways shift of position with respect to the beam of radiation. Besides, most beakers (and test tubes, etc.) are not quite round, and so the path length will vary if the vessel is rotated even though great care is taken to ensure that the beam passes through its center. Consequently it is very desirable to be able to leave the titration vessel entirely undisturbed throughout the titration.

thing is also true of many other factors which must be very carefully controlled in a direct spectrophotometric analysis.

The experimental execution of a spectrophotometric titration is very simple. A suitable clear-walled vessel containing the solution to be titrated is placed in the light path of the spectrophotometer, a wavelength appropriate to the particular titration is selected, and the absorbance reading is adjusted to some convenient value by means of the sensitivity and slit-width controls. A measured volume of reagent is added, the solution is stirred, and the absorbance is read again. This is done at several points before the end point, and at several more points after the end point. As in conductometric and amperometric titrations, the end point is found graphically.

Both the wavelength and the initial absorbance setting depend on the absorption spectra of the various chemical species involved in the titration. For example, a sulfuric acid solution of ferrous iron might be titrated with standard permanganate at a wavelength of 525 mμ, which is the wavelength of maximum absorption for permanganate ion. Manganous, ferrous, and ferric ions do not absorb at this wavelength. So the absorbance will remain equal to the value originally set on the instrument (which might very conveniently be zero) until the end point is passed. If $C_{MnO_4^-}$ is the concentration of the standard permanganate used and $C_{Fe^{++}}$ the initial concentration of the unknown ferrous solution, both in moles per liter, and if at some point on the titration curve $V_{MnO_4^-}$ ml. of permanganate has been added to $V_{Fe^{++}}$ ml. of the ferrous solution, the concentration of permanganate ion in the solution will be virtually zero if the end point has not yet been reached. If the end point has been passed, the concentration of excess permanganate will be

$$c = \frac{V_{MnO_4^-}C_{MnO_4^-} - \frac{1}{5}V_{Fe^{++}}C_{Fe^{++}}}{V_{MnO_4^-} + V_{Fe^{++}}}$$

The factor $\frac{1}{5}$ arises from the stoichiometry of the reaction between permanganate and ferrous ion. Taking into account the fact that the absorbance of the excess permanganate is proportional to its concentration, and rearranging, we secure

$$A\left(\frac{V_{MnO_4^-} + V_{Fe^{++}}}{V_{Fe^{++}}}\right) = \epsilon b\left[\left(\frac{C_{MnO_4^-}}{V_{Fe^{++}}}\right)V_{MnO_4^-} - \frac{1}{5}C_{Fe^{++}}\right]$$

This equation shows that the absorbance, after correction for dilution in the same way as in conductometric and amperometric titrations, increases linearly with the volume of permanganate beyond the end point. So the titration curve, which is a plot of the absorbance, corrected for dilution, against the volume of reagent, will be _/-shaped, and the point of intersection of the two straight-line portions will be the equivalence point of the titration.

Note that we have implicitly assumed that Beer's law is obeyed under the conditions of the titration. If it is not, the increase of the absorbance beyond the end point will not be linear, and this will adversely affect not only the ease of making the extrapolation required but also the accuracy and precision of the result. Consequently it is essential even in a spectrophotometric titration to keep in mind the various factors discussed in Secs. 8-7 and 8-9 which may lead to deviations from Beer's law. On the other hand, the importance of any such deviation depends on its magnitude, and a small deviation may be quite devoid of perceptible effect on the result of a spectrophotometric titration even though it would be relatively serious in direct spectrophotometry. Of course the same sort of consideration applies to other titration methods as well. In a conductometric weak acid–strong base titration, for example, the variation of ionic strength as the titration proceeds may be sufficient to cause a variation of several per cent in the equivalent conductance of every ion involved, but the error thus caused in the location of the end point is inappreciable. In effect this amounts to a deliberate use of the phenomenon discussed in the footnote on page 278 to conceal and overcome the variation of the equivalent conductance (or absorptivity) with concentration.

Obviously neither the shape of this titration curve nor the location of the end point will be affected by changing the arbitrarily chosen "value" of the absorbance of the initial solution; varying this will merely shift the entire curve up or down. One may feel more comfortable if the spectrophotometer is adjusted to read zero absorbance with the titration cell filled with water before the titration is begun, but this is really quite unnecessary.

Of course different kinds of titration curves will be secured in different situations. If only the substance being titrated absorbs at the wavelength used, the absorbance will decrease as the titration proceeds, and will become zero after the equivalence point. In this case it is often convenient to set the initial absorbance to some relatively high value, such as 1.0 or 2.0, depending on the concentration of the solution being titrated. Titration curves of this kind are _ -shaped. On the other hand, if one of the products of the reaction is the only absorbing species at the wavelength selected, a /‾-shaped curve will be obtained, and there are several other more or less obvious possibilities.

Just as in conductometric and amperometric titrations, and for exactly the same reasons, little or no attention is paid to points very near the end point. On the contrary, the end point is found by extrapolating the linear segments of the titration curve to their point of intersection. This allows spectrophotometric titrations, like the other titration methods, to be used even when the titration reaction has a rather low equilibrium constant.

The optimum concentration of a solution which is to be analyzed by a spectrophotometric titration depends on the absorptivity of the absorbing species involved, and is usually of the order of 10^{-4} or 10^{-5} M; if the absorptivity of the absorbing species is very high it should be possible to carry out a titration with good accuracy even if the unknown is as dilute as 10^{-6} M. This is well below the lower limit of the range of feasibility of either conductometric or amperometric titrations. Furthermore, spectrophotometric titrations are often applicable in cases where no good electrometric system is available for the indication of the end point. (Of course the converse is also frequently true.)

Many substances which are difficult or even impossible to convert to forms suitable for determination by direct spectrophotometry can easily be determined by a spectrophotometric titration with an absorbing reagent. For example, to determine +3 arsenic by direct spectrophotometry we might oxidize the arsenic to the +5 state, add excess molybdate to form a heteropoly arsenomolybdic acid, and reduce this with stannous chloride to give a so-called "heteropoly blue." (The success of this procedure depends on the fact that the molybdenum atoms in the heteropoly complex ions are much more rapidly reduced by the stannous chloride than are the molybdenum atoms of the excess molybdate.) However, both +5 arsenic and phosphorus would certainly interfere seriously. It would be much simpler to titrate the +3 arsenic solution with standard permanganate at the wavelength of maximum absorption for permanganate ion.

An even simpler technique, which is closely related to a spectrophotometric titration but is considerably more rapid, would require merely the addition of a carefully measured amount of permanganate just large enough to give a definite excess, followed by the direct spectrophotometric determination of the concentration of the excess permanganate. This would be less accurate and precise than a true spectrophotometric titration, because the errors in wavelength setting, optical path length, slit width, and the like would not be canceled out as they would be in the titration. But it would be considerably more accurate than an ordinary direct spectrophotometric analysis because these errors would affect only the determination of the relatively small excess of reagent.

By the use of an appropriate indicator it is often possible to perform a spectrophotometric titration even when none of the substances directly involved in the titration exhibits any measurable absorption. Acid-base titrations can be made with the usual indicators, often with far better precision and accuracy than can be secured when the end point is located visually (*cf.* Expt. IV-16).

Some indicator methods are much more subtle. For example, both bismuth and cupric ions form chelates with ethylenediaminetetraacetate, but the bismuth chelate is by far the more stable of the two. Hence the bismuth chelate forms first when a solution containing both bismuth and cupric ions is titrated with standard ethylenediaminetetraacetate. When virtually all of the bismuth has been chelated, the cupric chelate will begin to form. The bismuth chelate absorbs only in the ultraviolet (its wavelength of maximum absorption is 265 mμ), whereas the wavelength of maximum absorption for the copper chelate is 745 mμ. When the course of the titration is followed spectrophotometrically with the wavelength set at 745 mμ, the absorbance remains unchanged as the bismuth chelate is formed; then it increases with increasing concentration of the cupric chelate. Thus the cupric ion serves as an indicator for the titration of the bismuth. This is advantageous when the spectrophotometer available is suitable only for measurements in the visible region; otherwise bismuth could naturally be titrated directly at 265 mμ. The titration at 745 mμ permits the simultaneous determination of bismuth and copper, for the absorbance becomes constant when the equivalence point of the copper titration is passed. The complete titration curve is _/‾‾-shaped and has two end points, the first corresponding to the quantity of bismuth present and the second to the sum of the quantities of bismuth and copper.

In spectrophotometric as well as in conductometric and amperometric titrations, the concentration of the titrant should be considerably greater than that of the solution being titrated in order to minimize the dilution correction which must be applied.

8-14. Spectrophotometric Studies of Ionic Equilibria. Spectrophotometric procedures have been used to study many different kinds of ionic equilibria. In this section we shall describe some of the techniques most commonly employed.

A type of equilibrium which is perhaps more obviously amenable to spectrophotometric study than any other is the dissociation of an acid-base indicator. The potentiometric determination of the dissociation constant of an indicator, using the procedure described in Sec. 4-10, is often rather difficult because of the very low solubilities of many indicators in water. However, the absorbance of a 10^{-4} or even 10^{-5} F solution of an indicator is easily measured. The effect of pH on the spectrum of a two-color indicator was mentioned in Sec. 8-10. Two wavelengths of maximum absorption are found: one corresponds to the acid form and the other to the alkaline form of the indicator, and unless there is a side equilibrium involving one of these forms there will be an isosbestic point somewhere between these two wavelengths.

As the pH of a solution containing a fixed formal concentration of indicator is raised, the absorbance of the acid form will decrease while that of the alkaline form will increase. The relation among pH, pK_a, and the concentrations of the acid and alkaline forms is

$$pK_a = pH - \log \frac{[In^-]}{[HIn]} \tag{8-15}$$

We begin by preparing two solutions, each containing a known formal concentration of the indicator. One of these should be so acidic that essentially all of the indicator is converted to HIn, and the other should be so alkaline that essentially all of the indicator is converted to In^-. The wavelength of maximum absorption for each of these forms is located, and the absorbance of each solution is measured at both of these wavelengths. From these data the molar absorptivity of each form at each wavelength is easily calculated. Now we dissolve some of the indicator in a well-buffered solution whose pH is accurately known and is within about 0.5 unit of pK_a for the indicator. From the absorbance of this solution at each of the two wavelengths of maximum absorption, the values of $[In^-]$ and $[HIn]$ are found by the technique described in Sec. 8-10. Introducing these values, together with the measured pH, into Eq. (8-15) gives the formal dissociation constant of the indicator at the ionic strength of the particular buffer employed (*cf.* Table 4-1). If we wish to know the thermodynamic dissociation constant we need merely measure K_a at a number of ionic strengths and extrapolate to infinite dilution.

Having thus secured the dissociation constant of the indicator, we could use it to measure the dissociation constants of other acids or bases which do not absorb in the visible portion of the spectrum. For example, to determine the dissociation constant of acetic acid we might prepare a solution containing known formal concentrations of acetic acid and sodium acetate, add some of the indicator, and measure the resulting ratio $[In^-]/[HIn]$ spectrophotometrically. For obvious reasons it would be necessary to use an indicator whose dissociation constant is not far from that of acetic acid, and the acetic acid–acetate solution should be so well buffered that its pH is not appreciably affected by the addition of the indicator solution. For acetic acid this procedure would certainly have no advantages over the potentiometric one. On the other hand, it permits the extension of the acidity scale into both the strongly acid and strongly alkaline regions, where the liquid-junction potentials included in the potentiometric measurements would be so large as to make the results quite uncertain. Furthermore, the spectrophotometric method is virtually the only one available for the establishment of reliable acidity scales in nonaqueous media.

It is evident that the procedure described for studying the dissociation equilibrium of an indicator can be applied to the equilibrium between any pair of substances of which at least one absorbs in an accessible portion of the spectrum. Thus, for example, spectrophotometric methods can be applied to the investigation of keto-enol equilibria.

In order to study the equilibrium involved in the formation of a complex ion, it is necessary to determine the equilibrium concentration of at least one of the species present in a mixture containing known formal concentrations of the reactants, and this is often relatively easy to accomplish by spectrophotometric methods. Although the absorbance of such a mixture is usually not a linear function of the *total* concentration of metal present (Sec. 8-9), it is generally proportional to the concentration of the species which is actually responsible for the absorption, provided that due attention is paid to the sources of error previously discussed.

Perhaps the simplest of the spectrophotometric techniques which have been used for the study of complex-formation equilibria is the *molar ratio method.*[1] A series of solutions is prepared which contain equal formal concentrations of a metal ion but different formal concentrations of the complexing agent. The ratio of these concentrations should usually vary from about 0.1 to 10 or 20. The absorbance of each solution is then measured at a wavelength where the complex ion absorbs but the aquometal ion does not. These absorbances are proportional to the equilibrium concentrations of the complex ion in the solutions, and a plot of the absorbance against the ratio of the number of moles of ligand to the number of moles of metal ion (which is the same as the ratio of the corresponding total or formal concentrations) will resemble Fig. 8-10. An entirely similar graph can be obtained from the results of a spectrophotometric titration if one plots the absorbance, corrected for dilution, against the ratio of the numbers of moles of the ligand and metal ion. The extent of the curvature in the vicinity of the end point depends, of course, on the degree of dissociation of the complex. However, the stoichiometric formula of the complex can be found by extrapolating the straight-line portions of the graph, which is to say that the point at which these lines intersect corresponds directly to the ratio of ligand to metal ion in the complex. This procedure works very well for weakly dissociated complexes. But if the dissociation constant of the complex is too large the molar ratio plot will become a smooth continuous curve and it will be impossible to locate the stoichiometric point. In such cases better results can often be secured by the slope-ratio or continuous-variations methods, which will be described in later paragraphs.

Within a certain rather restricted range, however, the curvature around the "end point" of a molar ratio plot can be turned to good advantage and used for the calculation of the dissociation constant of the complex.

[1] *Cf.* A. S. Meyer, Jr., and G. H. Ayres, *J. Am. Chem. Soc.*, **79**, 49 (1957).

Let us suppose that the metal ion M and the ligand X are found to form a 1:1 complex MX. Suppose that the total concentration of M in the solution is C F; then that of X at the end point will also be C F. If the reaction between M and X at this point were complete, the concentration of MX would be C, and its absorbance would be the same as the absorb-

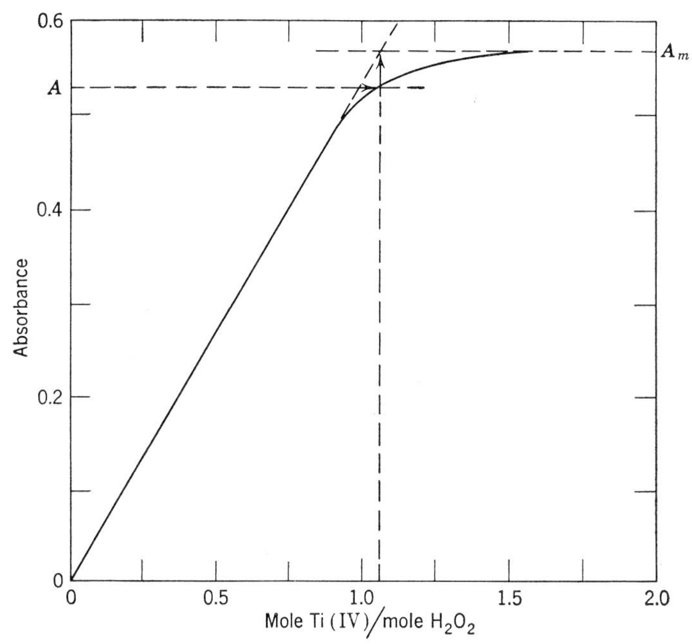

FIG. 8-10. Molar ratio plot for solutions containing 1.01 mM titanium(IV) and varying concentrations of hydrogen peroxide in 0.1 F perchloric acid. The absorbances were measured at 410 mμ, which is the wavelength of maximum absorption for the titanium–peroxide complex. (*W. P. Ferren, M.S. Thesis, Polytechnic Institute of Brooklyn,* 1957.)

ance A_m measured in the presence of so large an excess of X that the dissociation of the complex is virtually completely repressed. Actually, however, the formation of the complex is not quite complete at the "end point," and so the absorbance A at that point is smaller than A_m. These values are indicated in Fig. 8-10. We have, therefore,

$$[\mathrm{MX}] = \frac{A}{A_m} C$$

$$[\mathrm{M}] = [\mathrm{X}] = C - [\mathrm{MX}] = C\left(1 - \frac{A}{A_m}\right)$$

from which the dissociation constant of the complex is

$$K = \frac{[1 - (A/A_m)]^2}{A/A_m} C \tag{8-16}$$

The derivations of the corresponding equations for complexes of different formulas are left to the student.

Since A/A_m depends on C, because the degree of dissociation of the complex increases with increasing dilution, it is advisable to determine K at a number of values of C. The agreement among the values of K thus found gives a valuable indication of the reliability of their mean and helps to guard against the possibility of overlooking a nonabsorbing complex. The best values are generally secured when A/A_m is between about 0.7 and 0.9. If A/A_m is greater than 0.9, the uncertainty in the numerator of the right-hand side of Eq. (8-16) becomes intolerably large, whereas a value of A/A_m smaller than about 0.7 could only result from a plot whose curvature was so extensive that the "end point" could hardly be located with complete certainty. These difficulties with a highly dissociated complex may be partly overcome by increasing C and decreasing the optical path length, but there is no such simple way of dealing with a complex so strong that A/A_m is nearly 1 even in very dilute solutions.

Although the precision which can be attained in the determination of K is greatest at the stoichiometric equivalence point, data secured on either side of this point may also be used. Let us suppose that a number of mixtures containing 10^{-4} F metal ion, M, and varying concentrations of ligand, X, are prepared, and that the absorbance of each is measured at a wavelength where MX absorbs but M and X do not. Let us further suppose that the "end point" of the molar ratio plot is found to occur when the concentration of X is $2 \times 10^{-4} F$, that the absorbance approaches a limiting value of 0.600 in the presence of a large excess of X, and that the absorbance of a solution containing 2.5×10^{-4} F X is 0.450. From these suppositions it follows immediately that the formula of the complex is MX_2. In the solution containing 2.5×10^{-4} F X, the absorbance data give

$$[MX_2] = \frac{0.450}{0.600} \times 10^{-4} = 7.5 \times 10^{-5}$$

At the same time we can write two conservation equations

$$[M] = 1 \times 10^{-4} - [MX_2] = 2.5 \times 10^{-5}$$
$$[X] = 2.5 \times 10^{-4} - 2[MX_2] = 1.0 \times 10^{-4}$$

from which K is found by simple arithmetic.

It is interesting to note that a very similar amperometric method has been used occasionally. A solution of a metal ion is titrated with a ligand solution, and the diffusion current is measured at a potential which is on the plateau of the wave of the simple metal ion but well before the start of the wave of the complex. What is measured here is essentially the concentration of the simple metal ion not yet reacted, but K is readily calculated from this with the aid of the conservation equations for M and X. In this procedure the diffusion current which is measured at any point on the titration curve results from the reduction of nearly all of the simple metal ions in the thin layer of solution surrounding the indicator electrode. This displaces the complex-formation equilibrium, and therefore some of the complex ions in the diffusion layer dissociate to give more simple metal ions which are reduced in their turn. So the concentration of simple metal ion obtained by the amperometric measurement is always too high. The magnitude of the error is often very small, because many complex metal ions dissociate only relatively slowly. But it is easy to see that the spectrophotometric procedure is better than the amperometric one, because the former is not dependent on the validity of any assumption about the rate of dissociation of the complex.

The formula of a complex ion can also be found by the so-called *slope-ratio method.* This can often be used where the molar ratio technique fails, because it is based on absorbance measurements made with solutions containing large excesses of either the metal ion or the ligand to repress the dissociation of the complex, and yet it does not depend on any long uncertain extrapolation. Consider the complex ion formed by the reaction

$$m\,\mathrm{M} + n\,\mathrm{X} = \mathrm{M}_m\mathrm{X}_n$$

As before, we shall assume that it is possible to measure the absorbance of the complex without interference from M or X. If a small amount of M is added to a solution containing a very large excess of X, virtually every one of the M ions added will be complexed, by virtue of the common-ion effect, and we shall have

$$\Delta[\mathrm{M}_m\mathrm{X}_n] = \frac{\Delta C_\mathrm{M}}{m}$$

where ΔC_M is the number of moles of M added to each liter of the solution (*i.e.*, the increase in the formal concentration of M produced by the addition). Let us define the quantity S_M as the slope of a plot of the absorbance *vs.* the formal concentration of M in the presence of a very large constant excess of X. It will be given by the equation

$$S_\mathrm{M} = \frac{\Delta A}{\Delta C_\mathrm{M}} = \epsilon_c \frac{b}{m}$$

where ϵ_c is the molar absorptivity of the complex ion. Similarly, if a small increment of X is added to a solution containing an overwhelming excess of M, the slope of a plot of the absorbance against the formal concentration of X will be

$$S_\mathrm{X} = \frac{\Delta A}{\Delta C_\mathrm{X}} = \epsilon_c \frac{b}{n}$$

From these equations it follows at once that

$$\frac{S_M}{S_X} = \frac{n}{m}$$

from which the formula of the complex is easily deduced.

This method is valid whenever two conditions are satisfied: there must be only one complex formed, and it must obey Beer's law. If a series of successive complexes is produced, misleading results will be secured, for the value of S_M will correspond to the highest of these complexes (*i.e.*, the one containing the highest ratio of X to M), whereas the value of S_X obtained will be that for the lowest complex.

One of the most generally applicable and widely used techniques for elucidating the formula of a complex ion is the *method of continuous variations*. Suppose that a metal ion M and a ligand X react to form a complex MX_n. We prepare a series of solutions in which the *sum* of the formal concentrations of M and X is the same while their *ratio* varies. Representing these formal concentrations by C_M and C_X, respectively, we have

$$C_M + C_X = C$$

Let us define a quantity f such that, in any one of the series of solutions,

$$C_X = fC$$

and

$$C_M = (1 - f)C$$

In any of the mixtures of M and X we have

$$[M] = (1 - f)C - [MX_n]$$
$$[X] = fC - n[MX_n]$$

Obviously the concentration of the complex will change as f is varied: it will be zero when f is either zero or unity, and will have a maximum value at some intermediate point. If the absorbance is measured at a wavelength where the complex absorbs but M and X do not, the value of f at the point of maximum absorbance will naturally also be the value of f which corresponds to the maximum concentration of MX_n. This in turn will be the value of f at the point where $d[MX_n]/df$ is zero. Since

$$[MX_n] = \frac{\{(1 - f)C - [MX_n]\}\{fC - n[MX_n]\}^n}{K}$$

where K is the dissociation constant of the complex, it is easily shown that

$$K\frac{d[MX_n]}{df} = \{fC - n[MX_n]\}^n \left\{-C - \frac{d[MX_n]}{df}\right\} + \{(1 - f)C - [MX_n]\}\left\{C - n\frac{d[MX_n]}{df}\right\} n\{fC - n[MX_n]\}^{n-1}$$

Vosburgh & Cooper, JACS 63, 437 (1941).

Now, on introducing the qualification that $d[MX_n]/df = 0$, this alarming equation collapses into

$$n = \frac{f}{1 - f}$$

Accordingly we plot the absorbances of our solutions against f, the mole fraction of the ligand in the mixture. For a 1:1 complex MX, the absorbance will pass through a maximum at $f = 0.5$; if the complex has the formula MX_2, the maximum will occur at $f = 0.67$; if its formula is M_2X, the maximum will be found at $f = 0.33$; and so on.

The success of this method depends on the same conditions as those which govern the success of the slope-ratio method. The sharpness of the maximum provides a valuable indication of the stability of the complex. If this is only very slightly dissociated, the graph will consist of two straight lines, one passing through zero absorbance at $f = 0$ and the other passing through zero absorbance at $f = 1$, and the point of intersection of these lines will be quite easy to locate with good accuracy. For a moderately dissociated complex there will be considerable curvature resulting from the incompleteness of the reaction around the stoichiometric point. Provided that the complex is neither too weak nor too strong, this curvature can be used to estimate its dissociation constant by a procedure very similar to that employed in dealing with molar ratio data.

Some trouble, however, may be encountered in dealing with an extremely weak complex, for then the curve will be quite flat in the vicinity of the maximum. Of course one may draw one straight line through the points at low values of f, where the dissociation is largely repressed by the excess metal ion present, and another through the points at high values of f, where the dissociation is again largely repressed by the excess ligand. The point of intersection of these two lines then gives the stoichiometric value of f. Even with this technique one might expect difficulty in distinguishing between a 1:3 and a 1:4 complex, whose maxima would occur at $f = 0.75$ and $f = 0.80$, respectively, but fortunately this problem is very rare. The continuous variations method is one of the best available for the study of even very weak complexes.

To guard against the possibility that the solutions prepared actually contain two (or more) complexes, some of which are overlooked because they happen not to absorb at the particular wavelength used in the absorbance measurements, it is customary to carry out the continuous variations measurements at several different wavelengths and also at several different values of C. If neither of these changes has any effect on either the shape of the graph (apart from changing all of the values by a constant factor) or the value of f at the maximum, it is safe to conclude that only a single complex is formed.

Another spectrophotometric technique for finding the dissociation constant of a complex might be called the *matching absorbance method.* Two solutions are prepared which contain different M/X ratios, and the absorbance of each is measured at a wavelength where only the complex absorbs. The solution of higher absorbance is then diluted until its absorbance is exactly the same as that of the other solution. From the known formal concentrations of M and X in the original solutions and the extent to which one was diluted we can calculate the formal concentrations of M and X in the two equally absorbing solutions finally prepared. Let us call these concentrations $C_{\text{M},A}$, $C_{\text{X},A}$, and so on. If one of the methods previously described has shown that the complex has the formula MX, the conservation equations for M and X give

$$\begin{aligned}
[\text{M}]_A &= C_{\text{M},A} - [\text{MX}]_A \\
[\text{X}]_A &= C_{\text{X},A} - [\text{MX}]_A \\
[\text{M}]_B &= C_{\text{M},B} - [\text{MX}]_B \\
[\text{X}]_B &= C_{\text{X},B} - [\text{MX}]_B
\end{aligned}$$

When the absorbances of solutions A and B are the same, evidently the concentration of MX in each is equal to some (unknown) value c. Then

$$K = \frac{(C_{\text{M},A} - c)(C_{\text{X},A} - c)}{c} = \frac{(C_{\text{M},B} - c)(C_{\text{X},B} - c)}{c}$$

By eliminating c we can compute K; the algebraic details are left to the student. The absorbance matching is best done by using the solution which originally contains the smaller concentration of the complex to make the zero absorbance setting of the spectrophotometer. The cell containing this solution is then replaced by another of equal path length containing a known volume of the other solution, which is diluted until the spectrophotometer again indicates zero absorbance.

The method is well suited to the study of very weak complexes. It yields the most precise results when the solutions are prepared in such a way that both $C_{\text{M},A}C_{\text{X},A} - C_{\text{M},B}C_{\text{X},B}$ and $(C_{\text{M},A} + C_{\text{X},A}) - (C_{\text{M},B} + C_{\text{X},B})$ are widely different from zero. In general, this will be impossible when K is small, and in fact complexes for which this is the case are almost always more amenable to study by electrometric than by photometric methods. On the other hand, very weak complexes are inherently extremely difficult to study by electrometric techniques, and so the electrometric and photometric methods serve as valuable complements to each other.

PROBLEMS

The answers to the starred problems will be found on page 403.

*1. 1-Nitrobutane has a single absorption band in the visible and near ultraviolet;

its wavelength of maximum absorption in dilute ethanolic solutions is 277 mμ. What is the energy, in kilocalories per mole, required to bring about the electronic transition involved? The index of refraction of ethanol is 1.362.

2. One of the "benzenoid" bands on a spectrum of benzene in ethanolic solutions has its absorbance maximum at a wavelength of 248 mμ. At what wavelength should the corresponding point appear on a spectrum of a solution of benzene in cyclohexane if the indices of refraction of ethanol and cyclohexane are 1.362 and 1.429, respectively?

***3.** A 1-cm. layer of a certain potassium permanganate solution has a transmittance of 0.800 at 525 mμ. What would be the transmittance of a 5-cm. layer of the same solution under the same conditions?

4. A number of solutions of "dithizone"

(diphenylthiocarbazone, $S{=}C\begin{matrix}\diagup NH{-}NH{-}C_6H_5 \\ \diagdown N{=}N{-}C_6H_5\end{matrix}$)

in chloroform are prepared and found to have the following transmittances at 610 mμ in 1-cm. cells:

Concentration, M	*T, %*
1.00×10^{-6}	90.7
3.50×10^{-6}	70.3
1.17×10^{-5}	31.0
2.11×10^{-5}	11.9

a. Is Beer's law obeyed under these conditions?
b. If so, what is the molar absorptivity of dithizone in chloroform solutions?

***5.** The molar absorptivity of acetone in ethanolic solutions is 2.15×10^4 at 362 mμ. What will be the absorbance of a 2-cm. layer of a solution containing 1 mg. of acetone per liter of ethanol, measured at 362 mμ?

6. The following table gives the absorbances of a number of standard solutions of nickel(II) in a concentrated thiocyanate medium under certain experimental conditions:

Concentration of nickel, F	*Absorbance*
2.00×10^{-4}	0.033
5.00×10^{-4}	0.081
1.25×10^{-3}	0.195
3.12×10^{-3}	0.440
7.81×10^{-3}	0.855

a. Is Beer's law obeyed by these data?
b. What is the concentration of nickel in a solution whose absorbance, measured under the same experimental conditions, is 0.300?

7. Show that the limiting value of the slope of an absorbance–concentration curve obtained with a wide-band filter photometer as c approaches zero is equal to the constant value of the slope of a similar curve secured with a spectrophotometer operated at extremely narrow effective band widths.

8. A sulfuric acid solution of cupric sulfate was analyzed for cupric ion by the following procedure. Exactly 5 ml. of the unknown was transferred to an absorption cell, and its absorbance was found to be 0.123 at the wavelength of maximum absorption for cupric ion. Exactly 1 ml. of a standard 0.01000 F cupric sulfate solution was added, and the absorbance of the mixture was found to be 0.204 at the same wavelength. Another 1-ml. portion of the standard solution was added, and the absorbance of this mixture was found to be 0.262. What was the concentration of the unknown solution?

***9.** A color-forming reagent was added to a certain unknown at exactly 12:15 P.M., and the absorbance of the resulting mixture was measured at a number of times thereafter. What would the absorbance have been at 12:15 if the reaction responsible for the formation of the absorbing species had been instantaneous? State explicitly any assumption made regarding the kinetics of the decomposition of the absorbing species.

Time	*A*	*Time*	*A*
12:20	0.153	1:30	0.480
12:30	0.282	1:50	0.469
12:45	0.395	2:30	0.416
1:10	0.469	4:15	0.353

10. Prove the statement made on page 283 that the minimum relative error in c occurs at a transmittance of 36.8 per cent, and find the minimum value of the ratio of the relative error in c to the absolute photometric error.

11. It is proposed to determine the concentrations of vanadium(V) in a number of samples by adding enough sulfuric acid to make its concentration about 1 F, followed by an excess of hydrogen peroxide. The wavelength of maximum absorption for the yellow peroxyvanadic acid formed by this treatment is 455 mμ, and the absorbance at that wavelength of a solution known to contain 8.0 mg. of vanadium per liter is 0.55. Within what range should the concentrations of vanadium in the treated solutions fall in order that maximum precision will be secured in the analyses?

12. The molar absorptivity of ascorbic acid is 395 at its wavelength of maximum absorption, 265 mμ. A certain unknown solution of ascorbic acid had an absorbance of 0.583 in a 1-cm. cell at 265 mμ. However, it was known to contain an impurity which also absorbed at 265 mμ, and air was therefore bubbled through it until the ascorbic acid had been quantitatively oxidized. The oxidation product of the ascorbic acid does not absorb at 265 mμ. The impurity was not affected by this treatment. If the absorbance of the oxidized solution was 0.097, what was the concentration of ascorbic acid initially present?

***13.** A spectrophotometer was adjusted to read zero transmittance with the phototube in darkness, then to read zero absorbance with a 1.113×10^{-4} F solution of o-phenanthroline ferrous sulfate

N N Fe 3 ++ $SO_4^{=}$

in the light path. The cell was emptied and refilled with a 1.276×10^{-4} F solution of o-phenanthroline ferrous sulfate; the spectrophotometer now indicated an absorbance of 0.264. Finally the cell was filled with an unknown solution of o-phenanthroline ferrous sulfate and, without changing any of the instrument settings, the indicated absorbance was found to be 0.399. From other data it was known that o-phenanthroline ferrous sulfate obeys Beer's law under these conditions. Find the concentration of the unknown solution.

14. An absorption cell whose optical path length was roughly 1 cm. was filled with a 1.512×10^{-3} F solution of potassium permanganate. A spectrophotometer was adjusted to read zero transmittance with this cell in the optical path and with the wavelength dial set at 525 mμ, which is the wavelength of maximum absorption for permanganate ion. The cell was emptied and refilled with a 1.378×10^{-3} F permanganate solution, and this was used to set the 100 per cent transmittance reading of the instrument. Finally the cell was filled with an unknown permanganate solution, and the measured transmittance of this solution at 525 mμ was found to be 0.679. What was the concentration of the unknown permanganate solution?

***15.** The wavelength of maximum absorption for the acidic form of methyl red is 528 mμ, and that for the alkaline form is 400 mμ. A 1.22×10^{-3} F solution of methyl red in 0.1 F hydrochloric acid has absorbances of 0.077 and 1.738 at 400 mμ and 528 mμ, respectively; a 1.09×10^{-3} F solution of methyl red in 0.1 F sodium bicarbonate (pH = *ca.* 8.8) has absorbances of 0.753 and 0.000 at these wavelengths. A small amount of the indicator is dissolved in a buffer containing 0.1 F potassium acetate and an unknown and immaterial concentration of acetic acid. The pH of the resulting solution is 4.31 and its absorbances are 0.166 and 1.401 at 400 mμ and 528 mμ. The same absorption cell was used in all of the absorbance measurements. What is the formal acid dissociation constant of methyl red at an ionic strength of 0.1?

16. Three solutions of a certain indicator, HInd, in aqueous sodium hydroxide are prepared, and their absorbances are measured under such conditions that the absorption is entirely due to the indicator anion, Ind^-. Their compositions and measured absorbances are given below. Compute the formal dissociation constant of HInd at $\mu = 0.2$.

Composition	*Absorbance*
1×10^{-5} F HInd in 0.1 F NaOH–0.1 F $NaClO_4$	0.302
2×10^{-5} F HInd in 0.1 F NaOH–0.1 F $NaClO_4$	0.604
1×10^{-5} F HInd in 0.2 F NaOH	0.565

17. A newly synthesized organic acid was known to be monobasic and to have a formula weight of 297. Exactly 10 mg. of the pure acid was weighed into the absorption cell used in securing the data of Problem 15. It was dissolved in 5 ml. of 0.1 F sodium perchlorate; then exactly 0.150 ml. of 0.1000 F sodium hydroxide and one drop of a dilute methyl red solution were added. After thorough mixing, the absorbances of the resulting solution were measured and found to be 0.464 at 400 mμ and 0.927 at 528 mμ. What is the formal dissociation constant of the acid at an ionic strength of 0.1? (Use the answer to Problem 15 in the calculations.)

***18.** A certain metal ion M is reported to form a 1:1 complex with a ligand X. At the wavelength of maximum absorption by the complex, where neither M nor X exhibits measurable absorption, a solution containing 1.00 F M and 1.00×10^{-4} F X has an absorbance exactly equal to that of a solution containing 5.00×10^{-3} F M and 5.00×10^{-3} F X. What is the dissociation constant of the complex?

19. Some time after the data given in Problem 18 were secured, it began to seem doubtful that the formula of the complex really was MX, and the following measurements were made in an attempt to resolve this question. Four solutions containing known formal concentrations of M and X were prepared, and their absorbances were measured at the wavelength of maximum absorption by the complex. The compositions and absorbances of the solutions were as given in the following table. Unless these data show that the complex is indeed MX, find its actual dissociation constant under the conditions of Problem 18. These new measurements were made with a 5-cm. absorption cell. Find the molar absorptivity of the complex.

[M], *F*	[X], *F*	*A*
1.00	2.33×10^{-4}	0.127
1.00	5.91×10^{-4}	0.334
3.89×10^{-4}	0.98	0.443
5.21×10^{-4}	0.98	0.593

20. Ferric and mercuric ions are known to form chelates having the formulas FeX_3^{+++} and HgX_2^{++} with a certain chelating agent. A solution containing 1.00×10^{-3} *F* X and 1.00×10^{-3} *F* ferric iron is prepared and found to have an absorbance of 0.740 at a wavelength where the ferric chelate absorbs but where the mercuric chelate, ferric and mercuric ions, and the chelating agent are all nonabsorbing. Another solution, containing 1.00×10^{-3} *F* X, 1.00×10^{-3} *F* ferric iron, and 1.00×10^{-3} *F* mercuric mercury, is found to have an absorbance of 0.530 under the same conditions. The ferric chelate is known to obey Beer's law and to have a dissociation constant of 3.1×10^{-20}. What is the dissociation constant of the mercuric chelate?

CHAPTER 9

THE ABSORPTION OF INFRARED RADIATION

by Robert P. Bauman

9-1. Introduction. Infrared spectroscopy[1] may be considered to be an extension of the experimental methods and the general theory already encountered in Chap. 8. The principal differences lie in the variations of experimental technique, which are required, for example, by the necessarily different optical materials, sources, and detectors, and in the different molecular mechanism for the absorption, which involves vibrational and rotational activation of the molecules rather than electronic activation. In marked contrast, however, to the visible and near ultraviolet regions, all substances show absorption in the infrared region, and in all but a few cases, primarily alkali halides, this absorption includes one or more bands within the region commonly examined by commercial spectrometers.

It will be remembered that absorption in the ultraviolet or visible is nearly always dependent on the existence of unsaturation of some form in the absorbing compound or on the presence of certain metal ions, so that relatively few compounds absorb and even these few often show only a single broad band. The interpretation of infrared spectra is made more difficult by the very large number of absorption bands commonly observed and by the virtually unlimited possibilities which would seem to exist for compounds that might be contributing to a given absorption pattern. At the same time, the large number of bands in the infrared region provides a greatly increased number of data which may be brought to bear on the identification problem. In fact, so many data are available, in the form of frequencies and intensities of the absorption bands, that the identification has been shown to be unique for all substances studied thus far. To take full advantage of this wealth of information a number of projects are being actively pursued to apply high-speed com-

[1] The terms "spectroscopy" (or "absorption spectroscopy"), "spectrometry," "spectrophotometry," and "absorptimetry" are equivalent in meaning for the present purposes, as are the terms "spectrometer" and "spectrophotometer." In using one pair of terms in Chap. 8 and another pair here, we have followed the usage currently most prevalent among analytical chemists.

puters to the task of analyzing unknown spectra, qualitatively and quantitatively, for the presence of pure compounds whose spectra have been previously obtained and supplied to the computer's memory capacity.

Fortunately for the practicing spectroscopist and the chemist who cannot wait several years to have his sample analyzed, the high-powered methods of digital computers or card-sorters are not required for most identifications. Recognizing that any vibration of a molecule will involve the entire molecule (hence the uniqueness of the spectrum), it is still true that, to a very good approximation, certain bands within the spectrum may be assigned to particular types of chemical bonds or to specific functional groups. For the chemist who knows what he put into his reaction flask and, perhaps, the C/H/O ratio of his product, the determination within a few minutes that carbonyl, hydroxyl, amino, aliphatic C—H, ortho-, meta-, or para-substituted aromatic, or ether groups, for example, are present or absent will usually be the necessary key to determining the nature of the unknown compound. With an informed guess as to the specific compound present, the spectrum of this unknown may be compared with the spectra of appropriate known compounds. A suitable match of absorption spectra is a much better indication of the nature of the unknown than is identity of melting or boiling point or any other single datum which may be obtained. Unlike the latter, the infrared spectrum is generally insensitive to minor components (below 1 per cent), and even though impurities do produce small effects these will not often increase the difficulty of identifying the main component.

The very near infrared, or "overtone" region meets the visible region at about 12500 cm.$^{-1}$ (0.8 μ) and extends to about 4000 cm.$^{-1}$ (2.5 μ). The "fundamental" region of the infrared, which is by far the most frequently observed range, extends from the very near infrared to about 650 or 700 cm.$^{-1}$ (15 or 16 μ) or even as far as 200 or 300 cm.$^{-1}$ (30 to 50 μ). The "far infrared," which tends to be pushed continually toward lower frequencies by the development of better optical materials, extends from this nebulous boundary to about 25 cm.$^{-1}$ (400 μ) or even less. Beyond this, and now overlapping it to some extent, is the microwave region, which is merely an extension of the far infrared from the theoretical viewpoint but which involves drastically different experimental techniques. The far infrared and microwave regions yield information about the rotations of molecules in the gas phase, or, in a few cases, about low-energy crystal lattice vibrations. The present discussion will be restricted to those regions which can be explored with commercial infrared instruments and which arise primarily from vibrational modes of the molecules.

9-2. Mechanism of Absorption. From laws of classical mechanics it is easy to demonstrate that a vibrating system can efficiently absorb energy from a series of impulses only if the impulses strike the system with a frequency that is very near the natural frequency of the vibrator. Quantum mechanics treats the problem in a very different manner, but comes to the same conclusion. A molecule, or to a good approximation a bond within a molecule, may absorb radiant energy, therefore, if the natural frequency of vibration of the molecule is the same as the frequency of the radiation, and if the vibration which is to be stimulated will produce a change in dipole moment. The latter condition is nearly always satisfied unless it is made impossible by the symmetry of the molecule, but the magnitude of the change in dipole moment may in some cases be quite small, producing only weak absorption bands. It is also possible, because of deviations of any molecule from the ideal behavior of a simple harmonic oscillator, for the molecule to absorb radiation of a frequency equal to the sum of two vibrational frequencies. These "combination" bands are usually relatively weak, and will not be specifically considered here, but they contribute to the complexity of the observed spectrum.

The vibrational frequency of a molecule or functional group is determined by the geometrical arrangement of all of the atoms in the molecule, by the masses of the atoms involved in the motion, and by the strength, at the equilibrium position, of the bond or bonds which will be affected. For a diatomic molecule the wavenumber of the absorption, expressed in reciprocal centimeters, is given by

$$\sigma = \frac{1}{2\pi c} \sqrt{\frac{k}{m}} \qquad (9\text{-}1)$$

where k expresses the strength of the bond and is called its *force constant*, c is the speed of light, and m is the *reduced mass* of the two atoms:

$$\frac{1}{m} = \frac{1}{m_1} + \frac{1}{m_2}$$

The (vacuum) wavenumber σ is the frequency divided by the speed of light, or is the reciprocal of the vacuum wavelength.

Polyatomic molecules follow the same trend. Motions involving hydrogen atoms are thus found at much higher frequencies, or wavenumbers, than are motions involving heavier atoms. Motions involving double bonds absorb at higher frequencies than those involving single bonds, and triple bonds give rise to absorption of still higher frequencies. Bending motions, such as the opening and closing of the H—C—H angle in a CH_2 or CH_3 group or the bending of a carbon chain, produce lower frequency absorption, extending even beyond the range of most spectrometers. Figure 9-1 shows the spectra of phenol and nylon, with many

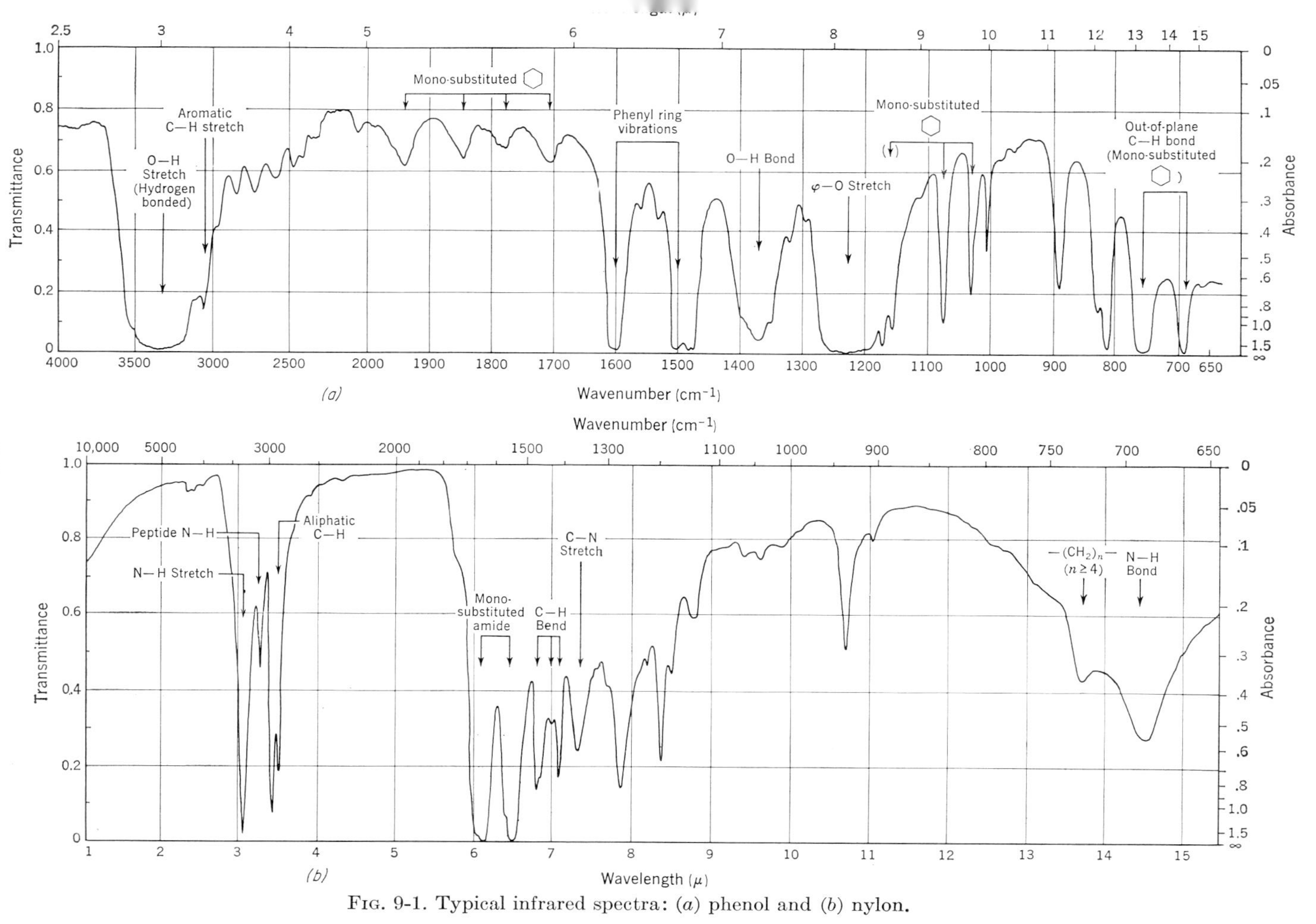

Fig. 9-1. Typical infrared spectra: (a) phenol and (b) nylon.

of the bands labeled to show their origin. It is usually not possible to assign all of the observed bands to particular molecular motions except for reasonably small or symmetrical molecules which have been thoroughly studied.

Molecules in the gaseous state cannot rotate with any arbitrary angular velocity, but are restricted to certain discrete rotational velocities.[1] In condensed states, neighboring molecules so disturb these rotational energy levels that they become merged into a continuum (except in the case of H_2, where the levels are so widely spaced that they are not merged in either liquid or solid). Rotational energy states do not affect the spectra of solids and liquids, therefore, but they do play a fundamental role in determining the shapes and intensities of absorption bands of gases. At room temperature, the molecules of a gas are distributed among the various possible rotational levels according to the Boltzmann distribution law. When radiant energy is incident on a molecule it will, in general, not only excite a particular vibrational mode but will also increase or decrease the rotational energy of the molecule (always to an adjacent rotational level). What would be observed, therefore, if the spectrum were to be examined with an instrument capable of showing the full details of the absorption band, would be a series of sharp lines which would not fall in quite the same place since the difference in energy between, for example, the tenth and eleventh rotational energy levels will be greater than that between the ninth and tenth.

If it is possible for the molecule to absorb energy to start vibrating (or to vibrate more energetically) without a concurrent rotational energy change, we may expect a sharp absorption band (Q branch) corresponding to the vibrational frequency. But at higher frequencies there will be the sharp bands which correspond to the sum of the changes of vibrational and rotational energies (R branch), and at lower frequencies there will be the sharp bands which correspond to the vibrational frequency minus the changes in rotational energy (P branch). Observed with a real spectrometer, these bands become merged into three comparatively broad bands (the P, Q, and R branches) or, especially for heavy molecules, even into two bands or only a single band. Figure 9-2 shows this effect. The exact structure of the rotation-vibration bands is thus not important to the casual spectroscopist except that the over-all or apparent intensity of a band will depend rather markedly on the pressure of the gas because of this fine structure. The effect of this "pressure broadening" will be discussed in Sec. 9-6.

9-3. Instrumentation. The design of infrared spectrometers differs little from the design of visible-ultraviolet spectrometers, although the optical materials which are used, as well as the sources and detectors,

[1] The justification for this statement may be found in texts on quantum mechanics.

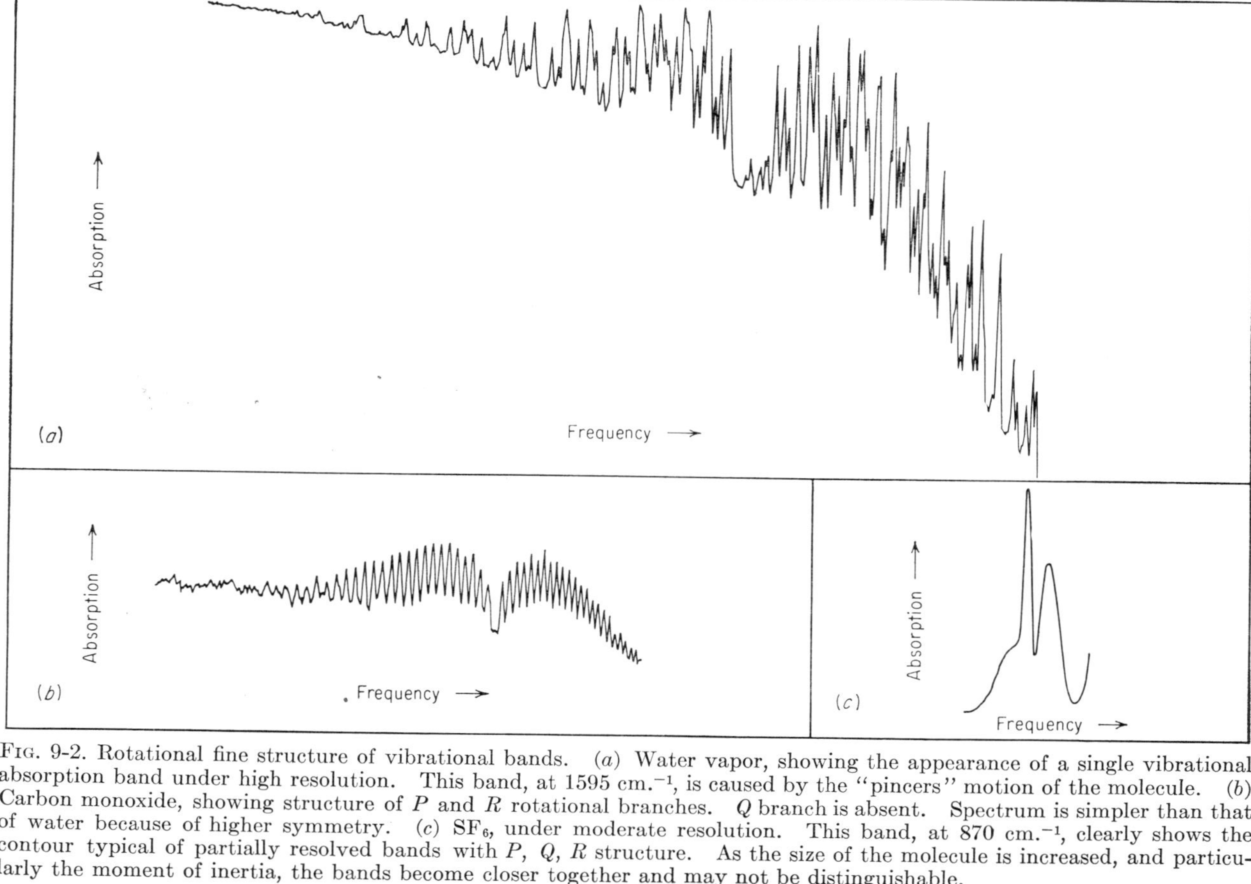

FIG. 9-2. Rotational fine structure of vibrational bands. (*a*) Water vapor, showing the appearance of a single vibrational absorption band under high resolution. This band, at 1595 cm.$^{-1}$, is caused by the "pincers" motion of the molecule. (*b*) Carbon monoxide, showing structure of *P* and *R* rotational branches. *Q* branch is absent. Spectrum is simpler than that of water because of higher symmetry. (*c*) SF_6, under moderate resolution. This band, at 870 cm.$^{-1}$, clearly shows the contour typical of partially resolved bands with *P*, *Q*, *R* structure. As the size of the molecule is increased, and particularly the moment of inertia, the bands become closer together and may not be distinguishable.

must of necessity be different. For the very near infrared even these differences are quite minor. Photomultiplier tubes are rather insensitive in this region, but the photoconductive cells, such as the lead sulfide cell, are quite satisfactory. Quartz and silica are transparent in this region, and in part of it even show better dispersion than in the visible. Over most of the region, however, the dispersion of quartz or silica is not fully satisfactory, so that the use of a double monochromator or a very large prism is advisable for most studies in the region. One commercial spectrometer solves the problem of decreased silica dispersion by employing silica and grating monochromators in sequence; in the very near infrared the grating carries the major load, whereas in the ultraviolet (between 250 and 185 mμ) the silica prism is the more effective dispersing agent.

In the fundamental region of the infrared, below 3600 cm.$^{-1}$, glass and silica are completely unsatisfactory because they absorb very strongly. It is necessary, therefore, to use an ionic crystal that will have vibrations of rather low frequency. Lithium fluoride is excellent to about 1700 cm.$^{-1}$, calcium fluoride to 1200 cm.$^{-1}$, sodium chloride to 650 cm.$^{-1}$, potassium bromide to 425 cm.$^{-1}$, cesium bromide or thallium bromide–iodide ("KRS-5") to 300 cm.$^{-1}$, and cesium iodide to 200 cm.$^{-1}$. It will be observed that, as predicted by Eq. (9-1), increasing the masses of the ions decreases the frequency of the absorption band.

Unfortunately it is not possible to use a single prism over the entire region, even though the prism material might be transparent. The dispersive power of a prism arises from its ability to differentiate among frequencies, which means that the refractive index must be different for adjacent frequencies. The change of refractive index with frequency $dn/d\nu$ becomes appreciable only in the vicinity of an absorption band; hence we have the paradox that a prism becomes best as a dispersing agent when it becomes worst as a transmitting medium. The choice of a prism must always involve a compromise between these opposing factors, for each prism has a region of optimum efficiency just above its absorption cutoff, and ideally one should use several prisms to scan the infrared region, as shown in Fig. 9-3. In practice, however, most analytical spectroscopists employ sodium chloride for the entire region from 4000 to 650 cm.$^{-1}$. If better performance is desired for a particular problem they use a lithium or calcium fluoride prism (or, more recently, a grating) for this task. Cesium bromide is the best choice below 650 cm.$^{-1}$.

Sodium chloride must be protected from high concentrations of water vapor, to which potassium and cesium bromides are even more susceptible. Air conditioning of the laboratory is highly desirable, both to control the humidity and to maintain constant temperature. If only sodium chloride is employed, this is much less necessary, especially since it is now common practice to thermostat the monochromator at a level above room temperature. In practice one finds that sodium chloride is surprisingly rugged, and cells with sodium chloride windows may be handled routinely for several years in the laboratory without excessive fogging of the surface if the relative humidity is maintained near or below 50 per cent. Fogging of a prism is far more serious than fogging of cell windows, however, since cell windows can easily be repolished by hand methods but a prism cannot.

Mirror optics are strongly preferred in the infrared, as in the ultraviolet, except for points in the optical system where focus is not critical.

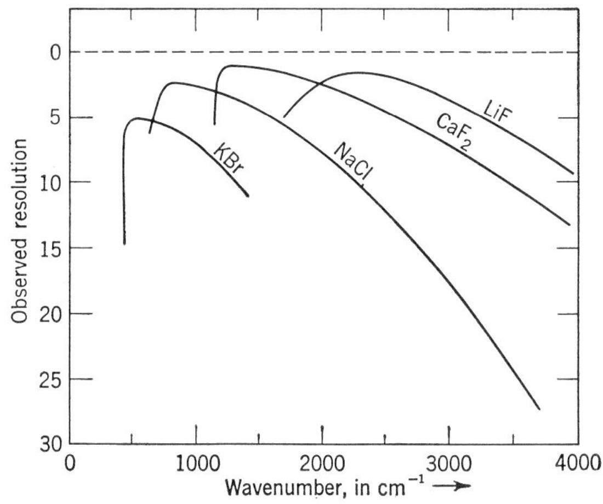

FIG. 9-3. Observed spectral resolution for various prism materials as a function of frequency [based on data of Gore, McDonald, Williams, and White, *J. Opt. Soc. Amer.*, **37**, 23 (1947)]. Optimum performance is secured in the region where the resolution, expressed in cm.$^{-1}$, has its smallest numerical values.

Potassium bromide lenses are occasionally used for this application. All instruments now on the market employ basically the same optical system, the Littrow mount, which reflects the beam from a plane mirror behind the prism and returns it through the prism a second time. The prism thus acts twice upon the beam, doubling the dispersion produced. A recent design, known as the "double-pass" system, goes a step farther by returning the beam through the prism again, producing a total of four passes through the prism with an attendant improvement in resolution. These systems are shown in Fig. 9-4.

The inherent complexity of infrared spectra compared to most ultraviolet and visible absorption patterns strongly encouraged the early development of recording instruments, since point-by-point plotting is extremely time-consuming. Most instruments sold today are also of the

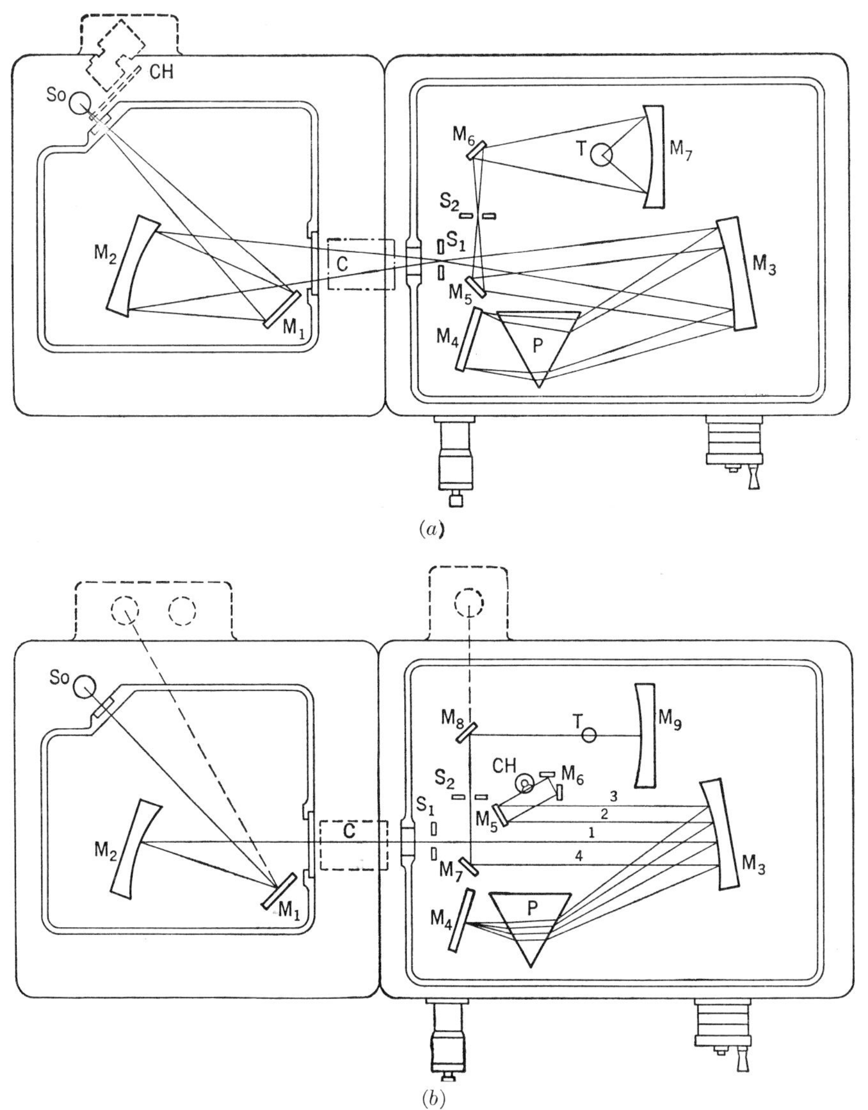

FIG. 9-4. Optical diagram of a single-beam instrument. (*a*) Conventional Littrow mount. The beam from the entrance slit S_1 is collimated by the parabolic mirror M_3 and sent through the prism. The Littrow mirror M_4 returns the beam through the prism to M_3. It is then reflected by M_5 onto the exit slit S_2 and finally by plane and elliptical mirrors onto the thermocouple detector. The sample is placed adjacent to S_1, between the Globar source S_0 and the monochromator housing. (*b*) The same monochromator modified to the "double-pass" system. After passing through the prism, to the Littrow mirror, and back to the parabolic mirror, the beam passes between mirrors M_5 and M_6, then returns to M_3, the prism and Littrow mirror, before striking M_7 and passing on to the exit slit. A rotating segment chops the beam between M_5 and M_6; since the radiation which follows the path shown in (*a*) is unchopped, the signal arising from this is not amplified and is thus eliminated from the recording. (*Courtesy of the Perkin-Elmer Corporation, Norwalk, Conn.*)

double-beam type described in Sec. 8-8, in which the signal transmitted by the sample is compared with that transmitted by the reference material (which is usually nothing more than the laboratory atmosphere). Variations of source output, detector sensitivity, and monochromator transmittance are eliminated from the record in this design. A schematic diagram of a typical instrument is shown in Fig. 9-5.

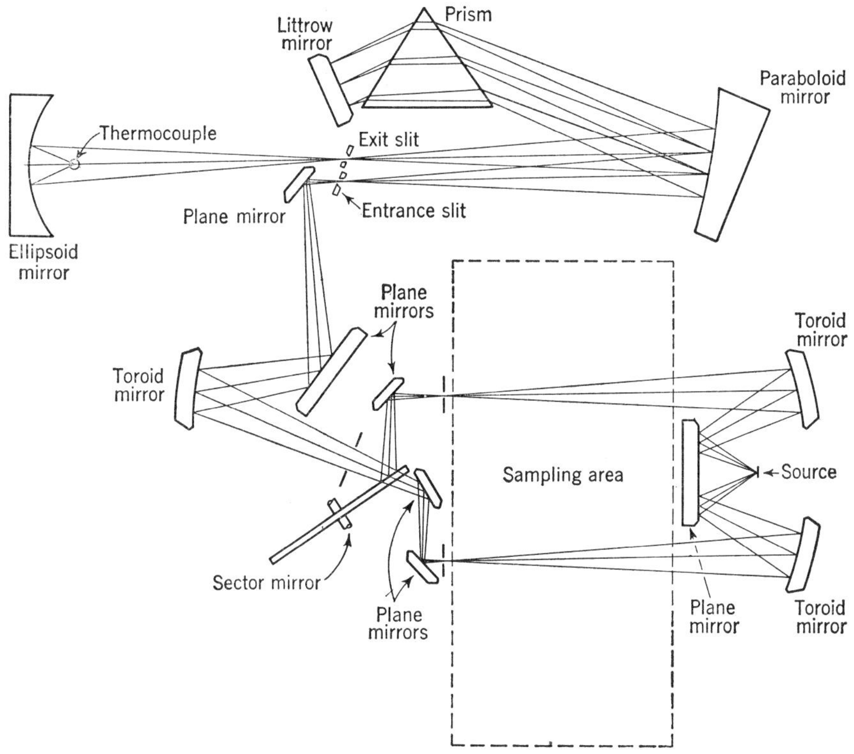

FIG. 9-5. Optical diagram of a double-beam instrument. Radiation from the source is collected by plane and toroidal mirrors and passed through the sampling area. The rotating sector alternately passes the sample beam or reflects the reference beam onto another toroidal mirror which focuses the radiation onto the entrance slit. The two beams are thus combined (though separated in time) before passing into the monochromator. The monochromator is very similar to the single-beam, single-pass monochromator of Fig. 9-4*a*. (*Courtesy of the Perkin-Elmer Corporation, Norwalk, Conn.*)

Two problems of instrumentation—sources and detectors—continue to plague the infrared spectroscopist. The best source of infrared radiant energy that has been found is an approximate "black-body" radiator, an object which has an energy output similar to that shown in Fig. 9-6. The total radiation (given by the area under the curve) is proportional to the fourth power of the absolute temperature, but the frequency corresponding to the maximum of the curve is proportional to temperature and is already into the very near infrared by the time the total energy output is significant ("red heat"). Increasing the temperature will therefore markedly increase the intensity of very near infrared and visible frequencies, but will have very little effect on the lower, more desired, frequencies. The primary effect of a temperature increase is to increase the amount of unwanted radiant energy, a certain fraction of which will find its way through the monochromator as stray light, producing a

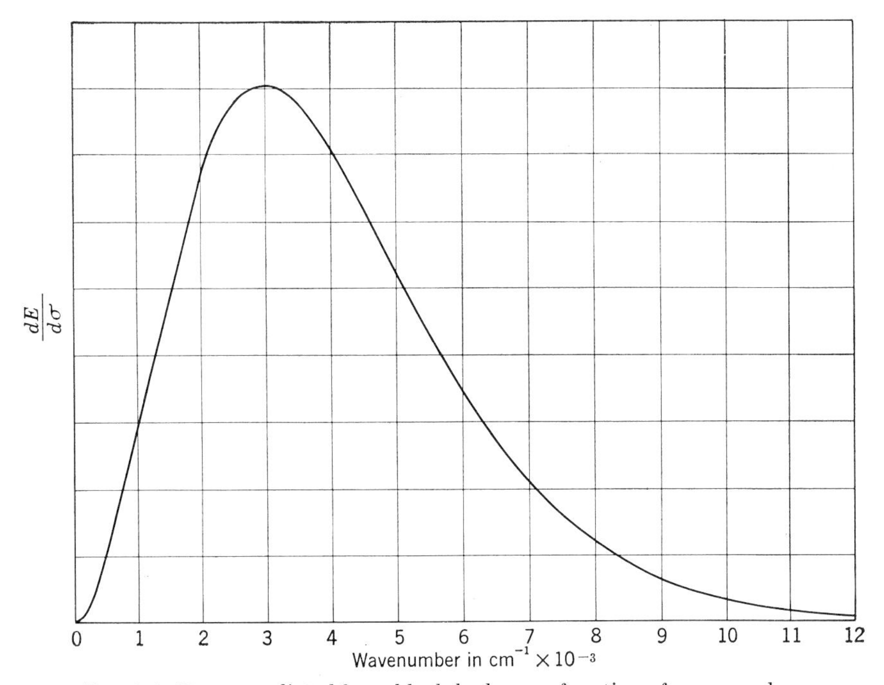

FIG. 9-6. Energy radiated by a black body as a function of wavenumber.

spurious signal. A silicon carbide rod, known as a Globar, and a cylinder of rare earth oxides, the Nernst glower, are the sources commonly found in commercial infrared spectrometers.

Throughout most of the infrared region it is necessary to use detectors which must be considered quite unsatisfactory in comparison with the detectors available for other regions of the spectrum. Several thermal detectors, which depend on the heating effect of radiant energy, are available, including the thermocouple or thermopile, the bolometer, and the Golay pneumatic detector. These have essentially uniform responses for all frequencies from the far ultraviolet to the far infrared, measured in terms of detector signal per watt of incident power. This means, of course, that the response measured in terms of detector signal per incident photon will be proportional to the frequency, and is thus low for infrared frequencies. In the ultraviolet and visible regions the photomultiplier is so vastly superior (roughly a factor of 10^5 in sensitivity) that one would not consider building a thermocouple into a spectrometer for this region. Similarly, in the very near infrared the photoconductive cell is the only possibility seriously considered. But in the infrared region below the range of the photoconductive cells there is nothing better than the thermal detectors. There is very little difference in efficiency among these; the thermocouple is most commonly employed.

To prevent the very weak signal which originates at the detector from being lost in the stray signals picked up by the wires leading away from the thermocouple, it is common practice to place a preamplifier as close to the thermocouple as is reasonably possible. The amplified signal is then fed to the main amplifier, and then to the recorder. Amplifier design principles are, of course, independent of the type of radiant energy which is being measured.

9-4. Sample Handling and Technique. The most striking of the differences between ultraviolet-visible and infrared spectroscopy is the problem of sample handling. Spectra in the former region are nearly always obtained in glass or silica cells, usually of 1 cm. thickness, with the sample dissolved in any of a large number of solvents—including water, alcohols, saturated hydrocarbons, or, for some regions, acetone, benzene, or chlorinated hydrocarbons. Once one has prepared the specific compound or complex to be investigated, it is seldom a problem to know how to handle the sample. The only real difficulty is encountered in dealing with either insoluble powders or fluorescent or photochemically unstable materials. In the infrared much of this pattern is completely reversed. There is no solvent which is nonabsorbing throughout even the "rock salt region," 4000 to 650 cm.$^{-1}$, nor is there any rugged material comparable to glass or silica which is transparent over this region. But powders, which would scatter strongly in the visible region and especially in the ultraviolet, can be used directly in the infrared region if certain precautions are taken to minimize the scattering. Because of the low energies of infrared photons, fluorescence and decomposition are very rarely, if ever, encountered.

Samples which are liquid at room temperature are usually scanned in their pure form. The cells must be quite thin, of the order of 0.05 mm. or less. Strongly absorbing pure liquids may be observed qualitatively by placing them between rock salt plates, with or without spacers, to give even shorter path lengths. If the sample is solid but is readily melted, it may be heated and scanned as a liquid between plates. Solidification of the sample between the plates should be avoided; the solid may often give an excellent spectrum, but orientation effects may change the positions and intensities of the bands.

Carbon disulfide and carbon tetrachloride, because of their simplicity, heavy atoms, and high symmetries, are the most transparent solvents for the infrared, and are the most commonly employed. Although each has regions of absorption, these regions overlap very little, so that the combination of the two solvents is sufficient to allow the entire range to be examined. Samples which are not soluble in these compounds must be dissolved in a somewhat larger number of solvents, such as chloroform, tetrachloroethylene, methylene chloride or bromide, or bromoform in order to observe the full range from 4000 to 650 cm.$^{-1}$. To minimize solvent absorption, high concentrations of solute are preferred, and cells having thicknesses between 1 mm. and 0.05 mm. are most common.

The amount of scattering by a fine powder is proportional to the fourth power of the frequency, and depends on the size of the particles and the difference in refractive index at each of the surfaces. Bathing the powder in a medium of comparable refractive index will therefore greatly reduce the amount of scattering. For finely ground powders dispersed in a medium such as mineral oil or potassium bromide the scattering at infrared frequencies may be quite negligible. The thick slurry produced by mulling a compound with a refined mineral oil, such as Nujol, is commonly referred to as a Nujol mull. The chief disadvantage of the mull technique is that the absorption by the hydrocarbon oil tends to mask the presence of absorption by C—H bonds of the sample. This is usually no real disadvantage, since the chemist nearly always knows whether or not his samples contain carbon and hydrogen. Masking of olefinic C—H or the methylene/methyl ratio may, however, represent a more serious loss of information. A much less obvious disadvantage of the method is the fairly high degree of skill required to make mulls that show low scattering losses and sharp absorption bands, although this skill is readily acquired by practice. If it is desired to examine the C—H stretch region around 3000 cm.$^{-1}$, other mulling agents, especially perfluorocarbons and hexachlorobutadiene, may be preferred.

A much more recently developed method, suitable for either the infrared or the ultraviolet and visible, is the potassium bromide pellet method. If a sample is very finely pulverized and mixed with potassium bromide powder, the mixture may be pressed under vacuum (to prevent occlusion of air) with a moderate-sized hand-operated press into a clear pellet, about 1 to 2 cm. in diameter, depending on the die, and a millimeter or two in thickness. Special apparatus is required, and the total time demanded for the preparation of a sample by this method is significantly higher than by the mull technique, but the potassium bromide matrix contributes no absorption throughout the entire spectral region from

400 cm.$^{-1}$ to 47500 cm.$^{-1}$ (210 mμ). Since the mass and area of the pellet and the concentration of the sample are known, the method provides an excellent possibility for quantitative analysis (*cf.* Sec. 9-6).

Aside from the time element, there are certain other disadvantages of the pellet method which limit its utility. If the concentration of sample in the potassium bromide is much greater than 1 per cent the pellets are frequently not clear and tend to stick to the die. In a few instances, compounds have been shown to react with the potassium bromide. It is very difficult to avoid some pickup of water during the grinding and mixing of the sample with the potassium bromide, and since determination of O—H is usually more important than that of C—H this is an important consideration. The most serious difficulties, however, are a consequence of the uncertainties in the appearance of the spectrum caused by the process of producing the pellet. Several workers have found that the appearance of a spectrum may change quite markedly from one pellet to another, owing quite probably to the existence of various polymorphic forms of the sample and to changes induced by the local pressure and temperature effects. In spite of these limitations, the pellet method has proved to be very helpful, especially for hygroscopic compounds that do not mix well with mineral oil. At present it seems safe to say that the majority of experienced workers use both the mull and pellet techniques, with emphasis on the former for routine analyses. In studies of structure it is advisable to avoid powders altogether when possible.

Water is a particularly unsatisfactory solvent for infrared spectroscopy. It attacks the alkali halides that are most often employed as cell windows, and it is an extremely strong absorber throughout nearly the entire infrared region. In spite of these difficulties, the advantages and importance of water as a solvent for polar compounds and for biological materials have caused an increasingly great interest during the last few years in obtaining infrared spectra of aqueous solutions. Deuterium oxide, because of the different atomic masses involved, absorbs at somewhat different frequencies and has been found to be of great help in examining regions where ordinary water absorbs. Several alternative cell window materials (especially silver chloride, calcium fluoride, and "KRS-5") are satisfactory for work with aqueous solutions. Instrumental problems do remain, however. A discussion of these will help to illustrate the problems which exist—usually in a less acute form—in nearly all infrared spectroscopic investigations.

An obvious apparent circumvention of the absorption due to the water is to employ a double-beam spectrometer, so that by placing an equal amount of water in each beam this absorption will be subtracted out by the instrument. Within limits this is quite satisfactory. It should, however, be kept in mind that it is not possible to obtain a spectrum of any sample on a double-beam instrument that cannot also be obtained on a single-beam instrument! The real advantage of the double-beam instrument is that when there is sufficient radiant power transmitted by both the sample and reference cells so that the transmittance of each could be reliably determined, the instrument will automatically perform the task of subtracting these transmittance (or absorbance) values. For solutions which absorb strongly it is necessary to alter the instrument settings in order to obtain a reliable measure of the transmittance. This may be done by opening the slit, with a corresponding loss in resolution; by increasing the gain, but only to the point at which the "noise" becomes objectionable; and by running at a lower speed in order that higher damping, and hence higher amplifier gain and higher noise level at the amplifier, may be tolerated. Alternatively, it may be possible to increase the concentration of the sample in the absorbing solvent and then decrease the total thickness of solution.

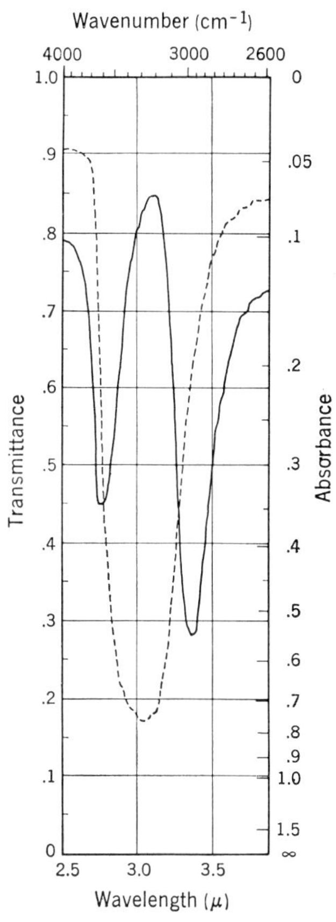

FIG. 9-7. Example of erroneous spectrum arising from lack of sufficient energy in reference beam. The water band, at 3300 cm.$^{-1}$, appears as shown by dashed curve when sample transmits appreciably and reference sample is thinner. When thickness of both sample and reference layers is increased, the amplifier unbalance causes apparent transmittance maximum in vicinity of band center.

A serious problem, especially for those without experience, is knowing whether a particular curve, obtained on a double-beam spectrometer, is to be considered reliable or whether the instrument was drifting during some portion of the run. Whenever possible, the performance of the instrument over any region in which the reference sample absorbs should be checked at the time the curve is obtained. This may be readily accomplished by inserting an opaque object into the sample beam, with the sample and reference cells in place, and noting whether the pen responds properly. If it is at all sluggish, corrective measures, as outlined above, should be taken. Increasing the source output should be avoided, however, since this may introduce stray radiation as the primary signal and the instrument will then measure the absorption of the sample in the frequency range of the stray radiation (in the very near infrared) rather than at the frequency for which the spectrometer is set.

Some of the most common indications of improper performance are a square corner, a moderately long flat top to an absorption band that is not at the zero per cent line, or occasionally an inverted band in the center of a band which is known to be very strong. Figure 9-7 shows a specific example of a strong absorption band that appears to be split into two bands. The apparent splitting is due to insufficient energy in the reference beam coupled with a condition of amplifier unbalance.

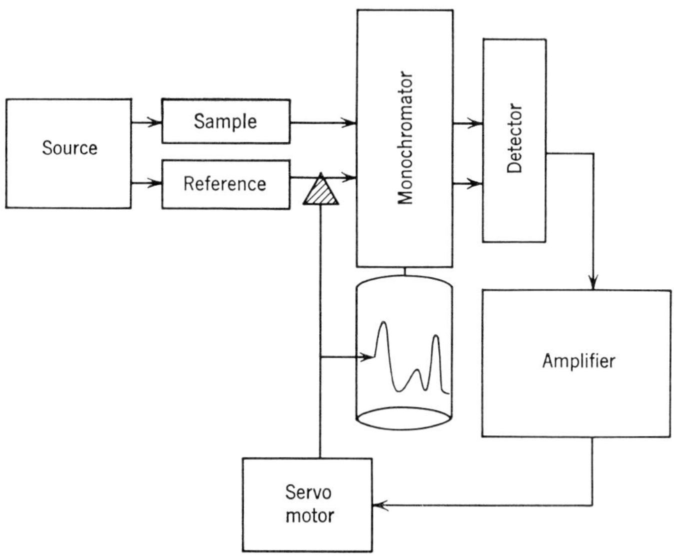

FIG. 9-8. Schematic diagram of a double-beam spectrometer. The unbalance signal at the detector is amplified and fed to a phase-sensitive servo motor which drives the optical wedge into or out of the sample beam to reestablish a balance between the signals. The recorder pen is mechanically linked to the optical wedge.

Figure 9-5 showed the full optical diagram of a double-beam instrument; Fig. 9-8 shows a block diagram of the same system. Radiant energy from the source is split into two equal beams, one of which passes through the sample and the other through the reference compartment. These are then recombined and passed through the monochromator, but they remain separated in time and impinge alternately on the detector. If the two signals are the same, there is no unbalance signal; but if, for example, the sample absorbs more strongly than the reference material, the unbalance signal which arises at the detector is amplified and fed to a servo motor which drives an optical wedge into the reference beam to decrease the transmission of that beam. The pen is mechanically linked to the wedge, and therefore moves across the recorder chart to show high transmittance when the wedge is out of the reference beam, low transmittance when the wedge is driven into the reference beam.

It should be clear that if a piece of cardboard were inserted into each beam, the signal at the detector would show no unbalance, regardless of the position of the wedge and pen. The instrument is then faithfully reporting what it sees no matter where the pen may be. If the amplifier is not precisely balanced under these conditions, the pen will drift in the direction corresponding to the signal arising in the amplifier. Of course one does not run spectra with cardboard in each beam, but it may be desirable to run spectra with water or some other solvent in each beam, and all solvents are opaque over certain regions. The curve traced out by the spectrometer does not tell anything at all about the absorption characteristics of the sample under these conditions. Atmospheric absorption, by water vapor and carbon dioxide, may cause similar dead spots over small portions of the spectrum, but these may be avoided by flushing the instrument housing with dry nitrogen or dry carbon dioxide–free air.

9-5. Qualitative Analysis. The basis of qualitative analysis employing infrared spectra has already been outlined. For the student who is especially interested in this invaluable technique there are now several excellent references available which are listed on page 407. The present discussion will attempt primarily to indicate the general possibilities and limitations of the method, with the hope of making it clear whether further investigation is warranted for a particular type of problem.

We may imagine two vibrating bodies, or two pendulums, which are attached to the same support. When one of them is set in motion, we know from classical mechanics that, if the frequency of its motion is the same as the natural frequency of the other pendulum, the condition of resonance exists and the vibrational energy of the first will be quite rapidly transferred to the second. The first will become motionless; then it will regain the energy from the second, and so forth. The steady-state vibrations of the system will involve motion of both halves, either in phase or 180° out of phase. The out-of-phase motion will have a higher frequency than that of either half alone; the in-phase motion will have a lower frequency than the isolated halves. If the natural frequencies of the two pendulums are quite different, however, the motion of the first will have little or no effect on the second.

Similar considerations will apply to molecular vibrations. An O—H bond will be quite different from a C—C bond, so that setting the O—H bond in motion will have little or no effect on the C—C bond, or on any of the other adjacent bonds. The frequency of the O—H bond vibration should therefore be characteristic solely of the O—H bond, with no significant difference in this frequency whether the hydroxyl group is attached to an aliphatic chain, an aromatic ring, or even a sulfur, phosphorus, or nitrogen atom (except in so far as the latter might really change the strength of the O—H bond). We would, on the other hand, expect that the C—C vibration would be significantly changed by the presence of adjacent C—C bonds, as in the series ethane–propane–butane–pentane. Hence we should not expect to find a single frequency which will be characteristic of C—C single bonds. In contrast, an olefinic bond, C═C, will nearly always be isolated from similar bonds, and thus we can expect to find a frequency which is characteristic of the C═C unit.

Bending motions of molecular bonds involve smaller restoring forces than do the stretching vibrations, and are therefore found at lower frequencies. Whereas the stretching of an O—H bond may be quite independent of its environment (excluding cases of hydrogen bonding), it seems reasonable that the ease of "wagging" of this group will depend fairly significantly upon its neighbors, and this is observed. Although we could readily predict, at least *a posteriori*, that many of the stretching vibrations would be characteristic and many of the bending motions would not, it does seem rather surprising that many of the bending frequencies are the same within rather large groups of molecules. For example, the hydrogen out-of-plane bending motions in olefinic or aromatic compounds are highly characteristic of the substitution pattern around the double bond or ring, but quite insensitive to the nature of the nonhydrogen substituents or the over-all size of the molecule.

A convenient summary of much of this information is presented in Fig. 9-9 in a form that is due to N. B. Colthup. The positions of the horizontal lines indicate the frequency ranges within which one may expect to find absorption bands due to the group indicated. A rough indication of the intensity (s, m, or w) is given for most of the bands, and in certain cases additional information, such as the fact that a band tends to be broad or is an overtone (2ν), may also be given.

The effective and accurate use of the Colthup chart or other sources of similar information requires a considerable amount of experience, since band shapes and intensities are frequently as significant as their positions. In the higher frequency region, between 1600 and 3700 cm.$^{-1}$, it is reasonably safe to conclude that any band of moderate or strong intensity should be identifiable by reference to the chart, although more than one possibility may be indicated. Below 1600 cm.$^{-1}$, however, there is no longer a 1:1 correspondence between characteristic frequencies, as given on the chart, and observed frequencies, since many of the latter are not characteristic of functional groups but rather of the particular molecule. This is partly because there are so many bands in this region, for most molecules, that one has no simple way of correlating the bands

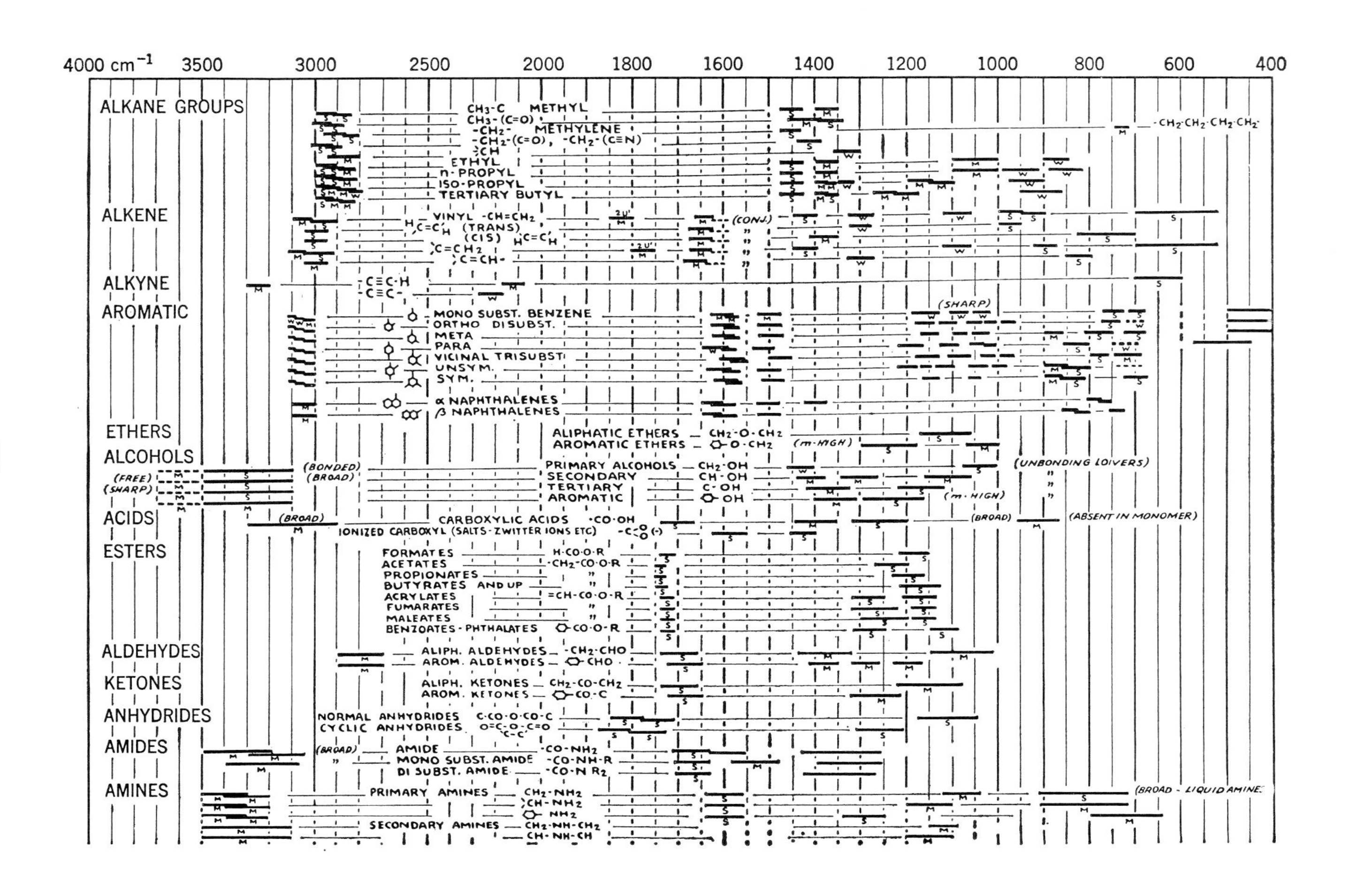
4000 cm^{-1}
3500
3000
2500
2000
1800
1600
1400
1200
1000
800
600
400
ALKANE GROUPS
CH3-C METHYL
CH3-(C=O)
-CH2- METHYLENE
-CH2-(C=O), -CH2-(C≡N)
CH
ETHYL
n-PROPYL
ISO-PROPYL
TERTIARY BUTYL
-CH2·CH2·CH2·CH2-
ALKENE
VINYL -CH=CH2
(TRANS)
(CIS)
C=CH2
C=CH-
(CONJ.)
ALKYNE
-C≡C-H
-C≡C-
AROMATIC
MONO SUBST. BENZENE
ORTHO DISUBST.
META
PARA
VICINAL TRISUBST.
UNSYM.
SYM.
α NAPHTHALENES
β NAPHTHALENES
(SHARP)
ETHERS
ALIPHATIC ETHERS CH2·O·CH2
AROMATIC ETHERS O·CH2
(m·HIGH)
ALCOHOLS
(FREE)
(SHARP)
(BONDED)
(BROAD)
PRIMARY ALCOHOLS CH2·OH
SECONDARY CH·OH
TERTIARY C·OH
AROMATIC OH
(UNBONDING LOIVERS)
ACIDS
(BROAD)
CARBOXYLIC ACIDS ·CO·OH
IONIZED CARBOXYL (SALTS-ZWITTER IONS ETC)
(ABSENT IN MONOMER)
ESTERS
FORMATES H·CO·O·R
ACETATES ·CH2·CO·O·R
PROPIONATES
BUTYRATES AND UP
ACRYLATES =CH·CO·O·R
FUMARATES
MALEATES
BENZOATES - PHTHALATES CO·O·R
ALDEHYDES
ALIPH. ALDEHYDES -CH2·CHO
AROM. ALDEHYDES CHO
KETONES
ALIPH. KETONES CH2·CO·CH2
AROM. KETONES CO·C
ANHYDRIDES
NORMAL ANHYDRIDES C·CO·O·CO·C
CYCLIC ANHYDRIDES O=C·O·C=O
AMIDES
AMIDE ·CO·NH2
MONO SUBST. AMIDE ·CO·NH·R
DI SUBST. AMIDE ·CO·N R2
AMINES
PRIMARY AMINES CH2·NH2
CH·NH2
NH2
SECONDARY AMINES CH2·NH·CH2
CH·NH·CH
(BROAD - LIQUID AMINE)

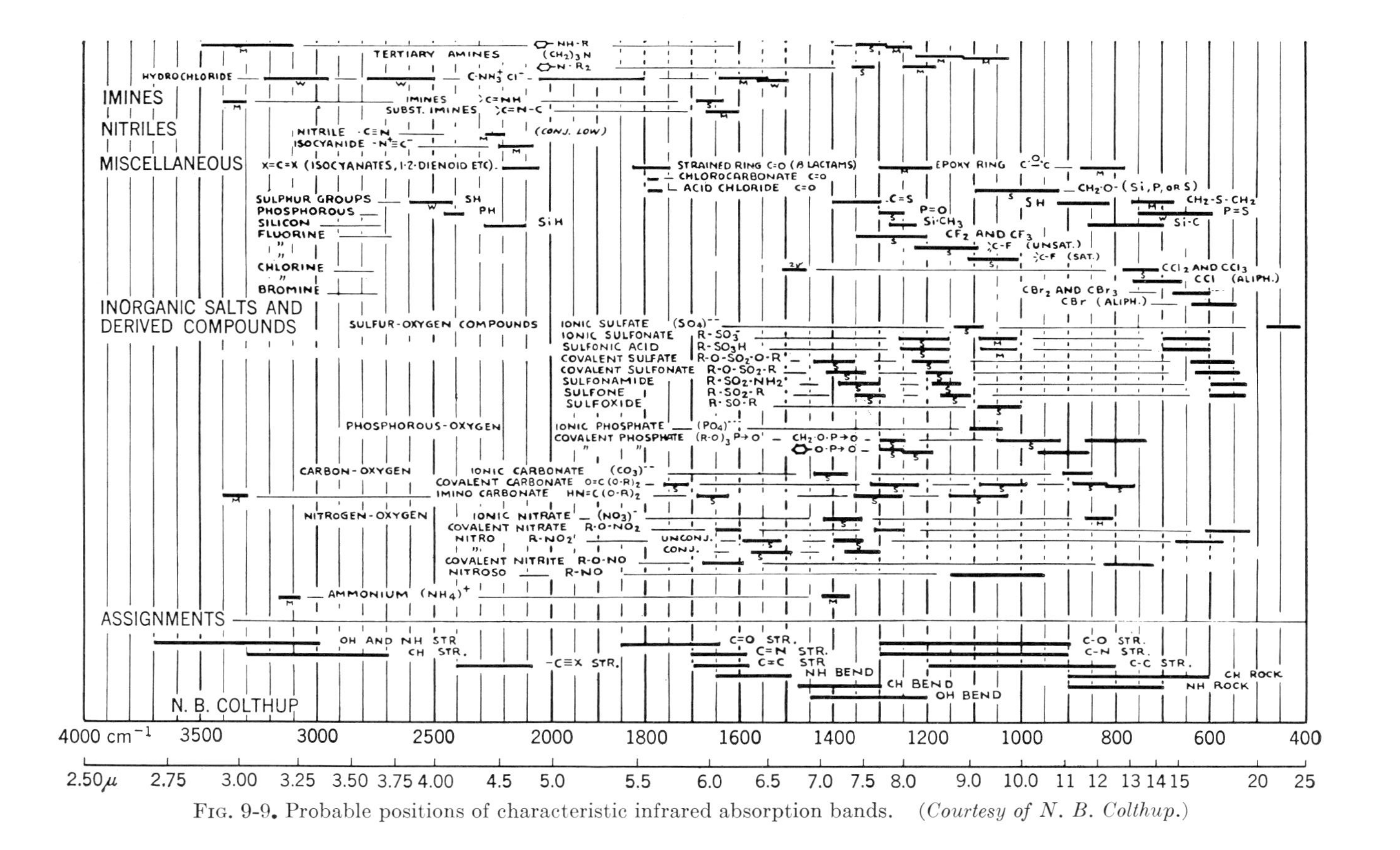

FIG. 9-9. Probable positions of characteristic infrared absorption bands. (*Courtesy of N. B. Colthup.*)

with particular vibrations, and partly because many bending vibrations occur in the low frequency region and the positions of these absorption bands are more susceptible to unpredictable changes. Primarily, however, the dependence of the low-frequency absorption pattern on the particular molecule may be attributed to the fact that there are many vibrations of approximately the same frequency and these will interact significantly with each other. The low-frequency region has quite appropriately been called the "fingerprint region" because it is this portion of the spectrum which is largely responsible for the uniqueness of the spectrum of an individual compound.

When it has been determined that a particular band in the spectrum is due to a specific functional group, such as C=O or O—H, further information may often be gleaned by observing the position of the band more precisely. For example, an O—H group which is not hydrogen bonded will appear as a weak but sharp band at 3600 cm.$^{-1}$; hydrogen bonding will greatly increase its intensity and move it down to 3300 cm.$^{-1}$. If the hydrogen bonding is especially strong, however, as in solid organic acids, the band may move even farther and become much broader. The position and shape of the band is therefore an indication of the extent of hydrogen bonding in the sample. The position of carbonyl absorption is only slightly different for aldehydes, ketones, esters, or organic acids, but it also varies slightly depending on whether these are aliphatic or aromatic compounds, and the position of the carbonyl absorption in anhydrides is quite noticeably distinct from the ketone, aldehyde, ester, or free acid position. Conjugation of the carbonyl group with an olefinic group, or the presence of an adjacent C—N bond, as in an amide, will significantly alter the nature and strength of the carbonyl bond and will accordingly cause a relatively large frequency shift (50 or 100 cm.$^{-1}$). Hydrogen bonding by the carbonyl or attachment of the oxygen to a strained ring will also change the position of the absorption band. There may, of course, be compensating changes by two or more factors. The infrared spectrum is frequently a highly useful probe into the molecule and its environment, providing information about the natures of the bonds in a molecule and the intramolecular or intermolecular interactions.

Certain bands are particularly helpful in the identification of an unknown material, for by their presence or absence one may often determine the general nature of the substance almost at a glance. Figure 9-1 showed the spectra of phenol and nylon; Fig. 9-10 shows the spectra of two additional compounds which for the moment we may consider to be unknown. Although the approach to the identification of these spectra may differ somewhat from one spectroscopist to another, the general method will be much like the following.

In Fig. 9-1*a* (page 307), the most striking features are the strong band at roughly 3400 $cm.^{-1}$, highly indicative of O—H, and the apparent absence of C—H absorption at 3000 $cm.^{-1}$. More careful examination, however, shows some evidence of a weak band, at about 3100 $cm.^{-1}$, that is characteristic of hydrogen next to an olefinic bond or attached to an aromatic ring. At 1500 and 1600 $cm.^{-1}$ we observe sharp bands that are usually indicative of an aromatic ring. We immediately jump to the conclusion, therefore, that this is probably an aromatic compound with hydroxyl substitution (and no alkyl substitution). We next look for indications of the positions of substitution and for confirming evidence. The strong bands at 700 and 755 $cm.^{-1}$ suggest mono- or meta-substitution; the pattern of four bands between 1700 and 2000 $cm.^{-1}$ even more strongly indicates mono-substitution. Finally, an aromatic alcohol should have a strong band between 1150 and 1300 $cm.^{-1}$, and this checks very nicely. Comparison with a known spectrum of phenol completes the identification.

We may assume that the analyst who is given a spectrum such as that of Fig. 9-1*b* would know that the sample was a polymer. He could then immediately note that there is aliphatic C—H from the bands at 2850 and 2900 $cm.^{-1}$. There is also evidence of O—H or N—H from the strong band at 3200 $cm.^{-1}$. If this were O—H it would be quite strongly hydrogen bonded and would quite probably be broader than the observed band; hence N—H is suggested. The strong bands at 1550 and 1620 $cm.^{-1}$ have the intensity and width to be expected of a band due to carbonyl. The lower band is too low for carbonyl and even the upper is too low for ordinary carbonyl, but the combination is typical of amides. The C—N bond is strengthened and the C═O bond weakened by delocalization of the electrons in this group, leading to the observed positions of absorption. The N—H bend also comes in this region, and the 1550 $cm.^{-1}$ band is due to a vibration which involves both the stretching of the C—N bond and the deformation of the C—N—H angle. At this point one would look at the spectrum of nylon for comparison and find the match quite satisfactory.

The sample of Fig. 9-10*a* is apparently a hydrocarbon, as may be quickly determined from the evidence for C—H, at 3000 $cm.^{-1}$ and between 1350 and 1500 $cm.^{-1}$ (bending vibrations), and by the lack of O—H, N—H, C═O, C—Cl, or other distinguishing features. From the sharpness and general scarcity of absorption bands one might guess that the compound is of fairly low molecular weight. The only strong clue to the exact nature of the hydrocarbon is given by the bands at about 2600 $cm.^{-1}$. Very few samples have bands in this region, but cyclic hydrocarbons frequently do. Comparison with the spectra of cyclopentane and cyclohexane would show that the latter is the correct compound. In the absence of a sufficient clue to the specific hydrocarbon involved, one could determine this most satisfactorily by machine sorting of index cards, as described below, or of course by comparison of other physical properties such as boiling point.

The polymer whose spectrum is shown in Fig. 9-10*b* is clearly composed of aromatic and aliphatic components, as shown by the aliphatic C—H absorption between 2850 and 3000 $cm.^{-1}$ and the aromatic (or possibly olefinic) C—H just above 3000 $cm.^{-1}$. The very sharp bands at 1500 and 1600 $cm.^{-1}$ confirm the presence of an aromatic ring. The very strong bands at 700 and 760 $cm.^{-1}$ and the pattern between 1700 and 2000 $cm.^{-1}$ indicate quite clearly that the ring is singly substituted. One might guess, therefore, that the sample is polystyrene, and this is the case. A comparison of the phenol and polystyrene curves will show the remarkable constancy of certain of the "characteristic" bands, including the hydrogen bending vibrations near 700 $cm.^{-1}$ and the combination and overtone bands near 1700 and 2000 $cm.^{-1}$ as well as the stretching vibrations mentioned. There are additional bands between 1000 and 1200 $cm.^{-1}$ which are also helpful in determining the position of substitution, although there is often interference from other bands in this region. The high-frequency group, 1700 to 2000 $cm.^{-1}$, is quite specific; the bands at 700 and 760 $cm.^{-1}$ cannot be considered nearly as reliable because small amounts of impurities may introduce strong absorption in this region. Not only aromatic compounds, but also chlorine compounds such as carbon tetrachloride, chloroform, etc., show strong bands in this region. These are frequently trapped in polymer films and could lead to confusion in interpretation if not recognized.

The four spectra discussed illustrate only a few of the many clues to structure which are provided by infrared absorption spectra. The problems at the end of this chapter offer some further examples, and several of the references (especially 49 and 50) listed in the Suggestions for Further Study (page 407) discuss many more specific applications of infrared spectroscopy to qualitative analysis and structure determination.

In most cases the interpretation of the spectrum on the basis of characteristic frequencies will not be sufficient to permit positive identification of a total unknown. It is very helpful to be able to compare the unknown spectrum with known spectra to confirm or disprove the assigned structure. Fortunately, the number of spectra which are available for comparison is rapidly increasing, and several collections of rather large numbers of these are available. The American Petroleum Institute has published the infrared spectra of a large number of compounds, primarily hydrocarbons but also many other materials. These can be secured[1] at nominal cost in loose-leaf form. Some of the earlier spectra are of poor quality, but useful nevertheless, and the purity of the

[1] From American Petroleum Institute Research Project 44, Carnegie Institute of Technology, Pittsburgh 13, Pa.

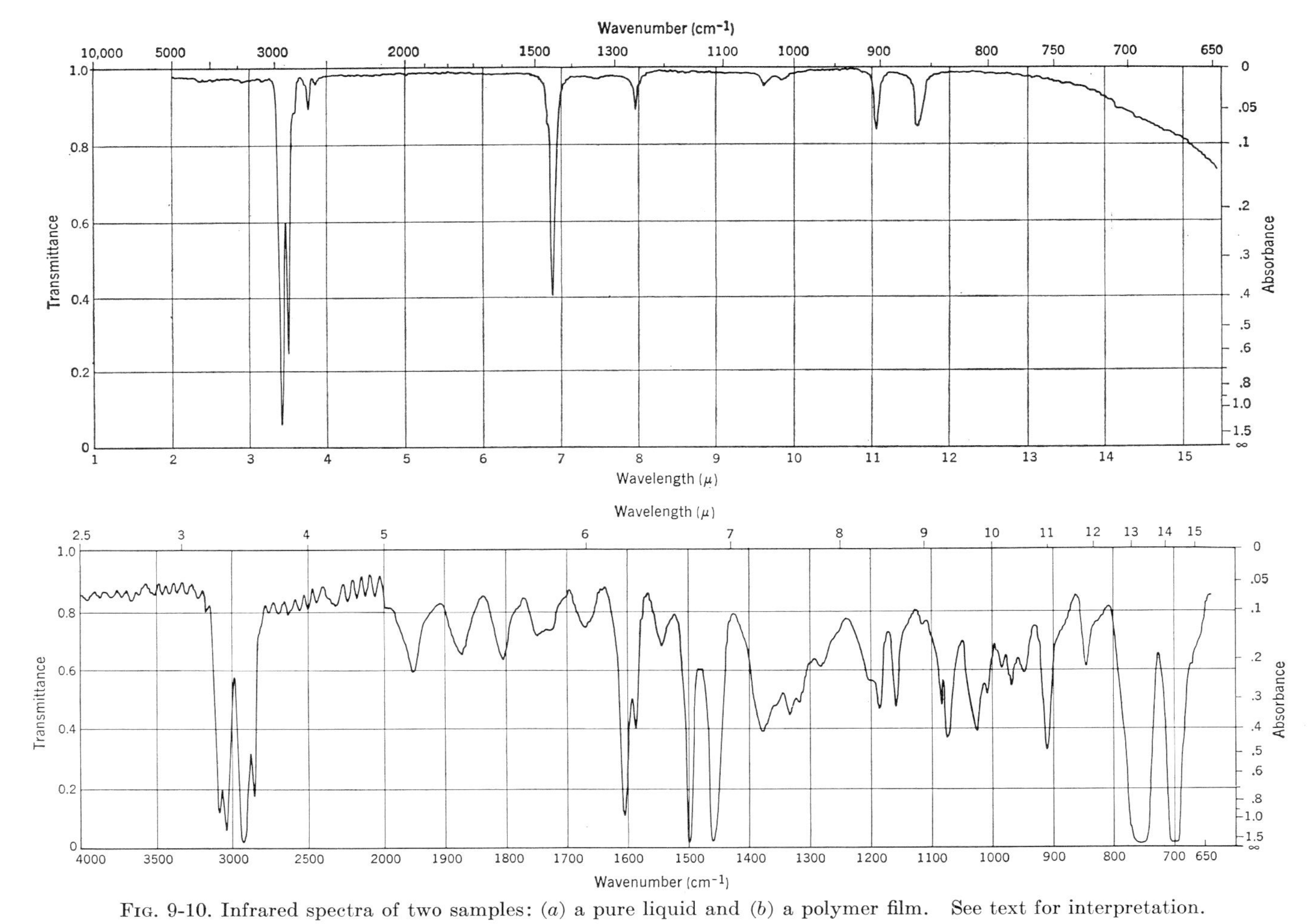

Fig. 9-10. Infrared spectra of two samples: (*a*) a pure liquid and (*b*) a polymer film. See text for interpretation.

compounds is generally quite satisfactory. Collections of spectra on edge-punched cards have been issued by the National Bureau of Standards, in cooperation with the National Research Council, and more recently by Butterworths Scientific Publications, in London. A very large number of spectra have been issued, in various forms, by the Sadtler laboratories in Philadelphia.

Within these four collections there are now several thousand spectra, so that even if these were on hand it would be an arduous task to sort through them. The American Society for Testing Materials undertook some time ago to provide an index to these collections, and also to the spectra reported in the literature, on cards which may be rapidly handled by electrical sorters. The positions of the absorption bands, as well as supplementary information about the compound, are punched into these cards, and thus in principle one may identify an unknown simply by sorting for those compounds which have the same absorption bands. There are limitations to the method, not the least of which is the requirement that the particular compound sought must have been included in the set of index cards, but the system has proved to be invaluable in many laboratories. As was mentioned earlier, even more elaborate instrumentation for automatic analysis of spectra is in the design stage, and this may greatly increase the effectiveness of infrared methods, particularly in fields involving complex mixtures, such as the biological sciences.

9-6. Quantitative Analysis. The principles of quantitative analysis in the infrared are essentially identical with those for the ultraviolet-visible region. There is little difference between the inherent accuracies of the methods used in the two regions, but there are variations of technique and procedure which depend upon the differences in instrumentation and sample handling. Each of these has been considered briefly in earlier sections. We shall consider here the specific implications of these differences in the procedures of quantitative analysis.

All double-beam instruments designed for use in the ultraviolet and visible are built upon the "electronic-null" principle. The balance of signals in such instruments is achieved by a potentiometer circuit. The pen or transmittance dial is connected to a moving contact on the potentiometer coil, so that the motion of the pen or transmittance knob serves to balance the instrument. In this design the full intensity of the reference beam is allowed to strike the detector at all times, providing an adequate unbalance signal to the amplifier no matter how strongly the sample absorbs. In the optical-null instruments employed in the infrared, it will be remembered, strong absorption by the sample causes equivalent attenuation of the reference beam by an optical wedge. If the reference material is nonabsorbing, the instrument can only be satisfied by driving the pen toward lower transmittance, but the closer the pen gets to the zero per cent line the smaller is the signal upon which the instrument operates. Eventually, some place in the vicinity of 1 per cent transmittance, there is no longer sufficient power received by the detector to operate the servo motor which drives the pen.

The effect of the optical-null design on the zero per cent line may be readily observed by blocking off the sample beam with an opaque object. If the shutter is inserted very rapidly, the pen will drive to its extreme position and remain there. If, on the other hand, the shutter is moved very slowly into the beam, the pen will move slowly down scale and eventually come to rest at a position noticeably short of the limiting position. If the pen has driven, under the inertia of its own mechanism, to the limiting position, it will be sluggish in returning from this position when the shutter is moved sufficiently to allow only a small amount of radiant energy through. Several methods have been employed to partially correct this situation. A small a-c signal applied to the pen will cause small oscillations about the equilibrium position, and a small up-scale signal, obtained by slightly unbalancing the amplifier circuit, will encourage the pen to move off the limiting position. It should be clear that these methods are of value in obtaining a more reproducible "zero per cent" line, and may improve the qualitative appearance of the absorption band, but they are scarcely adequate for determining the true position of the zero per cent line for quantitative purposes.

Fortunately, the uncertainty in the zero per cent line is quite small, and it becomes smaller as the instruments become more sensitive to small signal levels. The effect of the uncertainty is made even smaller by limiting analyses to bands which do not absorb too strongly. The error in absorbance ΔA due to an error δ in measuring the position of the zero per cent line is given by

$$\begin{aligned}\Delta A &= \log \frac{P_o + \delta}{P + \delta} - \log \frac{P_o}{P} \\ &= \log \frac{1 + (\delta/P_o)}{1 + (\delta/P)}\end{aligned}$$

The error in position δ will be small compared to P_o, and if P is not too small, δ/P will also be small compared to unity. Single-beam instruments are not subject to this uncertainty.

A similar difficulty exists in determining the position of the 100 per cent transmittance line, although for quite different reasons. Infrared cells are nearly always very thin (1 mm. or considerably less) and are usually made of sodium chloride or potassium bromide plates. These will change in transmittance as a result of fogging by water vapor and may also change in thickness if solvents are employed which can dissolve small amounts of these salts. (One cause of change in cell thickness, especially with potassium bromide, is the condensation of water vapor on the interior surfaces when a volatile solvent, such as chloroform, evaporates and cools the cell. This may be largely avoided by flushing with dry gas before applying a vacuum to the cell. A vacuum is necessary, however, for efficient removal of the solvent.) Since it is virtually impossible to produce and maintain accurately matched cells, and since at a high concentration of solute even matched cells will not provide equal thicknesses of solvent, there is always a serious problem, in either single-beam or double-beam operation of how to determine the 100 per cent transmittance level. In practice, therefore, this line is chosen somewhat arbitrarily.

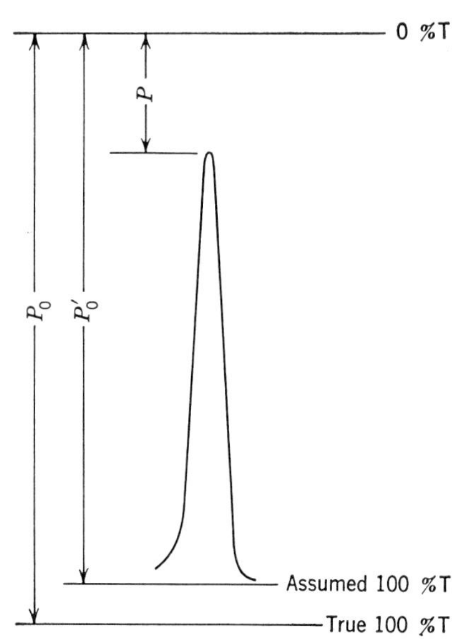

FIG. 9-11. Uncertainty in location of 100 per cent transmittance line. An incorrect or arbitrary choice of the 100 per cent line will not affect the linearity of the Beer's law plot, but at zero concentration the measured absorbance will not be zero.

To measure the transmittance of a sample, we must measure the radiant powers of the beams transmitted by the sample and by the reference. On a recorded spectrum that is linear in transmittance, this requires that we measure the distance on the chart from the zero per cent line to the absorption curve of the sample and also the distance from the zero per cent line to the 100 per cent line (at exactly the same frequency, of course). We may let P_o represent the distance between the zero per cent line and the true 100 per cent line for the sample and cell being scanned, and P_o' represent the distance between the zero per cent line and some other arbitrarily chosen line near the true 100 per cent line, as illustrated by Fig. 9-11. The absorbance of the sample is given by

$$\begin{aligned} A &= \log \frac{P_o}{P} = \log \frac{P_o'}{P} \frac{P_o}{P_o'} \\ &= \log \frac{P_o'}{P} + \log \frac{P_o}{P_o'} \end{aligned}$$

Letting

$$A' = \log \frac{P_o'}{P}$$

we see that

$$A' = A + \text{constant}$$

Thus, if Beer's law is followed,

$$A' = abc + \text{constant}$$

An incorrect choice of the 100 per cent line will not destroy the linear dependence of (apparent) absorbance on concentration, but will simply change the intercept of a plot of absorbance *vs.* concentration.

Because of uncertainty in determining the 100 per cent line, the "hill and valley" base-line method is very often used in infrared measurements. On a plot of absorbance *vs.* frequency, three frequencies are chosen which will usually (though not necessarily) correspond to a maximum and the adjacent minima of absorption. A straight line is drawn between the two outside points of the curve (Fig. 9-12), and the absorbance at the central point is measured between the curve and the arbitrarily drawn base line. If the background absorption (*i.e.*, all absorption not due to the compound being determined) varies linearly over the frequency interval chosen, the absorbance measured in this way will vary linearly with the concentration of the sample, independently of any changes in the amount of background absorption.

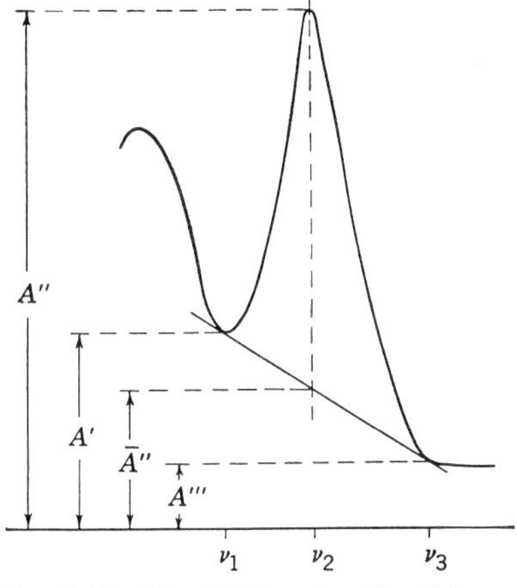

FIG. 9-12. The "hill and valley" base-line method. Three frequencies are chosen and the values of A', A'', A''', and $\bar{A}''$ found as shown. This will give the correct absorbance if (and only if) $\bar{A}''$ represents the intensity that would actually be transmitted by the sample in the absence of the compound being determined.

Three important limitations on this method should be kept in mind:

1. All measurements must be made at precisely the same frequencies, if they are to be comparable, even though these frequencies may not correspond to maxima or minima in the spectrum of a particular sample under study. The errors introduced by small variations in positions of measurement are smallest if the points do correspond to maxima or minima.

2. If other substances have absorption bands in the region chosen, the assumption that the background is linear is likely to be unjustified and the application of the technique will give the wrong answers. This is particularly true if the band chosen for measurement lies on the side of a band due to another substance. The procedure should be limited to cases in which the nature of the background absorbance is known.

3. The proof of the validity of the "hill and valley" method depends upon the assumption that the absorbance of the background varies linearly over the interval chosen; and therefore the method cannot be applied in the same way to plots which are linear in transmittance. It is possible to employ the same equations, but it is necessary to find the point $\bar{A}''$ algebraically rather than graphically. Since this point is on the straight line between A' and A''', it is given by

$$\bar{A}'' = A' - \frac{\nu_2 - \nu_1}{\nu_3 - \nu_1}(A' - A''')$$

If Beer's law is obeyed by the sample, it can be shown that

$$A'' - \bar{A}'' = \bar{a}bc$$

where $\bar{a}$ is an effective absorptivity that depends upon the absorptivity values of the sample at the three frequencies.

Single-component analyses in the infrared and ultraviolet-visible regions are too similar to warrant a repetition of the discussion of Sec. 8-10.

Multicomponent systems which obey Beer's law[1] can be handled quite easily in the infrared, for the large number of bands permits a relatively large number of experimental points to be applied to the calculation. Three- and four-component mixtures are handled quite routinely, and as many as ten components can be treated under favorable circumstances. Any such problem involving more than two unknowns can be most easily solved by matrix algebra.

Until recently the presence of "stray light" was a serious problem in infrared spectroscopy. Stray light is the term applied to any radiant energy, not within the frequency range for which the monochromator is adjusted, that gets through the monochromator and contributes to the measured signal. A small amount of this "false energy" has been shown to arise from minor imperfections which remain on the optical surfaces. These scatter a small fraction of the beam slightly off its path and cause an unpredicted broadening of the frequency range passed by the monochromator. Most of the stray radiation in infrared instruments, however, arises from radiation that is rejected at some point in the optical path within the monochromator but manages, by multiple reflections, to find its way back into the optical path. The amount of this is so small that it could be neglected if the source were of comparable intensity for all regions of the spectrum. But since the source produces much more radiant power in the very near infrared than in the far infrared (*cf.* Fig. 9-6), the amount of scattered very near infrared radiation may,

[1] Those which do not are less easily treated. Sometimes they may be solved approximately as a series of single-component analyses with small corrections applied to take into account the absorption due to other components.

in some cases, be comparable to or even greater than the total amount of far infrared radiation passed by the monochromator.

There are basically two methods of overcoming the problems of stray light. It may be removed by interposing a filter or a second monochromator (or a second pass through the same monochromator), or one may simply ignore the stray signal by chopping the beam with a blade that is transparent to the very near infrared radiation. In the latter case the stray signal, since it is not interrupted, will not be amplified even though it strikes the detector. Present commercial instruments have incorporated one or more methods for eliminating the effect of stray radiation and it has accordingly become a minor problem, especially above 600 cm.$^{-1}$. The amount of stray radiation increases strongly with decreasing frequency, and may become serious in some instruments in the potassium bromide or cesium bromide regions.

An approximate correction for stray light may be made by taking the signal measured at the center of a very strong absorption band as the zero per cent transmittance level. The difference between the signal measured at such a point and that measured with an opaque shutter is an indication of the total amount of stray light at that frequency. The exact measurement of stray light is in general difficult, if not impossible, because the sample employed will absorb varying amounts of the stray radiation, thus changing the stray light level as the sample concentration is changed. In accurate quantitative work it is necessary to eliminate stray light by the methods discussed above.

The quantitative analysis of a solid sample is somewhat more difficult than the same analysis performed on liquids or solutions. Unless the solid powder is distributed quite uniformly there will be significant deviations from Beer's law and there may be serious problems of reproducibility. Analyses have been successfully carried out on Nujol mulls by adding a known concentration of a reference material which serves to measure the thickness of the sample. Calcium carbonate is often a suitable internal standard for this method.

The potassium bromide pellet technique is much more promising, since the sample thickness may be accurately determined by weighing the final pellet. Those who have carefully tested the method have found that the results depend quite markedly on the way in which the sample is prepared, and especially on how finely the material is ground before the pellet is pressed. Mortar and pestle grinding, whether by hand or by an automatic mixer, is usually not adequate. Absorptivities obtained with mortar and pestle grinding have been shown in some cases to be low by 50 per cent compared to more effective pulverizing methods, such as freeze drying. Various types of vibratory or oscillatory mixers, such as the dental shakers, vary in effectiveness from very poor to very good.

Grinding under a volatile solvent is frequently helpful, but increases the likelihood of picking up moisture from the atmosphere. At present there is no general agreement among workers in the field concerning the best method for routine grinding of samples.

The determination of the absorptivity of a gas is quite straightforward if one is satisfied with moderate accuracy. It is found, however, that as the total pressure of gas in the cell is increased (with foreign gas if desired) the absorptivity will increase. Thus a gas absorbs infrared radiation more effectively if it is in a short cell at high pressure than if it is in a long cell at low pressure, and it absorbs more effectively if another gas is added to increase the total pressure. The effect of the higher pressures is to increase the frequency of collisions between gas molecules and hence to disturb the rotational energy levels of the molecule (we may recall the similar effect of condensing the gas to a liquid or solid which was mentioned in Sec. 9-2). The very narrow absorption bands, which make up the branches of the vibrational band, become broader and tend to merge. The area under these bands is unchanged, but the average transmittance over a frequency interval covering several such bands is decreased by the more uniform distribution of absorption with frequency. For example, if the narrow bands transmit 20 per cent ($A = 0.70$) of the radiation of some frequencies, while all of the radiation in an equivalent frequency range is transmitted, the observed transmittance will be 60 per cent ($T = 0.20/2 + 1.00/2 = 0.60$, $A = 0.22$). If the bands are broadened by the addition of a nonabsorbing gas so that they show an absorbance of 0.35 ($= 0.70/2$) over the entire frequency interval, the observed transmittance will be 46 per cent, even though the amount of the absorbing gas in the beam is unchanged.

The effect described should be strongly reminiscent of the difficulties of quantitative analysis of powders. The gas may be considered to be uniformly distributed in space, but it acts on only part of the beam, that is, on certain frequencies only. In the case of a powder dispersed, for example, in a potassium bromide pellet, the powder will absorb all frequencies in the beam about equally well, but only part of the beam actually strikes the sample particles. The remainder passes through unaffected, because of the inhomogeneous spatial distribution.

The change in absorptivity with pressure varies from one gas to another and may depend on the natures of the other gases present. Some experimenters have found it necessary to increase the total pressure to about 200 atm. to find a maximum absorptivity. In most practical analyses it would suffice to use a reproducible total pressure.

The analysis of gases in plant streams can frequently be performed effectively and economically by means of infrared spectroscopy. Some instruments built for this purpose are specially designed spectrometers similar in principle to those found in the laboratory, but most are non-dispersive analyzers. These contain no prism or grating, yet they show very high selectivity for the compound being determined and the equivalent of virtually infinite resolving power, under proper operating conditions. One such analyzer is shown in Fig. 9-13*a*. The radiant energy from the source is split into two beams, which pass through separate gas cells and strike separate detectors. The detectors are chambers filled with the pure gas which is to be determined in the plant stream. The absorption of radiant energy raises the temperature of the gas, and the resulting increase of pressure can be measured by virtue of the motion of

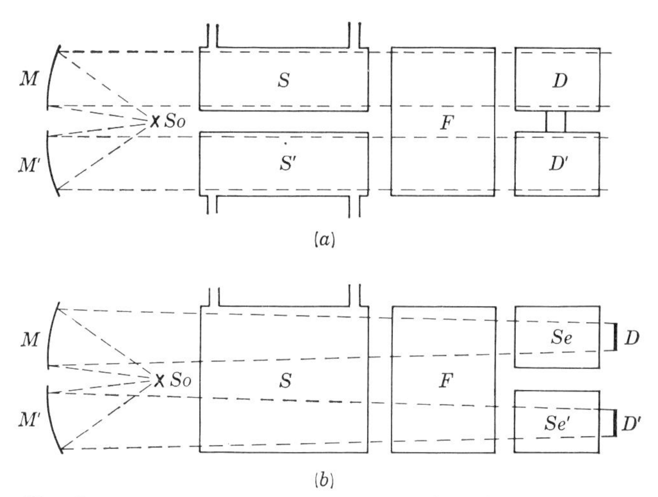

FIG. 9-13. Nondispersive gas analyzers. (*a*) Positive filter type. D and D' are detectors filled with the gas being determined. The sample is placed in S; a reference sample may be placed in S'. F is a filter cell to eliminate interferences. (*b*) Negative filter type. *Se* contains the gas being determined; *Se'* contains a nonabsorbing gas. D and D' are opaque detectors. The sample gas is contained in S. F is again a filter cell.

a thin diaphragm which serves as one plate of a condenser. Since the detector can absorb only those frequencies which can be absorbed by the component to be determined, the instrument selectively measures the transmittance of that compound. Provided that the pressures of the gas in the detector and sample chambers are suitably matched, the nondispersive analyzer will show a frequency discrimination that could be matched only by a dispersive analyzer with infinite resolving power.

The gas mixture to be analyzed is placed in one of the cells, or is allowed to flow through the cell continuously. If there is another component in the mixture which would also absorb part of the radiant energy being measured, filter cells, containing the interfering gas, may be added in both beams. The filter cell then absorbs all of the frequencies of overlapping absorption and the analyzer operates on the remaining regions of the spectrum. The difference in radiant power absorbed by the two detectors is clearly a measure of the amount of energy absorbed by the stream gases, and in particular by the compound in this stream to which the analyzer has been sensitized. If it is desired to determine two or more compounds simultaneously, additional detectors can be added in sequence, each removing from the beam only a portion of the total energy. An alternative design (Fig. 9-13*b*) places a cell containing the gas to be determined in front of one of two opaque detectors. Both beams pass through the sample being analyzed, and the difference between the energies striking the two detectors serves as an indication of the amount of the compound present. An excellent brief review of the operation of gas analyzers is given in Reference 46 (*cf.* page 407).

PROBLEMS

The answers to the starred problems will be found on page 403.

***1.** The energy emitted by a "black body" (a source which is in thermal equilibrium with its own radiation) is distributed with wavenumber according to the equation

$$\frac{dE}{d\sigma} = \frac{8\pi\sigma^2 Vhc\sigma}{e^{hc\sigma/kT} - 1}$$

where V is the volume in which the energy is measured, k is Boltzmann's constant (the gas constant), h is Planck's constant, c is the speed of light, and T is the absolute temperature.

a. Calculate the ratio of the energy per unit volume in an interval of one reciprocal centimeter at 5000 cm.$^{-1}$ to that at 600 cm.$^{-1}$ for a black-body source at 1200°C.
b. Compare this with the ratio for a source at 1800°C.
c. What percentage increase in energy at 600 cm.$^{-1}$ is obtained by increasing the source temperature from 1200 to 1800°C.?

2. The total emission of a black body, over all frequencies, was found empirically by Stefan, and shown (by means of an argument employing a Carnot cycle) by Boltzmann, to vary with the fourth power of the absolute temperature:

$$E = aT^4$$

where a is 7.57×10^{-15} erg cm.$^{-3}$ deg.$^{-4}$.

a. Find the ratio of the total energy per unit volume emitted by a black-body source at 1200°C. to that emitted by the same source between 600 and 605 cm.$^{-1}$.
b. What fraction of the unwanted radiant energy may be allowed to get through the spectrometer and still maintain the stray light level at 0.1 per cent of the energy available at 600 to 605 cm.$^{-1}$?
c. How will this factor change if the source temperature is increased to 1800°C.?

3. For a homopolar diatomic molecule A—A the restoring force will be proportional to the distance Δr the bond is stretched:

$$f = -k\,\Delta r$$

where k is the "force constant" of the bond. This leads to simple harmonic motion, in which the displacement of each atom from equilibrium is given by

$$x = \frac{\Delta r}{2} = x^0 \cos (2\pi\nu t)$$

By equating force to mass times acceleration, d^2x/dt^2, show that the frequency of the motion ν is given by Eq. (9-1).

*4. The observed vibrational frequency (neglecting the correction for anharmonicity) of hydrogen chloride is 2886 cm.$^{-1}$.

a. Find the force constant for the hydrogen chloride molecule.

b. If the molecule absorbs one photon of this energy, find the amplitude of the vibrational motion from the condition that the energy of the vibratory motion is equal to the potential energy

$$V = \int_0^x kx\,dx$$

at the maximum displacement. Express this amplitude as a percentage of the equilibrium distance, 1.275 Å.

5. To a good approximation one may consider the motions of a hydrogen atom attached to a hydrocarbon chain to be independent of the remainder of the molecule. The effective mass is therefore simply that of the hydrogen atom, and Eq. (9-1) may be applied. Aliphatic hydrocarbons show C—H stretching vibrations at approximately 2850 and 2950 cm.$^{-1}$. Where would these vibrations be observed if the compound were deuterated?

*6. Deuteration should have no other effect on a carbon–carbon vibration than that due to the increase in mass of the C—H unit. In a molecule such as benzene, therefore, the result of deuteration should be roughly the same as that of substitution of C^{13}.

a. What percentage shift in the frequency of a C—C vibration in benzene would be expected on deuteration?

b. Included among the vibrational frequencies of benzene are 606, 849, 992, 1178, 1596 (split by interaction with another band), 3047, and 3062 cm.$^{-1}$. The corresponding bands occur at 577, 661, 945, 867, 1559, 2264, and 2292 cm.$^{-1}$ for deuterobenzene, C_6D_6. On the basis of this evidence alone, which bands would you assign to vibrations involving primarily motions of the hydrogen atoms and which to vibrations involving primarily motions of the carbon atoms?

7. Assume that a certain spectrometer has a noise level that is independent of the power striking the detector. For a particular set of experimental conditions this is ± 0.5 per cent transmittance. What will be the average size of the ripple due to noise, expressed in absorbance units, at absorbance values of (*a*) 0.0, (*b*) 1.0, and (*c*) 2.0?

*8. A phototube or photomultiplier tube will show a noise level (expressed in energy or transmittance) that is approximately proportional to the square root of the radiant power striking the tube.

a. If the noise level of a spectrometer employing a photomultiplier tube is ± 0.5 per cent transmittance at 100 per cent transmittance, and this is assumed to be entirely due to the photomultiplier, what noise level would be anticipated at 1 per cent transmittance?

b. What noise level, expressed in absorbance units, would be expected at absorbance values of (1) 0.0, (2) 1.0, and (3) 2.0?

*9. From the spectrum shown in Fig. 9-14, identify the compound in terms of the functional groups present.

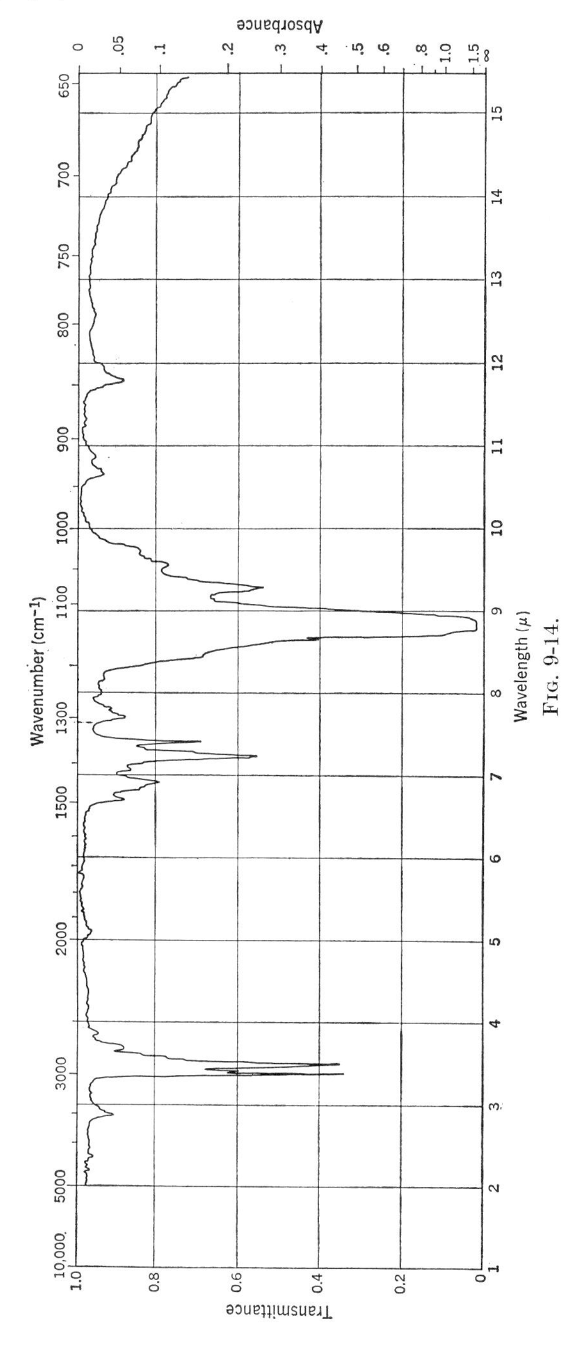

FIG. 9-14.

10. From the spectrum shown in Fig. 9-15, identify the compound in terms of the functional groups present.

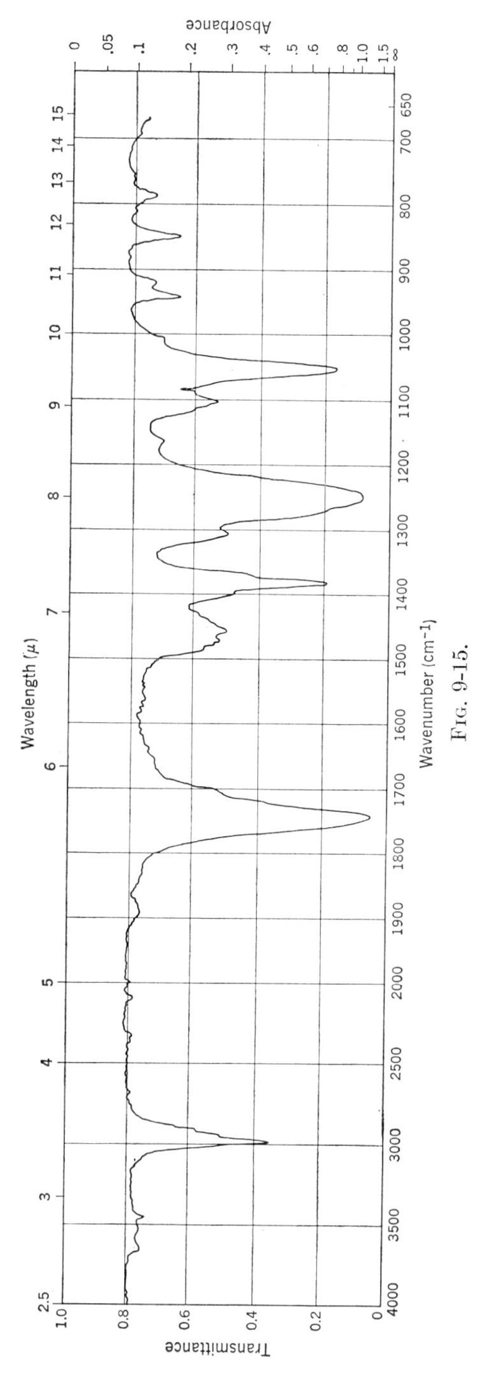

Fig. 9-15.

***11.** Identify the compound whose spectrum is shown in Fig. 9-16.

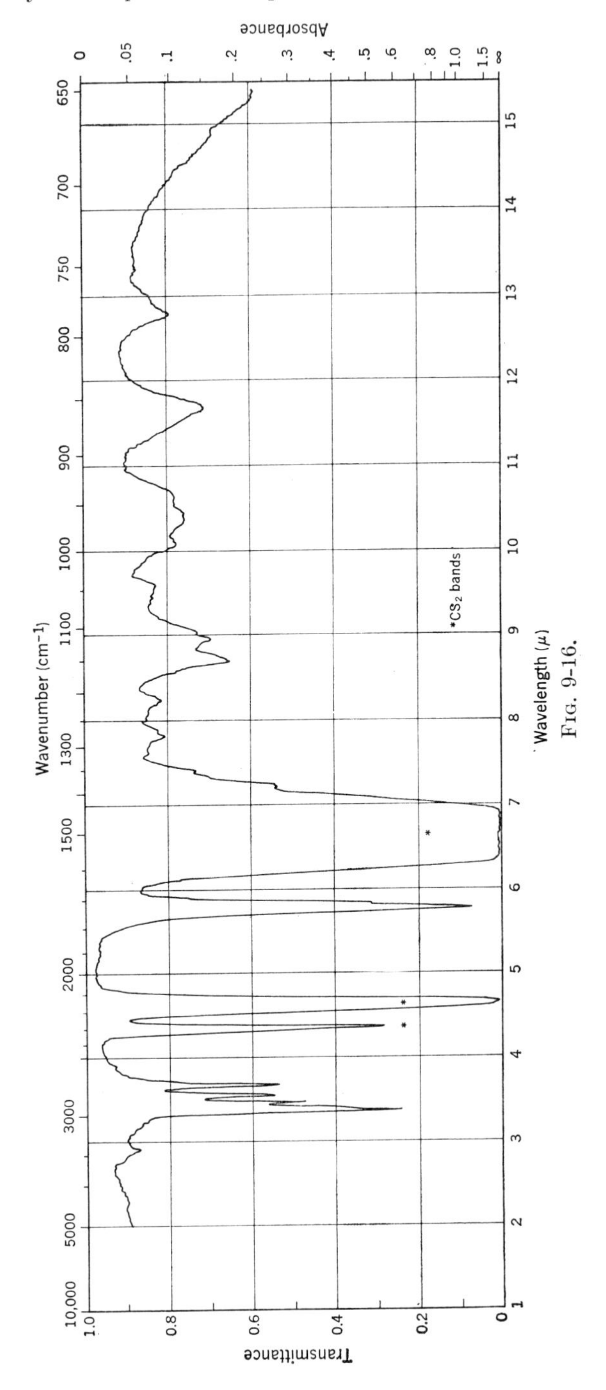

FIG. 9-16.

12. Identify the compound whose spectrum is shown in Fig. 9-17.

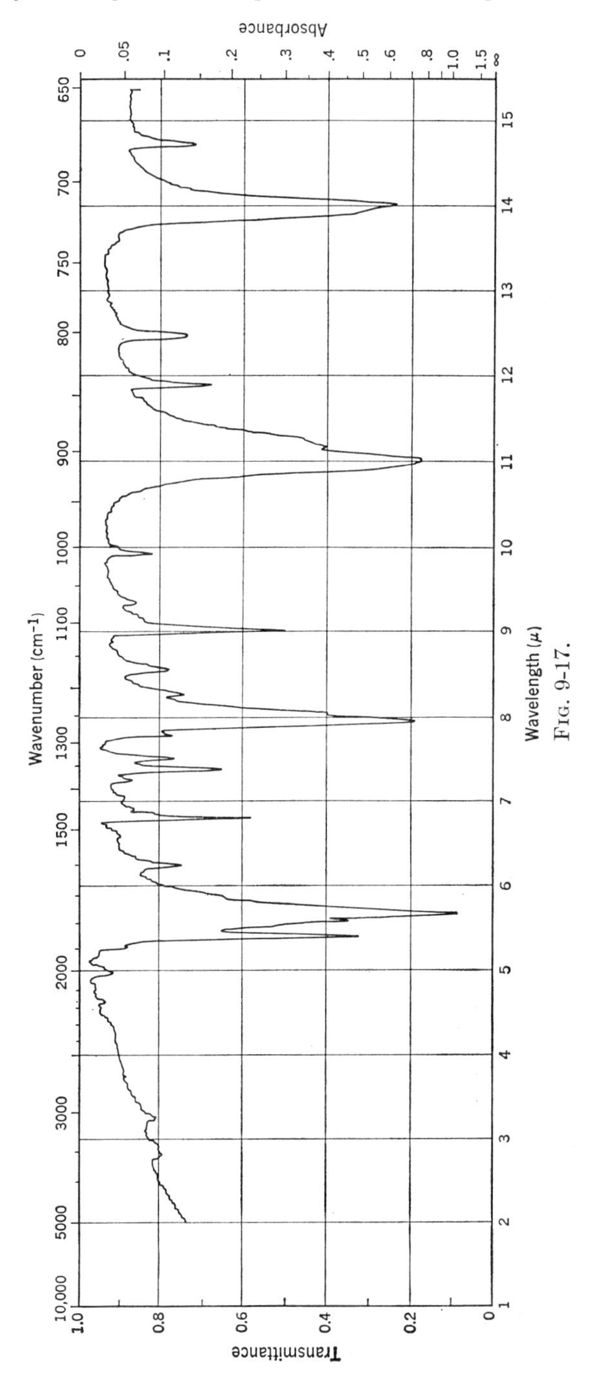

FIG. 9-17.

***13.** Does the compound whose spectrum is shown in Fig. 9-18 have the structure *A* or the structure *B*?

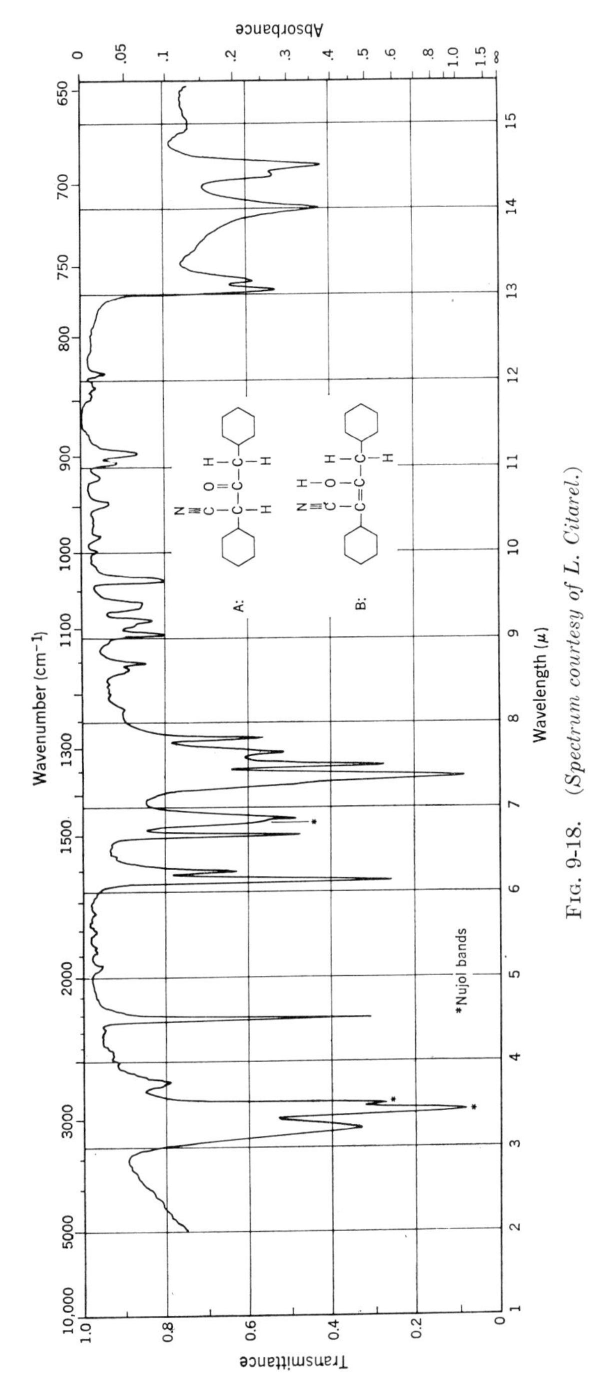

Fig. 9-18. *(Spectrum courtesy of L. Citarel.)*

14. Is the reaction product whose spectrum is shown in Fig. 9-19 the compound C or the compound D?

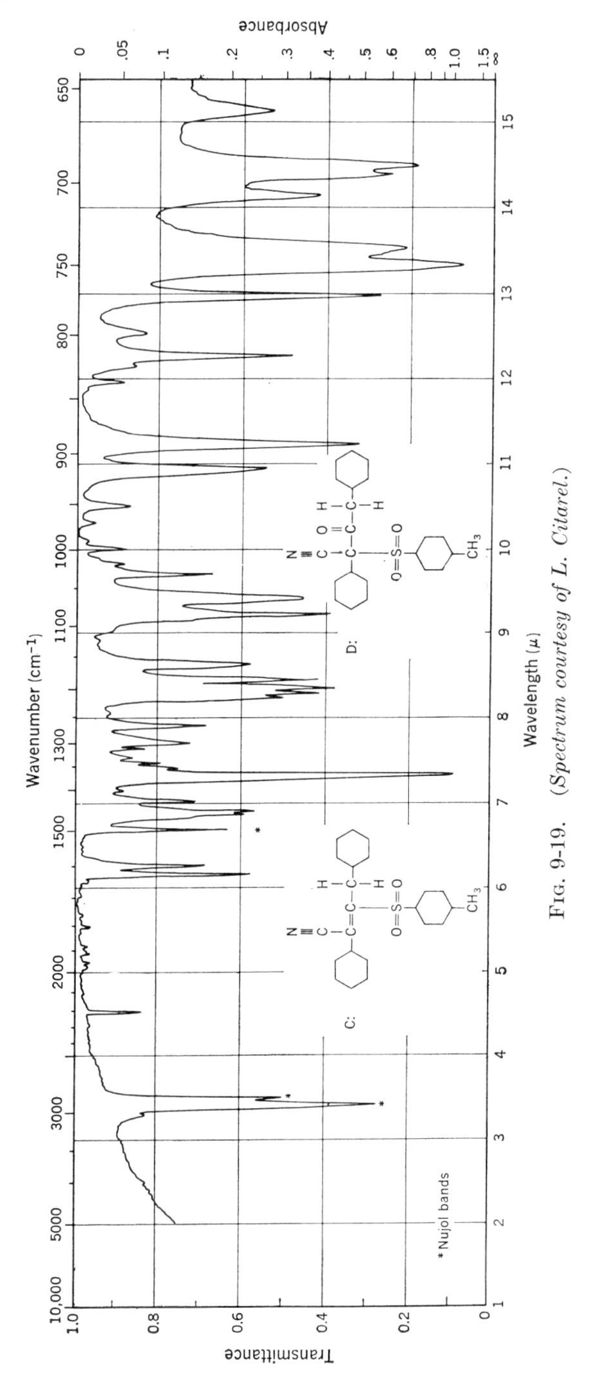

FIG. 9-19. *(Spectrum courtesy of L. Citarel.)*

*15. The following table gives the measured values of the transmittance of a certain compound at various concentrations.

a. Over what range of concentrations is Beer's law obeyed to an accuracy of ± 1 per cent?
b. Calculate the absorptivity, in liters per mole-cm., in the region of low concentrations. The path length of the cell was 0.54 mm.

Concentration, M	*Transmittance, %*
0.12	95.3
0.51	81.6
1.03	66.3
1.47	55.9
2.10	45.2
2.95	35.0

16. The following table gives the measured values of the transmittance of a certain compound at various concentrations over a range within which Beer's law is known to be followed. The path length of the cell was 0.079 mm.

a. Calculate the absorptivity algebraically, using several different points for the calculation.
b. Plot the absorbance against the concentration and estimate the absorptivity from the apparent slope of the line.
c. Calculate the slope of an absorbance–concentration plot by employing the method of least squares. (If we are willing to rely on our choice of the 100 per cent line to the extent of requiring that the curve go through the origin, so that $y = mx$, and if x is known more accurately than y, then minimizing the sum of the squares of the deviations gives the equation $m = \Sigma x_i y_i / \Sigma x_i^2$.)

Concentration, M	*Transmittance, %*
0.053	89.0
0.112	82.0
0.314	57.5
0.608	32.0
0.714	26.0
0.923	17.5

*17. Quantitative measurements on the carbonyl band of a certain ketone in solution show an apparent deviation from Beer's law at high concentrations. What simple method can be employed to determine whether this is a "dilution effect" or a result of instrumental limitations?

18. A solid powder is dispersed in a transparent medium in preparation for infrared measurements. Observation with a microscope reveals that the powder covers only 75 per cent of the total sample area. A certain amount of radiant energy will therefore always be able to pass through the sample. Assuming that the powder is uniformly distributed over the portion of the area which it covers, plot the expected value of the observed absorbance of the sample against the true absorbance of the powder as the latter varies due to changes in either the composition of the powder or the frequency of the incident radiation.

CHAPTER 10

RADIOCHEMICAL METHODS

10-1. Introduction. Every analytical method discussed elsewhere in this book depends directly on the *chemical* properties of the substance in question, which is to say on the behavior of the electrons surrounding the nuclei of the atoms. Thus, the oxidation of ferrous ion is a process in which one of these electrons is removed, and the corresponding formal potential is a measure of the tightness with which the electron is bound. The yellow color of ferric ion is a measure of the spacing of the energy levels of the remaining electrons, together with more subtle properties of these electrons which determine the extent to which radiation of certain frequencies is absorbed. In truly chemical work the nucleus of the atom manifests itself only indirectly through its fixed positive charge. This characteristic, of course, determines the entire chemistry of an atom but is *per se* not a matter of concern to the chemist. Not even the mass of a nucleus can be considered a chemical property. Ordinary oxygen contains nuclei which differ by $12\frac{1}{2}$ per cent in mass but which have identical chemical behavior. Truly nuclear properties can, however, be made the basis of methods for detecting the presence and measuring the relative abundance of an element. The simplest of these methods, in principle if not in practice, gives direct measures of nuclear masses. This method requires the elaborate and expensive modern mass spectrometer, and we shall not discuss it in this book.[1] Methods far less simple in principle but much easier in application, which we shall now discuss, are made to depend on the fact that an atomic nucleus is a complicated structure which can exist in different energy levels of its own. If one of these levels is excited, the resulting nucleus is unstable and is said to be radioactive. The unstable nucleus spontaneously returns to a stable state, and its excess energy is emitted in the form of easily detectable radiation; the unstable nucleus is said to undergo radioactive decay. Such nuclear transitions are unaffected by the state of chemical combination of the atom, so that radiochemical measurements can usually

[1] The interested student is advised to consult V. H. Dibeler, "Analytical Mass Spectrometry," in J. Mitchell, Jr. (ed.), "Organic Analysis," vol. 3, Interscience Publishers, Inc., New York, 1956.

be made after no, or at most very trivial, chemical separations. Investigations which would otherwise be forbiddingly difficult are frequently very much simplified by the use of radiochemical techniques.

In the following pages we shall, with no pretensions whatever to completeness, present some of those parts of nuclear physics which are of importance to the analytical chemist and describe the methods he uses in the solution of his problems.

We should interject at once a word of warning. Truly chemical methods of analysis are, in favorable cases, capable of almost unlimited accuracy. One can almost always pick and choose among such methods and arrive at a procedure which is good to, say, ± 0.05 per cent—if, of course, he is willing to forget the cost in time and care. In radiochemical work, however, the final measurement is subject to inherent, unavoidable, and relatively large statistical fluctuations. The measurement always consists of a count of individual events on an atomic scale, and the rate at which these events occur is necessarily erratic. It is true that the fluctuations can usually be made of less and less effect by the apparently simple expedient of taking larger and larger numbers of counts. However, the measurement of high counts with correspondingly high accuracy depends on the reliability of very complicated electronic circuits. It becomes an engineering feat of no small magnitude to push the reliability of counting techniques to better than 0.5 per cent. It will be seen that most radiochemical results depend on the ratio of two counts. Generally one cannot hope for better than ± 0.5 per cent accuracy in his final result even in favorable cases. If the method in use depends on a difference of counts, the final result may be subject to very much larger uncertainties. In this respect the radiochemical technique suffers in comparison with the ease of securing an accuracy of ± 0.1 per cent in the difference between a pair of weighings. However, most chemical questions can be settled with far less than the ultimate practical accuracy of the analytical balance, and the ease and speed of radiochemical work often far outweigh the lack of accuracy which must be accepted along with it.

10-2. Nuclear Reactions and Radiations. Before the discovery of artificial radioactivity, the applications in chemistry of radioactive tracers were highly limited. Only a few of the heavier elements have naturally occurring activities of any practical value to the chemist. In reality only the chemistry of lead, through its isotope RaD, was easy to investigate by this means. The method was extended with some uncertainty to the study of elements chemically similar to lead, but much is lost if one cannot rely on the chemical identity of the tracer and the element traced. Since 1934, when Curie and Joliot published their discovery of artificially induced radioactivity, truly isotopic tracers for almost every element have been found. Many of these have sufficiently long lives and emit radiation sufficiently easily measurable to be of value to the practical chemist.

Although the chemist is usually quite content to let someone else prepare for him the tracers he uses, it is necessary for him to understand the processes employed in these preparations. In particular, he must be able to recognize situations in which his results may be obscured by the presence of radioactive impurities. These may be present as a result of "side reactions" in the original preparative process, or they may be due to daughter activities of the tracer he uses.

The qualitative characteristics of the nuclear reactions by which artificial radioactivity is produced can be understood in terms of the Bohr-Wheeler model for the nucleus. The atomic nucleus is considered to be composed of neutrons and protons held together by very strong forces at present little understood. These assemblages of *nucleons* can exist in various energy states. (Any relatively massive constituent of a nucleus is termed a nucleon. A single type of nucleus, specified by its mass and charge, is termed a *nuclide.*) If by some means a nucleus is forced into an excited state, the excess energy will at first be distributed more or less uniformly among the nucleons, which may be pictured as being in a state of rapid and random motion within a skin of powerful surface forces. Eventually—that is, after some 10^{-12} to 10^{-14} second—either a single nucleon acquires enough energy to break through the surface skin, or else an internal transition takes place and energy is lost by the emission of a γ ray. In either case, having lost much energy, the nucleus settles back into a stable or, at any rate, less disturbed state. If a stable nucleus is produced, the chemist is not interested. If a disturbed nucleus is formed which much later decomposes further to emit additional radiation, the chemist may be very much interested indeed.

Perhaps the most important means by which a nucleus is lifted into an energetic state is through the absorption of a neutron of low energy and the nearly immediate emission of a γ ray. This process is referred to as the (n,γ) reaction. The neutron, being an uncharged particle, is unaffected by the intense coulombic force near the positively charged nucleus. Of all the massive elementary particles the neutron therefore has the best chance of striking a nucleus. To produce appreciable effects it is obviously necessary to provide an abundance of neutrons. This condition is easily fulfilled in the modern nuclear reactor, where the flux of neutrons is often 10^{12} to 10^{14} per square centimeter per second.

Another obvious necessity for hits on a target as small as an atomic nucleus is that the the projectile be as large as possible. The diameter of the nuclear target is of the order of 10^{-12} cm., and so its cross-sectional area is of the order of 10^{-24} cm.^{2} (= 1 *barn*). Nothing can be done to alter the size of a given nuclear target. It is otherwise with the neutron. From the quantum mechanical point of view the effective extension of a moving particle is quite vague and can in any case be altered by changing its velocity. According to quantum mechanics the size of a neutron is measured by its de Broglie wavelength λ, which is given in terms of Planck's constant h, its mass m, and its velocity v by

$$\lambda = \frac{h}{mv}$$

Thus the cross-sectional area of the projectile is *inversely* proportional to the square of its velocity. One has only to slow the neutron down to get a better chance for a hit. If energetic neutrons are allowed to bounce about in nonabsorbing matter and so lose their energy before they approach absorbing nuclei, a very much larger proportion of them produce nuclear reactions. It turns out that a bouncing particle (*i.e.*, one undergoing elastic collisions) loses on the average the greatest proportion of its energy to particles of a mass equal to its own. A good neutron moderator should therefore contain as large as possible a proportion of very light atoms—hydrogen is best—but these must not be capable of reacting directly with the neutrons. Paraffin will serve. Ordinary water, heavy water, and very pure carbon are in commercial use as neutron moderators. The effect was first discovered because of the presence of a wooden packing case. After 50 or 100 collisions in the moderator, the fast neutrons originally produced by some nuclear reaction are reduced to the ambient thermal energy corresponding to velocities (near room temperature) of 10^5 to 10^6 cm. per sec. and accordingly a cross section of the order of 10^8 barns. For purposes of comparison it may be noted that the cross section of a hydrogen *atom* is also 10^8 barns.

Now the *effective* cross section of a nucleus for the capture of such a thermal neutron varies over an enormous range from nuclide to nuclide. Thus, for He^4 it is zero; for Xe^{135} it is 3.5×10^6 barns, the largest known value. Isotopic nuclei of the same element differ greatly in ability to capture neutrons. Tables of these capture cross sections are available and make possible estimates of the yield of an (n,γ) reaction in a specified flux of neutrons. This is just what the chemist requires in estimating the feasibility of preparing a proposed tracer.

Some typical (n,γ) reactions leading to the formation of useful tracers are: Sb^{121} (n,γ) Sb^{122}; Ca^{44} (n,γ) Ca^{45}; Co^{59} (n,γ) Co^{60}; Fe^{54} (n,γ) Fe^{55}; and Zn^{64} (n,γ) Zn^{65}.

Reactions initiated by the collision of highly energetic charged particles with atomic nuclei are also of great importance for the preparation of tracers. A very high kinetic energy indeed is necessary to force a positively charged particle through the electrostatic potential barrier of an atomic nucleus. The collision cross section of a high-speed particle is necessarily small; hence intense ion currents are required to produce appreciable amounts of reaction. Very massive (and very expensive) equipment is required for the purpose. It is fortunate for the chemist that the physicist needs and can acquire such apparatus for other purposes; if this were not so, the supply of tracers would certainly be very much more limited. The cyclotron is the instrument best adapted to the production of tracers *via* high-energy nuclear bombardment. Many cyclotrons are now in operation throughout the world, and several prepare radioactive tracers on a commercial or semicommercial basis.

A very large number of nuclear excitations can be brought about by high-energy deuterons through the (d,p) reaction. Apparently the relative ease with which a deuteron can be forced into an atomic nucleus is due to a polarization effect: the deuteron approaches the nucleus neutron-end-foremost, and the neutron penetrates the nucleus before the proton suffers the full effect of the coulombic or electrostatic repulsion. The nucleus strips off the neutron and rejects the proton. In this reaction, as in the (n,γ) reaction, the charge on the nucleus remains the same while the mass increases by one unit; a radioactive isotope of the bombarded element is produced.

Two reactions which are brought about by high-energy neutrons, the (n,p) and (n,α) reactions, are of particular importance. In these reactions the charge on the nucleus of the bombarded element changes. Hence a chemical separation of the synthetic isotope becomes possible, and it is possible to prepare radioactive material of very high specific activity. Two such reactions are: ${}_{16}S^{32}$ (n,p) ${}_{15}P^{32}$ and ${}_{13}Al^{27}$ (n,α) ${}_{11}Na^{24}$. The subscripts at the left denote atomic numbers; the superscripts at the right, nuclear masses. For the same reason the (d,n) reaction involving the escape of a neutron on deuteron bombardment is also of importance: ${}_{26}Fe^{54}$ (d,n) ${}_{27}Co^{55}$.

Most of the many other types of nuclear reactions are of interest to the chemist largely because of their nuisance value; they are frequently the source of radiochemical impurities in his tracer preparations. A few of these are the $(d,2n)$, (p,γ), (p,n), (p,α), and (α,n) reactions. There are many others. Evidently when material is exposed to the heterogeneous radiation in a nuclear reactor or near the target of a cyclotron, the chemist must consider all possibilities. He must usually undertake for himself the radiochemical purification of his material. This is particularly necessary when dealing with tracers from the most prolific source of all, the fission of uranium.

When a thermal neutron penetrates the nucleus of an atom of U^{235}, there is a high probability that the excess energy will be dissipated by a splitting of the nucleus into two large fragments, accompanied by the liberation of several more neutrons. In U^{235}, as in other heavy elements, the ratio of neutrons to protons is higher than in the stable isotopes of the lighter elements. The nuclei of the fission fragments of uranium are then generally too rich in neutrons for stability and so are radioactive. They decay by β-ray emission to elements of higher atomic number. Many cases are known in which this decay takes place stepwise, passing through as many as six elements before a stable end product is reached. A large number of radioactive nuclides is to be found among the fission products, and an immense amount of work has gone into the identification and characterization of these substances. After appropriate separations and purifications, these by-products of the atomic power industry afford the chemist many possibilities for tracer work.

10-3. Properties and Availability of Tracers. By the spring of 1956 there were recognized 1058 artificially prepared unstable nuclides, 58 naturally occurring radioactive nuclides, and 272 stable nuclides. Relatively few of the nuclides are useful in chemical work. Some have life times which are too short; some emit radiation which is insufficiently penetrating to be measured easily; and some are simply too difficult to prepare in useful quantity. An excellent summary of the properties and interrelations of the nuclides is available in the chart prepared by the Knolls Atomic Power Laboratory of the General Electric Company.[1] This chart gives at a glance all of the nuclear information which is required in tracer work. One may, for example, assess at once the possibility of finding a given radioactive impurity in a tracer prepared by a known means.

The most practical source of information as to the availability of tracers is the *Catalog of Radioisotopes* of the Oak Ridge National Laboratory.[2] This booklet contains much useful information on the properties of the materials supplied. The catalog for 1956 lists 73 nuclides for 55 different elements.

10-4. Rate of Radioactive Decay. A single nucleus in a radioactive state decays when by chance one of its nucleons acquires sufficient energy and escapes from the nucleus, or when the circumstances become just right for the ejection of energy in the form of a γ ray. The time at which any particular nucleus will decay is completely unpredictable. On the other hand, many millions of energetically identical nuclei will on the average behave in a nearly exactly predictable fashion: a definite fraction

[1] This chart and an accompanying explanatory booklet may be obtained free on application to Department 2-119, General Electric Company, Schenectady 5, N.Y.

[2] Obtainable at a nominal charge from Union Carbide Nuclear Company, Oak Ridge National Laboratory, Radioisotopes Sales Department, P.O. Box P, Oak Ridge, Tenn. The sale and use of radioactive isotopes is subject to Federal regulation in the United States.

will decompose in a specified interval of time. If the time interval considered, dt, is short compared to the average life of the nuclei but still sufficiently long so that many decompositions occur within it, the fraction of the nuclei decomposing will be proportional to the length of the interval:

$$-\frac{dN}{N} = \lambda\, dt \tag{10-1}$$

The proportionality constant λ is characteristic of the nuclide and is unaffected by any outside influence. Therefore at any time t, after a time $t = 0$ when it is known that there were N_0 undecomposed nuclei present, the number of surviving nuclei is given by

$$N = N_0 e^{-\lambda t} \tag{10-2}$$

The number of decompositions which have taken place in this interval, t, is therefore

$$\Delta N = N_0 - N = N_0(1 - e^{-\lambda t}) \tag{10-3}$$

The decay constants for radioactive materials are almost always given as the times required for half disappearance. For $N/N_0 = \frac{1}{2}$, Eq. (10-2) becomes

$$\ln \tfrac{1}{2} = -\lambda t_{1/2} = -\ln 2$$

whence

$$\lambda = \frac{0.693}{t_{1/2}} \tag{10-4}$$

Frequently one must deal with an isotope which decays appreciably during the course of a chemical experiment, and sometimes even during the course of a measurement of its activity. Explicit formulas to take care of these eventualities may be written down at once. Suppose that an observation is made in the interval from $t - \Delta t/2$ to $t + \Delta t/2$, *i.e.*, that t is the half-way point of the count. Suppose, as is usually the case, that the number of counts observed is exactly proportional to ΔN, the number of decompositions taking place in Δt. (This proportionality is easy to ensure for a single determination. The difficulties arise in keeping the proportionality fixed from sample to sample.) Then the number of counts observed is

$$\begin{aligned} C &= k\,\Delta N = k[N_0 e^{-\lambda(t-\Delta t/2)} - N_0 e^{-\lambda(t+\Delta t/2)}] \\ &= kN_0 e^{-\lambda t}[e^{\lambda \Delta t/2} - e^{-\lambda \Delta t/2}] \\ &= 2kN_0 e^{-\lambda t} \sinh\left(\lambda \frac{\Delta t}{2}\right) \end{aligned} \tag{10-5}$$

This formula is exact, but, more often than not, unnecessarily so; and it is somewhat troublesome to use as it stands. For a counting interval which is short compared to the half-life of the nuclide, or, more exactly, if

$$\lambda \frac{\Delta t}{2} = 0.347 \frac{\Delta t}{t_{1/2}} \ll 1$$

or

$$\Delta t \ll 2.9 t_{1/2}$$

Eq. (10-5) can be simplified to

$$C = \lambda k N_0 e^{-\lambda \Delta t} \, \Delta t \tag{10-6}$$

It is usually convenient to express results in terms of the average counting rate R, which is equal to $C/\Delta t$. It is, in any case, seen from Eq. (10-5) that the measured activity decays with the time at which the count is taken just as does the total number of active nuclei.

The reader may easily show for himself that Δt must be less than $0.52 t_{1/2}$ if the approximate formula, Eq. (10-6), is to be in error by no more than 0.5 per cent. This condition is usually quite easily fulfilled; *i.e.*, it is usually possible to use a sufficiently active source so that an adequate number of counts is obtained in a time less than $0.52 t_{1/2}$.

If the entire time required for all the measurements is small compared to the half-life of the nuclide in question, which is to say if

$$e^{-0.693 t / t_{1/2}} \cong 1$$

then obviously the number of active nuclei present will change very little, and no correction for decay need be applied. This condition must, of course, be examined for each tracer and each experiment. If there is a choice of nuclides to use as the tracer, other things being equal, the longest-lived isotope should be used in order to keep decay corrections to a minimum. If corrections are necessary, it is best to determine the half-life of the nuclide in use with the apparatus at hand and as nearly as possible under the conditions of the chemical experiment. History seems to indicate that the accurate measurement of a half-life is not nearly so trivial an exercise as it would seem. A typical measurement of this type is shown in Fig. 10-1.

Since the decomposition of an atomic nucleus is a random event, the fluctuations in the measurements shown in Fig. 10-1 are just what one must expect in observations of this kind. Let us, however, confine our attention to a somewhat simpler case: the observation of a steady (*i.e.*, long-lived) source in which measurements are made over successive equal intervals of time. In spite of the "steady" character of the source, we record variable numbers of counts. It is important to know how large a deviation from a long-time average may reasonably be expected. The problem is one of probability theory and statistics. It is treated fully in many texts. Here we shall merely summarize the significant results.

Suppose that R is the long-time average counting rate, so that $\bar{n} = R\,\Delta t$ is the average number of counts observed in the interval Δt. The theory supplies us with a probability distribution function, $P(n,\Delta t)$ or $P(n,\bar{n})$, which we take as our measure of reasonable expectation. P is defined

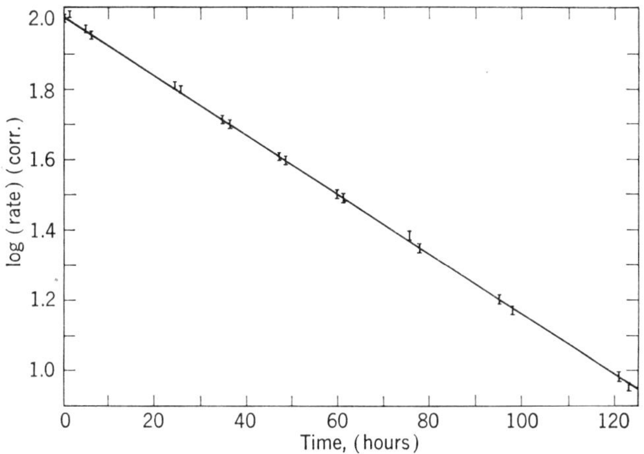

FIG. 10-1. Decay of Br^{82}. Radioactive bromine in potassium bromide was transferred to aluminum bromide by distilling a mixture of the salts, and an aqueous solution of the resulting aluminum bromide was placed in a jacketed counter. For each rate determination 81920 counts were taken. The lengths of the lines denoting the points are 10 times the values given by log $(C \pm \sqrt{C})$ and thus give a very pessimistic estimate of reliability. The straight line is drawn for $t_{1/2} = 35.9$ hr.

so that unity denotes certainty. $P(n,\bar{n})$ is the probability of observing n counts in an interval for which the average number of counts is $\bar{n}$. For the case of the steady source this function turns out to be the Poisson distribution:

$$P(n,\bar{n}) = e^{-\bar{n}} \frac{\bar{n}^n}{n!} \tag{10-7}$$

A plot of $P(n,10)$ is given in Fig. 10-2. The general characteristics of the probability function are immediately seen: the low probabilities for values of n much different from $\bar{n}$, and the nearly symmetrical shape of the curve near its maximum. This latter point expresses the reasonable expectation that, at least for large values of $\bar{n}$, small positive deviations from the average will be about as frequent as small negative deviations.

The formula for the Poisson distribution is a mathematical nuisance as it stands. (The presence of the factorial complicates calculations with it.) It can, however, be converted into a formula, good for large $\bar{n}$, which gives a distribution symmetrical about $n = \bar{n}$, and which fits the

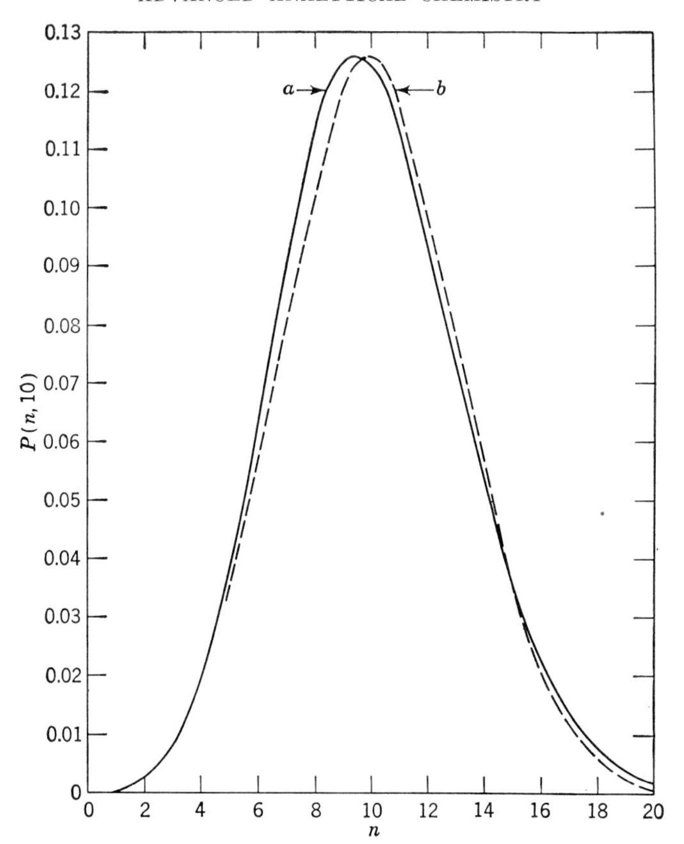

FIG. 10-2. Plots of (*a*) the Poisson and (*b*) the Gaussian probability functions for $\bar{n} = 10$. For larger values of $\bar{n}$ the discrepancy near the peak is not nearly so noticeable. Note that, strictly speaking, the "curves" consist only of discrete points at integral values of n.

true distribution well enough. If we use Stirling's approximation for the factorial ($n! = \sqrt{2\pi n}\, n^n e^{-n}$), write $\delta = \bar{n} - n$, and retain terms of order $\delta^2/\bar{n}$ in the exponent, we find

$$G(\delta) = \frac{1}{\sqrt{2\pi \bar{n}}} e^{-\delta^2/2\bar{n}} \tag{10-8}$$

This is the Gaussian distribution. It is also plotted in Fig. 10-2 for $\bar{n} = 10$ (not a very large value), where the character of the agreement is readily seen. The reader may exercise his calculus on this formula to show that the reasonable expectation for a value of δ between $-\infty$ and $+\infty$ is certainty:

$$\int_{\delta=-\infty}^{\delta=+\infty} G(\delta)\, d\delta = 1 \tag{10-9}$$

Statisticians generally specify the scatter of data in terms of an index σ^2, called the *dispersion* or *variance*, or else in terms of σ itself, which is called the *standard deviation*. These indices are simply the averages of our deviation $\delta = \bar{n} - n$ taken in such a fashion as to prevent negative values from canceling the effect of positive values:

$$\begin{aligned} \sigma^2 &= \overline{\delta^2} = \overline{(\bar{n} - n)^2} = \overline{(\bar{n}^2 - 2n\bar{n} + n^2)} \\ &= \overline{n^2} - \bar{n}^2 \end{aligned} \tag{10-10}$$

Using this expression for σ^2, we may easily calculate its value directly from a table of experimental data. We may also show what to expect for σ if the data obey a Gaussian distribution:

$$\begin{aligned} \sigma^2 &= \overline{\delta^2} = \frac{1}{\sqrt{2\pi\bar{n}}} \int_{-\infty}^{+\infty} e^{-\delta^2/2\bar{n}} \delta^2 \, d\delta \\ &= \bar{n} \end{aligned} \tag{10-11}$$

The standard deviation is therefore expected to be

$$\sigma = \sqrt{\bar{n}} = \sqrt{R \, \Delta t} \tag{10-12}$$

Now suppose that we have made a long series of counting-rate measurements. How many of these can be expected to differ from the average by as much as or more than the square root of the average count? The value of δ corresponding to this question is $\sqrt{\bar{n}}$. We need the value of $G(\delta)$ given by

$$\begin{aligned} G(\sqrt{\bar{n}}) &= \frac{1}{\sqrt{2\pi\bar{n}}} \int_{-\sqrt{\bar{n}}}^{+\sqrt{\bar{n}}} e^{-\delta^2/2\bar{n}} \, d\delta \\ &= \frac{1}{\sqrt{\pi}} \int_{-1/\sqrt{2}}^{+1/\sqrt{2}} e^{-u^2} \, du = 0.683 \end{aligned} \tag{10-13}$$

The numerical result is taken from a table of the probability integral.[1] Thus about three out of ten measurements in such a series may be expected to differ from the average by more than $\sqrt{\bar{n}}$. A test such as this is often applied to ensure that counting equipment is in proper running order.

We may turn this problem around to the point of view of the chemist planning an experiment. He might ask how many counts he should take so that the probability is 9 in 10 that the result will be accurate to ± 0.5 per cent. Here we specify $\delta/\bar{n} = 1/200$, and we require the value of $\bar{n}$ which satisfies the equation

$$\begin{aligned} 0.90 &= \frac{1}{\sqrt{2\pi\bar{n}}} \int_{-\bar{n}/200}^{+\bar{n}/200} e^{-\delta^2/2\bar{n}} \, d\delta \\ &= \frac{1}{\sqrt{\pi}} \int_{-\sqrt{\bar{n}}/200\sqrt{2}}^{+\sqrt{\bar{n}}/200\sqrt{2}} e^{-u^2} \, du \end{aligned} \tag{10-14}$$

[1] Such as that in B. O. Peirce, "A Short Table of Integrals," Ginn & Company, Boston, 1929.

Again using the table of the probability integral, we find

$$\sqrt{\bar{n}}/200\sqrt{2} = 1.163$$

or $\bar{n} = 1.08 \times 10^5$. Thus the chemist hoping to make a Geiger counter rival a balance in precision must record the time for some 1×10^5 counts in each measurement. With modern equipment this is not an impractical assignment.

One often hears in the laboratory the equivalent of the remark: "I took 40000 counts, so it's good to half a per cent." The remark is meaningful only up to the comma. It should have run something like "I took 40000 counts, and if I repeat the experiment nine times more, seven out of ten of the counts ought to be within half a per cent of the average." Or, after a little figuring with the table of the probability integral, the worker might have said, "I took 40000 counts, and I'm 99 per cent certain that the result is good to *thirteen* parts in a thousand." The reader is recommended to mull over these remarks and make similar calculations of his own. It is quite essential that the radiochemist have a full appreciation of the effect on his final conclusions of the necessarily random nature of the phenomena he observes.

Two other matters which arise in making radiochemical measurements must be considered. Every instrument used to determine an activity is subject to the effects of *background radiation*. Most of the bath of electromagnetic and other radiation in which we live is fortunately too feeble to affect an ionization chamber or a Geiger-Müller counter if the minimum precaution is taken of excluding visible and ultraviolet light. There is always present, however, some radiation—perhaps cosmic radiation—that is sufficiently energetic to penetrate even the thickest shield. Furthermore, all materials of construction, including those of a Geiger counter tube itself, contain minute traces of radioactive impurities. Even the most carefully shielded counter will therefore register a slow and hence highly erratic background. The rate of this background must be subtracted from the counting rate observed with a sample in place to obtain a true measure of the activity desired. The question of the accuracy of this difference immediately arises. We shall not discuss the somewhat intricate problem in probability here involved, but will merely quote a result which enables the operator to estimate the reliability of his measurements. If R is the rate sought and M and B the measured rates with and without sample in the counter, then for

$$R = M - B$$

the dispersion of the desired quantity, R, is the *sum* of the dispersions of the measured quantities:

$$\sigma_R^2 = \sigma_M^2 + \sigma_B^2 \tag{10-15}$$

Thus if $M = 40000$ and $B = 8000$, we find from Eqs. (10-12) and (10-15) that the standard deviation of R is $\sigma_R = \sqrt{48000} = 219$. We side-step the question of the probability to be assigned to the relative error in R, $^{219}/_{32000}$. It is obvious that B/M should be as small as possible. It is often possible in work with tracers to make B less than $\sqrt{M}$; it is then practically negligible.

The final matter to consider relates to the finite resolving time of all counting equipment. How fast can we count? Immediately after a counter receives an impulse there will be a short period of recovery during which the counter is insensitive. If a signal arrives during this period, it fails to register. Since the spacing between signals from a radioactive source is random, no matter how short the "dead time" of the counter or how low the average rate of arrival of counts, there will always be a certain probability that a particle will arrive during the dead time and fail to be counted. Evidently the fraction of particles thus missed increases with the average counting rate. Particularly in a chemical experiment in which an unknown, after much dilution of its activity, is to be compared with a relatively "hot" standard, these coincidence losses should always be taken into account.

If M is the observed rate of arrival of counts and τ is the dead time after a count, say in seconds, then $M\tau$ is the total dead time per second. If R is the true rate of arrival of counts, the relation between M and R is very nearly (provided that $M\tau$ is not too large)

$$M = R(1 - M\tau)$$

or

$$R = \frac{M}{1 - M\tau} \tag{10-16}$$

Geiger-Müller counters ordinarily have dead times between about 100 and 300 microseconds. Thus for a counting rate of 200 per sec. the correction amounts to $1/(1 - 200 \times 200 \times 10^{-6})$, so that the true counting rate is just over 4 per cent larger than that observed. If this correction happened to belong to a standard of activity which was being compared with an unknown counting only 50 per sec., omitting the corrections would introduce a 3 per cent error into the final ratio.

The determination of the dead time of a counter is best done directly. Considerable equipment is necessary, including an oscilloscope with a triggered and timed sweep and a generator supplying pairs of sharp pulses at variable and known spacing. If this equipment is not available, a fairly accurate estimate may be obtained by the technique of "stacked sources" described in Expt. IX-1 at the end of this text.

10-5. Equipment for Measuring Radioactivity. In chemical applications the two most practical instruments for the measurement of radioactivity are the Geiger counter and the scintillation counter. In modern practice the electronic circuits associated with these devices can scarcely be termed simple, although their functions are easy enough to describe. We shall omit all technical detail, for which the reader is referred to textbooks on experimental nucleonics or electronic engineering.[1]

The Geiger or Geiger-Müller counter tube itself is a special type of ionization chamber, constructed and operated in such a way that an initially small amount of ionization is rapidly and enormously multiplied.

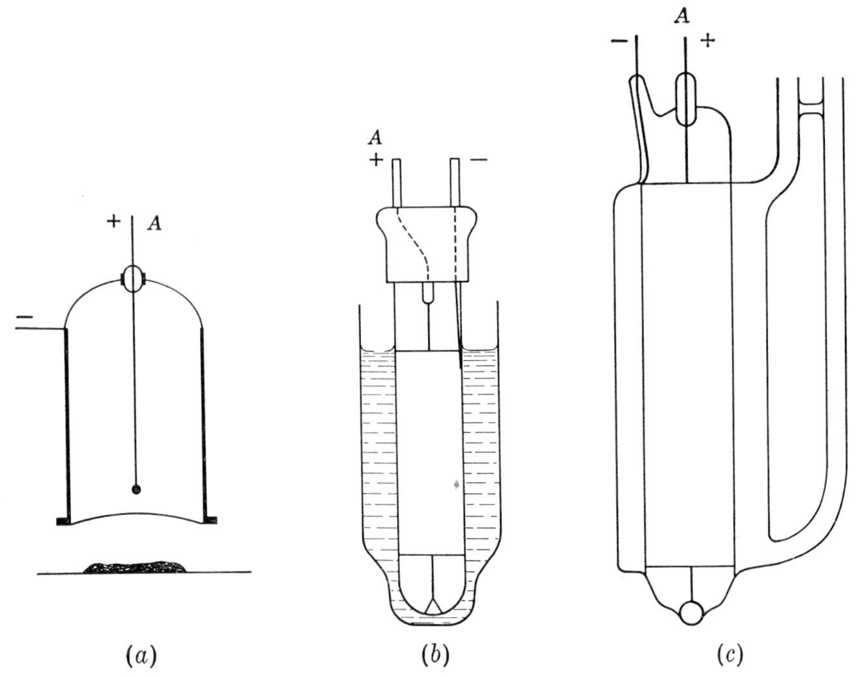

FIG. 10-3. Three types of Geiger counters. The end-window counter (*a*) is used principally for measuring the activities of solids. The dipping counter (*b*) and the jacketed counter (*c*) are used in work with solutions.

A few ion pairs from a beta particle, or a photoelectron ejected by a γ ray from the wall of the counter, trigger a discharge through the tube, and the accumulation of charge on an electrode, *A* in Fig. 10-3, produces an appreciable voltage pulse. To ensure a discharge of short duration the counter contains a small amount of some heavy gas (methane, alcohol, or a halogen) which dissipates the energy of the discharge and also prevents the emission of secondary electrons, which would cause spurious secondary discharges. The initial discharge in a Geiger counter is quite rapid; the relatively long recovery time after a discharge is the time

[1] See, for example, Reference 53, p. 408.

required for the migration of the heavy positive ions back to the outer electrode, where they are discharged. The ionization processes in the tube gradually destroy the heavy "quenching" gas (unless this is a halogen), and the tube eventually becomes useless.

The voltage pulse from the discharge may be further amplified and shaped. From the amplifier the pulse is fed into a scaling circuit. No mechanical register could possibly follow the rapid and erratic rate of counting which must be observed to obtain any accuracy in this work. The scaling circuit is an electronic divider. In one type each stage automatically passes on every other pulse; 12 stages divide by $2^{12} = 4096$. In another, each stage passes on every tenth pulse; four stages divide by 10^4, and the instrument very conveniently reads decimally. The circuits are generally so arranged that the output from the scaler can be used to operate a mechanical register. The timing can be done manually, with a stopwatch, but the more complete instruments usually have built-in electric clocks which start and stop with the count. These may be arranged so that the count is stopped either after a preset time or after a predetermined number of counts has accumulated. Each has its advantages. If the time is preset, the rate is directly proportional to the instrument reading; if the count is preset, the statistics remain constant.

Many forms of the Geiger counter are used. Three of these, particularly well adapted to chemical work, are shown in Fig. 10-3. Much care is necessary to ensure that a counter is placed reproducibly with respect to each of a series of samples. If it is necessary to measure the activity of a precipitate, an end-window counter, (*a*) in Fig. 10-3, is used. The precipitate must be spread uniformly over a well-defined area and positioned accurately below the window of the counter, every care being taken to ensure reproducible geometry. In addition to difficulties with positioning, the measurement of precipitates is afflicted with another uncertainty. Reproducible results are obtained only with "infinitely" thin or "infinitely" thick layers of precipitate. The radiation from the tracer is in part absorbed in the precipitate itself. Furthermore, radiation may be scattered into the counter from massive material behind the source. With a very thin layer of precipitate mounted on light (nonscattering) material, one may reasonably suppose that just half of the radiation proceeds upward toward the counter. The fraction of this radiation intercepted by the active volume of the counter depends on the area of the precipitate, its distance from the counter, and the thickness of the window of the counter. How thin a layer of precipitate may be classed as "infinitely" thin depends on the character of the radiation. Beta particles are absorbed in thin layers of matter which scarcely affect hard γ rays. For a γ emitter a sufficiently thin precipitate may be easy to prepare. But in dealing with a β emitter the precipitate should be so thick that radiation from its bottom layers cannot penetrate the upper surface; the precipitate should be "infinitely" thick. Under these conditions for a given precipitate a definite fraction of the radiation is seen by the counter, and the counting rate is proportional to the activity of the precipitate. This technique is particularly useful with the soft β radiation from C^{14} and S^{35}. Alpha radiation is so extensively absorbed by matter that special techniques are required for handling α emitters; these are fortunately of very limited chemical interest.

Precipitates may be collected by suction on small discs of filter paper. They should be throughly dried before measurement to prevent variable absorption of the radiation by evaporating water. If necessary a dry precipitate can be fixed in place by moistening it with a dilute solution of lacquer or plastic cement in some volatile solvent.

For samples which can be prepared in the form of solutions the Geiger tubes shown in (*b*) and (*c*), Fig. 10-3, are used. The simple dipping counter serves admirably for relatively rough work. The same volume of solution should be used for each sample and care taken to see that the counter is reproducibly placed, particularly as to depth of immersion. The jacketed counter tube, Fig. 10-3*c*, has many of the characteristics of a precision instrument. If the tube is always filled to a definite mark on the side arm, no problems of geometry arise. Self-absorption in the solutions must be considered if solutions of different densities are to be measured. When possible, standards of activity should be prepared in solutions nearly identical chemically with those to be measured. Little difficulty is experienced in getting reproducibility to ± 1 per cent with counters of this type.

In order that a Geiger counter be sensitive to β radiation it must be made with thin windows or thin walls. (These, incidentally, are very fragile.) The same counter is then rather insensitive to γ rays; relatively few photoelectrons are formed to initiate the discharge. Far more efficient counting of γ rays can be done with the scintillating crystal. This comparatively recent development can be considered as a very great refinement of the zinc sulfide screen used by the early workers in radioactivity for the detection and counting of α particles. Scintillators are of two general types. Organic phosphors, such as large crystals of anthracene or stilbene, are primarily useful for the measurement of β radiation. Large transparent crystals of salts containing heavy elements are most useful for γ rays. Crystals of sodium iodide, activated with a little thallium iodide, are in common use. Either type of crystal is mounted directly on the window of a photomultiplier tube (Sec. 8-8). Radiation absorbed in the crystal is converted into light which is scattered and reflected into the photomultiplier, there producing a voltage pulse.

In addition to its greater efficiency, this process is very much faster than the action of a Geiger counter. Counting at much higher speeds can be done.

The γ ray-sensitive salt crystal offers another important advantage. The γ rays of a single nuclide have sharply defined energies, just as do rays in the visible portion of the spectrum. In a large sodium iodide crystal the entire energy of a γ ray may be converted into photons of visible light, the number of these being proportional to the energy of the γ ray. If the voltages on the electrodes of the photomultiplier tube are carefully stabilized, and if the output pulse passes through a truly linear amplifier (in which the gain is independent of the height of the input pulse), then the height of the pulse from the amplifier will be proportional

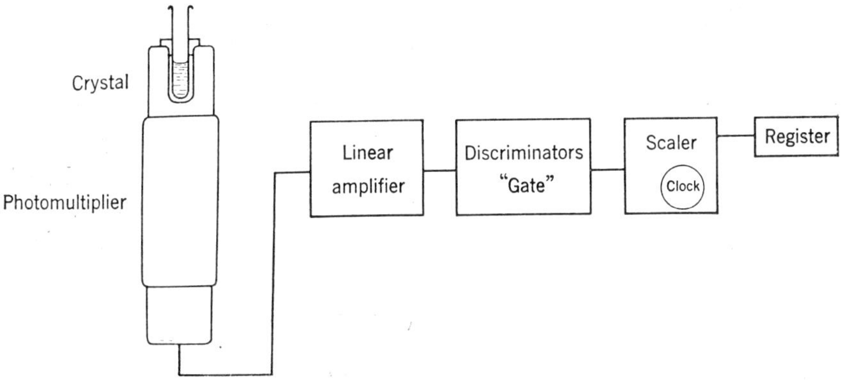

Fig. 10-4. Block diagram of a scintillation spectrometer. The instrument depicted is generally useful. Less elaborate equipment is necessary for simple counting.

to the energy of the γ ray. By the use of suitable discriminators in the circuit (see Fig. 10-4), the number of pulses with heights in a given narrow range can be measured. The operator may thus map the γ-ray spectrum of his material. On the other hand, he may choose to measure only the radiation within a certain narrow range of energies, and it then becomes quite feasible to measure the amount of a chosen radioactive nuclide in the presence of radioactive impurities. The technique is very important, but it is not a cure-all for lack of chemical purification; there are definite interferences. Those γ rays which are not absorbed completely in the crystal produce fewer photons and hence produce the effect of γ radiation of lower energy. For every sharp peak found in the spectrum there is a broader and lower peak at lower energy. This effect may obscure the significance of a low-energy peak for a different nuclide.

Crystals are made with a well for receiving a tube containing the radioactive material. The fraction of the radiation lost to the measurement is thereby greatly reduced. The operator must do a little preliminary experimentation to define the reproducibility of his arrangements. A satisfactory setup is shown in Fig. 10-4. A lucite bushing holds a small plastic test tube in a reproducible position. Identical volumes of different samples, preferably in the same test tube, may be compared very accurately.

10-6. Tracers and Traces. The unit of measurement for the amount of radioactivity is the defined *curie*, that amount of material which undergoes 3.700×10^{10} disintegrations per second. It is instructive to calculate for a typical case the actual weight of radioactive nuclide which might be taken in a chemical experiment. Let us suppose that we are investigating the coprecipitation of bromide with some other slightly soluble silver salt. We propose to collect a tenth of a gram of precipitate, redissolve it, introduce it into a jacketed Geiger counter, and measure its activity. With the equipment available we can count 200 per sec., and so we adjust the amount of tracer in the original solution to give approximately this result. Suppose that the entire amount of bromide in the precipitate is 1 mg. If our counter is 10 per cent efficient, what weight of the 35.9-hr. Br^{82} will the precipitate have to contain?

The sample is undergoing 2000 disintegrations per second, and so it contains $2000/(3.700 \times 10^{10}) = 5.4 \times 10^{-8}$ curie of radiobromine.

If N is the number of atoms of Br^{82} which produces just a curie of activity,

$$-\frac{dN}{dt} = \lambda N = 3.700 \times 10^{10}$$

but

$$\lambda = \frac{0.693}{t_{1/2}} = \frac{0.693}{35.9 \times 60 \times 60}$$

$$= 1.07 \times 10^{-5} \text{ per sec.}$$

so that

$$N = \frac{3.700 \times 10^{10}}{1.07 \times 10^{-5}} = 3.46 \times 10^{15}$$

and the weight of Br^{82} per curie is

$$\frac{3.46 \times 10^{15}}{6.02 \times 10^{23}} \times 82 \times 1000 = 4.71 \times 10^{-4} \text{ mg.}$$

The weight of Br^{82} in the milligram of bromine contained in the precipitate is

$$4.71 \times 10^{-4} \times 5.4 \times 10^{-8} = 2.54 \times 10^{-11} \text{ mg.}$$

Several things immediately become evident. In the first place, we certainly need not worry over the change in average atomic weight of our bromine because of the presence of the tracer. Less trivially, it appears that the application of the tracer technique is reliable only because relatively massive amounts of the natural element are present. It has frequently been found with traces that the chemistry of the trace appeared to be entirely different from that of the same element *en masse*. The trace might remain entirely in solution, or it might be entirely adsorbed on the walls of the vessel and never appear in the precipitate. Such effects must be particularly considered with a tracer producing a radioactive daughter on decay. The daughter trace may be present in a system containing no massive amount of the corresponding element and so behave in an anomalous fashion.

The highly specific behavior of traces may temporarily alter the activity of a solution prepared for measurement. As an example we may cite the case of Cs^{137}. This nuclide is an excellent tracer for cesium. It has the conveniently long half-life of 30 years. In its decay, however, it emits a β particle and forms not the stable Ba^{137} but rather an isomeric barium nucleus in an active state. This daughter in turn decays by a process known as internal conversion with the emission of a hard γ ray, the charge on the nucleus remaining fixed. The half-time for this process is 2.60 min. If a glass test tube containing a solution of cesium chloride "spiked" with Cs^{137} is quickly emptied and rinsed with water, it will be found to have retained a very considerable amount of radioactivity which decays with a period of 2.6 min. It is also found that the activity of the cesium chloride solution is not now constant, but that it grows with a period of 2.6 min. The daughter is being reborn in the solution. After a half hour virtually no activity is detected on the walls of the empty test tube, and no further increase in activity of the solution is observed. In the solution the rate of growth of activity has become equal to the rate of decay of Ba^{137}:

$$\lambda_{Cs} N_{Cs} = \lambda_{Ba} N_{Ba}$$

This is a case of what is known as *secular equilibrium*. (The term *secular* refers to the fact that the amount of Ba^{137} now changes only very slowly, according to the half-life of Cs^{137}.) The reader will find it instructive to show for himself that after a period of 10 half-lives of the short-lived nuclide (26 min. in the present case) the activity will be within 0.1 per cent of the value corresponding to equilibrium.

With these considerations in mind we may now proceed to describe situations in which the use of tracers can be a great aid to the analytical chemist.

10-7. Applications. A somewhat trivial use of the tracer technique arises from the necessity or desire to save time and work in an analysis which might otherwise be done by ordinary chemical methods. For example, in a study of the distribution of ferric iron between aqueous hydrochloric acid and some organic solvent one need only introduce Fe^{55} into the stock of ferric chloride and determine the relative counting rates of aqueous and organic solutions. Evidently, perhaps at the cost of some precision, a great deal of information can be obtained very rapidly by this means. Many examples of this nature might be cited. The technique is most useful in investigations requiring the analysis of large numbers of samples, especially if the chemical methods are difficult or unreliable.

A far less trivial application is the method of *isotopic dilution*, although few examples of this are to be found in inorganic analytical chemistry. Consider the situation in which it is possible to separate a small amount of a compound in a pure state, but impossible to recover it quantitatively. A quantitative determination may be obtained by adding to the unknown a weighed amount of the pure compound in question carrying a measured amount of tracer. When it is certain that the mixture is homogeneous, a portion of the compound is separated in the pure state. This sample of the compound will contain an amount of tracer determined by the dilution ratio of the experiment:

$$A_s = \frac{A_a N}{N + N_0} \tag{10-17}$$

where A_s and A_a are the specific activities of the material recovered and of the "spiked" material added, respectively, N is the number of moles of the pure compound added, and N_0 is the number of moles of the same compound originally present in the sample. The method has found important use in organic chemistry and, more especially, in biochemistry. Particularly in the latter field purifications may be relatively easy but quantitative separations are practically unheard of. A particularly striking application of the method was made by Calvin *et al.* in studying the mechanism of photosynthesis.[1]

Although inorganic substances are usually far less difficult to recover quantitatively than organic ones, the isotopic dilution method may still be very helpful in dealing with samples containing two or more elements whose chemical behaviors are very similar. This would be true, for example, of a sample containing a number of the lanthanide elements. By recently developed techniques depending on ion-exchange processes it is not too difficult to separate a small pure sample of a single rare earth from such a mixture, but a quantitative separation would be very difficult indeed. One might assay a mixture of rare earths for its praseodymium content by adding a known amount of a pure praseodymium preparation carrying the 19-hour Pr^{142} as tracer. By appropriate means a small amount of a pure praseodymium compound is separated, weighed, and its activity determined. Comparing the specific activities of the

[1] M. Calvin *et al.*, *J. Am. Chem. Soc.*, **72**, 1710, 5266 (1950).

added and recovered materials thus permits the calculation of the praseodymium content of the original material.

The study of coprecipitation, mentioned in Sec. 10-6, is very greatly facilitated by the use of radioactive tracers. The method is here to be recommended over any other. A direct and characteristic determination of the impurities in a precipitate is obtained. In this way, for example, it is easy to study the way in which the amount of potassium coprecipitated with calcium oxalate depends on the experimental conditions. Without tracers we would have to either rely on the necessarily very imprecise determination of the error in the weight of the precipitate caused by the coprecipitation, or carry out a difficult and tedious analysis of the precipitate for potassium.

The method has been extended with success to studies of the aging of precipitates. If fresh and aged precipitates of silver bromide are placed in bromide solutions containing Br^{82}, very great differences are found in the rates of the exchanges between the bromide of the precipitates and that in the solutions. An aged precipitate will absorb 10 per cent of its final activity within a minute or so, but will require 20 or 30 hours to reach isotopic equilibrium with the solution. Evidently exchange at the surface of the particle is rapid, while equilibrium is eventually reached only through the slow diffusion of the tracer into the interior of the particle.

Tracer techniques may be applied directly to the determination of the solubilities of very slightly soluble substances. The method is, for example, very useful with the organometallic chelate complexes. Thus, Co^{60} has been employed to show that the solubility of the α-nitroso-β-naphthol derivative is 1.5 mg. per l., while that of the β-nitroso-α-naphthol analogue is 0.17 mg. per l. The possibility of using tracers of high specific activity makes this application quite attractive.

10-8. Exchange Reactions. We shall conclude our discussion of the application of isotopic tracers in chemistry with a description of a phenomenon quite impossible to investigate by any other means. The implications in this connection may be very important in any application of tracers, particularly in organic chemistry.

By the use of isotopic tracers we can observe an atomic exchange process where no gross chemical reaction is taking place. This observation is direct proof of the dynamic nature of chemical equilibrium, if such proof were needed. It makes possible a study of chemical reactions under the simplest possible conditions and is thus of considerable theoretical significance. The existence, or nonexistence, of exchanges between different chemical species may evidently determine the usefulness of a tracer in any given situation.

Suppose, as an example, that we consider the processes taking place when carbon disulfide solutions of aluminum bromide and ethyl bromide are mixed. Chemically speaking, nothing happens, and with very little ingenuity the constituents of the mixture can be recovered unchanged. If, however, the aluminum bromide initially contained Br^{82}, the recovered ethyl bromide is found to be radioactive. Without implying anything at all as to the mechanism of the interaction which results in the exchange of bromine, we can write

$$Al_2Br_6^* + C_2H_5Br = Al_2Br_6 + C_2H_5Br^*$$

indicating merely that the ethyl bromide gains activity, denoted by the asterisk, at the expense of the aluminum bromide. Suppose a and b are the concentrations of *bromine* in forms Al_2Br_6 and C_2H_5Br, respectively, and let $R(a,b)$ be the rate of the exchange reaction, measured perhaps in moles per liter per second. Let θ and φ be the fractions of bromine atoms which are radioactive in Al_2Br_6 and in C_2H_5Br. If the aluminum bromide is separated and placed near a counter, the counting rate will be proportional to θa. If all the activity was initially introduced along with the aluminum bromide, then

$$\theta a + \varphi b = \theta_0 a, \qquad \text{a constant}$$

The rate of loss of activity from the aluminum bromide will be given by the rate of interchange times the fraction of the atoms lost which are radioactive times the fraction of atoms gained which are not radioactive. Then the net rate of loss of activity is

$$-\frac{d}{dt}(a\theta) = R\theta(1 - \varphi) - R(1 - \theta)\varphi$$

which immediately simplifies to give

$$-a\frac{d\theta}{dt} = R(\theta - \varphi) \tag{10-18}$$

Making use of the relation between θ and φ, we can at once integrate this equation and find

$$\frac{\theta}{\theta_0} = \frac{1}{a + b}\left[a + be^{-[(a+b)Rt/ab]}\right] \tag{10-19}$$

It is thus seen that the movement of radioactivity follows a first-order law whatever the kinetics of the reaction may be, provided only that the rate of the exchange can be expressed as a function of the concentrations. (More complicated cases can occur.)

Now the rates of exchange reactions vary widely. They may be practically instantaneous or so slow as to be negligible. (The reaction between aluminum bromide and ethyl bromide has a half-time of about thirty minutes at 0°.) In the use of tracers in analytical chemistry it is necessary for the chemist to assure himself that the reactions involved are so rapid that he may consider that he is dealing with a true equilibrium, and that side reactions are so slow that they may be disregarded. Intermediate cases involve all the complications of chemical kinetics and should be avoided by the analytical chemist at any cost. A great many exchange reactions have been studied in recent years, but it does not seem possible to give any very useful generalization to summarize the results. When a problem arises one must consult the original literature or, as more often than not proves necessary, carry out the required investigation oneself.

The study of isotopic exchange reactions has produced much information on the actual mechanism of chemical reactions and in some cases has produced definite proofs of structure for chemical species. Perhaps the most important example of the former is the unequivocal proof that the rate of the exchange of iodide ion with the iodine of *sec*-octyl iodide is directly related to the rate of the inversion at the asymmetric carbon. The mechanism of this Walden inversion is thereby elucidated. One of the clearest instances of structural information is the proof through the use of S^{35} of the nonequivalence of the sulfur atoms in the thiosulfate ion. Thiosulfate prepared from radioactive sulfur and ordinary sulfite ion fails to give radioactive sulfur dioxide on acid decomposition. The sulfur atoms retain their identities throughout; there is no exchange between the sulfur atoms even in the close association of the ion. It seems probable that the method has by no means been exploited to its limit, and that many interesting questions remain to be settled by its use.

10-9. Radioactivation Analysis. We shall conclude our survey of radiochemical methods with a mention of a relatively new and quite powerful method particularly applicable to the determination of traces: radioactivation analysis. When the chemist can command the use of a high-energy particle accelerator or a nuclear reactor, he can investigate the composition of materials *via* the character and behavior of the radiations they produce after suitable nuclear excitation. It is true that the requirement in equipment for both producing and observing this radiation is considerable, generally so much so that the method must be reserved for problems of considerable importance. For just this reason, however, a knowledge of the potentialities of the method is of importance to the analytical chemist.

The discussion in Secs. 10-2 and 10-4, together with a cursory glance at a chart of the radioisotopes, makes it clear that under favorable circumstances, *i.e.*, with not too complex mixtures, most of the elements can be characterized in terms of the nature and rate of decay of radiations from excited nuclides produced from them. In the simplest situations it is sufficient to determine the decay curve of these activities. When only a qualitative analysis is wanted a single determination of this curve on the unknown is sufficient. If quantitative estimates are needed, similar determinations on one or more standard samples of known composition will be required. In an excellent account of activation analysis by Boyd[1] an example of this type is given. The decay curve of a sample of rubidium carbonate after neutron irradiation is analyzed. The curve is shown to be composite and made up of contributions from four activities: (1) 18.6-day Rb^{86}, (2) 12.5-hr. K^{42}, (3) 3.1-hr. Cs^{134}, and (4) 18-min. Rb^{88}. Evidently then the total counting rate of the sample at any time is given by

$$R = \lambda_1 k_1 N_{01} e^{-\lambda_1 t} + \lambda_2 k_2 N_{02} e^{-\lambda_2 t} + \lambda_3 k_3 N_{03} e^{-\lambda_3 t} + \lambda_4 k_4 N_{04} e^{-\lambda\ t}$$

The constants k measure the efficiencies of the counting apparatus for the four different radiations. The N_0's are the amounts of the various active nuclides present at the end of the neutron irradiation. The numbers depend on the capture cross sections of the isotopes being activated as well as on the actual amounts of these present in the sample.

At times long compared to 12.5 hr. (perhaps 2 days) effectively only Rb^{86} remains in the sample and

$$\ln R_1 = \ln (\lambda_1 k_1 N_{01}) - \lambda_1 t$$

If this straight line is subtracted from the curve of log R *vs.* t, the resulting curve at times long compared to 3.1 hr. will be determined solely by K^{42}. One then subtracts

$$\ln R_2 = \ln (\lambda_2 k_2 N_{02}) - \lambda_2 t$$

to find a curve, determined at times long compared to 18 min., by the Cs^{134}. Thus by measuring the slopes of the straight portions of the successive curves the identity of the impurities is determined. The intercepts, *i.e.*, the ln $(\lambda k N_0)$, are determined by the relative proportions of these impurities. These are best interpreted in terms of results on standard samples irradiated and measured in an exactly similar fashion.

With the aid of a scintillation spectrometer and equipment to produce a record of the various α-ray peaks in the spectrum, much more detailed information can be obtained. By this means the method can be extended to give accurate quantitative determinations of very small traces. As little as 0.001γ (10^{-9} g.) of an element can often be detected, and the ordinary accuracy of the result is about ± 10 per cent or, in favorable cases, even somewhat less. Such methods have been applied to the analysis of semiconductors, such as silicon and germanium. Extremely small traces can produce enormous effects in the behavior of these materials in transistors. The interested reader is referred to the original literature[2] for an account of this application which also gives an excellent survey of the various problems involved.

[1] G. E. Boyd, *Anal. Chem.*, **21**, 335 (1949).
[2] G. H. Morrison and J. F. Cosgrove, *Anal. Chem.*, **27**, 810 (1955); **28**, 320 (1956).

PROBLEMS

The answers to the starred problems will be found on page 404.

***1.** The capture cross section for thermal neutrons of Cl^{35} is 40 barns. The isotopic abundance of Cl^{35} is 75.5 per cent. The half-time for the decay of Cl^{36} is 4.4×10^5 years. A 10-ml. sample of carbon tetrachloride is sealed in a quartz tube and placed in a nuclear reactor in a flux of slow neutrons of 10^{14} per cm.^{2} per sec. The tube is removed after 30 days of irradiation.

a. How many disintegrations per second of Cl^{36} are to be expected for the entire sample?
b. How many millicuries of Cl^{36} are produced?

2. The fission product $_{60}Nd^{147}$ decays by β emission to form $_{61}Pm^{147}$, which is itself radioactive and decays by β emission to $_{62}Sm^{147}$. (Sm^{147} occurs naturally. It is radioactive and decays by α emission with a half-life of 1.3×10^{11} years.) The half-lives of Nd^{147} and Pm^{147} are 11.6 days and 2.6 years, respectively. Plot curves showing (*a*) the decrease in activity of a sample of Nd^{147} initially free from promethium, (*b*) the increase in activity of the promethium, and (*c*) the gross activity of the sample. Is it possible to arrive at any type of radioactive equilibrium in this system?

***3.** Two solutions are prepared by diluting to exactly one liter portions of a stock solution of $Na^{22}Cl$. (A solution about 0.01 *F* in ordinary sodium chloride is used for the dilution. Why not pure water?) In the first case 5.00 ml. of the radioactive stock is taken; in the second case 10.00 ml. is taken. A jacketed Geiger counter, used with equipment designed for preset counts, with the first solution gives 65536 ($= 2^{16}$) counts in 643.0 sec. With the second solution 325.0 sec. is required for the same number of counts. What is the dead time of the counter? Estimate the reliability of the result.

4. A sample of silver bromide weighing 0.1000 g. is known to contain 1 millicurie of Br^{82}. The solubility product of silver bromide is 7.4×10^{-13}.

a. Calculate the number of disintegrations per second in 10 ml. of a saturated solution of the silver bromide.
b. If one employs a scintillation counter at 50 per cent efficiency, how long should one count to obtain a value for the solubility which, with 90 per cent certainty, is correct to ± 1 per cent?

***5.** It is desired to determine the amount of sodium associated with a portion of an ion-exchange resin in equilibrium with a solution 0.01 *F* in sodium chloride and 0.02 *F* in cesium chloride. A sample of this mixed solution is prepared containing a "spike" of Na^{22}. This solution is found to count 202.5 per sec. The resin is transferred to a flask containing 50.00 ml. of the labeled solution. After shaking for some hours to attain isotopic equilibrium, the solution is again measured *in the same Geiger counter* and found to give a rate of 92.0 per sec. How many millimoles of sodium are in the resin sample? All of the counting was done with totals of 90000 counts. Comment on the accuracy of the final result.

6. Carry out the integration which leads from Eq. (10-18) to Eq. (10-19).

7. *p*-Chlorophenol in benzene solution does not exchange chlorine with HCl. If *p*-chlorophenol is chlorinated by the addition of chlorine to a benzene solution containing radioactive HCl, it is found that the chlorinated product is radioactive. A solution contained initially 10.45 millimoles of radioactive HCl, and 2.85 millimoles of Cl (as Cl_2) was added. The ratio of the activity of the chlorinated product to the initial activity of the HCl was found to be 0.216. What can be said about the relative rates of the chlorination reaction and of the exchange between HCl and Cl_2?

CHAPTER 11

ION-EXCHANGE AND CHROMATOGRAPHIC METHODS

11-1. Introduction. The term *ion exchange* is generally understood to refer to an interaction resulting principally in a redistribution of ionic constituents between an insoluble solid and a solution. The solids considered in this connection are always polymers of high molecular weight which never of themselves disperse to form homogeneous solutions. Thus, in the somewhat limited sense in which we wish to consider the phenomenon, the formation of mixed crystals of varying composition when a saturated solution is evaporated does not constitute an ion-exchange process, although a redistribution of ions certainly takes place; the crystals will redissolve completely on diluting the solution. Over a hundred years ago, even before the recognition of the existence of ions, the ion-exchange process was observed in the displacement of the calcium associated with the clay minerals of the soil by ammonium salts. Some time later synthetic alumino-silicates with ion-exchange properties were applied in water softening, that is, in the replacement of the undesirable calcium in water by sodium from the synthetic silicate.

An essential feature of the ion-exchange process is its *reversibility*. When the sodium content of the water softener has been practically exhausted through nearly complete replacement by calcium, the exchanger may be regenerated by passing over it a strong solution of sodium chloride. In the case of the soil minerals, the reversibility of the process is intimately involved with the chemistry of plant nutrition and is thus an extremely important effect.

The early synthetic alumino-silicates failed to find wide chemical application, mainly because they are extensively decomposed in acidic solutions. About 1935 the British chemists B. A. Adams and E. L. Holmes first prepared synthetic ion exchangers of an entirely different nature, namely polymeric organic compounds having very high molecular weights. These resins have high chemical and mechanical stability.

They can be prepared either as acids or as bases, and thus we have both *cation exchangers* and *anion exchangers*. Extraordinary advances in all branches of chemistry concerned with separations have followed the introduction of these substances. Since they have proved to be of commercial value on quite a large scale, a wide variety of exchangers has been developed and is now easily available. Inorganic separations formerly quite impossible in a practical sense have become relatively easy; thus the preparation of pure salts of the rare earth elements is now being carried out on a commercial scale. Separations which are simple in principle but tediously long by classical methods can frequently be carried out in a matter of minutes with the aid of the appropriate ion exchanger. These facts are of great importance to the analytical chemist, who, however well he may be equipped with modern tools, must very frequently make separations before he can make a determination. A working knowledge of the theory and techniques of the ion-exchange method will be an invaluable part of his equipment.

11-2. Ion Exchangers, Resinous and Inorganic. The most important natural ion exchangers are inorganic in character. However, the chemist in the laboratory has thus far found much more use for the synthetic organic polymers, although useful ion exchangers can be prepared from natural carbonaceous materials. When coal is treated with sulfuric acid, sulfonic acid groups are fixed in the material and a useful cation exchanger results. In an early piece of work on ion exchange, unbleached sulfite paper pulp was successfully used as an exchanger for the removal of copper from solution. Generally speaking, however, the synthetic exchangers have properties which make them more attractive to the chemist.

A synthetic organic ion exchanger is an aggregate of chains of a polyelectrolyte, so cross-linked as to produce an insoluble matrix. These aggregates may be prepared either by the copolymerization with a cross-linking agent of molecules carrying functionally active groups, or by the introduction of the active groups after the formation of the resin. Thus, an ion exchanger with weak acid characteristics is readily made by the copolymerization of methacrylic acid with glycol bismethacrylate as shown at the top of page 372.

The polymethacrylate chains are linked together in proportion to the amount of glycol ester added to the reaction mixture. The result is a highly insoluble matrix with a flexible and open structure, into which solvent molecules as well as ions can easily penetrate. This particular polymer is easily prepared in various forms, in rods or sheets of almost any desired size. It is transparent and water-white; with it many of the features of the ion-exchange process can be demonstrated in a striking manner.

$$
\begin{array}{l}
n\mathrm{CH_2{=}C(CH_3){-}COOH} + m\mathrm{CH_2{=}C(CH_3){-}COOCH_2} \\
\hphantom{n\mathrm{CH_2{=}C(CH_3){-}COOH} + m}\qquad\qquad\qquad\qquad\quad | \qquad \rightarrow \\
\hphantom{n\mathrm{CH_2{=}C(CH_3){-}COOH} + m}\mathrm{CH_2{=}C(CH_3){-}COOCH_2}
\end{array}
$$

$$
\begin{array}{ccccccc}
 & \mathrm{COOH} & & \mathrm{COOH} & & \mathrm{CH_3} & \\
 & | & & | & & | & \\
\mathrm{-CH_2-} & \mathrm{C} & \mathrm{-CH_2-} & \mathrm{C} & \mathrm{-CH_2-} & \mathrm{C} & \mathrm{-CH_2-} \\
 & | & & | & & | & \\
 & \mathrm{CH_3} & & \mathrm{CH_3} & & \mathrm{C{=}O} & \\
 & & & & & | & \\
 & & & & & \mathrm{O} & \\
 & & & & & | & \\
 & & & & & \mathrm{CH_2} & \\
 & & & & & | & \\
 & & & & & \mathrm{CH_2} & \\
 & & & & & | & \\
 & & & & & \mathrm{O} & \\
 & & & & & | & \\
 & & & \mathrm{CH_3} & & \mathrm{C{=}O} & \\
 & & & | & & | & \\
 & & \mathrm{-CH_2-} & \mathrm{C} & \mathrm{-CH_2-} & \mathrm{C} & \mathrm{-CH_2-} \\
 & & & | & & | & \\
 & & & \mathrm{COOH} & & \mathrm{CH_3} &
\end{array}
$$

Another valuable type of ion exchanger, a strong acid in character, is prepared by the sulfonation of the resin obtained in the copolymerization of styrene and divinylbenzene:

$$
\begin{array}{l}
\qquad\qquad\quad \mathrm{H} \qquad\qquad\qquad\qquad\qquad\qquad\qquad\qquad\qquad\qquad\quad \mathrm{H} \\
\mathrm{-C-CH_2-C-CH_2-C-CH_2-C-CH_2-C-CH_2-C-CH_2-} \\
\quad \mathrm{H} \qquad\qquad\qquad\qquad\quad \mathrm{H} \qquad\qquad\quad \mathrm{H} \qquad\qquad\quad \mathrm{H} \\
\\
\qquad\qquad\qquad\qquad\qquad \mathrm{H} \qquad\qquad\quad \mathrm{H} \qquad\qquad\quad \mathrm{H} \qquad\qquad\qquad\qquad\qquad\quad \mathrm{H} \\
\mathrm{-CH_2-C-CH_2-C-CH_2-C-CH_2-C-CH_2-C-CH_2-C-} \\
\qquad\quad \mathrm{H} \qquad\qquad\qquad\qquad\qquad\qquad\qquad\qquad\qquad\qquad\quad \mathrm{H} \\
\\
\qquad\qquad\qquad\qquad \mathrm{-CH_2-C-CH_2-} \\
\qquad\qquad\qquad\qquad\qquad\quad \mathrm{H}
\end{array}
$$

This hydrocarbon is prepared in an emulsion copolymerization, from which it may be obtained as nearly spherical beads. These are treated with concentrated sulfuric acid and so converted into an insoluble substituted toluenesulfonic acid. In a preparation such as this, there is no guarantee that the functional sulfonic acid groups are all structurally equivalent; indeed, there is indirect evidence to show that they are not.

It is probable that this dissimilarity has some effect on the quantitative nature of the ion-exchange processes into which the resin enters; as a practical matter it is of no great consequence. The cross-linked sulfonated polystyrenes, aside from their behavior as strong acids, are chemically quite inert. They are among the most useful types of ion exchangers we have.

The physical properties of an ion-exchange resin are in large part determined by the extent of cross-linking. This cannot be determined directly in the resin itself; it is merely specified as the mole per cent of cross-linking agent in the mixture polymerized. Thus, "polystyrene sulfonic acid (5 per cent DVB)" refers to a resin containing nominally 1 mole in 20 of divinylbenzene. The true degree of cross-linking in the resin undoubtedly differs somewhat from the nominal value and is probably variable within a resin particle. Highly cross-linked resins are generally harder, more brittle, and more impervious than the lightly cross-linked materials. The preference of a resin for one ion over another, its *selectivity*, is much influenced by the degree of cross-linking. To a limited extent resins with prescribed properties can be prepared through control of the degree of cross-linking.

Anion exchange resins are likewise cross-linked high polymers. Their basic character is due to the presence of amino, substituted amino, or quaternary ammonium groups. The polymers containing quaternary ammonium groups are strong bases; those with amino groups are weakly basic in character.

Table E in the Appendix gives a list of ion exchangers easily available in the United States together with some of their characteristics and principal properties.

In addition to the clay minerals already mentioned, other natural silicates, the zeolites, have ion-exchange properties. The analytical chemist is, however, little interested in either of these substances. Because of their physical characteristics, the clays are troublesome to handle, and the zeolites are so slow in action as to be of no practical value. Both are decomposed by acids.

Two very recent and promising additions to the list of inorganic exchangers should be mentioned, although their potentialities have not yet been extensively investigated. A substance with many of the characteristics of an anion exchanger is obtained by adding ammonia to a solution of zirconium oxychloride and drying the resulting precipitate at a temperature below 300°C. Complex zirconium phosphates and tungstates prepared by the addition of phosphoric acid or sodium tungstate to solutions of zirconium oxychloride have characteristic cation exchange properties. Excellent separations of the alkali metals have been obtained with these materials. These zirconium compounds are stable in acid solutions. Many interesting separations should become possible through their use.

11-3. Description of the Ion-exchange System. The usefulness of an ion exchanger rests first, of course, on its specific chemical activity; its practicality will be determined largely by the speed with which it reacts. A high porosity on a molecular scale is an essential feature of a practical ion exchanger. Because of this porosity, an ion exchanger immersed in a solution absorbs not only the ions with which it reacts specifically but also the solvent and to some extent ions of opposite charge, the *nonexchange* ions. The interaction is quite complicated. A bead of a polysulfonate resin in an aqueous solution of sodium and calcium chlorides is depicted in a highly schematic fashion in Fig. 11-1. To the analytical chemist the most important feature of this system is the distribution of sodium and calcium between the exchanger and the solution. Some knowledge of the other aspects of the interaction may be essential in planning a separation, but applications of ion exchangers which do not depend on their specific functionality are rather rare.

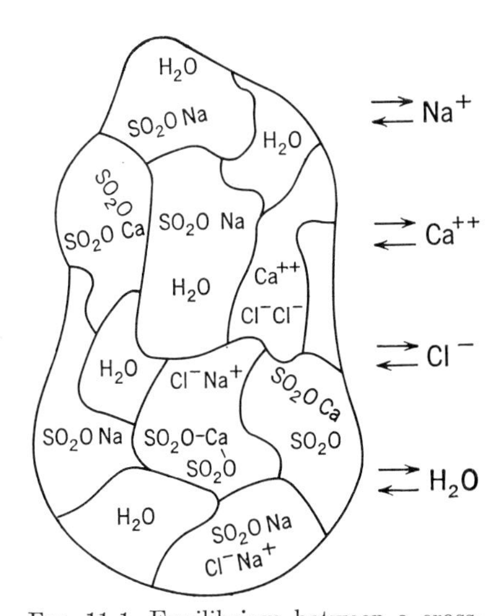

FIG. 11-1. Equilibrium between a cross-linked polysulfonic acid and an aqueous solution containing sodium and calcium chlorides. Generally the uptake by the resin of nonexchange ion, and equivalent counter ion, is less than is implied in the figure.

For a complete description of a cationic exchange process, the following facts must be known:

1. The amount of water in the resin.
2. The amount of anion in the resin.
3. The relative amounts of the two cations (and of hydrogen ion if the solution is significantly acidic) in the resin and in the solution.
4. The total amount of cation in the resin and the total concentration of the solution. The former involves a knowledge of the *exchange capacity* of the resin, *e.g.*, the number of sulfonate groups attached to a given mass of hydrocarbon. The total amount of cation in the resin is necessarily equal to the exchange capacity plus the anion content (all measured in equivalents).
5. The rates of diffusion of the ions into the resin structure. The rate of the exchange will be determined by the diffusion of ions either within the

exchanger or through a static film of solution surrounding the exchanger. We shall return to this point later.

The absorption of the solvent has one very practical aspect. The degree of cross-linking determines the stiffness of the resin, which in turn determines the extent of swelling due to an increase in water content. A dry resin which is lightly cross-linked will swell far more when immersed in water than a similar but much more extensively cross-linked resin. In most applications of ion exchange a solution is passed through a tube packed with exchanger. Changes in extent of swelling are quite enough to cause mechanical difficulties in the operation of these columns. A bed of exchanger may become so tightly packed as to stop the flow of fluid, and ion-exchange columns have even been known to burst because of the swelling of their contents.

The matter of the distribution of ions between solvent and exchanger is of prime importance to the chemist and must now be examined in greater detail.

11-4. Ion-exchange Equilibria. For present purposes it will be sufficient to disregard the effects due to the absorption of water and non-exchange ion and discuss the ionic equilibrium for a somewhat hypothetical exchanger in which these effects are absent. Since we shall use our results largely as a qualitative guide rather than to make quantitative calculations, this neglect will not be serious. With this limitation the ion-exchange equilibrium is analogous to that between a mixed crystal and a solution. If solid silver bromide is shaken with a nearly saturated solution of silver chloride, the resulting solid will be found to contain chloride, and a close examination will reveal that this chloride is incorporated in the crystal lattice with the bromide. This result is essentially more complicated than considerations based on the solubility product could imply. The methods of chemical thermodynamics, which serve to derive the form of the solubility product constant for a pure solid, give equilibrium constants involving variable activities in the solid phase when applied to either a mixed crystal or an ion exchanger.

For the simplified sodium–calcium equilibrium, we find for the reaction

$$2\mathrm{NaEx} + \mathrm{Ca}^{++} = \mathrm{CaEx}_2 + 2\mathrm{Na}^+$$

$$K = \frac{(\mathrm{CaEx}_2)[\mathrm{Na}^+]^2}{(\mathrm{NaEx})^2[\mathrm{Ca}^{++}]} \frac{g_{\mathrm{CaEx}_2}}{g^2_{\mathrm{NaEx}}} \frac{f^2_{\mathrm{Na}^+}}{f_{\mathrm{Ca}^{++}}} \tag{11-1}$$

Here the brackets denote, as usual, concentrations in the solution, and we use parentheses to indicate some convenient measure of the composition of the exchanger. The solid composition might conveniently be measured as a fraction of the constant exchange capacity, expressed perhaps as milliequivalents per gram of exchanger in the dry hydrogen form. We would then have $(\mathrm{NaEx}) + (\mathrm{CaEx}_2) = 1$. Evidently only for ions of like charge are the concentration units immaterial. The symbol g denotes an activity coefficient in the resin phase.

Although Eq. (11-1) seems to imply the necessity for individual ion-activity coefficients (*cf.* Sec. 2-6), this is actually not the case. The ratio needed is determinate:

$$\frac{f^2_{Na^+}}{f_{Ca^{++}}} = \frac{f^2_{Na^+}f^2_{Cl^-}}{f_{Ca^{++}}f^2_{Cl^-}} = \frac{(f_\pm)^2_{NaCl}}{(f_\pm)_{CaCl_2}}$$

Unfortunately, without directions for finding the activity coefficients, both the g's and f's, Eq. (11-1) is nearly useless. The problem is not trivial. The f's refer to activity coefficients in the mixed solution, and not many measurements have been made of these. In most cases the best we can do is to rely on the Debye-Hückel theory, which will fail us in any but quite dilute solutions. Very few measurements indeed have been made of activity coefficients in the exchanger phase (the g's), and here no established theoretical recourse is available. Within the exchanger the ions are part of a concentrated solution of a peculiar type, one in which all the ions of one sign are fixed in position. If Dowex 50 (a commercial polystyrene sulfonic acid) is brought to equilibrium with a solution of sodium chloride, it is found that the "inside solution" has a concentration of about 6 moles per kilogram of imbibed water. No theory of electrolytic solutions is competent to handle such a high concentration.

In view of all the uncertainties in the computation of true activities, the most useful approach is to correlate ion-exchange results in terms of an equilibrium concentration quotient. We must remember that this quantity cannot be expected to remain more than approximately constant as the composition of the exchanger is altered. This approach is, of course, no different from that generally adopted in the semiquantitative discussion of most of the equilibria of interest to the analytical chemist. No harm is done thereby; the theories are indispensable in roughing out plans for an experiment, but of course one must always be alert to the fact that the treatment thus secured is necessarily somewhat inexact.

The equilibrium quotient for two ions is usually denoted by the symbol $K_A{}^B$. Thus, for the reaction

$$m\text{AEx}_n + n\text{B}^{\pm m} = n\text{BEx}_m + m\text{A}^{\pm n}$$

we have

$$K_A{}^B = \frac{(\text{BEx}_m)^n[\text{A}^{\pm n}]^m}{(\text{AEx}_n)^m[\text{B}^{\pm m}]^n} \tag{11-2}$$

The dependence of $K_A{}^B$ on the composition of the exchanger varier greatly with the nature of the exchanging ions and of the exchanges itself. Several examples are given in Figs. 11-2 to 11-4. A knowledge of the equilibrium quotient permits an assessment of the feasibility of a given separation. This may be illustrated with a simple yet frequently encountered situation.

If an ion exchanger is used in the separation of a trace constituent—perhaps an impurity which must be removed or a valuable component which must be concentrated—the equilibrium quotient assumes the simple form of a distribution ratio; the concentrations in both exchanger and solution of the major component remain essentially constant. Thus, it is found that traces of ferric iron in concentrated hydrochloric acid are selectively held by a strong base *anion* exchanger such as Dowex 1. Evidently the iron is present in the form of a negatively charged chloro complex, probably $FeCl_4^-$. We expect then to find a nearly constant distribution ratio

$$\frac{(FeCl_4^-)}{[FeCl_4^-]} = K_{Cl^-}^{FeCl_4^-} \frac{(Cl^-)}{[Cl^-]}$$

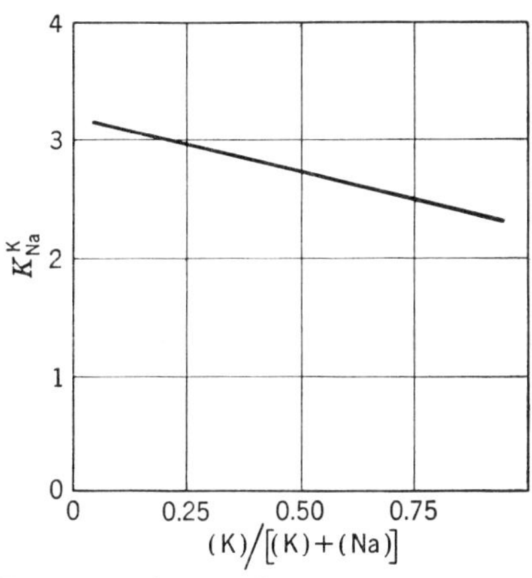

FIG. 11-2. The equilibrium quotient $K_{Na}{}^{K}$; for a resin containing dipicrylamine residues. In basic solution the nitro groups are in their tautomeric acid form and show a marked selectivity for potassium over sodium. Note the small change in equilibrium quotient with composition of the exchanger. [*From D. Woermann, K. F. Bonhoeffer, and F. Helfferich, Z. physik. Chem.*, **8**, 271 (1956).]

Of course, if this ratio is computed from experimental data given in terms of total iron content of solution and exchanger, it cannot be expected to remain even approximately constant. The proportion of iron in the form of the chloro complex will certainly vary markedly with the composition of the solution. The value of the distribution ratio

$$D = \frac{(Fe^{III})}{[Fe^{III}]}$$

in terms of total iron is, however, just what is wanted for estimating the feasibility of a separation. For Dowex 1 in 10 F hydrochloric acid the distribution coefficient D for ferric iron is about 3.2×10^4, with (Fe^{III}) measured in moles per liter of resin. As the reader may readily show for himself, if a volume V of impure hydrochloric acid is treated with a volume v of resin in the chloride form, a fraction $V/(Dv + V)$ of the total iron present will remain in the acid. Thus, a liter of hydrochloric acid treated successively with two 10-ml. portions of Dowex 1 will have its

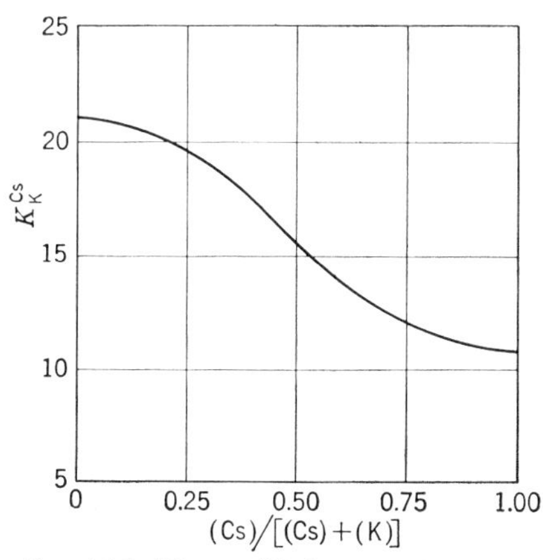

FIG. 11-3. The equilibrium quotient $K_K{}^{Cs}$ for the exchange on a montmorillonite clay. Note the extreme selectivity of the clay for cesium and the marked variation of the equilibrium quotient with exchanger composition. [*From J. A. Faucher and H. C. Thomas, J. Chem. Phys.*, **22,** 261 (1954).]

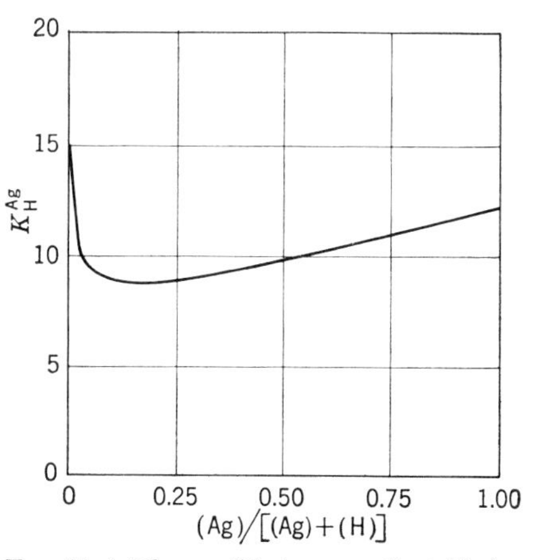

FIG. 11-4. The equilibrium quotient $K_H{}^{Ag}$ for the exchange on Dowex 50. Note the extraordinary increase in selectivity for silver at low silver content. The silver ion has been found to behave in an anomalous fashion in many exchange applications.

ferric iron content reduced by a factor of 10^5. Of course it is not to be expected that every separation will be so effective. In the present case this advantageous situation is spoiled completely merely by diluting the acid. In 2 F hydrochloric acid, for example, D is only 10, and two treatments as described would leave 80 per cent of the iron in the acid. To reduce the ferric iron content of a liter of 2 F acid by a factor of 10^5, no less than 125 successive treatments with fresh 10-ml. portions of resin would be required. A more practical procedure than batchwise treatment must be found. A very simple expedient serves this purpose. If the exchanger is packed into a tube and the acid passed slowly over it, each layer of exchanger comes nearly to equilibrium with the acid in contact with it, and by the time the first of the acid reaches the end of the column, it will have undergone many equilibrations with fresh exchanger. In a single operation, using an appropriate amount of resin, it is easy to prepare any required amount of very highly purified acid.

In the following section we shall discuss the theory of *ion-exchange chromatography* in enough detail so that the reader can design for himself experiments which will easily produce otherwise very difficult separations.

11-5. Elements of Chromatographic Theory. A detailed theoretical study of the flow of a fluid through a bed of granular solid with which it reacts chemically is quite an involved task. Such a complete study is fortunately not essential for an intelligent and productive use of the chromatographic technique. Most of the guidance needed to plan an exchange or a separation can be obtained from a discussion of a rather highly idealized case. The effects of the finer points omitted in the idealization can be assessed qualitatively and need give no serious trouble in practice.

The principal simplification which we shall make in the discussion to follow will be to consider that each thin layer of exchanger, taken transverse to the direction of flow of the fluid, is in true chemical equilibrium with the solution passing over it. Evidently this can be only an approximation. Sorption processes proceed at a finite rate and equilibrium with a moving fluid can certainly never be quite reached if the composition of the solution is changing, as it must be in any case of interest. For a reason which will later become apparent, however, it turns out that the approximation is often quite good, particularly in cases where sharp separations are indicated by the equilibrium quotients involved.

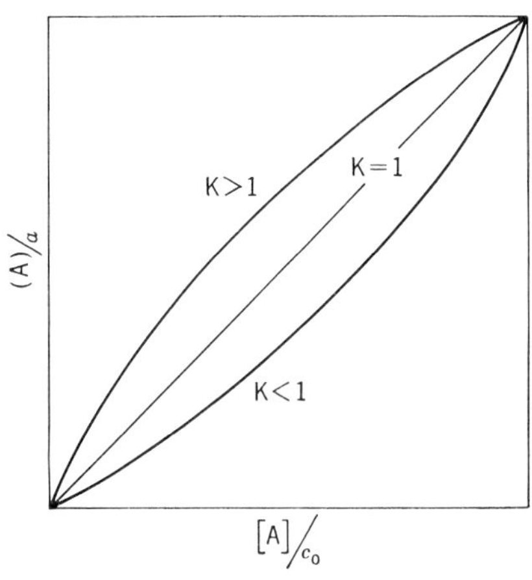

FIG. 11-5. Exchange isotherms. The figure shows the relation between exchanger composition and solution composition for various values of K. For $K > 1$ the resin is *selective* for A.

In the following discussion we do not want to limit ourselves to a case as simple as that mentioned in connection with the removal of ferric iron from a hydrochloric acid solution; that is, we shall not confine our attention to a consideration of trace components. We shall, however, continue to neglect the effect of imbibed nonexchange ions and the relatively small effects due to changes to the solvent content of the exchanger. For simplicity of presentation we shall consider only the case of a pair of exchanging ions of like charge. (The reader should repeat the arguments for more complex cases.) The equilibrium quotient for the exchange reaction may then be written

$$K = \frac{(\mathrm{A})\{c_0 - [\mathrm{A}]\}}{[\mathrm{A}]\{a - (\mathrm{A})\}} \tag{11-3}$$

where a is the constant capacity of the exchanger and c_0 is the constant total concentration of the solution, both in molar units. Thus

$$(\mathrm{A}) = \frac{(Ka/c_0)[\mathrm{A}]}{1 + \left(\frac{K-1}{c_0}\right)[\mathrm{A}]} \tag{11-4}$$

When A is preferentially held by the resin, K is greater than unity, and the exchange isotherm is a curve concave toward the [A] axis, as shown in Fig. 11-5. If perchance $K = 1$ (*i.e.*, the resin shows no preference for one ion over the other), a linear isotherm results, and the ion is distributed between resin and solution in the ratio

$$\frac{(\mathrm{A})}{[\mathrm{A}]} = \frac{a}{c_0}$$

This case is similar to that for a trace component. In the case of the trace, however, the value of the equilibrium quotient is all-important in determining the extent of adsorption:

$$\frac{(\mathrm{A})}{[\mathrm{A}]} = K\left(\frac{a}{c_0}\right)$$

In exchange adsorption[1] of ions of like charge there is no need to consider in detail the situation for which K is less than unity; we may simply transfer our attention to the other ion and revert to the case of an isotherm concave toward the axis of concentration in solution. One case or the other will be of interest depending on what one is trying to accomplish. If we wish to remove a trace of iron from hydrochloric acid we must work in the region of large K. If we wish to determine the iron content of the acid, the iron first removed must be recovered from the resin with a solvent which favors the desorption.

In discussing the behavior of a chromatographic column, we need not limit ourselves to a consideration of exchange adsorption alone. The motion of a chemical species along a column of sorbent will shortly be seen to depend primarily on the shape of the concentration isotherm and only in a relatively unimportant way on the mechanism which brings about the sorption. No ions are involved in the adsorption of anthracene

[1] At this point it may be well to interject a remark on the terminology applied to processes which result in the transfer of chemical species between phases. We shall use the generic term *sorption* for all such purposes. Thus, the adsorption of hydrogen by platinum black, the absorption of water by a sponge, and ion exchange between a resin and a solution are all processes of sorption. The more specific term *adsorption* implies effects dependent on the existence of a surface of separation between the phases. Distinctions frequently become very vague (and even useless) on close analysis, and a completely self-consistent terminology is nearly impossible. Thus, ion exchange with mica is known to take place only on the surface, and for mica the surface can be defined with extraordinary precision. Ion exchange on mica is also known to be due to just the same chemical causes which produce ion exchange within the crystals of the clay mineral attapulgite. The lack of definition of the "surface" of attapulgite is immaterial. The same gross effects are produced in the two cases. As we have seen, the macroscopic description of the ion-exchange process can be given in the form of an exchange equilibrium constant. This relation can be considered as a special case of an adsorption isotherm. We shall frequently refer to the process of *exchange adsorption* in contexts where the existence of this relation is of importance.

by alumina from a hydrocarbon solvent, but the behavior of the column is in many respects indistinguishable from a similar operation involving ion exchange from an aqueous solution. To avoid committing ourselves to unimportant details, let us write the equation for the isotherm as an unspecified relation between q, the concentration on the sorbent, and c, the concentration in solution:

$$q = f(c)$$

The minimal condition containing the essential statement that the chemical species A moves along the column of sorbent is the obvious necessity of conservation of mass: all of A which enters a layer of the column must either remain there or move out. Suppose that a uniformly packed column contains a mass m of sorbent per unit volume of packed bed, and that the free volume is everywhere α per unit volume of bed. [If the density of the sorbent is s, then $(m/s) + \alpha = 1$.] If we consider a slice of thickness dx through the column perpendicular to the average direction of flow of the fluid, the diagram of Fig. 11-6 will make evident the validity of the relation

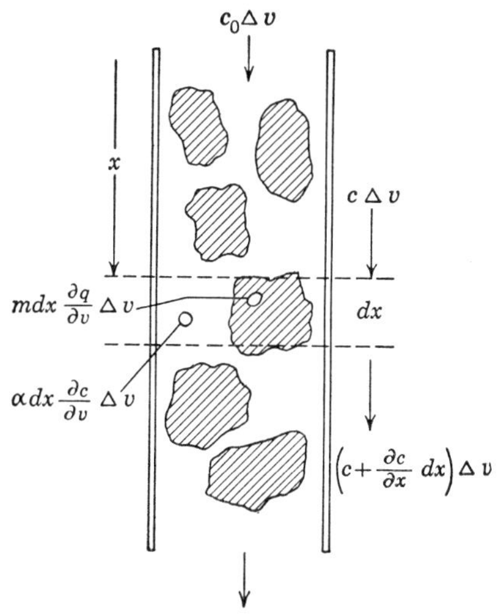

FIG. 11-6. Material balance in a chromatographic column. Δv ml. of solution at initial concentration c_0 is passed into the column. After passing through x cm. of column the concentration is c. All the material flowing into the slice of thickness dx passes through, is adsorbed, or produces a change in concentration in the solution within dx. The figure is drawn for a column of unit cross section of fractional free volume α and packed so that every unit volume of the bed contains m g. of adsorbent.

$$\left(\frac{\partial c}{\partial x}\right)_v + \alpha \left(\frac{\partial c}{\partial v}\right)_x + m \left(\frac{\partial q}{\partial v}\right)_x = 0 \qquad (11\text{-}5)$$

Here v is the volume of fluid passed into the column *per unit cross section of column;* q is conveniently measured in millimoles per gram of sorbent and c in millimoles per milliliter of solution. This relation merely states that the excess of solute flowing into a region of column over that carried downstream is distributed between sorbent and solution in this region. No account is taken of a diffusive flow of solute downstream, which is frequently of small importance but is one of the things which must be remembered when it is found that the theory based on Eq. (11-5) does not quite exactly fit the facts.

If we now introduce the assumption of instantaneous local equilibrium by putting $q = f(c)$, we find the differential equation

$$\left(\frac{\partial c}{\partial x}\right)_v = -[a + mf'(c)]\left(\frac{\partial c}{\partial v}\right)_x$$

or, using a well-known property of partial derivatives,

$$\left(\frac{\partial x}{\partial v}\right)_c = \frac{1}{a + mf'(c)} \tag{11-6}$$

The meaning of this equation is made clear by the following consideration. It may evidently be written for a fixed value of c as follows:

$$\frac{\Delta x}{\Delta t} = \frac{1}{a + mf'(c)}\frac{\Delta v}{\Delta t}$$

and it is thus apparent that the rate of motion down the column of a zone of concentration c is proportional to the volume rate of flow of the solution and that the proportionality constant, frequently denoted by R, is determined by the manner in which the column is packed (a and m), by the shape of the adsorption isotherm $[f'(c)]$, and by the concentration in question. For a given column the adsorption isotherm is the determining factor in its operation. If the isotherm is concave toward the c axis, then $f'(c_1) > f'(c_2)$ for $c_1 < c_2$, and as a result the high concentrations tend to overtake the low concentrations in the motion down the bed of sorbent. The consequences of this simple fact are important, as will immediately become apparent. In large part all of the extraordinary successes of the chromatographic technique rest upon it. Let us illustrate this fundamental chromatographic principle with a simple example.

Suppose that we are required to find in a great hurry the concentration of a certain solution of sodium chloride. We have available a supply of Dowex 50 in the hydrogen form. We propose to pass a known volume of the sodium chloride solution through a column packed with the resin and to titrate the hydrochloric acid emerging from the column with a standard solution of sodium hydroxide. It is known that the exchange isotherm for the replacement of hydrogen by sodium on the resin available is given by

$$(Na^+) = \frac{54[Na^+]}{1 + 54[Na^+]}$$

in which (Na^+) is measured as a fraction of the capacity of the resin. For Dowex 50 (8 per cent DVB) this capacity is about 1.5 meq. per milliliter of packed bed (*not* per milliliter of resin). It is proposed to pass 50 ml. of approximately 0.1 F sodium chloride into a 20-ml. column packed with exchanger. When the first small portion of solution comes into contact with the bed, a narrow zone at the top of the column will contain sodium chloride at all concentrations from zero to 0.1 F. As more solution is added the question immediately arises as to whether the regions of low sodium concentration will leak from the bottom of the column before the entire 50 ml. has passed into the top. To answer this question we proceed as follows. The derivative of the isotherm is

$$f'([\mathrm{Na}^+]) = \frac{54}{(1 + 54[\mathrm{Na}^+])^2}$$

so that the slope at zero concentration is $f'(0) = 54$, as is otherwise apparent. The displacement of the zone of near-zero concentration may now be calculated in the units convenient for the problem, namely in terms of a fraction of the total column capacity, which is $1.5 \times 20 = 30$ meq. To measure distance along the column in terms of milliequivalents of capacity ξ we write, using a as the capacity of the resin per gram,

$$\Delta\xi = maA(\Delta x) = \frac{1}{(a/ma) + [f'(c)/a]}\Delta V \tag{11-7}$$

in which now $f'(c)/a$ is the slope of the isotherm in units of fraction-of-capacity, a/ma is the free volume per milliequivalent of resin, and ΔV is the total volume poured into the column of cross-sectional area A (and *not* the volume per unit area). The free volume per milliequivalent of resin for packed Dowex 50 is close to 0.3, and $f'(c)/a = 54$, so that $\Delta\xi = 0.9$. Thus the zone of near-zero concentrations would traverse, if undisturbed, less than one-thirtieth of the total length of the column. The same computation applied to the zone of high concentration leads us into a difficulty. Here $f'(0.1)/a = 54/(6.4)^2 = 1.3$ and

$$\Delta\xi = \frac{50}{1.6} = 31 \text{ meq.}$$

This is, of course, ridiculous since only 5 meq. has been fed into the column. What actually happens is that the zone of high concentration moves $31/0.9 = 34$ times as fast as the zone of low concentration, but only until it overtakes the latter. At this stage a sharp concentration front is formed, which now moves only as fast as material is supplied to the top of the column, while the concentration within the expanding zone remains constant at $c = 0.1$ and $q/a = f(0.1)/a = 0.844$. No contradiction of our fundamental differential equation is involved; the equation is evidently identically satisfied by fixed values of c and q. By the time that the entire 50 ml. has passed into the top of the column, the region of saturation is given by

$$\xi \left(\frac{a}{ma}\right) \times 0.1 + \xi \left(\frac{f(0.1)}{a}\right) = 50 \times 0.1$$

$$\xi = \frac{5}{0.03 + 0.844} = 5.72$$

Approximately one-fifth of the column will be occupied by sodium, and most of the equivalent hydrogen ion will have passed into the effluent as hydrochloric acid.

To complete the design of a satisfactory procedure for quantitatively converting sodium chloride into hydrochloric acid, we must ask if it is possible to rinse the column with water and so collect in the effluent just 5.000 meq. of acid. Certainly when the 50 ml. of sodium chloride solution has just passed into the column, the interstices of the uppermost 5.72 meq. of the column are filled with unreacted sodium chloride, in amount $0.3 \times 0.1 \times 5.72 = 0.171$ meq. If we now pour water slowly into the top of the column, the free hydrochloric acid in the lower part of the column will be forced into the effluent, and the sodium chloride solution from the upper part will pass into unreacted exchanger and so be converted into hydrochloric acid. A detailed calculation of the behavior of the system becomes somewhat involved at this stage and is in any case of doubtful utility, since the small changes in question are frequently much influenced by disturbances not taken into account in the simple theory we are using. Evidently the displacement of hydrogen ion from an additional 0.171 meq. of resin will release the required amount of acid. If no dilution of the chromatogram occurred, a volume of $0.171/(0.844 \times 0.1) = 2$ ml. of water would suffice. This small volume, however, would not serve to force all of the acid still contained in the column into the effluent. We would expect in any case to have to pass at least one "column volume" ($30 \times 0.3 = 9$ ml.) into the column to produce the desired effect. In practice more is almost always necessary.

Our calculations have indicated the feasibility of the experiment. Recourse must now be had to the laboratory to check these estimates and to modify the details so as to produce the desired accuracy.

We may summarize this discussion of what might be called ideal chromatography by repeating the conditions of motion for the various parts of a zone of sorbed material, such as might, for example, be formed in a column of Dowex 50 by introducing first a small amount of salt solution and then partially eluting the zone of salt by passing into the column a solution of hydrochloric acid of the same concentration. (This situation is, of course, different from that described in the preceding paragraphs.) The fate of the sorbed sodium, as represented by the equilibrium concentrations of sodium to be found in the various parts of the zone, might resemble that depicted in Fig. 11-7. The concentration of hydrochloric acid used in the elution is c_0. The motion of the sharp front A is determined by

$$\Delta x\{ac_0 + mf(c_0)\} = c_0\,\Delta V \tag{11-8}$$

The motion of the region of varying concentration B is given by

$$\Delta x\{a + mf'(c)\} = \Delta V \tag{11-9}$$

It is seen that the point B', representing the rear of the zone of maximum concentration, for an isotherm curved toward the c axis moves forward more rapidly than the sharp front. For in this case we have $f(c_0)/c_0 > f'(c_0)$ (draw a sketch!), and so

$$\frac{1}{a + m[f(c_0)/c_0]} < \frac{1}{a + mf'(c_0)}$$

The trailing region of lower concentrations eventually coalesces with the sharp front. Although the front remains sharp (on the basis of the simple theory), after a time it becomes progressively more dilute and the whole chromatogram becomes more diffuse as it passes down the column.

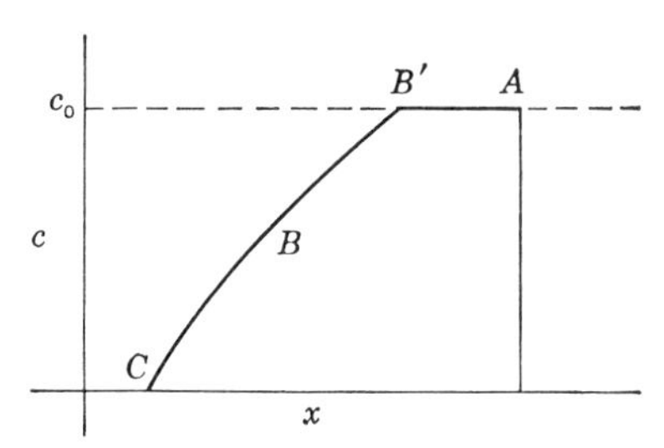

FIG. 11-7. A moving zone of exchanging material in a chromatographic column. The figure is based on the idealized theory. In practice the front at A is always more or less diffuse, and the boundary CBB' is usually longer, trailing off slowly to zero concentration.

The effect just described may often be overcome by using as an eluent a solution of an ion more strongly sorbed than those involved in the chromatogram. In this case two exchange equilibria are involved; one produces a sharp leading boundary in the chromatogram while the other produces a sharp trailing boundary. The zone of sorption of interest is, so to speak, pushed forward as a slug through the column. The method favors the collection of the substance of interest in less diluted form and often makes possible separations otherwise impossible because of mixing in the trailing zone.

It is possible to give a theoretical discussion of the separation of substances which interfere with each other on the sorbent, but the theory is quite complicated and subject to many uncertainties. Furthermore, very few of the mixed adsorption isotherms have been measured; in nearly every new case the fundamental data required are lacking. We shall therefore content ourselves with illustrating the principles as applied to trace constituents, where mutual interferences are insignificant. Actually many important instances arise in separating trace amounts of chemical substances. The growing use of nuclear fission is supplying us with large amounts of complex mixtures of the radioactive fragments of the fission reaction. Many of these substances have practical uses and are often separated from the mixtures by the methods we are discussing.

For a pair of substances at concentrations so low that the exchange isotherms with the "bulk" ion of the exchanger can be considered to be linear and also so low that there is no mutual interaction on the exchanger, our theory predicts that the zones of sorption will move independently and at constant and characteristic rates through the column. (The low concentrations required for this simplification are just those met with the fission fragments.) For the linear isotherm, $q = kc$, all concentrations of the same substance move at the same rate, and there is no self-sharpening effect in the zone of sorption. Any disturbance which tends to smear out the zone remains uncorrected and usually becomes progressively worse.

The source of some of these disturbances is immediately obvious. If a column is poorly packed, with grossly nonuniform regions of sorbent, it will "channel"; the flow will not be uniform across every section, and very large disturbances will result. These can easily be avoided by proper care in the laboratory. No matter how carefully a column is packed, however, the necessary inhomogeneity of the bed of solid permeated by a moving solution produces what might be called microscopic channeling, which is unavoidable. A given filament of solution may be broken into two on contact with a granule of sorbent and one portion may find its way down the column relatively faster than the other. The net effect of such a process is the dispersion of an initially sharp front. If the effects of the sorption are not such as to correct this, the dispersion necessarily becomes more and more apparent as time goes on. A qualitatively similar effect is caused by downstream molecular diffusion. Furthermore, the effects of the finite rates of the sorption processes also produce diffuse fronts. All of these phenomena, in one degree or another, are always present in any actual experiment. The result is that boundaries of sorption zones, which should be sharp according to the simple theory, are always more or less diffuse. Indeed, only in very carefully controlled experiments with slow flow rates, rapid sorption reactions, and favorably curved isotherms can really sharp boundaries be obtained. Fortunately, useful chromatographic separations are not by any means wholly dependent on the ability to produce sharp boundaries. In the case of trace constituents, where the boundaries are always diffuse, excellent separations are still obtainable if the average rates of motion of the zones are sufficiently different.

Figure 11-8 depicts the course of an experiment in which an initially sharp zone of a trace adsorption is moved down the column by an appropriate eluent. (In this case several trace elements are present.) The nonequilibrium effects and other uncorrected disturbances gradually distort the shape of the zone, both the leading and following edges becoming more diffuse. The moving zone, if initially sufficiently narrow, eventually assumes a nearly symmetrical bell shape. More detailed theories of the chromatographic process give a qualitatively accurate description of this effect. A very useful result of the simple theory

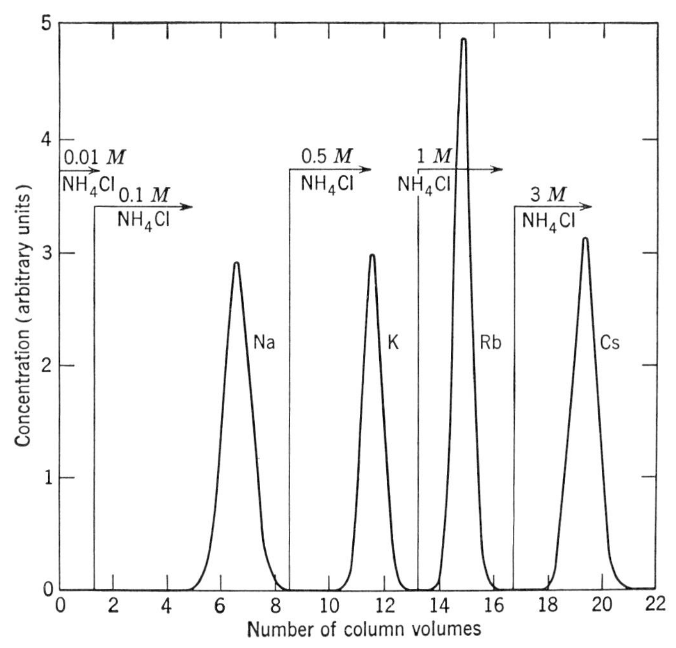

FIG. 11-8. Separation of the alkali metals on a column of zirconium tungstate. In this separation a mixture of the alkalis has been placed on the column and eluted with solutions containing successively higher concentrations of ammonium chloride. Note the excellent separations obtained even in the difficult case of the Cs-Rb-K mixture. [*From K. A. Kraus, T. A. Carlson, and J. S. Johnson, Nature,* **177,** 1129 (1956).]

remains true for the motion of the distorted zones: the rate of motion of the maximum of concentration is given by the formula deduced from the simple theory, namely

$$\frac{\Delta x}{\Delta V} = \frac{1}{a + mk}$$

for $q = kc$. In so far as the constituents of a mixture do not interfere on the sorbent, this formula is applicable to each of the constituents with the appropriate value of k.

In an actual case the moving zone continues to spread, with two adverse effects. First, the separated material is collected in a diluted solution. More seriously, there is an increase in the possibility of cross-contamination due to an overlapping of the zones. This difficulty can often be avoided by an appropriate choice of eluent. The elution is started with a solution into which only one component is appreciably desorbed. When this constituent is moved well along the column, or possibly entirely through it, the eluent is changed and another separation is effected. An excellent illustration of this is seen in Fig. 11-8, which shows the progress of a separation of the alkali metals on a column of precipitated zirconium tungstate. The elutions are done with solutions of ammonium chloride of different concentrations. Evidently the tenacity with which the alkali ions adhere to this exchanger increases as we move down the periodic table. The experiment orders the strengths of the sorptions with respect to ammonium ion. We have here an example of a general technique which gives fundamental information as to the behavior of ions in exchange-adsorption reactions in addition to providing the kind of information needed in planning separations.

11-6. Experimental Techniques. The equipment necessary for the application of ion exchange in the laboratory is very simple. Only in separations of several substances is there any necessity for complicated devices to regulate flow and collect samples. Generally the applications fall into one of three groups; purifications on a large scale, ionic replacements for analytical convenience, and chromatographic separations on a small scale incidental to other investigations. The principles involved are always the same; the object in view determines the character of the experiment.

The most extensive practical use of ion exchange is for the purification of water. Equipment especially designed for this purpose may be had for any scale of operation from purifying enormous quantities of boiler feedwater down to the production of a few liters for laboratory purposes. There is no question here of recovering the entire input to the exchanger and, at least in the laboratory, a large excess of exchanger is used so as to obtain as pure a product as possible. The usual practice is to employ a "mixed bed," consisting of intimately mixed anion and cation exchangers. The cation exchanger converts salts to the corresponding acids, which are in effect converted to water by the anion exchanger. The process is very effective and produces water of a quality comparable to that obtainable from a good laboratory still.

The purification of hydrochloric acid on an anion exchanger, which was mentioned in Sec. 11-4, points to a process which might well be turned to the purification of the chlorides of any of the alkalis, alkaline earths, aluminum, or the rare earths. None of these elements forms a chloro complex which is adsorbed by Dowex 1, whereas every other metal in the periodic table, with the single exception of nickel, is sorbed at some concentration of hydrochloric acid.[1] It must not be supposed, however, that the ion-exchange process is a cure-all for purifications. Here, as in any other method of purification, the best results are obtained when there are gross qualitative differences between the behaviors of the substances being separated. But even when no such differences can be found, purifications can still be made to depend on quantitative differences in equilibrium quotients, as in the separation of the alkalis depicted in Fig. 11-8.

Ionic replacements for analytical purposes are usually quite simply carried out. Here the entire input to the column must be recovered; the experimental arrangements are designed to make this easy. Most frequently metallic ions are replaced by hydrogen ion and anions by hydroxyl, so that simple acid-base titrations can be made. It is far simpler, for example, to titrate the hydrochloric acid equivalent to an unknown amount of calcium than to carry out a gravimetric determination *via* a precipitation of the oxalate. Similarly, the chemist who had to analyze an unknown solution of sodium perchlorate might profitably contemplate its conversion into either perchloric acid (by passage through a column containing a cation exchanger in the hydrogen form) or, say, sodium chloride (with the aid of the chloride form of an anion exchanger), followed by a titration with standard base in the first case or with standard silver nitrate in the other.

Ion-exchange chromatography designed to produce separations of similar ions is a much more difficult operation to carry out properly. Much depends on having a well-prepared column; "channeling" through a poor bed of resin might completely spoil the results. As has been pointed out, some dispersion of a concentration front always occurs, but the effects of this microscopic channeling are small compared to the enormous disturbances which result from gross inhomogeneities in the bed. A common example is the leakage along the walls of a column made with a small tube packed with large beads of exchanger. In all cases the diameter of the bead should be less than one-tenth that of the tube. The lower limit of particle size is determined by the resistance of the bed to the flow of fluid through it. In difficult separations a close approach to equilibrium is essential, and very small particles are desirable to ensure rapid reaction with the solution. Since in these cases low flow rates must be used, the resistance of the bed is unimportant. Usually, however, one does not employ beads smaller than about 0.08 mm. in diameter; particles of this size are just held on a 200-mesh screen.

[1] See, for example, K. Kraus, "Anion Exchange Studies of the Fission Products," Proceedings of the International Conference at Geneva, August, 1955, 7, Session 9B1 (United Nations, 1956). A summary of the work of Kraus and his collaborators is reproduced in *Ann. Rev. Phys. Chem.*, **7**, 157 (1956).

In order to prepare a well-packed column a supply of exchanger of narrow size range is desirable. Very fine particles should be removed by wet-sieving. The tube intended for the column is filled with water and the resin is sprinkled in, a little at a time, while the tube is vigorously tapped. If small portions of the resin are taken, little size segregation takes place, and a nearly uniformly packed bed is obtained.

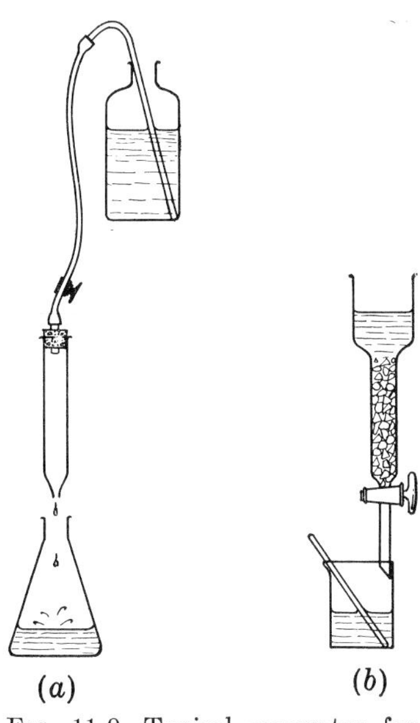

FIG. 11-9. Typical apparatus for ion-exchange chromatography: (*a*) for large-scale purifications, (*b*) for analytical replacements and small-scale purifications where quantitative recovery is necessary.

Two typical kinds of laboratory-scale columns are illustrated in Fig. 11-9. For the purification of large amounts of material the arrangement of Fig. 11-9*a* is useful. If the siphon has sufficient head, the flow rate can be well enough controlled with a screw pinch clamp. Volume changes of the resin during an experiment usually cause a compaction of the bed, making it necessary to readjust the rate of flow from time to time. Not much in a general way can be said about the selection of a proper flow rate for an ion-exchange experiment without going into a detailed analysis of the various rate processes which determine the behavior of a column. Sometimes very satisfactory results can be obtained at high rates of flow (up to 10 cm. per sec.); in other applications it may be necessary to use flow rates as low as 0.02 cm. per sec. The reader interested in a detailed analysis of column performance is referred to an excellent discussion of this subject by E. Glueckauf.[1]

For a simple replacement reaction, which must be carried out quantitatively with a limited volume of solution, the apparatus of Fig. 11-9*b* is quite satisfactory. The enlarged top of the tube makes quantitative transfer easy.

The term *chromatography* stems from the original use of the method by the Russian botanist Tswett for the separation of natural pigments. When such separations are made on a column of a white adsorbent, such as magnesium carbonate, very striking zones of color may be obtained, and it is easy to judge the progress of the operation. In the ion-exchange process we generally have no such aid; the products are colorless more

[1] E. Glueckauf, *Trans. Faraday Soc.*, **51**, 34 (1955).

often than not, and in any case the highly colored resins obscure any effects which might otherwise be observed. It is therefore necessary to follow the course of a separation by the tedious but sure process of analyzing samples of the effluent. (The work may be made very much less tedious if radioactive tracers can be applied.) If the separation is a matter of routine and the zones containing the products are sufficiently widely separated, it may be quite satisfactory to collect large samples by hand. In exploratory work, where the character of the results is necessarily uncertain, some type of automatic fraction collector is most desirable. These devices usually consist of a reel carrying a large number of test tubes and a driving mechanism to position the tubes serially under the column. Samples may be collected at equal intervals of time, or a photoelectric drop counter may be so arranged as to change tubes at approximately equal volumes of effluent. In either case the actual sample sizes are best determined by weighing; no reliance should be placed on either the constancy of the flow rate or the infallibility of the drop counter.

11-7. Partition Chromatography. We should not conclude this discussion of ion-exchange chromatography without at least a mention of two closely related means of separating chemical substances: liquid and gas partition chromatography. In so far as the discussion of Sec. 11-5 relates to the behavior of an ion-exchange column for the case of a linear isotherm, it may be taken over bodily to describe the partition of a dissolved substance between two fluid phases, one of which is held fixed while the other moves past it. The moving phase may be either a gas or a liquid; the fixed phase is always a liquid held on some inert solid. The applicability of the chromatographic method is enormously expanded by these simple devices. Much of the early work in adsorption chromatography was rendered suspect because of the tendency of delicate molecules to decompose or rearrange in the adsorption process. When the distribution between phases is brought about by a mere dissolution in an inert solvent, difficulties due to the instability of the molecule are largely avoided. In addition, the method is applicable to a far larger class of compounds; specific adsorption processes are not required for success.

Liquid partition chromatography takes on many forms. The immobile liquid may perhaps be held in a column of inert silica gel, or on alumina or starch. An extremely productive variation of the technique, known as *paper chromatography,* makes use of paper for the support and utilizes capillary flow to produce the effects. Many important separations of natural products obtainable only in microgram amounts have been made in this way. Under given conditions very characteristic rates of motion of the zones of distribution are obtained. Otherwise hopelessly complicated separations and identifications can be made. The reader interested in pursuing this subject may profitably consult the monograph by Cassidy (Reference 56 in the Suggestions for Further Study, page 408).

The most recent addition to the array of chromatographic techniques is *gas chromatography*, which involves the distribution of a compound between an immobile liquid and a moving gas. Here the distribution is determined by the Henry's law coefficient relating the partial pressure in the gas to the concentration in the liquid. Sensitive devices register small changes in some property of the effluent gas, such as its density, coefficient of thermal conductivity, or absorbance at some suitable wavelength, and so note the arrival of a zone containing a given molecular species. The Henry's law coefficient determines the time of travel through the tube of fixed liquid, and under fixed experimental conditions this serves to determine which of the constituents of a mixture is emerging from the tube at any given instant. The method has been applied with great success to the separation of extremely complex mixtures, such as those obtained in the fractionation of petroleum, and is rapidly becoming a versatile and useful analytical tool.

PROBLEMS

The answers to the starred problems will be found on page 404.

1. Dowex 1 adsorbs lead from hydrochloric acid solutions most strongly in the region 1 to 3 F. Above 8 F there is little or no adsorption. Cupric copper, on the other hand, is but slightly, if at all, adsorbed at hydrochloric acid concentrations below 3 F; its distribution coefficient rises to a value near 10 between 6 and 12 F hydrochloric acid. Devise a procedure for separating lead from copper.

2. The stoichiometric equilibrium quotient for the exchange

$$\mathrm{CsEx} + \mathrm{Na}^+ = \mathrm{NaEx} + \mathrm{Cs}^+$$

on a certain carboxylic acid exchanger is $K_{\mathrm{Cs}}^{\mathrm{Na}} = 1.55$. Plot a curve showing how the fraction of the capacity of the resin used by sodium varies with the composition of the solution when the sum of the concentrations of sodium and cesium in the solution is kept constant. Comment on the use of this exchanger as a means of separating sodium and cesium.

***3.** A column of the exchanger mentioned in Problem 2 is packed so that $\alpha = 0.5$ and $m = 0.75$. The exchanger is initially saturated with cesium and is known to have a capacity of 3.1 meq. per g. A solution 0.1 F in sodium chloride is passed slowly into the column. How fast does the front of the zone containing sodium move with respect to the linear rate of flow of the fluid?

4. A soil is known to consist of sand and other inert minerals, together with about 10 per cent of montmorillonite clay in the potassium form. The clay has a capacity of 1.0 meq. per g. Calculate the weight of the soil necessary to remove completely 1000 curies of Cs^{137} from a solution. The half-life of Cs^{137} is 30 years.

***5.** With the aid of Fig. 11-3, calculate the radioactivity of a solution initially containing 1 curie of $Cs^{137}NO_3$ after it has been equilibrated with 100 g. of the soil described in Problem 4.

6. Amberlite IRC-50 derives its activity from slightly ionized carboxylic acid groups. When packed in a column its free volume is about 40 per cent. On transformation from the hydrogen form to the sodium form this resin swells 100 per cent. Its capacity is 10.0 meq. per g. of dry resin. The average density of the bulk material is 0.69 at 50 per cent moisture. Devise and give the details of a method for removing small amounts of acid from sodium chloride solutions.

CHAPTER 12

THE DEVELOPMENT OF AN ANALYTICAL METHOD

Now that we have completed our discussion of the most important of the tools of modern analytical chemistry, it seems fitting to recall what was said in Chap. 1 about the concerns of the analytical chemist. One of the most important of these—to many analytical chemists it is even *the* most important—is the development of an analytical method, a procedure which the analyst can use to obtain quantitative information about the composition of a sample. There are a great many circumstances which may necessitate the development of a new method. The mushrooming growth of modern technology forces us to be concerned about the presence and effects of small amounts of substances which formerly could have been shrugged off as "traces." New types of materials may be impossible to handle by older methods because of interferences from constituents never before encountered along with the substance to be determined. The previously existing methods may be incapable of giving results of the desired accuracy, or they may be excessively tedious and time-consuming when time is of the essence. In any of these events a new analytical method must be developed, and it is the purpose of this chapter to describe the way in which this is done. We cannot hope to equip the student with the experience and factual knowledge which must be his before methods for the determination of cesium in cerium or of erbium in terbium will spring full-fledged from his brain. But we shall try to outline the philosophy and technique of developing such methods, for this is a vital part of the skeleton around which the experience and knowledge must be built during the analytical chemist's professional career.

Let it first of all be said that the considerations which may dictate a choice among the possible methods of carrying out any given analysis are both numerous and complex, so much so, in fact, that in actual practice it is by no means as uncommon as it ought to be to find such choices being made wrongly or illogically. There are, for example, many laboratories which employ other techniques to carry out analyses which could be better and more quickly done by polarography, simply because they do not own a polarograph; from the point of view of the scientist or even the alert accountant this is merely irrational and scarcely deserves discussion. More subtle, however, is the fact that the technique to which a chemist most readily turns is the one with which he is most familiar. The chemist who has had many years' experience with spectrophotometry, who is intimately acquainted with every detail of spectrophotometric theory and practice, and who has often been successful in developing elegant procedures for spectrophotometric analysis will quite naturally tend to try to apply spectrophotometry to almost any new problem with which he is confronted. In fact, he may do so even when it is clear to the dispassionate observer that the difficulties which can reasonably be anticipated in the spectrophotometric approach are far more substantial than those likely to be encountered in an attempt to do the very same thing by a potentiometric titration.

It has to be admitted that this sort of thinking often reflects no more than a wholly justifiable confidence in one's ability to surmount almost any kind of difficulty, and when we consider the enormous versatility of most of the individual techniques discussed in the preceding chapters it does not seem at all surprising that almost any of them can somehow be used in solving (or at least contributing to the solution of) any particular problem. In addition, even when the path he has selected is not necessarily the easiest one, the chemist who has resourcefully and imaginatively overcome difficulties which would have felled a lesser man tends, perfectly understandably, to bask in the admiration of his scientific colleagues.

However, especially for the beginner who has had neither the opportunity to develop an irresistible affection for any one technique nor enough experience in any single field to enable him to embark upon a virtuoso display of his powers, it is important to recognize that there are certain general principles which may render one approach distinctly preferable to another in any particular case. In Sec. 5-10 we outlined the thought processes that one might go through in trying to decide between a potentiometric and a conductometric method for determining the solubility of lead thiocyanate in water. Now, however, the number of techniques at our disposal has transcended the limitations of any specific illustrative example, and we shall have to use more general terms to illustrate the manner in which one of these techniques should be chosen.

Let us suppose that a large number of nearly identical samples have to be analyzed for a single constituent, and that the analytical chemist must design a method of carrying out the analysis so that it can be routinely performed by an analyst. This is the usual function of an analytical chemist in an industrial laboratory.[1] We shall assume that he

[1] The analytical chemist who teaches in a college or university may occasionally do

knows the identity of the substance which is to be determined (if not, he must carry out a qualitative examination of the material before beginning to think about a quantitative procedure), the approximate range of concentrations at which it is likely to be present, and the order of magnitude of the accuracy which must be secured in the analysis. In addition, he must know the identity and approximate concentration of each of the other substances which may be present in the samples. This is the minimum amount of information necessary for the rational development of an analytical method. It should be obvious that the determination of 0.01 per cent of lead in a sample of potassium chromate with an accuracy of ± 1 per cent is a completely different problem from the determination of 1 per cent of lead in an aqueous lead nitrate solution containing a little nitric acid with an accuracy of ± 0.1 per cent.

Having thus secured a statement of the problem at hand, one should naturally turn to the literature to make sure that someone else has not already solved the same problem, or perhaps one very similar to it. If one has to determine 0.005 per cent of bismuth in electrolytic copper, and found in the literature a method which could be used without any modification whatever for the determination of bismuth in the presence of a hundred or a thousand times as much copper, this would probably provide an excellent starting point, for only relatively trivial changes might be needed to adapt this method to the problem at hand. However, in this situation one would pay very little attention to a published method for the determination of bismuth which could only be applied in the presence of no more than an equal amount of copper, for it would probably take less time to design an analytical procedure capable of achieving the desired end directly than to adapt the published method to the task or to design a procedure for separating the trace of bismuth from so large a fraction of the copper present.

In this initial search of the literature it is better not to spend much time in studying papers dealing with problems which are only distantly related to the one of interest. In the example just mentioned, it is rather unlikely that a published method for the determination of traces of bismuth in the presence of a large excess of lead would be of any very great assistance

---

the same thing, but because he is not vitally concerned with specific applications of analytical techniques he is not likely to do it very often. Most of the academic analytical chemist's research activity is devoted to studying questions of a less immediate nature, such as the theory of a new analytical technique, the design or modification of an instrument, or the basic chemistry underlying some existing or imagined analytical method. When he does take the time to solve some specific problem of analysis, it is usually either to test the correctness of his conclusions about the pertinent basic chemistry or to demonstrate the application of a new instrument or technique.

in the development of the desired procedure. Of course this should not be taken to imply that the analytical chemist (or any other chemist) does not study every paper in his field that he can get his hands on, whether it deals with a specific problem in which he is interested at the moment or not. In doing so, however, his primary objective is not to secure detailed factual information, because this is so easy to look up at a time when it is really needed that feats of memory are invariably rather poorly rewarded. On the contrary, his reading should be directed toward the acquisition of ideas and principles which may be useful to him at a later date and perhaps even in a completely different context. Without exaggeration we may say that to any scientist one idea is worth a thousand facts.

Assuming that one cannot find in the literature any procedure which seems likely to be readily adaptable to the immediate problem, the first thing to do is to write a set of very brief outlines—one or perhaps two for each classical or instrumental technique—of methods which could be used for the determination of the desired substance if nothing else were present in the sample. In doing this one naturally has to take into account the concentration of the material expected to be present in the sample, the accuracy with which it must be determined, and the inherent sensitivity and accuracy of each of the techniques. Lead ion in moderate concentrations can easily be determined with good accuracy by a conductometric titration with standard oxalate, but it would be a waste of time to consider a conductometric titration for the determination of lead in 10^{-6} M solutions. If the analyses must be accurate to within a few tenths of a per cent, one will certainly tend to reject such techniques as direct potentiometry and direct spectrophotometry, but will turn instead to the electrolytic and titration methods. On the other hand, when an accuracy of about ± 5 per cent will suffice, the greater inherent accuracy of the titration procedures will be unnecessary and so one will be inclined to prefer one of the quicker direct methods.

During this first step it will generally be helpful to make a rather cursory search of the literature, not in quest of specific experimental details, but merely to make sure that no general type of method has been overlooked. In this process *Chemical Abstracts* tends to be more useful than the original literature, because the much more detailed information furnished by the latter cannot yet be put to use, and for the same reason it is usually still more convenient to consult a reasonably up-to-date monograph on analytical methods and procedures.

Next one should consult a handbook[1] or, better, a collection of recent

[1] It is unfortunately true that a substantial fraction of the factual data pertaining to modern instrumental techniques is not yet to be found in handbooks, and that an about equally substantial fraction of the information which does appear in the hand-

monographs on the various instrumental techniques to secure factual information bearing on each of the methods which were outlined in the preceding step. Suppose, for example, that the sample is a fairly strongly alkaline solution of sodium tartrate containing about 1 mM cadmium, whose concentration must be determined to within ± 5 per cent. At first glance this system might seem to be ideally suited to an analysis by direct polarography. However, inspection of a table of polarographic data will reveal that the cadmium wave in this supporting electrolyte is abnormally small and rather poorly defined, and that its height is not proportional to the cadmium concentration. Further inspection of the table might suggest acidifying the solution or wet-ashing the sample with sulfuric acid to destroy the tartrate. In this way the chemist may be led to make some modifications in his original outline, to start in again with a different way of applying the same instrumental technique, or even to discard the technique entirely without further ado.

At this point one will usually be left with just three or four possible methods, and now each of these should be written out in detail to facilitate the next step. This should result in several rather comprehensive sets of directions, any one of which could be used by the analyst to perform the desired analysis if he were confronted with a sample identical in every way with the actual samples but free from any of the impurities contained in the latter. These directions should include rough sketches of titration curves, approximate values of the pH (or indicator electrode potential, diffusion current, absorbance, etc.) at the equivalence point, some estimate of the instrumental and chemical errors likely to be inherent in each procedure, and in particular they must contain exact statements of the experimental conditions under which the final measurement or titration is to be carried out. It is during this step that all of the pertinent equilibrium constants, standard or formal potentials, diffusion current constants, absorptivities, and other similar data are put to use.

Then one considers how each of the other substances expected to be present in the sample would behave if it were carried through the same procedure all by itself, starting with the one which is present at the highest concentration and proceeding to the most dilute. Let us suppose that we are faced with the problem of determining, with an accuracy of ± 0.2 per cent or so, the concentration of ferrous iron in a sample which contains roughly 0.1 F ferrous iron and 0.1 F stannous tin in 1 F hydrochloric acid. The preceding step might have furnished us with, among

books has been superseded by newer and better experimental measurements. It is only fair to state that this situation is the virtually inevitable consequence of the rapid strides of present-day research on instrumental techniques. However, the student should realize that few if any practicing chemists share the common student belief that a handbook is the repository of the whole truth and nothing but the truth.

other things, a fairly complete set of directions for the potentiometric titration of ferrous iron with standard permanganate. Now we would realize that the oxidation of chloride ion by permanganate would consume some of the reagent if a pure 1 *F* hydrochloric acid solution were "titrated" with permanganate. Obviously this would vitiate the procedure, and in attempting to salvage it we would probably turn to the use of ceric sulfate instead of the permanganate. Having thus eliminated the interference of the hydrochloric acid, we must now examine the effect of this change in the procedure on the determination of the iron. In 1 *F* hydrochloric acid the formal potentials of the ferric–ferrous and ceric–cerous couples are +0.70 and +1.28 v., respectively. These are not as far apart as the formal potentials of the ferric–ferrous and permanganate–manganous ion couples, and consequently the slope of the titration curve around the equivalence point will be decreased by substituting ceric sulfate for the permanganate. However, plotting the titration curve secured with ceric ion indicates that there should still be no difficulty in securing results of the desired degree of accuracy.

Now we would turn to the effect of the stannous ion in the sample. When a 1 *F* hydrochloric acid solution containing 0.1 *F* stannous tin is titrated with ceric sulfate, the stannous ion is naturally oxidized. This fact would eliminate a possible alternative procedure consisting of a photometric titration of the sample with standard ceric solution using *o*-phenanthroline ferrous sulfate as a redox indicator. However, we should find that the formal potential of the stannic–stannous couple in 1 *F* hydrochloric acid, +0.14 v., is so much less positive than the formal potential of the ferric–ferrous couple that a differential potentiometric titration would be perfectly feasible. So we would merely have to expand the directions for the potentiometric titration to provide for the location of the equivalence point of the oxidation of the stannous ion and the measurement of the volume of reagent required to get from this point to the second end point of the titration.

Unless one is very lucky, every one of the procedures which has survived this far will be found to involve some interference from one or more of the constituents of the sample. If this is indeed the case, one must begin to think about some sort of prior chemical treatment of the sample. This may involve merely an oxidation or reduction of the interfering substance before beginning the analysis proper. For example, the interference of excess cupric ion in the determination of lead by direct polarography can be eliminated by adding cyanide to the supporting electrolyte to reduce the copper to $Cu(CN)_2^-$, which is not reducible at a dropping mercury electrode. In more severe cases it may require an actual physical separation, using such techniques as precipitation, volatilization, extraction with an immiscible solvent, or electrolytic reduction to a solid product (such as a metal) which can be mechanically separated from the solution. The details of the numerous kinds of separations used in chemical analysis cannot be discussed in the space available here. However, a good general rule is that the separation and the subsequent analysis should be as closely interrelated as possible. We may illustrate this by considering the determination of a trace of lead in a sample containing relatively large amounts of cadmium, cobalt, arsenic, copper, and barium. This curious mixture would doubtless defy any attempt to develop a simple direct method of analysis, so that some sort of separation would become essential. One way of separating lead from these other metals involves extracting it from an alkaline solution containing cyanide into a solution of dithizone in chloroform; the lead passes into the chloroform layer as lead dithizonate, which is intensely colored and could be determined by direct spectrophotometry. At this point we could certainly shake the chloroform solution of the lead dithizonate with aqueous hydrochloric acid to reextract the lead into an aqueous phase, and then analyze the resulting hydrochloric acid solution of the lead by direct polarography. On the other hand, this would be a remarkably inelegant way of squandering the analyst's time, and it would be much simpler and better to make the chloroform solution up to known volume and analyze it spectrophotometrically. In Sec. 7-9 we mentioned the fact that the solution resulting from the removal of one constituent by controlled-potential electrolysis is ordinarily ideally suited to polarographic analysis (or amperometric titration), and there are quite a few other examples of analytical techniques which are so closely related to separation procedures that the one is virtually a logical consequence of the other.

When, therefore, it becomes necessary to incorporate a separation into the analytical scheme, the chemist must go back and reexamine his entire proposed procedure. He may wish to modify it drastically or even to discard it entirely, for the material which is isolated by the separation is, after all, an entirely new sample which might perhaps best be analyzed in a manner quite different from the original sample. The separation may have resulted in the elimination of many constituents of the sample whose presence had previously led the chemist to change his first ideas considerably. Sometimes an excess of a reagent added during the separation procedure would interfere seriously if the analysis were carried out by a method which seemed perfectly adequate before the necessity for the separation was perceived.

In this way a more and more detailed set of experimental directions will gradually take form. During all of this time the chemist will have been sitting at his desk or else in the library finding out how other chemists have solved problems more or less similar to the one with which he must cope. Eventually, however, he comes to a point at which he is convinced that his method will work *if* something about which he can find no information in the literature is true, or *if* an apparently doubtful statement in the literature is false, or *if* . . . . Then he goes into the laboratory and begins to enjoy life again.

Some time later he will have convinced himself of the validity of his approach. Then he will usually have to make some measurements designed to find out the exact optimum conditions for carrying out the entire procedure: the optimum range of pH or of the concentration of some complexing agent, the wavelength of maximum absorption, the potential at which the diffusion current should be measured, and so on. He now has merely to convert the results of his labors into an exact and detailed description of every step in the method from beginning to end, use the method himself for a few analyses—preferably of known samples, partly to make sure that the method does indeed give the right answer, and partly to permit any lingering "bugs" in the method to be gotten out without spoiling any samples of real importance—and finally hand it over to the analyst who will be responsible for its routine application in the future.

As the years and the problems pass by, the chemist finds himself telescoping the process just outlined. He will automatically and more or less subconsciously reject techniques which have not proved useful for similar problems in the past, and by the same token he will often turn straight to the single technique which was once used with success in a closely related situation. As he gains more experience, he will begin to include the interfering substances in his first thoughts rather than introduce them at a later step. Occasionally he will at the first attempt design a procedure which needs only to be investigated and confirmed in the laboratory, and eventually he may do this more often than not. As his experience, knowledge, and self-confidence increase, as the imaginative resourcefulness of his solutions to difficult problems grows and becomes known to his colleagues, he comes to be a worthy inheritor of the magnificent tradition of the science of analytical chemistry.

ANSWERS TO PROBLEMS

Most of the answers which follow contain one significant figure beyond the number warranted by the data given. In comparing his results with the values given here, the student should keep in mind the possibility that a discrepancy, even if only in the last decimal place, may reflect an error in his entire approach to the problem. In such a case he should not succumb to the temptation to overlook the error because it is small, for an approximation unwittingly made may be fairly satisfactory in the particular problem at hand but ludicrously wrong in another which seems quite similar.

Chapter 2

1. 6.285
3. 3.907
5. 3.855×10^{-11}
7. 8.630
9. $[Ag^+] = 3.94 \times 10^{-7}$; $[Ag(NH_3)_2^+] = 4.58 \times 10^{-4}$; $[Br^-] = 1.02 \times 10^{-6}$; $[Cl^-] = 4.57 \times 10^{-4}$; $[H^+] = 2.47 \times 10^{-11}$; $[NH_3] = 9.08 \times 10^{-3}$; $[NH_4^+] = [OH^-] = 4.04 \times 10^{-4}$
11. $AgCl_2^-$; $K_c = 3.33 \times 10^{-6}$
13. 4.1×10^{-26}
15. 9.22
17. $[Ag^+] = 8.80 \times 10^{-4}$; $[CO_3^{=}] = 7.93 \times 10^{-6}$; $[HCO_3^-] = 8.80 \times 10^{-4}$; $[Cl^-] = 2.05 \times 10^{-7}$; $[H^+] = 4.88 \times 10^{-9}$; $[OH^-] = 2.05 \times 10^{-6}$

Chapter 3

1. *a.* $Ag = Ag^+ + e$; $Fe^{+++} + e = Fe^{++}$
b. $Ag|Ag^+||Fe^{+++}, Fe^{++}|Pt$
c. -0.028 v.
d. Ag +, Pt −
e. 0.34
6. *a.* $Ag + I^- = AgI + e$; $AgCl + e = Ag + Cl^-$
b. $Ag|AgI(s), I^-||Cl^-, AgCl(s)|Ag$
c. $+0.373$ v.
d. Ag|AgI −, AgCl|Ag +
e. 2.10×10^6
11. *a.* $H_2 = 2H^+ + 2e$; $Cl_2 + 2e = 2Cl^-$; $H_2 + Cl_2 = 2H^+ + 2Cl^-$
b. 1.4599 v.
16. *a.* $2Ag + SO_4^{=} = Ag_2SO_4 + 2e$; $PbSO_4 + 2e = Pb + SO_4^{=}$; $2Ag + PbSO_4 = Ag_2SO_4 + Pb$
b. -1.009 v.
19. $+0.153$ v.

24. -0.136 v.
26. 2.89
28. 0.0097
30. Re(VI); $+0.180$ v. *vs.* S.C.E.
34. *a.* -0.267 v.
b. $+443\%$

Chapter 4

1. 3.09
5. 0.0037 ml.
9. 34.48 ml.
13. -0.050%
17. 1.51×10^{-13}
21. $7.4 \times 10^{-13} \leq K_a \leq 1.1 \times 10^{-12}$
25. 7.013
29. Dibasic
31. 1.36×10^{3} seconds

Chapter 5

1. 1.1291 cm.^{-1}
4. 2.2754×10^{-3} mho/cm.
9. $2.73 \times 10^{-8}\ F$
11. 3211 ohms
15. $+1.55\%$/degree

Chapter 6

1. $[\text{Ti(IV)}] = 1.858 \times 10^{-3}\ F$; $[\text{Ti(III)}] = 1.42 \times 10^{-4}\ F$
5. 1.086×10^{-5} cm.^2/second
9. 6.177 microamperes
13. Os(III)
17. *a.* $\text{Cr(VI)} + 3e = \text{Cr(III)}$
b. -0.852 v. *vs.* S.C.E.
c. No
21. 0.0840 mM

Chapter 7

1. 0.425%
4. 1.527 v.
6. 80.51 ml.
8. -0.535 ± 0.016 v. *vs.* S.C.E.
11. 41.60 mg.

Chapter 8

1. 75.73 kcal./mole
3. 0.3277
5. 0.740
9. 0.604
13. $1.359 \times 10^{-4}\ M$
15. 6.33×10^{-6}
18. 0.320

Chapter 9

1. *a.* 3.51
b. 9.65
c. 54
4. *a.* 4.79×10^{5} dynes/sec.^2
b. 12%

6. *a.* 1.02

b. $^{849}/_{661}$, $^{1178}/_{867}$, $^{3047}/_{2264}$, $^{3062}/_{2292}$ H; $^{606}/_{577}$, $^{992}/_{945}$, $^{1596}/_{1559}$ C

8. *a.* 0.05% transmittance

b. (1) ± 0.0022; (2) ± 0.0069; (3) ± 0.022

9. Aliphatic C—H (2850–3000 cm.$^{-1}$); C—O (1130 cm.$^{-1}$). Suggests aliphatic ether (no O—H or C═O). Compound is $(C_2H_5)_2O$.

11. C_6H_5COOH

13. Structure B. Hydroxyl (3000 cm.$^{-1}$) is strongly bonded to N. Carbonyl (1720 cm.$^{-1}$) is absent. C≡N is lowered (2200 cm.$^{-1}$) by conjugation.

15. *a.* Up to about 1.5 *M*

b. 3.21

17. Vary the length of the cell and determine whether the deviation depends on the concentration value or on the absorbance value.

Chapter 10

1. *a.* 4.15×10^6

b. 0.112

3. 107 ± 100 microseconds

5. 0.60

Chapter 11

3. $^1/_{46}$ as fast

5. 50 microcuries

SUGGESTIONS FOR FURTHER STUDY

This is a list of the sources which, in our opinion, can be most profitably consulted by the student for further information concerning each of the nine instrumental techniques discussed in the text. Under each heading we have arranged the references in approximate order of increasing difficulty for the beginner, so that the undergraduate can consult the first book or two in each group with small difficulty and much profit, while the graduate student can easily select the more advanced or more specialized works for detailed study.

CHAPTER 1: THE NATURE OF ANALYTICAL CHEMISTRY

1. Wilson, E. B., Jr.: "An Introduction to Scientific Research," McGraw-Hill Book Company, Inc., New York, 1952.

CHAPTER 3: POTENTIOMETRIC MEASUREMENTS[1]

2. Delahay, P.: "New Instrumental Methods in Electrochemistry," chap. 1, Interscience Publishers, Inc., New York, 1954.
3. Lingane, J. J.: "Electroanalytical Chemistry," chaps. 2, 3, 6, 7, and 8, Interscience Publishers, Inc., New York, 1953.
4. MacInnes, D. A.: "The Principles of Electrochemistry," chaps. 5–10, 13, 14, 16, and 17, Reinhold Publishing Corporation, New York, 1939.
5. Dole, M.: "Principles of Experimental and Theoretical Electrochemistry," chaps. 14–19, McGraw-Hill Book Company, Inc., New York, 1935.
6. Kolthoff, I. M., and N. H. Furman: "Potentiometric Titrations," 2d ed., John Wiley & Sons, Inc., New York, 1931.
7. Clark, W. M.: "The Determination of Hydrogen Ions," 3d ed., chaps. 13–17, The Williams & Wilkins Company, Baltimore, 1928.
8. Kortüm, G., and J. O'M. Bockris: "Textbook of Electrochemistry," chap. 7 and secs. 1 and 3 of chap. 8, Elsevier Publishing Company, New York, 1951.
9. Wawzonek, S., and C. Tanford: Potentiometry, in A. Weissberger (ed.), "Physical Methods of Organic Chemistry" (vol. I of "Technique of Organic Chemistry"), 3d ed., Interscience Publishers, Inc., New York, in press.
10. Latimer, W. M.: "The Oxidation States of the Elements and Their Potentials in Aqueous Solutions," 2d ed., Prentice-Hall, Inc., Englewood Cliffs, N.J., 1952.

[1] Most of these books were written or published before the promulgation of the Stockholm Convention in 1953, and therefore employ the now outdated American sign convention: *cf.* footnote on p. 19.

CHAPTER 4: THE MEASUREMENT OF pH

11. Bates, R. G.: "Electrometric pH Determinations," John Wiley & Sons, Inc., New York, 1954.
12. Ref. 3, chaps. 4 and 5.
13. Ref. 4, chaps. 11 and 15.
14. Ref. 5, chaps. 20, 21, 24, and 25.
15. Ref. 6, chap. 8.
16. Ref. 7.
17. Ref. 8, chap. 8, sec. 2, and chap. 9.
18. Fritz, J. S.: "Acid-Base Titrations in Nonaqueous Solvents," G. F. Smith Chemical Co., Columbus, Ohio, 1952.
19. Ricci, J. E.: "Hydrogen Ion Concentration. New Concepts in a Systematic Treatment," Princeton University Press, Princeton, N.J., 1952.

CHAPTER 5: CONDUCTOMETRIC MEASUREMENTS

20. Ref. 3, chap. 9.
21. Ref. 4, chaps. 3, 4, and 18–20.
22. Ref. 5, chaps. 4–10.
23. Britton, H. T. S.: "Conductometric Analysis," D. Van Nostrand Company, Inc., Princeton, N.J., 1934.
24. Ref. 8, chap. 2, secs. 7 and 8, chaps. 5 and 6.
25. Shedlovsky, T.: Conductometry, in A. Weissberger (ed.), "Physical Methods of Organic Chemistry" (vol. I of "Technique of Organic Chemistry"), 3d ed., Interscience Publishers, Inc., New York, in press.
26. Reilley, C. N.: High-frequency Methods, chap. 15 in Ref. 2.

CHAPTER 6: POLAROGRAPHY AND AMPEROMETRIC TITRATIONS

27. Meites, L.: "Polarographic Techniques," Interscience Publishers, Inc., New York, 1955.
28. Elving, P. J.: Application of Polarography to Organic Analysis, in J. Mitchell, Jr., *et al.* (eds.), "Organic Analysis," vol. 2, Interscience Publishers, Inc., New York, 1954.
29. Kolthoff, I. M., and J. J. Lingane: "Polarography," 2d ed., Interscience Publishers, Inc., New York, 1952.
30. Müller, O. H.: Polarography, in A. Weissberger (ed.), "Physical Methods of Organic Chemistry" (vol. I of "Technique of Organic Chemistry"), 3d ed., Interscience Publishers, Inc., New York, in press.
31. Ref. 2, chaps. 2–5.
32. Milner, G. W. C.: "The Principles and Applications of Polarography and Other Electroanalytical Processes," Longmans, Green & Co., Inc., New York, 1957.

CHAPTER 7: ELECTROLYTIC METHODS

33. Ref. 3, chaps. 10–18.
34. Ref. 27, chap. 9.
35. Meites, L.: Electrolytic and Coulometric Methods, in A. Weissberger (ed.),

"Physical Methods of Organic Chemistry" (vol. I of "Technique of Organic Chemistry"), 3d ed., Interscience Publishers, Inc., New York, in press.
36. Ref. 2, chaps. 11–14, 18, and 19.
37. Cooke, W. D.: Coulometric Methods, in J. Mitchell, Jr., *et al.* (eds.), "Organic Analysis," vol. 2, Interscience Publishers, Inc., New York, 1954.

CHAPTER 8: THE ABSORPTION OF VISIBLE AND ULTRAVIOLET LIGHT

38. Sandell, E. B.: "Colorimetric Determination of Traces of Metals," 2d ed., Interscience Publishers, Inc., New York, 1950.
39. Hiskey, C. F.: Absorption Spectroscopy, and J. F. Scott: Ultraviolet Absorption Spectrophotometry, chaps. 3 and 4, in G. Oster and A. W. Pollister (eds.), "Physical Techniques in Biological Research," Academic Press, Inc., New York, 1955.
40. Mellon, M. G. (ed.): "Analytical Absorption Spectroscopy," John Wiley & Sons, Inc., New York, 1950.
41. Gillam, A., and E. S. Stern: "An Introduction to Electronic Absorption Spectroscopy in Organic Chemistry," Edward Arnold & Co., London, 1954.
42. Duncan, A. B. F., and F. A. Matsen: Electronic Spectra in the Visible and Ultraviolet, in W. West (ed.), "Chemical Applications of Spectroscopy" (vol. 9 of "Technique of Organic Chemistry," edited by A. Weissberger), Interscience Publishers, Inc., New York, 1955.
43. West, W.: Spectroscopy and Spectrophotometry, and Colorimetry, Photometric Analysis, and Fluorimetry, in A. Weissberger (ed.), "Physical Methods of Organic Chemistry" (vol. I of "Technique of Organic Chemistry"), 3d ed., Interscience Publishers, Inc., New York, in press.
44. Koller, L. R.: "Ultraviolet Radiation," John Wiley & Sons, Inc., New York, 1952.
45. Snell, F. D., and C. T. Snell: "Colorimetric Methods of Analysis," 3d ed., D. Van Nostrand Company, Inc., Princeton, N.J., vol. I (Theory, Instruments, pH), 1948; vol. II (Inorganic Determinations), 1949; vol. III (Organic Compounds I), 1953; vol. IV (Organic Compounds II), 1954.

CHAPTER 9: THE ABSORPTION OF INFRARED RADIATION

46. Williams, V. Z.: Infrared Instrumentation and Techniques, *Rev. Sci. Instr.*, **19**, 135–178 (1948).
47. Miller, F. A.: Applications of Infrared and Ultraviolet Spectra to Organic Chemistry, in H. Gilman *et al.* (eds.), "Organic Chemistry," vol. III, John Wiley & Sons, Inc., New York, 1953.
48. Lord, R. C., R. S. McDonald, and F. A. Miller: Notes on the Practice of Infrared Spectroscopy, *J. Opt. Soc. Amer.*, **42**, 149–159 (1952).
49. Bellamy, L. J.: "The Infrared Spectra of Complex Molecules," John Wiley & Sons, Inc., New York, 1954.
50. Jones, R. N., and C. Sandorfy: Infrared and Raman Spectroscopy: Applications, in W. West (ed.), "Chemical Applications of Spectroscopy" (vol. IX of "Technique of Organic Chemistry," edited by A. Weissberger), Interscience Publishers, Inc., New York, 1955.
51. Herzberg, G.: "Molecular Spectra and Molecular Structure," vol. II, "Infrared and Raman Spectra of Polyatomic Molecules," D. Van Nostrand Company, Inc., Princeton, N.J., 1945.

CHAPTER 10: RADIOCHEMICAL METHODS

52. Reynolds, S. A.: Analytical Radiochemistry, *Record Chem. Progr.*, **16**, 99 (1955).
53. Willard, J. E.: Some Applications of Radioactive Tracers to Physical Chemical Problems, *Record Chem. Progr.*, **12**, 163 (1951).
54. Friedlander, G., and J. W. Kennedy: "Nuclear and Radiochemistry," John Wiley & Sons, Inc., New York, 1955.
55. Bleuler, E., and G. J. Goldsmith: "Experimental Nucleonics," Rinehart & Company, Inc., New York, 1952.

CHAPTER 11: CHROMATOGRAPHIC METHODS

56. Cassidy, H. G.: Fundamentals of Chromatography, in A. Weissberger (ed.), "Technique of Organic Chemistry," vol. X, Interscience Publishers, Inc., New York, 1957.
57. Samuelson, R. D.: "Ion Exchangers in Analytical Chemistry," John Wiley & Sons, Inc., New York, 1953.
58. Brimley, R. C., and F. C. Barrett: "Practical Chromatography," Reinhold Publishing Corp., New York, 1953.
59. Strain, H. H.: Chromatographic Adsorption Analysis, in B. L. Clarke and I. M. Kolthoff (eds.), "Chemical Analysis," vol. II, Interscience Publishers, Inc., New York, 1945.
60. Pollard, F. H., and J. F. W. McOmie: "Chromatographic Methods of Inorganic Analysis," Academic Press, Inc., New York, 1953.
61. Phillips, C.: "Gas Chromatography," Academic Press, Inc., New York, 1956.

APPENDIX

Table A. Standard and Formal Potentials
The data are referred to the normal hydrogen electrode at 25°C.

Half-reaction	E^0, v.	$E^{0\prime}$, v., in		
		1 F HCl	1 F $HClO_4$	Other media
$Na^+ + e = Na$	−2.714			
$Al^{+++} + 3e = Al$	−1.66			
$Mn^{++} + 2e = Mn$	−1.18			
$Cr^{++} + 2e = Cr$	−0.91			
$2H_2O + 2e = H_2 + 2OH^-$	−0.828			
$Cd(OH)_2 + 2e = Cd + 2OH^-$	−0.809			
$Zn^{++} + 2e = Zn$	−0.763			
$TlI + e = Tl + I^-$	−0.74			
$TlCl + e = Tl + Cl^-$	−0.557	−0.551		
$Fe^{++} + 2e = Fe$	−0.440			
$Cr^{+++} + e = Cr^{++}$	−0.41			
$Cd^{++} + 2e = Cd$	−0.403			
$PbI_2 + 2e = Pb + 2I^-$	−0.365			
$PbSO_4 + 2e = Pb + SO_4^=$	−0.356	....	....	−0.29 (1 F H_2SO_4)
$Tl^+ + e = Tl$	−0.3363	....	−0.33	
$V^{+++} + e = V^{++}$	−0.255	....	−0.21	
$N_2 + 5H^+ + 4e = N_2H_5^+$	−0.23			
$Cu_2I_2 + 2e = 2Cu + 2I^-$	−0.185			
$AgI + e = Ag + I^-$	−0.151	....	....	−0.137 (1 F KI)
$Pb^{++} + 2e = Pb$	−0.126	....	−0.14	−0.32 (1 F NaOAc)
$2H^+ + 2e = H_2$	±0.0000	+0.005	+0.005	
$UO_2^{++} + e = UO_2^+$	+0.05			
$S_4O_6^= + 2e = 2S_2O_3^=$	+0.08			
$HgO + H_2O + 2e = Hg + 2OH^-$	+0.098			
$Sb_2O_3 + 6H^+ + 6e = Sb + 3H_2O$	+0.152			
$Cu^{++} + e = Cu^+$	+0.153			
$Sn^{+4} + 2e = Sn^{++}$	+0.154	+0.14		
$SO_4^= + 2H^+ + 2e = SO_3^= + H_2O$	+0.17	....	....	+0.07 (1 F H_2SO_4)
$AgCl + e = Ag + Cl^-$	+0.2222			
$Hg_2Cl_2 + 2e = 2Hg + 2Cl^-$	+0.2676			
$UO_2^{++} + 4H^+ + 2e = U^{+4} + 2H_2O$	+0.334			
$Cu^{++} + 2e = Cu$	+0.337			
$Ag_2O + H_2O + 2e = 2Ag + 2OH^-$	+0.344			
$Fe(CN)_6^{-3} + e = Fe(CN)_6^{-4}$	+0.36	+0.71	+0.72	+0.56 (0.1 F HCl)
$VO^{++} + 2H^+ + e = V^{+++} + H_2O$	+0.361	....	....	+0.360 (1 F H_2SO_4)
$NiO_2 + 2H_2O + 2e = Ni(OH)_2 + 2OH^-$	+0.49			
$I_2 + 2e = 2I^-$	+0.5355			
$I_3^- + 2e = 3I^-$	+0.536			
$2Cu^{++} + 2Cl^- + 2e = Cu_2Cl_2$	+0.538	+0.45		
$H_3AsO_4 + 2H^+ + 2e = HAsO_2 + H_2O$	+0.559	+0.577	+0.577	
$MnO_4^- + e = MnO_4^=$	+0.564			
$MnO_4^= + 2H_2O + 2e = MnO_2 + 4OH^-$	+0.60			
$UO_2^+ + 4H^+ + e = U^{+4} + 2H_2O$	+0.62			
$AgOAc + e = Ag + OAc^-$	+0.643			
$Ag_2SO_4 + 2e = 2Ag + SO_4^=$	+0.653			
$ClO_2^- + H_2O + 2e = ClO^- + 2OH^-$	+0.66			
$O_2 + 2H^+ + 2e = H_2O_2$	+0.682			
$Q + 2H^+ + 2e = H_2Q$	+0.6994	+0.696	+0.696	
$Fe^{+++} + e = Fe^{++}$	+0.771	+0.700	+0.732	+0.68 (1 F H_2SO_4)
$Hg_2^{++} + 2e = 2Hg$	+0.789	....	+0.776	
$Ag^+ + e = Ag$	+0.7991	....	+0.792	+0.77 (1 F H_2SO_4)
$2Cu^{++} + 2I^- + 2e = Cu_2I_2$	+0.86			
$ClO^- + H_2O + 2e = Cl^- + 2OH^-$	+0.89			
$2Hg^{++} + 2e = Hg_2^{++}$	+0.920	....	+0.907	
$VO_2^+ + 2H^+ + e = VO^{++} + H_2O$	+1.00	+1.02	+1.02	
$2ICl_2^- + 2e = I_2 + 4Cl^-$	+1.06	+1.06		
$Br_2 + 2e = 2Br^-$	+1.0652			
$ClO_4^- + 2H^+ + 2e = ClO_3^- + H_2O$	+1.19			
$2IO_3^- + 12H^+ + 10e = I_2 + 6H_2O$	+1.195			
$O_2 + 4H^+ + 4e = 2H_2O$	+1.229			
$MnO_2 + 4H^+ + 2e = Mn^{++} + 4H_2O$	+1.23	....	+1.24	
$Cl_2 + 2e = 2Cl^-$	+1.3595			
$MnO_4^- + 8H^+ + 5e = Mn^{++} + 4H_2O$	+1.51			
$2BrO_3^- + 12H^+ + 10e = Br_2 + 6H_2O$	+1.52			
$Ce^{+4} + e = Ce^{+3}$	+1.61	+1.28	+1.70	+1.44 (1 F H_2SO_4)
$NiO_2 + 4H^+ + 2e = Ni^{++} + 2H_2O$	+1.68			
$MnO_4^- + 4H^+ + 3e = MnO_2 + 2H_2O$	+1.695			
$H_2O_2 + 2H^+ + 2e = 2H_2O$	+1.77			
$S_2O_8^= + 2e = 2SO_4^=$	+2.01			

TABLE B. POTENTIALS OF SOME COMMON REFERENCE ELECTRODES

Electrode	*Potential at 25°C., volts vs. N.H.E.*
$Hg|Hg_2SO_4(s)$, $H_2SO_4(1\ F)$	+0.682
$Hg|Hg_2SO_4(s)$, $K_2SO_4(s)$	+0.65
$Hg|Hg_2Cl_2(s)$, $KCl(0.1\ F)$	+0.335
$Hg|Hg_2Cl_2(s)$, $KCl(1\ F)$ ("N.C.E.")	+0.282
$Hg|Hg_2Cl_2(s)$, $KCl(s)$ ("S.C.E.")	+0.246
$Ag|AgCl(s)$, $KCl(0.1\ F)$	+0.290
$Ag|AgCl(s)$, $KCl\ (1\ F)$	+0.237
$Ag|AgCl(s)$, $KCl(s)$	+0.201
$Hg|HgO(s)$, $NaOH(1\ F)$	+0.140

TABLE C. IONIC EQUIVALENT CONDUCTANCES AT INFINITE DILUTION

The data refer to solutions at 25°C.

Ion	λ^0, *mho-cm.²/ g. equivalent*	*Ion*	λ^0, *mho-cm.²/ g. equivalent*
Ag^+	61.92	$Fe(CN)_6^{-4}$	110.5
$B(C_6H_5)_4^-$	21	$Fe(CN)_6^{-3}$	100
Ba^{++}	63.64	H^+	349.82
Br^-	78.48	I^-	76.80
BrO_3^-	55.8	IO_3^-	40.75
$HCOO^-$	54.6	IO_4^-	54.53
HCO_3^-	44.48	K^+	73.52
$CO_3^=$	69.3	Li^+	38.69
$HC_2O_4^-$	40.2	Mg^{++}	53.06
$C_2O_4^=$	74.2	NH_4^+	73.4
CH_3COO^-	40.9	NO_3^-	71.44
$C_6H_5COO^-$	32.3	Na^+	50.11
Ca^{++}	59.50	OH^-	198
Cl^-	76.34	Pb^{++}	73
ClO_3^-	64.58	$SO_4^=$	80.00
ClO_4^-	67.31	Sr^{++}	59.46
$Co(NH_3)_6^{+++}$	102.3	Tl^+	74.7
Cu^{++}	54	UO_2^{++}	51
F^-	55	Zn^{++}	52.8

Table D. Polarographic Half-wave Potentials

All values are referred to the saturated calomel electrode at 25°C. The symbol > 0 indicates that the wave starts from zero applied potential, merging with the anodic wave due to oxidation of the mercury. NR indicates that the substance in question gives no wave. A half-wave potential enclosed in parentheses corresponds to an anodic (oxidation) wave. The appearance of a w following the half-wave potential indicates that the wave is well-defined and that its height is easily and accurately measurable, whereas i means that the wave is ill-defined and that its diffusion current cannot be measured accurately. The Roman numeral in parentheses gives the oxidation state of the product of the half-reaction which occurs at the dropping electrode when this is known with reasonable certainty, and an R following this signifies that the oxidation or reduction proceeds with thermodynamic reversibility.

Element and oxidation state	Supporting electrolyte				
	1 *F* NH_3, 1 *F* NH_4Cl	1 *F* KCl	7.3 *F* H_3PO_4	1 *F* NaOH	1 *F* KSCN
Ag(I)	>0 (0)				
Al(III)	....	−1.75 w (0)	NR	NR	
As(III)	−1.46 w (0) −1.64 w (−III)	....	−0.46 w (0) −0.71 i (−III)	(−0.27) w (V)	−0.68 w (0) −0.09 i (−III) −1.56 i
As(V)	NR	NR	NR	NR	NR
Bi(III)	....	−0.09 w (0,R)*	−0.15 w (0)	−0.6 (0)	>0 w (0)
Cd(II)	−0.81 w (0,R)	−0.64 w (0,R)	−0.71 w (0,R)	−0.76 (0)	−0.65 w (0,R)
Co(II)	−1.29 w (0)	−1.20 w (0)	−1.20 i (0)	−1.43 (0)	−1.06 (0)
Cr(III)	−1.43 (II) −1.71 (0)	−0.61 (II) −0.85 (II) −1.47 (0)	−1.02 w (II)	....	−1.05 i (II)
Cr(VI)	−0.2 (III) −1.6 (0)	−0.3 (III) −1.0 (III) −1.55 (II) −1.8 (0)	>0 i (III)	−0.85 w (III)	>0 (III) −0.46 (III) −0.95 (II)

Table D. Polarographic Half-wave Potentials (*Continued*)

Element and oxidation state	Supporting electrolyte				
	1 *F* NH_3, 1 *F* NH_4Cl	1 *F* KCl	7.3 *F* H_3PO_4	1 *F* NaOH	1 *F* KSCN
Cu(II)	−0.24 w (I,R) −0.51 w (0,R)	+0.04 (I) −0.22 w (0)	−0.09 w (0,R)	−0.41 w (0)	>0 w −0.54 w (0)
Fe(II)	(−0.34) (III)	−1.3 (0)	. . . .	(−0.9) (III,R) −1.46 (II)	−1.52 i (0)
Fe(III)	. . . .	>0 (II)	+0.06 (II,R)	−0.9 (II,R)	>0 w (II)
Mn(II)	−1.66 w (0)	−1.51 w (0)	. . . .	−1.70 (0)	−1.54 w (0)
Mo(VI)	−1.71 w (V?)	. . . .	±0.0 i −0.49 w	. . . .	NR
Ni(II)	−1.10 w (0)	−1.1 (0)	−1.18 i (0)	. . . .	−0.68 w (0)
Pb(II)	. . . .	−0.44 w (0,R)	−0.53 w (0,R)	−0.76 w (0,R)	−0.44 w (0,R)
Sb(III)	. . . .	−0.15 w (0,R) *	−0.29 w (0)	(−0.45) w (V)	>0 w (0)
Sb(V)	NR	. . . .	NR	. . . .	NR
Sn(II)	. . . .	(−0.1) i (IV) −0.47 (0)	−0.58 (0)	(−0.73) w (IV) −1.22 w (0)	−0.46 w (0,R)
Sn(IV)	. . . .	−0.1 (II)	−0.65 i	NR	−0.5 i (0)
Tl(I)	−0.48 w (0,R)	−0.48 w (0,R)	−0.63 w (0,R)	−0.48 w (0,R)	−0.52 w (0,R)
U(VI)	−0.8 (V) −1.4 (III)	−0.2 (IV + V) −0.9 (III)	−0.12 −0.58	−0.95 w	−0.24 i (V,R) −0.54 w (IV + V) −1.21 w (IV)
V(IV)	(−0.32) w (V) −1.28 w (II)	Depends on pH	−0.6 i −0.93 i (II)	(−0.43) w (V)	Depends on pH
V(V)	−0.96 w (IV) −1.26 w (II)	Depends on pH	>0 (IV) −0.54 i −0.91 (II)	−1.7 i (II)	Depends on pH
Zn(II)	−1.35 w (0)	−1.00 w (0,R)	−1.13 i (0)	−1.53 w (0,R)	−1.06 w (0,R)

* In 1 *F* HCl.

TABLE E. ION-EXCHANGE RESINS

Trade name	Type	Cross-linking	Comments
Strong acids:			
Amberlite IR-120	Polystyrene nuclear sulfonic acid	....	4.3 meq./g. dry resin
Dowex	Polystyrene nuclear sulfonic acid	1–16% DVB	5.0 meq./g. dry H resin
Duolite C-3, C-10	Phenolic matrix methylene sulfonic acid		
Duolite C-20, C-25	Polystyrene nuclear sulfonic acid		
Acids of intermediate strength:			
Duolite C-62	Hydrocarbon phosphonous acid	....	One H^+ per group
Duolite C-63	Hydrocarbon phosphonic acid	....	Dibasic exchange group
Duolite C-65	Phosphoric acid		
Weak acids:			
Amberlite IRC-50	Carboxylic acid	4–6% DVB	10.0 meq./g. dry resin
Duolite CS-101	Carboxylic acid	....	Acrylic resin
Strong bases:			
Amberlite IRA-400, -410	Polystyrene quaternary amine	3–5% DVB	3.0 meq./g. dry resin
Amberlite IRA-401	Polystyrene quaternary amine	<2% DVB	3.0 meq./g. dry resin
Dowex 1	Polystyrene trimethyl benzyl amine	1–10% DVB	3.5 meq./g. dry resin
Dowex 2	Polystyrene dimethyl ethanol benzyl amine	1–10% DVB	3.5 meq./g. dry resin
Duolite A-40, -42	Quaternary ammonium		
Duolite A-101, -102	Quaternary ammonium	....	Highly porous
Bases of intermediate strength:			
Duolite A-30	Mixed polyalkylene amines	Alkylene bridging	
Duolite A-41, -43	Tertiary and quaternary amines		
Weak bases:			
Amberlite IR-45	Polystyrene polyamine	3–5% DVB	5.0 meq./g. dry resin
Dowex 3	Polystyrene polyamine	Not defined	5.5 meq./g. dry resin
Duolite A-2, -7	Primary, secondary, tertiary amines		
Duolite A-6	Tertiary amines		
Duolite A-14	Polystyrene secondary, tertiary amines		
Duolite A-114	Secondary, tertiary amines	....	High porosity

PRACTICAL APPLICATIONS OF INSTRUMENTAL TECHNIQUES

The directions for laboratory experiments which constitute the remainder of this text are divided into two groups, the first comprising the experiments numbered from I through VIII, and the second consisting of Expts. IX and X. No doubt every instructor who makes assignments from this material will be influenced in his choice by his own personal experiences, interests, and predilections, as well as by such factors as the amount of laboratory time and the quantity and nature of the equipment available. However, it seems desirable to provide a brief description of the experiments and of the way in which the authors have employed them in their own courses.

Each of the sets of experiments numbered I, II, III, and IV actually includes from 12 to 20 or more experiments dealing with the determination of the aqueous solubility of a simple inorganic salt. Within these sets, the experiments numbered -1 and -2 deal with the preparation and analysis of a sample of the solid salt and the preparation of a pair of saturated solutions of the product. Experiments -3 and -4 involve direct potentiometry and potentiometric titrations, respectively, and must be preceded by Expt. V, which describes the assembly of apparatus for potentiometric measurements. Experiments -5, -6, and -7 deal with pH measurements, direct conductometry, and conductometric titrations. These should be followed by Expt. VI, which describes the assembly of apparatus for polarographic measurements, and then by the experiments numbered -8, -9, and -10, which involve the execution of polarographic procedures. The electrolytic techniques discussed in Chap. 7 of the text are illustrated by Expts. -11, VII, -12, VIII, and -13. Visual colorimetry, direct spectrophotometry, and spectrophotometric titrations are covered in Expts. -14 through -16.

Not every one of these techniques is represented in every set of experiments. For example, the thallous chloride system being studied in Expt. II is not amenable to pH measurements, and so Expt. II-5 does not exist. Similarly, there are no experiments numbered II-11, II-14, IV-9, IV-11, IV-13, and IV-14. On the other hand, Expt. I-5 consists of six parts, which may advantageously be divided among the various groups of students so that their data may be discussed and compared outside the laboratory, and there are many other examples of alternative ways of applying the same technique to a given problem.

Even when only one of each of these alternatives is assigned, the work involved in any of these sets of experiments is more than sufficient to consume six hours of laboratory time each week for 16 weeks. This gives the instructor the opportunity to omit some of the individual experiments entirely or to assign them to the best students as extra credit work.

The groups of experiments numbered IX and X deal with the radiochemical and chromatographic procedures which are discussed in Chaps. 10 and 11. It did not seem possible to integrate these techniques with the others in Expts. I to IV without giving the student a somewhat unrealistic idea of the customary methods of applying them, and so they have been segregated in this fashion. We have been in the habit of assigning one experiment from each of these two sets to be done after the completion of the work of Expts. I to IV.

It is desirable to require each student to write a comprehensive report on every experiment, excepting a very few (such as the experiments numbered -1 in sets I to IV) which obviously require no report. These reports should serve to summarize the data obtained and the way in which these data can be interpreted and put to use, especially those suggested by the questions and directions included in the experiments. It is probably worth mentioning that not every experiment included here "works" in the sense of giving the results expected by the student or listed in a handbook. It has been our experience that this often stimulates the student to an appreciation of the accuracy and limitations of an experimental technique and of the factors which enter into the design of an experiment in a way that the mere routine verification of preconceived ideas or duplication of literature values rarely does.

EXPERIMENT I-1

Preparation of Lead Iodide

In a 2000-ml. beaker heat a solution of 33.2 g. (0.2 mole) of potassium iodide in 1000 ml. of water nearly to boiling. While stirring continuously, *slowly* add a solution of 33.1 g. of lead nitrate in 1000 ml. of water. Cover with a watch glass and let stand for several days, stirring occasionally if possible.

At the end of this time decant the supernatant liquid and wash the precipitate several times with a 1:100 solution of perchloric acid, then several times with water alone. Transfer the solid to a fine- or medium-porosity sintered-Pyrex Buchner-type filtering funnel, wash several times more with water, and suck as dry as possible. Spread the moist solid on a large watch glass to air-dry. Place three glass rods on the watch glass to form a triangle, and invert a second watch glass on the triangle. This arrangement permits free circulation of air while greatly decreasing contamination by dust. Let stand for about a week with occasional stirring.

EXPERIMENT I-2

Analysis of Lead Iodide

A. Determination of Water. Weigh about a gram of air-dried lead iodide into a tared weighing bottle and determine the loss in weight on drying to constant weight at 105 to 120°C.

B. Determination of Lead. Weigh about 1 g. of the iodide into a 250-ml. beaker or erlenmeyer flask, add 25 ml. of 6 *F* nitric acid, and evaporate to dryness. To the residue add 10 ml. of 6 *F* nitric acid and 25 ml. of 3 *F* sulfuric acid and evaporate to fumes of sulfur trioxide. *Cool,* add 50 ml. of water (caution!), and let stand on the steam bath or hot plate for at least an hour. At the end of this time transfer the precipitate to a medium- or fine-porosity sintered-Pyrex filtering crucible and wash several times with 0.05 *F* sulfuric acid, then twice with water, once with 95 per cent ethanol, and once with ether (caution, flame hazard!). Suck dry on the aspirator, then remove the crucible from the suction flask and let it stand in the air for a few minutes. Then heat to constant weight at 105°C.

C. Determination of Iodide. Weigh 0.8 to 1.0 g. of the iodide into a 1000-ml. beaker and add about 800 ml. of water. Heat almost to boiling and stir until the salt has completely dissolved, then add 50.00 ml. of 0.1000 *F* silver nitrate from a buret or volumetric pipet. Let cool to room temperature, add 10 ml. of 3 *F* sulfuric acid and 5 ml. of saturated ferric ammonium sulfate, and titrate with standard potassium thiocyanate solution. It is helpful to use comparison solutions in locating the end point.

Calculate the percentages of water, lead, and iodide in the sample and compare these with the expected values for PbI_2.

In preparation for later experiments, place about 1900 ml. of conductivity water or of the best deionized or distilled water available in each of two 2000-ml. glass-stoppered Pyrex bottles, and add about 2 g. of lead iodide to each. Warm one of the flasks to 40 to 50°C. (not quite all of the salt should dissolve), and then let both of the flasks stand, with occasional shaking, at room temperature or in a water thermostat at 25°C. if one is available. The flasks should be clearly marked so that they can be told apart later on.

EXPERIMENT I-3*a*

Determination of the Solubility of Lead Iodide by Direct Potentiometry with a Lead Electrode

Clean and slightly etch the surface of a ¼-inch-diameter stick of the purest available metallic lead by immersing all but a centimeter or two of it in 1 *F* nitric acid for about 30 sec. Wash it thoroughly with distilled water and keep the etched portion of the electrode covered with water slightly acidified with nitric acid except when it is actually in use.

Secure a reference electrode and a potassium nitrate salt bridge from the instructor, or prepare them in accordance with his directions.[1] Place two 150-ml. beakers side by side on the laboratory bench. Fill one about half full with saturated potassium nitrate, and immerse the tip of the reference electrode in this solution. In the other beaker place a mixture of 50.0 ml. of 0.1 *F* lead nitrate and 1.0 ml. of 0.1 *F* perchloric acid, connect the two solutions by means of the salt bridge, and connect the reference electrode to the potentiometer circuit. (Will the lead electrode be positive or negative?)

Standardize the potentiometer against the standard cell. Rinse the lead electrode with distilled water and immediately place it in the lead nitrate solution. (The lead electrode can be most easily manipulated if it is placed in the hole of a rubber stopper, which can then be held in a buret clamp mounted above the beaker of solution.) Connect it to the potentiometer and measure the potential of the cell. After the potentiometer has been balanced, stir the solution briefly and turn the potentiometer knob to a value roughly 0.05 v. distant; then repeat the measurement to make sure that reproducible values are being secured.

Remove the lead electrode from the solution, rinse it with water, and replace it in the very dilute nitric acid. Discard the solution and replace it with another prepared by diluting 30.0 ml. of 0.1 *F* lead nitrate and 2.0 ml. of 0.1 *F* perchloric acid to 100.0 ml. Repeat the measurements described in the preceding paragraph, and proceed in the same way to measure the potential of the lead electrode in 0.002 *F* perchloric acid containing successively 0.01, 0.003, 0.001, 0.0003, and 0.0001 *F* lead ion.

The utmost care in keeping the surface of the lead electrode scrupulously clean is essential to the success of these measurements.

Finally add 1.0 ml. of 0.1 *F* perchloric acid to 50.0 ml. of each of the saturated lead iodide solutions[2] prepared in Expt. I-2, and measure the potential of the lead electrode in each of these mixtures.

[1] Directions for the preparation of one type of silver–silver chloride–saturated potassium chloride reference electrode may be found in Expt. V.

[2] Unless the lead iodide solution is perfectly clear and free from suspended solid salt, filter about 5 ml. of it through a very fine quantitative filter paper. Discard this,

Disassemble the cell and store the reference electrode and the salt bridge according to the instructor's directions.

Plot the potential of the indicator electrode against log $[Pb^{++}]$. How well does the slope of the resulting straight line agree with the theoretical value, 0.02957 v. at 25°C.? Is better agreement secured by introducing the activity coefficient of lead ion calculated from the Debye-Hückel equation? Extrapolate the data to log $[Pb^{++}] = 0$, and use the value of E_{ind} thus obtained and the potential of the silver–silver chloride–saturated potassium chloride reference electrode (+0.201 v. *vs.* N.H.E.) to calculate the standard potential of the lead ion–lead half-reaction. Compare this value with the literature value, −0.126 v. Account for any difference between these values. Attempt to secure a more accurate value of E^0 by plotting $E_{ind} - 0.02957 \log [Pb^{++}]$ against the square root of the ionic strength and extrapolating to infinite dilution. Does the resulting value seem to be more reliable than the one obtained previously? Why or why not?

Use the plot of E_{ind} *vs.* $[Pb^{++}]$ to find the concentration of lead ion in each of the saturated lead iodide solutions, and calculate the solubility product of lead iodide. What do you think is a fair estimate of the uncertainty in K secured from these data?

EXPERIMENT I-3*b*

Determination of the Solubility of Lead Iodide by Direct Potentiometry with a Silver–Silver Iodide Electrode

Prepare a silver wire electrode by coiling and etching a piece of silver wire according to the directions in Expt. V. Secure or prepare a reference electrode and 100 ml. of each of a series of potassium iodide solutions containing 0.1, 0.03, 0.01, 0.003, 0.001, 0.0003, and 0.0001 M iodide.

Place the silver electrode and the reference electrode in a 150-ml. beaker, add 50 ml. of 0.1 F potassium iodide and a few milligrams of pure solid silver iodide,[1] and stir for several minutes. Meanwhile decide which of the two electrodes will be positive, make the appropriate connections to the potentiometer, and standardize the potentiometer against the standard cell. Measure the potential of the silver electrode, stir for several minutes more, and measure the potential again after turning the potentiometer knob to a value about 0.05 v. away from the initial

---

filter about 60 ml. of the solution through the same paper, and use a 50-ml. aliquot of this.

[1] If no pure solid silver iodide is available, add 0.001 F silver nitrate *dropwise*, stirring constantly, until a definite but barely detectable permanent turbidity is just produced.

balance point. Repeat, if necessary, until a potential which is reproducible to ± 0.5 mv. is secured.

Then discard the 0.1 *M* iodide solution, rinse the beaker and electrodes thoroughly, and replace it with 50 ml. of 0.03 *M* iodide. Again add a few milligrams of silver iodide and measure the potential of the cell as directed in the preceding paragraph.

When the potential of the silver–silver iodide electrode has been measured in each of the known iodide solutions, measure it in each of the saturated lead iodide solutions prepared in Expt. I-2.

Finally disassemble the cell and store the indicator and reference electrodes according to the instructor's directions.

Plot the potential of the indicator electrode against log $[I^-]$. How well does the slope of the best straight line through the experimental points agree with the theoretical value? If the potentials of the two electrodes can be measured with equal precision, will the silver–silver iodide electrode give more or less accurate information concerning the composition of a saturated lead iodide solution than the lead electrode? Which electrode would you use to secure information concerning the lead–iodide complexes present in solutions containing both potassium iodide and lead iodide?

Extrapolate the plot of E_{ind} *vs.* log $[I^-]$ to log $[I^-] = 0$, and from the value of E_{ind} thus obtained, together with the potential of the reference electrode employed, calculate the standard potential of the silver–silver iodide electrode. From this compute the solubility product of silver iodide, and compare your values for these quantities with the literature values. On your graph locate the potential at log $[I^-] = 0$ which corresponds to the literature value of $E^0_{AgI,Ag}$. Does this point represent a reasonable extrapolation of your data?

Now construct a plot of $E_{ind} + 0.05915 \log [I^-]$ against the square root of the ionic strength and extrapolate this to infinite dilution. How does the length of this extrapolation compare with the length of the extrapolation of the plot of E_{ind} *vs.* log $[I^-]$? Which plot gives the more reliable value of the standard potential of the silver–silver iodide electrode?

Use the plot of E_{ind} *vs.* log $[I^-]$ to find the concentration of iodide ion in each of the saturated lead iodide solutions, and calculate the solubility product of lead iodide. What do you think is a fair estimate of the error in the value of K secured from these data?

Lead iodide is known to dissolve in solutions containing excess iodide. What assumption concerning the stabilities of the lead iodide complex ions is made in calculating the solubility product of lead iodide from the data secured in this experiment? How could you test the validity of this assumption experimentally?

EXPERIMENT I-3*c*

Determination of the Solubility of Lead Iodide by Direct Potentiometry with an Iodine–Iodide Electrode

Carry out the experimental measurements described in Expt. I-3*b*, but use a platinum wire indicator electrode instead of the silver electrode, and add five drops of freshly prepared 10 per cent tincture of iodine (1 g. of pure iodine dissolved in 9 ml. of ethanol) to each solution instead of the solid silver iodide.

Because each of these solutions will be saturated with iodine, we may take the activity of iodine as unity and write for the potential of the iodine–iodide electrode, $Pt|I_2(s), I^-(c)$,

$$E = E^0 - \frac{0.05915}{2} \log [I^-]^2$$

where the literature value of E^0 is $+0.5355$ v. *vs.* N.H.E. When iodine is added to an iodide solution, the following reaction takes place:

$$I_2(s) + I^- = I_3^-$$

The equilibrium constant of this reaction is

$$K = \frac{[I_3^-]}{[I^-]} = 0.94$$

If the original solution contains c moles of iodide per liter, then after the addition of excess iodine we shall have

$$[I^-] + [I_3^-] = 1.94[I^-] = c$$

and so the potential of the iodine–iodide electrode will be

$$\begin{aligned} E &= E^0 - \frac{0.05915}{2} \log \left(\frac{c}{1.94}\right)^2 \\ &= E^0 + 0.017 - 0.05915 \log c \end{aligned}$$

Treat the data by the method described in Expt. I-3*b*.

Compare the precision and accuracy of the results secured with the three indicator electrodes used in the different sections of this experiment.

EXPERIMENT I-4*a*

Potentiometric Titration of Iodide with Lead Ion

Crush 35 to 40 g. of pure lead nitrate in a mortar, transfer it to a weighing bottle, and dry at 120°C. for two hours or more. Cool to room temperature and weigh 33.12 g. into a 150-ml. beaker. Dissolve in water and transfer quantitatively to a 1000-ml. volumetric flask. Add 1 ml. of concentrated nitric acid (or, better, 72 per cent perchloric acid), and dilute to the mark with water.

Weigh about 0.6 g. of pure potassium iodide into a 250-ml. beaker, dissolve it in 100 ml. of water, and add a few milligrams of pure solid silver iodide. Place a helical silver wire indicator electrode in the solution, and connect this to one wire leading from the potentiometric assembly. Also place in the solution the tip of an agar–1 F potassium nitrate salt bridge; immerse the other end of this bridge in a small beaker containing saturated potassium chloride or nitrate, and in this small beaker immerse the tip of the silver–silver chloride electrode prepared in Expt. V. This arrangement gives electrical contact between the iodide solution and the reference electrode without any danger of contaminating the former with chloride or the latter with nitrate.

If a stirring motor or a magnetic stirrer is available, use it to keep the iodide solution in slow but constant motion.

Measure the potential of the cell, wait a minute or two, and repeat the measurement to ensure that a steady value has been obtained. (If the potentiometer cannot be balanced, it may be necessary to interchange the wires leading to the indicator and reference electrodes.) Add the standard lead nitrate solution in 1- or 2-ml. aliquots from a 50-ml. buret; after each addition wait a minute or two before measuring the potential. When the slope of the curve appears to be starting to increase, begin adding smaller increments of the lead nitrate; when it is certain that the end point has been passed, start adding successively larger portions of the lead nitrate solution. Continue until about 40 ml. of lead nitrate has been added.

Plot the titration curve and locate the end point. How accurately does this correspond to the stoichiometric prediction? How can you account for any discrepancy? From the data secured at each of several points before the end point calculate the standard potential of the cell $Ag|AgI(s),\ I^-(c)||KCl(s),\ AgCl(s)|Ag$, neglecting liquid-junction potentials.

Look up the standard potential of the $Ag|AgCl(s),\ Cl^-$ electrode. How is this related to the activity of chloride ion? If the activity of chloride ion in saturated potassium chloride is 2.50 M, what is the expected potential of the $Ag|AgCl(s),\ KCl(s)$ electrode? How does this compare with the literature value?

From your value for the standard potential of the cell $Ag|AgI(s),\ I^-(c)||KCl(s),\ AgCl(s)|Ag$ and the literature value of the potential of the reference electrode, calculate the standard potential of the half-reaction $AgI + e = Ag + I^-$. Combine this with the standard potential of the half-reaction $Ag^+ + e = Ag$ ($E^0 = +0.7991$ v.) to secure the solubility product of silver iodide. How do you account for any difference between your value and that found in a handbook?

What interpretation can you attach to the potentials measured at points beyond the equivalence point?

EXPERIMENT I-4*b*

Potentiometric Titration of Iodide with Silver Nitrate

Pipet exactly 100 ml. of one of the saturated lead iodide solutions into a 150-ml. beaker containing a magnetic stirring bar, and add 2 to 3 g. of solid barium nitrate. In the solution immerse a helical silver wire to serve as an indicator electrode, and one end of a 4 per cent agar–1 *F* potassium nitrate salt bridge. Immerse the other end of the salt bridge in a small beaker containing saturated potassium nitrate, and place the tip of the reference electrode prepared in Expt. V in this solution.

Carrying out the titration at 70°C. greatly improves the results obtained; it gives greater accuracy, a much more rapid attainment of steady potentials near the end point, and a considerably greater "break" at the end point. This is most easily accomplished by the use of a combination magnetic stirrer and hot plate. Consult the instructor for directions on this point.

From a 10-ml. microburet add a few drops of standard 0.1 *F* silver nitrate. Measure the potential of the silver electrode as soon as it has become constant, and record this and the buret reading. Continue adding the silver solution in 0.5-ml. portions until the potential of the indicator electrode reaches about −0.13 v. *vs.* the reference electrode being used; then add successive 0.1-ml. aliquots until the equivalence point (which occurs at a potential of about +0.05 v.) has definitely been passed.

Discard the solution, clean the titration cell thoroughly, and repeat the titration with a 100-ml. portion of the other saturated lead iodide solution.

Locate the end points of the titrations, using the procedure illustrated by Table 3-2. From these data calculate the concentrations of iodide in the saturated lead iodide solutions and the solubility product of lead iodide. How do these values compare with those obtained in Expt. I-3*b*? Which are more reliable, and why?

Calculate the formal potential of the Ag|AgI(*s*), I^- electrode by the method illustrated by Table 3-6. How well does this agree with the literature value of the standard potential of the electrode and with the data of Expt. I-3*b*? Suggest possible explanations for any differences.

Does the potential at the end point of this titration agree with the value calculated from the formal potentials involved? From your data, what is the value of log $[I^-]$ at the end point? The solubility product of silver iodide is approximately 10^{-16}; how can you account for any difference between the experimental value of log $[I^-]$ at the end point and the expected value, -8.0? [*Cf.* I. M. Kolthoff and J. J. Lingane, *J. Am. Chem. Soc.*, **58,** 1524 (1936).]

EXPERIMENT I-4*c*

Potentiometric Titration of Iodide with Ceric Sulfate

Transfer exactly 50 ml. of one of the saturated lead iodide solutions to a 250-ml. beaker and add 50 ml. of concentrated hydrochloric acid. Using a platinum wire as the indicator electrode and either an S.C.E. or the silver–silver chloride electrode prepared in Expt. V as the reference electrode, titrate with a standard 0.05 *F* ceric solution (prepared by diluting exactly 50 ml. of 0.1 *F* ceric sulfate to 100.0 ml. with roughly 2 *F* hydrochloric acid). Use a 10-ml. microburet, and continue the titration until a total of 10 ml. of the ceric solution has been added. Repeat with a 50-ml. aliquot of the other lead iodide solution.

Locate the end points of the two titrations and calculate the concentrations of iodide in the saturated lead iodide solutions. The iodide is oxidized according to the equation

$$I^- + 2Cl^- = ICl_2^- + 2e$$

What is the solubility product of lead iodide? Which of the methods used here and in Expt. I-4*b* gives the more reliable value of the iodide concentration? Why is this? What are the principal reasons for preferring either of these methods to the one used in Expt. I-4*a*? Which method would you choose for the analysis of a solution containing 10^{-5} *M* iodide?

Following the procedure illustrated by Table 3-6, find the formal potentials of the ICl_2^-, I^- and Ce^{IV}, Ce^{III} couples in 6 *F* hydrochloric acid. How does the precision of each set of values compare with the estimated precision of the potential measurements? Account for any significant discrepancy.

Compare the formal potential of the ceric–cerous couple in 6 *F* hydrochloric acid with the standard potential of the couple. Assuming that the entire difference between these values is due to complexation, calculate the ratio of the dissociation constants of the chloro-ceric and -cerous complex ions. Estimate the liquid-junction potential across a boundary between saturated potassium chloride and 6 *F* hydrochloric acid. In view of this estimate, what is the most probable value of this ratio of dissociation constants? Is the result of your calculations consistent with the literature value of the formal potential of the ceric–cerous couple in 1 *F* hydrochloric acid?

From your values of the formal potentials of the couples involved in this titration, calculate the potential of the indicator electrode at the equivalence point. By interpolation in your data (*cf.* Table 3-2) find the actual potential at the point you took as the end point in each titration. How well do these values agree?

What is the principal chemical cause of error in this procedure, and how could it be eliminated? Explain why this titration gives better results than a titration of iodide with ceric ion in dilute sulfuric acid.

EXPERIMENT I-5*a*

The Hydrolysis of Lead Ion

Prepare 100 ml. of an 0.10 F solution of lead nitrate in the best water available. The concentration of the lead nitrate should be known to within about ± 1 per cent. Dilute 25.0 ml. of this solution to 100.0 ml. to prepare an 0.025 F solution; dilute 25.0 ml. of this in turn to prepare an 0.00625 F solution; and so on until a series of lead nitrate solutions is available whose concentrations range from 0.1 to 2.4×10^{-5} F in successive factors of four.

Set the pH meter to read 3.57 with the electrodes immersed in saturated potassium hydrogen tartrate. Rinse the electrodes thoroughly and immerse them in a portion of the 0.1 F lead nitrate solution. When a steady reading is secured, discard the solution and replace it with another portion of the 0.1 F solution. If the pH now measured differs from the value finally secured with the first portion, continue in the same way until a constant pH is obtained. Then discard the lead nitrate solution, rinse the electrodes thoroughly with dilute nitric acid and then with water, and measure the pH of the saturated potassium hydrogen tartrate solution. If the value secured lies between 3.55 and 3.59, reset the meter to 3.57 and proceed to measure the pH of the 0.025 F lead nitrate solution. If not, reset the meter and repeat the measurements with the 0.1 F solution.

When the pH of each of the lead nitrate solutions has been measured, proceed to measure the pH of each of the saturated lead iodide solutions prepared in Expt. I-2.

For each of the lead nitrate solutions compute the equilibrium constant of the reaction $Pb^{++} + H_2O = Pb(OH)^+ + H^+$. What is the mean acid dissociation constant of lead ion? How does the effect of ionic strength on the various activity coefficients enter into this experiment? How could this be eliminated?

Plot the pH of a lead nitrate solution against the concentration of lead ion, and use this to estimate the concentrations of lead ion in the saturated lead iodide solutions. Calculate the solubility product of lead iodide.

What is the effect on this calculation of the formation of complex ions such as PbI^+ and PbI_3^-? What would be the effect of a trace of perchloric acid occluded in the lead iodide? How could you determine whether enough perchloric acid had been occluded to cause a significant error?

How would you proceed to calculate the solubility product of the salt PbA_2 from data like these if you knew the dissociation constant of the weak acid HA?

EXPERIMENT I-5*b*

Alkalimetric Titration of Lead Ion

Prepare an approximately 0.1 *F* solution of sodium hydroxide by diluting 5 ml. (use a calibrated medicine dropper or an automatic pipet) of 50 per cent sodium hydroxide to a liter with freshly boiled distilled water. Store the solution in a rubber-stoppered glass bottle or, better, in a polyethylene bottle with a tightly fitting screw cap. Standardize it by titrating weighed samples of potassium hydrogen phthalate or potassium hydrogen tartrate, using phenolphthalein as the indicator. (If preferred, the titration can be made potentiometrically according to the directions in Expt. I-5*d*.)

Accurately weigh 0.2 to 0.3 g. (0.7 to 0.9 millimole) of reagent grade lead nitrate into a 250-ml. beaker and add exactly 100 ml. of water. Set the pH meter to read 3.57 with the glass and calomel electrodes immersed in a saturated solution of potassium hydrogen tartrate; then rinse the electrodes thoroughly with water and immerse them in the lead nitrate solution.

Measure the pH of this solution (how does the result compare with the data obtained in Expt. I-5*a*?), then add 0.50 ml. of the standard 0.1 *F* sodium hydroxide solution from a 50-ml. buret, stir thoroughly, and measure the pH again. Continue the titration until 50 ml. of sodium hydroxide has been added, being sure to secure enough data to define the entire course of the titration curve. Note the volume of base required to produce the first permanent precipitate. Is the final solution perfectly clear? Discard the solution and rinse the beaker and the electrodes thoroughly with dilute nitric acid followed by water. Leave the electrodes standing in distilled water.

Plot the titration curve and locate the end point of the titration. How does the volume of base required to reach this point compare with the expected value if the precipitate is assumed to be $Pb(OH)_2$? How is this comparison affected by the adsorptive properties of hydrous lead oxide? Do you think that this titration could be developed into a useful method for the determination of lead?

Is there any indication of an end point corresponding to the reaction

$Pb^{++} + OH^- = PbOH^+$? If such an end point cannot be found, is it safe to conclude that $PbOH^+$ is not formed during the titration? What does a failure to observe two inflection points during the titration of a weak dibasic acid indicate about the value of K_1/K_2?

At each experimental point preceding the end point, calculate the number of millimoles of sodium hydroxide added to the solution. Divide this by the number of millimoles of base needed to reach the end point (*not* the calculated equivalence point), and take the resulting ratio as the fraction of the lead precipitated.

Calculate the concentration of lead left in solution; this is $[Pb^{++}] + [PbOH^+]$. From this and the result of Expt. I-5*a* calculate $[Pb^{++}]$, and from this and the measured pH calculate the solubility product of $Pb(OH)_2$. Look up the literature value of this constant. Suggest an explanation for any difference between the average of your values and the literature value. Do your values seem to vary systematically as more of the lead is precipitated? What assumption in the prescribed method for estimating $[Pb^{++}] + [PbOH^+]$ might be responsible for such behavior?

Devise a method for calculating the equilibrium constant of the reaction $Pb(OH)_2 + OH^- = Pb(OH)_3^-$ from the data secured at points past the end point of the precipitation reaction. [The formation of $Pb(OH)_4^=$ does not become appreciable until the pH becomes very high.] Carry out the calculation for several of the experimental points. From the resulting values and the measured solubility product of the hydrous oxide, calculate the dissociation constant of $Pb(OH)_3^-$. The literature value is 1×10^{-14}. To what may any difference be due?

EXPERIMENT I-5*c*

Analysis of a Saturated Lead Iodide Solution by Potentiometric Titration with Sodium Hydroxide

Pipet exactly 100 ml. of each of the saturated lead iodide solutions prepared in Expt. I-2 into a 250-ml. beaker. Titrate with standard 0.1 F sodium hydroxide using a 10-ml. microburet.

Locate the end points of the titration curves. From the volumes of sodium hydroxide used here and in Expt. I-5*b*, calculate the amount of lead ion in each of the samples of saturated lead iodide solution, and from these values compute the solubility product of lead iodide. Why is the use of a standardization against lead nitrate likely to give more accurate results than that of a standardization against, say, potassium hydrogen phthalate?

Compare the results of Expts. I-5*a* and I-5*c*. Which method seems to you to be likely to give the more reliable value? In the light of the results of Expt. I-3, does this expectation agree with fact? Propose reasons for any discrepancies among the results of these three experiments.

EXPERIMENT I-5*d*

Potentiometric Standardization of 0.1 *F* Sodium Hydroxide

Dry a few grams of pure potassium hydrogen phthalate or potassium hydrogen tartrate to constant weight at 110°C. Weigh out two 0.6- to 0.7-g. portions (about 3 millimoles) of the dry salt into 250-ml. beakers. Add 100 ml. of water and stir until the salt is dissolved. (Ignore any small residue which remains undissolved after a few minutes; it will go into solution as base is added.)

Measure the pH of the solution, add 5 ml. of the sodium hydroxide from a 50-ml. buret, and measure the pH again. Continue adding the base in 5-ml. portions until the slope of the titration curve begins to increase, then start decreasing the size of the successive aliquots to make sure that the end point is not overrun. Around the end point the base should be added in 0.10-ml. portions.

Titrate the other weighed sample of the standard in the same way.

Locate the end point of each titration and calculate the concentration of the sodium hydroxide. From one or two points on the nearly flat portion of each curve calculate the second dissociation constant of phthalic or tartaric acid. The literature values of K_2 are 3.0×10^{-6} for phthalic acid and 6.9×10^{-5} for tartaric acid. How well does your value agree with the accepted value? How large a discrepancy would be caused by an error of 0.05 unit in each of your pH measurements? Propose an explanation for any larger difference between the measured and literature values.

From the known amount of the primary standard used and the volumes of the solutions calculate the concentration of phthalate or tartrate ion present at the equivalence point of the titration. Use this and your value of K_2 to calculate the pH which should have been observed at the end point. By interpolation in the titration data find the pH actually observed at each end point, and compare these with the predicted value. What conclusion do you draw from the comparison?

EXPERIMENT I-5*e*

Alkalimetric Titration of Lead Ion in Citrate Medium

Dissolve 52.5 g. (0.15 mole) of crystalline sodium citrate ($Na_3Cit \cdot 5H_2O$) in 150 ml. of water.

Transfer 60 ml. of the citrate solution (reserve the rest for Expt. I-5*f*) to a 150-ml. beaker and insert the glass and calomel electrodes. Slowly and with stirring add dilute hydrochloric acid (1:50) until the pH of the solution becomes 6.0 ± 0.2. Transfer exactly 25 ml. of this to another beaker, add exactly 75 ml. of water, and titrate with standard 0.1 *F* sodium hydroxide. Try to obtain points at intervals no larger than 0.1 or 0.2 pH unit, and continue adding the sodium hydroxide until the pH becomes at least 11.5.

Discard the solution, rinse the electrodes very thoroughly, and reset the meter to read pH 3.57 with the electrodes immersed in saturated potassium hydrogen tartrate solution. Again rinse the electrodes thoroughly and replace them in the titration beaker.

Transfer exactly 25 ml. of the buffer of pH 6.0 to this beaker; add an accurately weighed amount [0.6 to 0.7 g. (1.8 to 2.1 millimoles)] of reagent grade lead nitrate and exactly 75 ml. of water. Stir until the lead nitrate has completely dissolved, then measure the pH of the mixture. Titrate with 0.1 *F* sodium hydroxide exactly as before.

Plot the two sets of points on the same sheet of graph paper (the larger the better) and draw the best curve through each set. Lay a straightedge on the graph paper parallel to the volume axis and measure the distance between the two curves at pH 6.0 (or whatever the initial pH of the dilute citrate buffer may have been). Convert this distance to the corresponding volume of sodium hydroxide; this gives the volume of base required to restore the pH to 6.0 after adding the amount of lead ion used in the experiment, and so it gives a measure of the extent of the reaction $Pb^{++} + Cit^{\equiv} + H_2O = PbOHCit^{=} + H^+$ when the pH is kept constant at 6.0. (The equation given is suggestive only; it represents the stoichiometry of the reaction but doubtless not the true constitution of the complex.)

Similarly measure the distance between the two curves at pH 6.1, 6.2, . . . , and so on up to the highest pH which appears on both curves. Plot the corresponding volumes of base against the pH on a new sheet of graph paper. This gives a titration curve for the lead–sodium hydroxide reaction. Find the volume of base required to reach the inflection point of this curve. This is the volume of base required to react with all of the lead—call it V^*. Secure the average value of V^* from the original data. From this calculate the number of hydroxyl ions which react with one lead ion in the titration reaction. How does the accuracy of this titration compare with that of the direct titration of lead with sodium hydroxide (Expt. I-5*b*)? What is responsible for the difference?

Measure the volume of base V consumed by the lead at each of several pH values on the rising portion of the graph just plotted. Show that the concentration of lead remaining unneutralized is proportional to $V^* - V$, while that of the complex ion formed is proportional to V. At a pH of, say, 8, is it reasonable to assume that the lead not yet titrated was present as Pb^{++}? (What was the appearance of the solution?) From these data calculate the corresponding values of the equilibrium constant $K = [PbOHCit^{=}][H^{+}]/[PbCit^{-}]$. What chemical reaction is responsible for the increasing consumption of base beyond the equivalence point?

EXPERIMENT I-5*f*

Analysis of a Saturated Lead Iodide Solution by Potentiometric Titration with Sodium Hydroxide in Citrate Medium

Dilute exactly 10 ml. of the 0.1 F sodium hydroxide used in Expt. I-5*e* to 100.0 ml. with freshly boiled distilled water.

Transfer exactly 25 ml. of the stock 1 F sodium citrate solution prepared in Expt. I-5*e* to a 250-ml. beaker and add exactly 100 ml. of water. Titrate with the 0.01 F sodium hydroxide just prepared, trying to secure points at intervals of 0.1 to 0.2 pH unit. Continue until 50 ml. of base has been used.

Discard the solution and replace it with a mixture of 25 ml. of the stock citrate solution and 100 ml. of one of the saturated lead iodide solutions prepared in Expt. I-2. Titrate this in exactly the same way as the "blank," then similarly titrate a portion of the other saturated lead iodide solution.

Plot all of the data on a large sheet of graph paper and measure V^* for each of the lead iodide titrations by the method described in Expt. I-5*e*. Calculate the concentration of lead ion in each of the saturated solutions of lead iodide, and compute the solubility product of lead iodide.

Would this method be applicable to the analysis of a 10^{-5} M solution of lead ion? Compare it in this respect with the procedure of Expt. I-5*c*. When applied to 10^{-3} M solutions, which procedure is more accurate? Why?

EXPERIMENT I-6

Determination of the Solubility of Lead Iodide by Direct Conductometry

Secure a dip-type conductance cell and a direct-reading a-c operated conductance bridge. If necessary, platinize the electrodes of the cell according to the directions in Secs. 4-3 and 5-13.

Place exactly 500 ml. of water in a clean dry 600-ml. beaker and measure the conductance of the water. Add 1.00 ml. of stock 0.100 N (0.0500 F) lead nitrate solution, mix very thoroughly, and measure the conductance again. Add another 1.00 ml. of the stock lead nitrate and repeat the measurement, then add 3.00 ml. more, then 5.00 ml., 10.0 ml., 10.0 ml., and 20.0 ml.

Discard the solution and repeat the measurements, first with a stock 0.100 N (0.100 F) solution of potassium nitrate, then with a stock 0.100 N (0.100 F) solution of potassium iodide. Then measure the conductance of each of the saturated lead iodide solutions prepared in Expt. I-2.

Finally measure the conductance of a standard potassium chloride solution supplied by the instructor. After making this measurement, rinse the cell very thoroughly with distilled water and leave it standing in a clean beaker with the electrodes covered with water.

Consult Table 5-1 to secure the specific resistance of the potassium chloride solution used, and calculate the cell constant of the cell from Eq. (5-1).

Calculate the equivalent conductance of each of the salt solutions used, and at each normality of salt calculate $\Lambda_{Pb(NO_3)_2} + \Lambda_{KI} - \Lambda_{KNO_3}$. Multiply this quantity by the normality to which it refers; this gives $\Lambda_{PbI_2}C$, which is equal to $1000(l/A)(1/R)$. Plot this against the normality. On the ordinate of this graph locate the value corresponding to each of the saturated lead iodide solutions, and read the normality of each solution off the abscissa scale.

Plot the equivalent conductance of each of the salts used against the square root of the normality of the salt. These should be straight lines, in accordance with Eq. (5-4). By extrapolating to infinite dilution, find the value of Λ^0 for each salt, and compare these with the values secured from Table C in the Appendix. Account for any discrepancies observed.

These straight lines may be represented by the following equations:

$$\begin{aligned} \Lambda_{Pb(NO_3)_2} &= \Lambda^0_{Pb(NO_3)_2} - a\sqrt{C} \\ \Lambda_{KI} &= \Lambda^0_{KI} - b\sqrt{C} \\ \Lambda_{KNO_3} &= \Lambda^0_{KNO_3} - c\sqrt{C} \end{aligned}$$

Since

$$\Lambda_{PbI_2} = \Lambda_{Pb(NO_3)_2} + \Lambda_{KI} - \Lambda_{KNO_3}$$

it is possible to find C algebraically by solving the equation

$$1000\left(\frac{l}{A}\right)\left(\frac{1}{RC}\right) = \Lambda^0_{Pb(NO_3)_2} + \Lambda^0_{KI} - \Lambda^0_{KNO_3} - (a + b - c)\sqrt{C}$$

where R is the measured resistance of the (C normal) lead iodide solution. Calculate the normality of the saturated lead iodide solution from this equation, and compare it with the value found graphically above.

Identify any sources of error in this experiment, and compare the results obtained with those secured in previous experiments. How well do your values for Λ^0 agree with those calculated from the equivalent ionic conductances listed in Table C of the Appendix?

EXPERIMENT I-7

Analysis of Saturated Lead Iodide Solutions by Conductometric Titration

Prepare 100.0 ml. of an approximately 0.03 F solution of sodium oxalate by weighing accurately 0.4 g. of the pure salt, dissolving it in water, and diluting to volume in a 100-ml. volumetric flask.

Using a dip-type conductance cell, accurately measure the resistance of a mixture of a 100.0-ml. aliquot of one of the saturated lead iodide solutions with a volume of water which is just sufficient to completely immerse the electrodes of the cell. From a 10-ml. microburet add 1.00 ml. of the sodium oxalate solution. Mix the solution thoroughly and observe whether any precipitate of lead oxalate has formed. Measure the resistance of the mixture and add another 1.00 ml. of the oxalate solution. Continue until 10 ml. of oxalate has been added. Then discard the solution and rinse the cell very thoroughly with distilled water. Titrate a 100.0-ml. aliquot of the other saturated lead iodide solution in exactly the same way. Finally wash the cell very thoroughly and leave it immersed in distilled water.

Plot the titration curves and calculate the concentrations of the saturated lead iodide solutions. Compare the resulting values for the solubility product of lead iodide with those secured in preceding experiments.

Explain the shapes of the curves obtained, especially any abnormal behavior of the portion of the curve preceding the end point.

Predict the shape of the curve which would have been obtained if silver sulfate had been used as the reagent. Lead chromate is considerably less soluble than lead oxalate; why might it have been dangerous to use sodium chromate in this experiment instead of sodium oxalate?

Describe a way in which the solubility product of lead iodide could be calculated from the results of a conductometric titration of potassium iodide with lead nitrate. What accuracy would you expect to be able to obtain from such a procedure?

EXPERIMENT I-8

Determination of the Solubility of Lead Iodide by Direct Polarography

Deaerate exactly 50 ml. of a solution containing 1 F acetic acid and 1 F sodium acetate in the solution compartment of an H cell, insert the dropping electrode, and adjust and check the apparatus according to the directions given in the fourth part of Expt. VI.

Adjust the "bridge" H (Expt. VI, Fig. 1) to apply about 0.9 v. across the cell circuit. Make sure that the d.e. is the *negative* electrode. Measure $E_{d.e.}$ and E_R. Repeat the measurements after adjusting the bridge to deliver 1.0 v., 1.1 v., etc., until E_R has begun to increase rapidly because of the onset of hydrogen evolution. This will occur at about $E_{d.e.} = -1.3$ v.

Then adjust the bridge to deliver about 0.8 v. and carry out the measurements at successively less negative values of $E_{d.e.}$. Consult Expt. III-8 for a description of the procedures to be adopted for the measurement of negative currents and positive potentials. Continue the measurements at more and more positive values of $E_{d.e.}$ until the (negative) current has begun to increase rapidly because of the oxidation of mercury. This will be at about $E_{d.e.} = +0.3$ v. (What effect will the presence of chloride ion in the agar bridge have on this value?)

After the entire residual current curve has been thus secured, add exactly 1.00 ml. of an 0.0200 F lead nitrate solution. Bubble nitrogen through the solution for 3 to 5 min., then divert the gas stream over the surface of the solution, remove any gas bubbles from the tip of the capillary, and repeat the measurements. Pay special attention to the steeply rising portion of the wave.

Add, in the order given, 2.00 ml. of 0.1 per cent Triton X-100 and two more 1.00-ml. portions of the 0.0200 F lead nitrate. Secure the polarogram of the solution after each addition (and deaeration).

Finally add two 1.00-ml. portions of 0.0200 F potassium iodide, deaerating the solution and securing the polarogram of the mixture (from about +0.3 to −0.3 v.) after each addition. Then discard the solution and thoroughly rinse both the cell and the capillary.

Place exactly 50 ml. of one of the saturated lead iodide solutions in a 100-ml. volumetric flask, add 13.6 g. (0.10 mole) of sodium acetate trihydrate, 6 ml. (0.1 mole) of glacial acetic acid, and 4.0 ml. of 0.1 per cent Triton X-100, and dilute to the mark with water. Shake until the sodium acetate has completely dissolved.

Rinse the solution compartment of the cell with one or two small portions of this solution, then pour about 50 ml. of the solution into the cell, deaerate it, and secure its polarogram. Repeat with a solution prepared in the same way from the other saturated lead iodide solution. *The height of the column of mercury above the capillary during these measurements must be exactly the same as in the measurements with the known lead and iodide solutions.* (Why?)

At the completion of the experiment refer to the last paragraph of Expt. VI.

Plot the residual current curve and the polarograms obtained with the stock lead and iodide solutions. What effects does the addition of the maximum suppressor have on the height and half-wave potential of the lead wave? What effect does the addition of a maximum suppressor have on the anodic iodide wave? [*Cf.* I. M. Kolthoff and C. S. Miller, *J. Am. Chem. Soc.*, **63,** 1405 (1941).]

Calculate the concentration of lead ion in each known solution, measure the diffusion current ($i_l - i_r$) of each lead wave, and calculate the value of i_d/C. Measure the diffusion current of the lead wave secured from each of the solutions prepared from the saturated lead iodide, and from these and the mean value of i_d/C for lead ion calculate the concentration of dissolved lead in a saturated lead iodide solution.

Repeat these calculations, using the heights of the iodide waves, to secure the concentration of iodide in a saturated lead iodide solution. Compute and report the solubility product of lead iodide.

What is the average value of $E_{1/2}$ for the lead waves? Why does this differ from the half-wave potential of lead ion, -0.397 v. *vs.* S.C.E., in a 1 F potassium nitrate supporting electrolyte? Does this experiment give a value for $[Pb^{++}]$ in the saturated lead iodide solution? If not, to what does the concentration you have calculated actually correspond? If S moles of lead iodide is dissolved in a liter of the saturated solution, and if this solution contains Pb^{++}, PbI^+, (undissociated) PbI_2, PbI_3^-, and $PbI_4^=$, write an equation for the total concentration of dissolved lead in such a solution. Write another equation for the total concentration of dissolved iodide. How are these concentrations related? Show how the solubility product of lead iodide, $[Pb^{++}][I^-]^2$, could be calculated from these equations together with any other necessary information.

Assuming that the accuracy and precision of the students' potentiometer are ± 0.5 mv., what is the smallest diffusion current that could certainly be detected with this apparatus? To what concentration of lead does this correspond under the conditions of this experiment? How could the circuit be modified to allow smaller currents to be detected? Is this a practical method of increasing the sensitivity of polarographic analysis?

What electrode reactions are responsible for the two waves? Are they reversible? How could you confirm your answers to these questions by further experiments?

The literature values for the half-wave potentials of these waves (in similar but not quite identical media) are -0.476 v. for lead in 1 F acetic acid–1 F sodium acetate–0.001 per cent gelatin, and -0.03 v. for 1 mM iodide (why does the iodide concentration have to be specified?) in 0.1 F potassium nitrate. How well do your values compare with these? If the diffusion current constant of lead in this supporting electrolyte is 2.9, what is the value of $m^{2/3}t^{1/6}$ for your capillary? (To what value of $E_{d.e.}$ does this refer?) Why is the diffusion current constant of lead in this medium smaller than in 1 F nitric acid (where $I_{Pb^{++}} = 3.67$)?

EXPERIMENT I-9*a*

Polarographic Study of the Lead Iodide Complexes

Weigh 83.0 g. (0.50 mole) of potassium iodide into a 100-ml. volumetric flask, add 4.0 ml. of 0.1 per cent Triton X-100 and 1.00 ml. of 0.100 *F* lead nitrate, then slowly add water until all of the potassium iodide has dissolved and the solution is just diluted to the mark. Both because the solution will be very nearly saturated and because potassium iodide has a large negative heat of solution, it is advisable to add only enough water to give a volume of about 95 ml. and then, after the solution has stood for a day with occasional shaking, to make the final adjustment of volume after most or all of the iodide has dissolved and the solution has returned to room temperature.

At the same time prepare another solution containing the same concentrations of lead nitrate and Triton X-100, but with 40.0 g. (0.50 mole) of ammonium nitrate instead of the potassium iodide. Let this solution also stand for a day to regain room temperature before making the final adjustment of volume.

Transfer exactly 50 ml. of the nitrate solution to an H cell, deaerate it, and obtain its polarogram. Then add successively 1.00, 1.00, 3.00, 5.00, and 15.0 ml. of the iodide solution, deaerating the mixture and repeating the measurements after each addition. It is not necessary to secure more than two or three points on either the residual current portion of the polarogram or the plateau, but care should be taken to secure enough points on the rising portion of the wave to permit evaluation of the slope of a plot of $E_{\text{d.e.}}$ *vs.* $\log i/(i_d - i)$.

Discard this solution, rinse the cell thoroughly, and pipet exactly 50 ml. of the iodide solution into the cell. Obtain its polarogram and also the polarograms of the solutions prepared by the successive addition of 1.00, 1.00, 3.00, 5.00, 15.0, and 20.0 ml. of the nitrate solution. (Consult the instructor for directions concerning the possible omission of any of these 13 polarograms.)

Discard the solution, rinse both the cell and the capillary thoroughly, and put the apparatus away according to the directions given in the last paragraph of Expt. VI.

Calculate the concentration of iodide present in each solution and measure the half-wave potential of each wave. Construct a plot of $E_{1/2}$ against the logarithm of the iodide concentration.

Are the waves reversible? What is the value of n for the electrode reaction? What would you expect to be the slopes of a plot of $E_{1/2}$ *vs.* $\log [I^-]$ for various values of p in the general formula, PbI_p^{2-p}, for the lead iodide complex? Is your plot a straight line? What conclusions do you draw from this?

The ammonium nitrate is used in this experiment to maintain a constant ionic strength. Look up the values of λ^0 for the ammonium, potassium, nitrate, and iodide ions. How do you think the liquid-junction potential at the boundary $KI(c)$, $NH_4NO_3(5\text{-}c)\|KCl(4.2\ F)$ changes with c? Why is ammonium nitrate preferable to sodium nitrate? Why could potassium nitrate not be used?

Interpolate on your graph to find the value of $E_{1/2}$ in a solution containing 1 F potassium iodide. Is it possible to use this value to predict the solubility of lead iodide in a solution containing 1 M iodide ion? From the slope of the curve at $\log [I^-] = 0$, which complex appears to predominate?

If the instructor so directs, consult *J. Am. Chem. Soc.*, **73,** 5321, 5323 (1951) for a detailed description of the method used for the treatment of data such as those secured here.

EXPERIMENT I-9*b*

The Formula and Dissociation Constant of the Lead(II) Hydroxide Complex

Place 1.00 ml. of 0.100 F lead nitrate in each of three 100-ml. volumetric flasks. To each add 4.0 ml. of 0.1 per cent Triton X-100, then dilute one to the mark with a 1 F sodium hydroxide solution whose concentration is known to within ± 2 per cent. Add 30.0 ml. of the sodium hydroxide to another solution and 10.0 ml. to the third, and dilute the solutions to the mark with water.

Transfer about 50 ml. of each solution in turn to a polarographic cell, and obtain the polarogram of each from about $E_{\text{d.e.}} = -0.3$ to $E_{\text{d.e.}} = -1.2$ v. *vs.* S.C.E. Use the same pressure of mercury above the capillary as was used in Expt. I-8. Discard the solutions, and follow the instructions given in the last paragraph of Expt. VI.

Measure the half-wave potential of each wave. Are the waves reversible (taking into account the fact that the only known oxidation states of lead in aqueous solutions are +4, +2, and 0)? Construct a plot of $E_{1/2}$ *vs.* $\log [OH^-]$. Is this a straight line? What conclusion do you draw from this regarding the number of complex ions present in significant amounts?

Measure the slope of this line, and with the aid of Eq. (6-16) find the value of p in the half-reaction

$$Pb(OH)_p^{2-p} + x\text{Hg} + 2e = Pb(Hg)_x + pOH^-$$

What is the formula of the complex ion which predominates in these solutions?

Extrapolate the line to log [OH⁻] = 0 and find the value of $E_{1/2}$ at that point. Note that this is very nearly the formal potential (*vs.* S.C.E.) of the couple shown in the preceding paragraph. Taking $E_{1/2}$ for simple lead ion in 1 F potassium nitrate as -0.397 v. *vs.* S.C.E., calculate the dissociation constant of the hydroxy complex. Compare the result with that obtained in Expt. I-5*b*. Which value of this constant is likely to be more accurate, and why?

How does the change in the liquid-junction potential affect the slope of your plot of $E_{1/2}$ *vs.* log [OH⁻]? Use the values given in Table 3-1 to correct the measured values of $E_{1/2}$ in 0.1 and 1 F sodium hydroxide. Does this bring the difference between them into better agreement with the prediction of Eq. (6-16)?

Compare the heights of the waves secured in this experiment with those obtained in Expt. I-8 for equal concentrations of dissolved lead. What is responsible for the difference? Using the value of $m^{2/3}t^{1/6}$ computed in the last paragraph of Expt. I-8, together with the values of i_d and C obtained here, calculate the diffusion current constant of lead in 1 F sodium hydroxide. Compare this with the literature value, 3.40.

EXPERIMENT I-10

Analyses of Saturated Lead Iodide Solutions by Amperometric Titrations

With a volumetric pipet transfer exactly 50 ml. of one of the saturated solutions of lead iodide to a polarographic cell, and add 2.0 ml. of 0.1 per cent Triton X-100 and 5.0 g. (0.05 mole) of potassium nitrate. Bubble nitrogen through the solution for about 5 min., then divert the nitrogen stream over the surface of the solution, insert the dropping electrode, and adjust the height of the mercury reservoir to give a drop time of about 3 sec.

Adjust the potential of the dropping electrode to approximately -1.0 v. *vs.* S.C.E. and measure the limiting current at this potential. Add 0.50 ml. of 0.005 F potassium dichromate from a 10-ml. microburet, bubble nitrogen through the solution for about a minute, then measure the limiting current again. Continue adding 0.5-ml. aliquots of the dichromate until a total of 10 ml. has been added.

Discard the solution, wash the cell and capillary thoroughly with dilute (0.5 to 1 F) nitric acid followed by water, and repeat the titration with a portion of the other saturated lead iodide solution. If the instructor so directs, carry out this titration at $E_{\text{d.e.}} = -0.2$ v.

Again clean the cell and pipet 50 ml. of one of the lead iodide solutions into it. Add 5 ml. of 72 per cent perchloric acid and 2.0 ml. of 0.1 per cent Triton X-100, deaerate the solution, and titrate at $E_{\text{d.e.}} = -0.7$ v. with 0.01 F mercurous nitrate (dissolved in 1 F perchloric acid) as the reagent. Repeat with 50 ml. of the other saturated lead iodide solution. At the end of the experiment refer to the last paragraph of Expt. VI.

For each of the titrations plot the limiting current, corrected for dilution, against the volume of reagent. Locate the end points and compute the concentrations of lead or iodide ion in the samples and the solubility product of lead iodide.

Explain the shape of each curve. What effects would be observed if the supporting electrolyte were omitted? If the maximum suppressor were omitted? How do the results of these titrations compare with the results of the titrations of Expt. I-7? In the titration of the lead, do your data indicate any interference from a reaction between dichromate and iodide? If so, how could this interference be eliminated?

EXPERIMENT I-11

Electrogravimetric Determination of Lead by Electrolysis at Constant Current

Weigh two 0.2-g. portions of the air-dried lead iodide prepared in Expt. I-1 into 250-ml. beakers, add 20 ml. of concentrated nitric acid, and evaporate nearly to dryness on a hot plate or steam bath. Repeat the evaporation with successive 20-ml. portions of nitric acid until iodine is no longer evolved and the solution is perfectly clear and colorless. Dissolve the moist residue from the last evaporation in a barely sufficient volume of water, add 5 ml. of 72 per cent perchloric acid,[1] and cautiously evaporate over a low flame to copious fumes of perchloric acid.

Cool the residue and take it up in 100 ml. of 4 F nitric acid. Immerse a platinum cathode and anode in the solution (the latter should have been previously cleaned with 6 F nitric acid, then ignited in the *oxidizing* cone of a good Meker flame and weighed), and adjust them so that they are as nearly concentric as possible. Connect the electrodes to the terminals of a d-c power supply, making sure that the tared anode is connected

[1] Perchloric acid is such a valuable reagent that we could hardly do without it in the chemical laboratory, and yet in the hands of an ignorant or careless worker it is a very hazardous substance indeed. Perchloric acid solutions which are *either* cold *or* dilute (or both) are not at all dangerous. But a hot concentrated solution of the acid is an extremely rapid and powerful oxidizing agent. If such a solution contains any organic matter *or any other oxidizable material* the resulting explosion may be exceedingly destructive. It is therefore absolutely essential to make perfectly sure that all oxidizable matter is completely oxidized before the appearance of perchloric acid fumes. The fuming itself must be done in a good hood provided with a sturdy safety shield, and the chemist must be provided with safety goggles. The fumes must not be allowed to come in contact with any organic vapors, wood, or any other readily oxidizable material. When the evaporation has proceeded to copious fumes of the acid, remove the flame and let the solution cool while it remains in the hood.

to the positive terminal. Not more than about three-quarters of the anode should be covered by the solution. Cover the beaker with a split watch glass if one is available.

Start the stirrer, if one is provided with the apparatus, and set the rheostat R to zero. Close the electrolysis circuit and read the current and the applied voltage, both of which should be zero. Increase the applied voltage in 0.2-v. increments up to about 1 v., then in 0.1-v. increments up to 2.5 v., measuring the current at each voltage. Then reset the applied voltage to a value at which the current is approximately 0.5 amp., and let the electrolysis proceed overnight.

At the end of this time measure the electrolysis current flowing through the cell, then stop the electrolysis by lowering the beaker away from the electrodes without interrupting the electrical circuit, meanwhile rinsing the anode with a *gentle* stream of water. Dry the anode at 125°C. for about 30 min., cool it in a desiccator, and weigh. The lead dioxide thus deposited is not quite pure, and the generally accepted figure for its lead content is 86.4 per cent rather than the theoretical value of 86.6 per cent.[1]

Dissolve the lead dioxide by immersing the anodes in warm 4 F nitric acid containing a little hydrogen peroxide, then rinse them with water and return them to the instructor.

Plot the current–voltage curve and find the decomposition potentials with respect to lead dioxide deposition and oxygen evolution. Compare these with the expected values, assuming that the overpotentials for the deposition of hydrogen and oxygen on smooth platinum are 0.1 and 0.4 v., respectively.[2] Explain any difference between your curve and Fig. 7-2*a*.

From the mean of the initial and final electrolysis currents at the applied voltage used in the electrolysis calculate the average current efficiency for the oxidation of lead ion. What would this have been if the electrolysis had been allowed to proceed for four hours more? What would it have been if the electrolysis had been carried out at an applied voltage just smaller than the decomposition potential with respect to oxygen evolution? Sketch the current–time curve you would expect in an electrolysis at the latter voltage. What are the significances of the slopes of the portions of the current–voltage curve at voltages lower and higher than the decomposition potential with respect to oxygen evolution?

Calculate the percentage of lead in the lead iodide and compare the results of this experiment with the data secured in Expt. I-2.

[1] *Cf.* H. J. Emeléus and J. S. Anderson, "Modern Aspects of Inorganic Chemistry," D. Van Nostrand Company, Inc., Princeton, N.J., 1938, pp. 465–466.

[2] The overpotential for oxygen evolution on an electrode coated with lead dioxide is unknown.

Why is it necessary to destroy the iodide before beginning the electrolysis? Why does lead not deposit on the cathode during this experiment? How would you proceed to secure a quantitative deposition of metallic lead on a solid cathode?

EXPERIMENT I-12

Determination of Lead by Coulometry at Controlled Potential

Accurately weigh 0.45 to 0.55 g. of solid lead iodide into a 125-ml. erlenmeyer flask, add 20 ml. of dilute (3 F) sulfuric acid and 5 ml. of concentrated nitric acid, and evaporate to copious fumes of sulfur trioxide, taking care to avoid spattering in the final stages of the evaporation. Cool to room temperature and *cautiously* add 10 ml. of water, followed by *one drop* of methyl red indicator, enough 6 F sodium hydroxide to just change the color of the indicator to yellow, and 20 ml. of 6 F sodium hydroxide in excess. The solution should be completely clear and free from any suspended particles of undissolved lead sulfate.

Prepare 100 ml. of 3 F sodium hydroxide and use this for filling the central and auxiliary electrode compartments of a cell for controlled-potential electrolysis. Transfer the lead solution quantitatively to the working electrode compartment of the cell, using as little water as possible for rinsing out the flask. Proceed as directed in Expt. VII, keeping the potential of the working electrode constant at -1.0 v. *vs.* S.C.E.

When the electrolysis has been completed, open the stopcock of the coulometer buret and raise the level of the electrolyte nearly to the top of the graduated portion of the buret. (Be sure to measure and record the final buret reading before doing this!) Immediately close the stopcock. Reset the potentiometer to read 0.50 v., interchange the connections from the storage batteries to the switch S' and the rheostat Rh, and, after reading the coulometer buret again, restart the electrolysis by closing S' and S''. During this controlled-potential electrolysis with a lead amalgam anode as the working electrode, the metallic lead deposited during the previous electrolysis will be reoxidized to the $+2$ state and will reenter the solution. This electrolysis will proceed much more rapidly than the first one. (Why?)

At the end of the second electrolysis empty the cell and leave it half filled with distilled water.

Look up the half-wave potential of lead in a sodium hydroxide supporting electrolyte, and use it to explain the selection of control potentials in this experiment. What error, if any, would have been caused by carrying out the first electrolysis at a working electrode potential of -0.80 v. *vs.* S.C.E.? Why were the connections from the cell to the potentiometer not reversed before beginning the second electrolysis?

Calculate the quantity of electricity consumed in each electrolysis, and suggest an explanation for any difference between the two values. What would have been the effect of the formation of a trace of iodate during the preparation of the sample?

Compute the percentage of lead in the lead iodide, and compare the result with the values secured in Expts. I-2*b* and I-11.

Outline a method by which pure thallous and lead chlorides could be quantitatively separated from a mixture of the two, using controlled-potential electrolyses at mercury working electrodes.

EXPERIMENT I-13

Determination of Iodide by Coulometry at Controlled Current

Transfer about 250 ml. of 2 F hydrochloric acid to the titration cell constructed in Expt. VIII. Insert the reference electrode and a helical platinum wire indicator electrode, and turn on the magnetic stirrer. Turn on the electrolysis circuit by means of switch S (Fig. 7-12) so that the current flows through R_3 instead of through the cell, and wait about 15 min. so that the resistors will come to thermal equilibrium. R_1 should be set at about 20000 ohms to give a current of about 4.5 ma. Make sure that the working generator electrode is the anode by checking the polarity of the connections to the cell.

Meanwhile standardize the potentiometer against the standard cell, connect it across the standard resistor R_2, and balance it approximately. This will save time during the exact measurement of the voltage drop across R_2 while the electrolysis is taking place.

At the end of the 15-min. warm-up period add 6 g. (0.04 mole) of sodium bromide dihydrate and exactly 10 ml. of one of the saturated lead iodide solutions. When the sodium bromide has completely dissolved, measure the potential of the platinum indicator electrode. Set the stop clock to zero and throw switch S in the opposite direction so that the electrolysis current passes through the cell.

Immediately measure the voltage drop across the standard resistor R_2, working as quickly as possible. At the end of about 100 sec. reverse switch S, record the reading of the stop clock, and again measure the potential of the indicator electrode. Reconnect the electrolysis circuit and wait about 50 sec. before disconnecting it and measuring the potential of the indicator electrode again. When this has changed to about 0.5 v. *vs.* the reference electrode being used, decrease the generation interval between successive potential measurements to 20 sec. Be sure that a steady value of the potential is obtained at each point.

The equivalence point occurs at a potential of about 0.6 v. When you are quite certain that it has been passed, measure the voltage drop across

R_2 again with the electrolysis current flowing through the cell. Then discard the solution, rinse the cell thoroughly, and repeat the "titration" with a 10-ml. portion of the other lead iodide solution. If the current is allowed to flow through R_3 between the "titrations" it will be unnecessary to allow a second warm-up period.

Plot the "titration" curves (potential of the indicator electrode *vs.* generation time) and locate the end points of the reactions. Compute the quantity of electricity required to reach each end point, calculating the current from the mean of the initial and final voltage drops across R_2, which should be the same within ± 0.1 per cent. The net reaction which occurs at the working generator anode is $I^- + 2Br^- = IBr_2^- + 2e$; beyond the equivalence point, excess free bromine is produced by the reaction $3Br^- = Br_3^- + 2e$. Calculate the concentration of iodide in each of the saturated lead iodide solutions, and compute the solubility product of lead iodide. What do you think is the probable error of this value? Did you see any evidence for the formation of I_3^- (or I_2Br^-) at any point during the titration?

What would have happened if the generator cathode had not been isolated from the solution? If the sodium bromide had not been added?

Outline a procedure for the potentiometric titration of iodide with a standard solution of a suitable reagent, employing the same net chemical reaction that occurred at the working generator anode in this experiment. Wooster, Farrington, and Swift[1] were able to determine 10 to 50 micrograms of iodide coulometrically with an average error of ± 2 per cent. Would this be feasible by your procedure?

EXPERIMENT I-14

Determination of Lead by Visual Colorimetry

Transfer exactly 1 ml. of each of the saturated lead iodide solutions to 100-ml. volumetric flasks, and in a third flask place exactly 5 ml. of a standard 0.001 *F* lead nitrate solution. To each flask add 10 ml. of a lead-free 1 *F* solution of potassium sodium tartrate which has been made slightly ammoniacal,[2] and dilute to the mark. Transfer exactly 25 ml.

[1] W. S. Wooster, P. S. Farrington, and E. H. Swift, *Anal. Chem.*, **21**, 1457 (1949).

[2] Dissolve 28 g. of reagent grade potassium sodium tartrate tetrahydrate in 50 ml. of water, add a little ammonia, then 1 ml. of 2 per cent sodium sulfide solution. Let settle for a few days and filter through a fine quantitative filter paper. Acidify the filtrate slightly with hydrochloric acid, boil to remove hydrogen sulfide, add a small excess of ammonia and dilute to 100 ml.

Alternatively, dissolve 28 g. of the salt in 50 ml. of water, transfer to the working electrode compartment of a controlled-potential electrolysis cell, and electrolyze for several hours at a mercury cathode whose potential is kept constant at -1.0 ± 0.2 v. *vs.* S.C.E. Drain off the solution without interrupting the electrolysis current, add a little ammonia, and dilute to 100 ml.

(or some other suitable exactly known volume) of the known lead solution to each of two clean dry colorimeter cups and place these in position on the instrument.

Raise the cells as far as possible; then turn on the lamp (if one is used) and adjust the mirror so that the two halves of the field of view are brightly and equally illuminated. Read the two verniers which give the distances between the bottoms of the plungers and the bottoms of the cells (*i.e.*, the lengths of the columns of solution); these should both be 0.0 (mm.) if the instrument is properly adjusted. Consult the instructor if this is not the case.

Now add exactly 1 ml. of 2 per cent sodium sulfide solution (freshly prepared from clear colorless crystals of sodium sulfide) to the solution in each cell. Mix by gently raising and lowering the cells; do not agitate any more than necessary to make the solutions homogeneous, for fear of flocculating the lead sulfide.

Lower the right-hand cell until its vernier reads 20.0 mm., then adjust the left-hand cell until the two halves of the field of view appear indistinguishable. Repeat three times, alternately raising and lowering the left-hand cell toward the point of balance. Then set the right-hand cell at 40.0 and repeat.

Discard the solution in the left-hand cell and rinse the cell thoroughly. Dry it with a roll of lens tissue or filter paper, and pipet in the same volume of one of the prepared lead iodide solutions as was used of the known lead solution. Add exactly 1 ml. of 2 per cent sodium sulfide and mix as above. Leaving the right-hand vernier set at 40.0, balance the instrument four times according to the directions in the preceding paragraph. Then reset the right-hand vernier to 20.0 and repeat.

Treat the other prepared lead iodide solution in exactly the same way.

Calculate the concentrations of the saturated lead iodide solutions, consulting Sec. 8-4 for a discussion of the method of treating the raw data. From these values compute the solubility product of lead iodide, and compare the result with the values obtained in preceding experiments.

What is the apparent precision of the measurements? What is the chief source of error in the final value? How would you modify the experiment to avoid the necessity of removing the lead present as an impurity in the tartrate solution? Lead sulfide is virtually insoluble in 0.1 F hydrochloric acid; why could this not be used instead of the tartrate?

What other metals can be determined by "colorimetric analyses"[1] of

[1] This procedure is not a true colorimetric analysis. At least some of the light entering the suspension is reflected or scattered by the particles of lead sulfide, rather than being absorbed, although these losses are much less significant here than in similar measurements with, say, white precipitates. Even with lead sulfide, however, there is a definitely detectable effect of variations in the size of the colloidal particles.

sulfide sols? How would you proceed to analyze a solution containing traces of both cupric and lead ions by such a procedure? How could the interference of copper be eliminated without resorting to a separation if only the amount of lead were of interest?

EXPERIMENT I-15

Spectrophotometric Determination of Lead

Prepare a 1×10^{-4} *M* solution of lead ion by diluting 1.00 ml. of standard 0.1 *F* lead nitrate to a liter with 0.05 *F* perchloric acid. Transfer 1.00 ml. of this solution to a clean dry glass-stoppered separatory funnel, 2.00 ml. to another, 3.00 ml. to a third, and 4.00 ml. to a fourth. Into each of two additional funnels measure 5.00 ml. of a solution prepared by diluting 5.00 ml. of each of the saturated lead iodide solutions to 100 ml. Use a seventh funnel to prepare a "blank" solution, containing no added lead, for use as a reference solution in the spectrophotometric measurements.

Add 1.00 ml. of a 20 per cent solution of hydroxylamine hydrochloride to each funnel, followed by 10.0 ml. of a solution prepared by mixing 500 ml. of water with 150 ml. of concentrated ammonia, adding 20 g. of reagent grade potassium cyanide, and diluting to one liter. Then add exactly 25 ml. of an 0.001 per cent solution (10 mg. per liter) of dithizone

$$
S{=}C\begin{matrix} \diagup NH{-}NH{-}C_6H_5 \\ \diagdown N{=}N{-}C_6H_5 \end{matrix}
$$

in reagent grade chloroform, stopper the funnel tightly, and shake for one minute. Let stand until the phases separate, and note the colors of the chloroform layers. If any of them is red,[1] add another 5.00 ml. of the dithizone solution to each funnel and shake again. Remove the stopper from the funnel containing the "blank" solution and draw off a few drops of the chloroform layer to rinse out the stem of the funnel. Discard this and fill an absorption cell with the remainder of the chloroform solution. Immediately stopper this and transfer it to the cell holder of the spectrophotometer.

Similarly transfer the chloroform layer from one of the funnels containing a known amount of lead to another cell and place this in the cell compartment of the spectrophotometer. Using the "blank" solution to make the 100 per cent transmittance setting at each wavelength in accordance with the operating instructions furnished with the spectro-

[1] Lead dithizonate is red, whereas dithizone itself in chloroform solutions is green.

photometer, measure the absorbance of the lead solution at 10-mμ intervals from 400 to 700 mμ. The greatest absorption by the lead dithizonate occurs at 510 mμ, whereas the excess dithizone absorbs most strongly at 610 mμ. Measure the absorbances of each of the remaining solutions at these two wavelengths.

From the data secured with the four known solutions plot the absorbance against the weight of lead in the sample. Do these plots indicate that Beer's law is obeyed at both wavelengths? If not, suggest a possible reason for the behavior observed. Draw the best curve through the experimental points, and use this to find the weight of lead present in each of the samples prepared from the lead iodide solutions. Calculate the solubility product of lead iodide, and compare the probable reliability of the resulting value with those of the values secured in preceding experiments.

Why was it necessary to dilute the lead iodide solutions before taking aliquots for the analysis? What was the purpose of adding the hydroxylamine hydrochloride?[1] What would have been the effect of a trace of zinc present as an impurity in the lead iodide, and why? In the light of your answer to the preceding question, explain the statement that the presence of tartrate or citrate ion in the sample usually does not interfere with the extraction of a metal dithizonate.

An excellent discussion of the theory and practice of the use of dithizone in the spectrophotometric determinations of various metals is given by E. B. Sandell, "Colorimetric Determinations of Traces of Metals," 2d ed., Interscience Publishers, Inc., New York, 1950, pp. 87–120.

EXPERIMENT I-16

Spectrophotometric Titration of Lead with Ethylenediaminetetraacetate

Assemble the apparatus for spectrophotometric titrations according to the instructor's directions. An ultraviolet spectrophotometer is required in this experiment.

Pipet exactly 5 ml.† of one of the saturated lead iodide solutions into a quartz or Vycor titration cell and add 50 ml. of water and 1 ml. of 0.1 *F* perchloric acid. Measure the absorbance of the mixture at a wavelength of 240 mμ; then add 0.50 ml. of standard 0.005 *F* disodium dihydrogen

[1] In principle it would be better to decompose the lead dithizonate present in the first extract from each of the lead iodide samples by shaking the entire chloroform layer with about 10 ml. of very dilute (0.01 *F*) hydrochloric acid, then reextracting the lead according to the procedure described above. Consult the instructor for directions on this point.

† It is assumed that the titration cell has a capacity of about 100 ml. If this is not the case, the volumes used must be adjusted accordingly.

ethylenediaminetetraacetate from a 10-ml. microburet, stir briefly, and measure the absorbance again. Continue the titration until a definite excess of the reagent has been added, then discard the solution and repeat the titration with an aliquot of the other lead iodide solution.

Plot the titration curves and locate the end points. Is it necessary to apply corrections for dilution? Calculate the concentrations of lead ion in the saturated lead iodide solutions and the solubility product of lead iodide. Compare your results with the values secured in preceding experiments, particularly Expts. I-7 and I-10.

What is the structure of the lead ethylenediaminetetraacetate complex? Why does this titration give a much sharper end point than would the titration of lead with, say, acetate ion, even assuming that a wavelength could be found at which a lead acetate complex would absorb light whereas the aquo-lead ion would not? (*Cf.* A. E. Martell and M. Calvin, "Chemistry of the Metal Chelate Compounds," Prentice-Hall, Inc., Englewood Cliffs, N.J., 1952, pp. 473–479.)

Outline a direct spectrophotometric method for the determination of lead as the ethylenediaminetetraacetate complex, and discuss its advantages and disadvantages relative to the procedure employed in this experiment.

EXPERIMENT II-1

Recrystallization of Thallous Chloride

To about 1500 ml. of water add 30 g. (0.126 mole) of commercially available, purified grade thallous chloride. Heat to boiling and filter while hot through a sintered-glass crucible or Buchner funnel with suction, collecting the filtrate in a 2-l. suction flask. Set aside for a day or two to permit crystallization and digestion, then filter the product on a clean sintered-glass crucible and wash it several times with small portions of distilled water. Dry at 110°C. for one hour.

EXPERIMENT II-2

Analysis of Thallous Chloride

A. Determination of Thallium. Weigh out several 0.25-g. samples of recrystallized thallous chloride into 250-ml. beakers and dissolve in 10 ml. of 6 *F* nitric acid with heating. Dilute to 50 ml. with distilled water, neutralize with 1 *F* potassium hydroxide, add 25 ml. in excess and then 25 ml. of 0.25 *F* potassium ferricyanide. Allow to digest for 18 hours, filter through a sintered-glass crucible, and wash the precipitate with hot water. Dry for 1 hour (no longer) at 200°C., and weigh as Tl_2O_3.

B. Determination of Chloride. Weigh out two 0.5-g. samples of recrystallized thallous chloride into 400-ml. beakers. Add 200 ml. of distilled water, heat to boiling, and stir until dissolved, breaking up any large clumps of the thallous chloride with the end of the stirring rod. When the salt has completely dissolved, add 5 ml. of 6 *F* nitric acid and 10 ml. of 0.3 *F* silver nitrate. Heat the mixture nearly but not quite to boiling, and let it stand in the dark overnight. The supernatant liquid should then be perfectly clear. Filter through a medium-porosity sintered-glass crucible, retaining as much as possible of the precipitate in the beaker, and wash with 0.1 *F* nitric acid solution by decantation. Redissolve the precipitate of silver chloride in 50 ml. of 1.0 *F* ammonia, added in small portions to ensure complete solution of the precipitate. Then wash the crucible with 50 ml. of 0.1 *F* ammonia added in small portions and combine this with the previous filtrate. Transfer the ammoniacal solution to a 250-ml. beaker, add 2 ml. of 0.3 *F* silver nitrate, and acidify *immediately* with 15 ml. of 6 *F* nitric acid. Make sure that an excess of acid is present. Heat to boiling and let digest as before. Prepare the crucible for filtration by washing out the remaining ammonia wash solution several times with distilled water followed by several small portions of 0.1 *F* nitric acid. Filter the silver chloride suspension and wash it as before with 0.1 *F* nitric acid solution. Dry at 110°C. for one hour and weigh as AgCl.

In preparation for later experiments place about 1900 ml. of conductivity water or of the best deionized or distilled water available in each of two 2000-ml. glass-stoppered Pyrex erlenmeyer flasks, and add about 8 g. of thallous chloride to each. Warm one of the flasks to 30 to 35°C. (not quite all of the salt should dissolve), and then let both of the flasks stand, with occasional shaking, at room temperature or in a water thermostat at 25°C. if one is available. The flasks should be clearly marked so that they may be later differentiated.

EXPERIMENT II-3

Determination of the Solubility Product of Thallous Chloride by Direct Potentiometry with a Silver–Silver Chloride Electrode

Prepare a silver wire indicator electrode by coiling and etching a piece of silver wire according to the directions in Expt. V. Also secure or prepare a silver–silver chloride–saturated potassium chloride reference electrode,[1] a 4 per cent agar–potassium nitrate salt bridge, and 100 ml. of each of a series of potassium chloride solutions containing 0.1, 0.03, 0.01, 0.003, 0.001, and 0.0003 *M* chloride.

[1] The instructor may prefer to use a commercial S.C.E. instead.

Place the indicator electrode in a 150-ml. beaker and pour about 50 ml. of the 0.1 M chloride solution into the beaker. Half fill another 150-ml. beaker with saturated potassium nitrate and connect the solutions in the two beakers with the salt bridge. Place the reference electrode in the beaker containing the potassium nitrate and connect the two electrodes to the potentiometer. (Which of the electrodes will be the positive electrode?)

Add a few milligrams of pure silver chloride to the chloride solution and stir for a few minutes. Meanwhile standardize the potentiometer against the standard cell. Measure the potential of the silver electrode, stir for a few minutes more, and measure the potential again. It is a good idea to turn the potentiometer knob to a value about 0.05 v. away from the first point of balance before beginning to measure the potential for the second time. Repeat, if necessary, until two successive measurements of the potential agree to ± 0.5 mv. or better.

Discard the solution and rinse the beaker and indicator electrode with several small portions of the 0.03 M chloride solution. Then proceed to measure the potential of the silver–silver chloride electrode in 0.03 M chloride, and so on through the remainder of the series of known solutions.

Finally carry out similar measurements with portions of the two saturated thallous chloride solutions prepared in Expt. II-2. Then disassemble the cell and store the indicator and reference electrodes according to the instructor's directions.

For the data secured with the silver–silver chloride electrode plot E_{ind} against log $[Cl^-]$. Does the slope of the best straight line through the experimental points agree with the theoretical slope, 0.05915 v.? Extrapolate the data to $E_{ind} = 0$ v. *vs.* the reference electrode employed and compare the corresponding value of $[Cl^-]$ with the concentration of a saturated potassium chloride solution. From the extrapolated value of E_{ind} at log $[Cl^-] = 0$, together with the literature value of the potential of the silver ion–silver electrode, calculate the solubility product of silver chloride. Take the potential of the silver–silver chloride–saturated potassium chloride reference electrode as $+0.201$ v. *vs.* N.H.E. To how large an error in the extrapolated cell potential does the error in the solubility product correspond?

Which of the experimental points received the greatest weight in these extrapolations? Explain why this procedure cannot be used for the accurate determination of standard potential data. Now plot $E_{ind} + 0.05915 \log [Cl^-]$ against the square root of the ionic strength, and extrapolate this to infinite dilution. Why do the points not fall on a perfectly horizontal straight line?

Use the plot of E_{ind} against log $[Cl^-]$ to find the concentration of chloride ion in each of the saturated thallous chloride solutions. Calculate the solubility product of thallous chloride.

How would you proceed to prepare a silver–silver chloride electrode whose potential would be accurate and reproducible to within ± 0.05 mv.?

EXPERIMENT II-4*a*

Analysis of a Saturated Thallous Chloride Solution by Potentiometric Titration with Silver Nitrate

Using a volumetric pipet, transfer exactly 100 ml. of one of the saturated thallous chloride solutions prepared in Expt. II-2 to a 250-ml. beaker, and add 50 ml. of water. Place a helical silver wire indicator electrode (see Expt. V for instructions) in the solution, and connect this to one of the wires leading from the potentiometric assembly. Also place in the solution the tip of an agar–1 F potassium nitrate salt bridge; immerse the other end of this bridge in a small beaker containing saturated potassium chloride or nitrate, and in this beaker immerse the tip of the silver–silver chloride electrode prepared in Expt. V. Connect the silver wire in the reference electrode to the other wire leading from the potentiometric assembly. Arrange a stirring motor or a magnetic stirrer, if one is available, to keep the solution in slow but constant motion.

Measure the potential of the cell; if the potentiometer cannot be balanced, it may be necessary to reverse the polarity of the wires leading from the cell. From a 50-ml. buret add 1 ml. of standard 0.1 F silver nitrate, stir the solution briefly, and record the potential as soon as it becomes steady. Continue adding measured volumes of the silver nitrate until about a 100 per cent excess has been added. Near the beginning of the titration, where the potential changes only very slowly, a milliliter or two of silver nitrate may be added at a time. Near the end point, however, the silver nitrate should be added in small uniform portions, as shown in Table 3-2.

Discard the solution and repeat the experiment with a 100-ml. portion of the other saturated solution of thallous chloride prepared in Expt. II-2.

Plot the titration curves and locate the end points. Calculate the molarities of the thallous chloride solutions and the solubility product of thallous chloride. From the potentials measured at several points before the end point calculate the standard potential of the cell $Ag|AgCl(s), Cl^-(c)||KCl(s), AgCl(s)|Ag$. From this and the literature value of the potential of the reference electrode, compute the standard potential of the half-reaction $AgCl + e = Ag + Cl^-$. Combine this with the standard potential of the half-reaction $Ag^+ + e = Ag$ ($E^0 = +0.7991$ v.) to secure the solubility product of silver chloride. How do you account for any discrepancy between your value and the literature value?

How well do the concentrations of chloride ion in the two saturated solutions agree? If sufficient time has not elapsed since these solutions were prepared, which of them should contain the higher concentration of dissolved salt?

EXPERIMENT II-4*b*

Potentiometric Titration of Thallous Ion with Potassium Bromate

Prepare 100 ml. of a standard 0.01 *F* potassium bromate solution by weighing out an appropriate amount of the pure dry salt, dissolving it in water, and diluting to volume in a 100-ml. volumetric flask.

Measure about 100 ml. of 3 *F* hydrochloric acid into a 250-ml. beaker and add exactly 50 ml. of one of the saturated thallous chloride solutions prepared in Expt. II-2. Place a helical platinum wire indicator electrode in the solution, and clamp a saturated calomel electrode or the silver–silver chloride electrode prepared in Expt. V in such a position that the tip of the electrode is immersed in the solution but does not interfere with the rotation of a magnetic stirring bar. Connect the indicator electrode to the positive lead from the potentiometer circuit and the reference electrode to the negative lead, and standardize the potentiometer against the standard cell.

Add 2 g. (0.015 mole) of sodium bromide dihydrate and measure the potential of the indicator electrode after the salt has completely dissolved. From a 50-ml. buret add 1 ml. of the standard bromate solution and record the potential as soon as it becomes steady. Continue adding 1- or 2-ml. portions of bromate until the end point seems to be approaching, then add the bromate in small uniform portions until the end point has been definitely passed, and finally add several 1- or 2-ml. portions of bromate in excess.

Repeat the titration with an aliquot of the other thallous chloride solution.

Plot the titration curves and locate the end points. Calculate the molarities of thallous ion in the solutions and the solubility product of thallous chloride. Suggest an explanation for any discrepancy between the mean value of [Tl^+] and the mean value of [Cl^-] found in Expt. II-4*a*. Which of these results do you think is more reliable? What is the most probable value of the solubility product of thallous chloride on the basis of these two experiments?

From the measured potential at each point between the beginning of the titration and the end point calculate the formal potential of the thallic–thallous couple in 2 *F* hydrochloric acid–0.1 *F* sodium bromide, using the method illustrated by Table 3-6. How does the precision of the values secured compare with your estimate of the precision of the measurements of the potentials? To what causes might any difference be due? Compare this formal potential with the standard potential of the half-reaction $Tl^{+++} + 2e = Tl^{+}$. What conclusions do you draw from the difference? Similarly, using the data secured in the presence of excess bromate, find the formal potential of the half-reaction $Br_3^- + 2e = 3Br^-$ and compare this with the standard potential of this half-reaction. (Why does this not give the formal potential of the half-reaction $BrO_3^- + 6H^+ + 6e = Br^- + 3H_2O$? How would you proceed to obtain a value for the formal potential of this half-reaction?)

From the two formal potentials you have calculated, compute the potential of the cell at the equivalence point of the titration. How well does this agree with the interpolated values of the potentials at the end points you have selected? Account for any discrepancy.

The titration curves secured in this experiment will be useful in carrying out Expt. II-13.

EXPERIMENT II-6

Determination of the Solubility of Thallous Chloride by Direct Conductometry

Secure a dip-type conductance cell and a direct-reading a-c operated conductance bridge. If necessary, platinize the electrodes of the cell according to the directions in Secs. 4-3 and 5-13.

Place exactly 500 ml. of water in a clean dry 800-ml. beaker and measure its conductance. Add 1.00 ml. of 0.100 *N* (0.100 *F*) thallous nitrate, stir thoroughly, and measure the conductance again. Add 1.00 ml. more of the thallous nitrate and measure the conductance again. Then add successively 3.00, 5.00, 10.0, 30.0, and 50.0 ml. of the thallous nitrate, measuring the conductance after each addition.

Discard the solution and repeat the measurements, using first 0.1 *F* potassium nitrate and then 0.1 *F* potassium chloride in place of the thallous nitrate. Then measure the conductances of each of the saturated thallous chloride solutions prepared in Expt. II-2.

Finally measure the conductance of a standard potassium chloride solution supplied by the instructor. Then rinse the cell very thoroughly and leave it standing in a clean beaker with the electrodes covered with water.

Consult Expt. I-6 for a discussion of the treatment of the data. Why is the graphical method superior to the analytical one in dealing with solutions as concentrated as saturated thallous chloride? What do you think of the validity for these solutions of the assumption that the equivalent conductances of chloride and nitrate ions are not affected by changing the cation present from potassium to thallium?

EXPERIMENT II-7

Analysis of Saturated Thallous Chloride Solutions by Conductometric Titration

Pipet exactly 50 ml. of one of the saturated solutions of thallous chloride into a beaker of just sufficient size to accommodate a dip-type conductance cell conveniently, and add enough distilled water to cover the electrodes of the cell. Stir thoroughly, measure the conductance of the solution, and titrate with standard 0.1 *F* potassium chromate from a 10-ml. microburet. Measure the conductance after the addition of each 0.5 ml. of reagent.

Discard the solution and repeat the titration with a 50-ml. aliquot of the other saturated thallous chloride solution. Then, in exactly the same way, titrate a 50-ml. aliquot of each solution with standard 0.1 *F* silver nitrate. At the end of the last titration wash the cell very thoroughly and leave it standing with the electrodes immersed in water.

Plot the titration curves, locate the end points, and compute the concentrations of thallous and chloride ions in the solutions. Are these equal within experimental error? If there is a difference between $[Tl^+]$ and $[Cl^-]$, how can the electroneutrality rule be satisfied? What effect must this have had on the conductances of the saturated solutions measured in Expt. II-6? Do the data secured in previous experiments seem to agree with the values of $[Tl^+]$ and $[Cl^-]$ secured here? If not, suggest a possible explanation. Calculate the solubility product of thallous chloride.

Explain the shapes of the curves obtained. Why could the titration of the thallium not be carried out in an acidic solution? What sort of curve would have been secured if lithium chromate had been used instead of potassium chromate? Why is lithium chromate not a practical reagent for the routine determination of thallium by conductometric titration?

EXPERIMENT II-8

Determination of the Solubility of Thallous Chloride by Direct Polarography

Deaerate exactly 50 ml. of 1 *F* potassium nitrate in the solution compartment of an H cell, insert the dropping electrode, and adjust and check the apparatus according to the directions given in the fourth section of Expt. VI.

Adjust the "bridge" *H* (Expt. VI, Fig. 1) to apply about 0.9 v. across the cell circuit. Make sure that the d.e. is the *negative* electrode. Measure $E_{\text{d.e.}}$ and E_R. Repeat the measurements after adjusting the bridge to deliver 1.1 v., 1.3 v., etc., until E_R has begun to increase rapidly because of the onset of hydrogen evolution. This will occur at about $E_{\text{d.e.}} = -1.9$ v.

Then adjust the bridge to deliver about 0.8 v. and carry out the measurements at successively less negative values of $E_{\text{d.e.}}$. Continue the measurements at more and more positive values of $E_{\text{d.e.}}$ until the (negative) current has begun to increase rapidly because of the oxidation of mercury. This will be at about $E_{\text{d.e.}} = +0.3$ v. (What effect will the presence of chloride ion in the agar bridge have on this value?)

After the entire residual current curve has been thus secured, add exactly 2.00 ml. of an 0.0100 F thallous chloride solution. Bubble nitrogen through the solution for 3 to 5 min., then divert the gas stream over the surface of the solution, remove any gas bubbles from the tip of the capillary, and repeat the measurements. Pay special attention to the steeply rising portion of the wave.

Add, in the order given, 2.00 ml. of 0.1 per cent Triton X-100 and two more 2.00-ml. portions of the 0.0100 F thallous chloride. Secure the polarogram of the solution after each addition (and deaeration). Then discard the solution and thoroughly rinse both the cell and the capillary.

Place exactly 10 ml. of one of the saturated thallous chloride solutions in a 100-ml. volumetric flask, add 10.1 g. (0.10 mole) of potassium nitrate and 4.0 ml. of 0.1 per cent Triton X-100, and dilute to the mark with water. Shake until the potassium nitrate has completely dissolved.

Rinse the solution compartment of the cell with one or two small portions of this solution, then pour about 50 ml. of the solution into the cell, deaerate it, and secure its polarogram. Repeat with a solution prepared in the same way from the other saturated thallous chloride solution. *The height of the column of mercury above the capillary during these measurements must be exactly the same as in the measurements with the known thallous solutions.* (Why?)

At the completion of the experiment refer to the last paragraph of Expt. VI.

Plot the residual current curve and the polarograms obtained with the solutions containing known concentrations of thallous ion. What effects does the addition of the maximum suppressor have on the height and half-wave potential of the wave?

Calculate the concentration of thallous ion in each of the solutions, measure the diffusion currents ($i_l - i_r$) of the waves, and calculate the values of i_d/C. Measure the diffusion current of the wave secured from each of the solutions prepared from the saturated thallous chloride, and from these values and the mean value of i_d/C find the concentration of dissolved thallium in a saturated thallous chloride solution.

What is the average value of $E_{1/2}$? Does this differ appreciably from the half-wave potential, -0.480 v. *vs.* S.C.E., of thallous ion in a 1 *F* potassium chloride supporting electrolyte? What conclusions do you draw from this concerning the extent to which thallous ion is complexed by chloride in dilute solutions? Is it safe to take the total concentration of dissolved thallium in the saturated thallous chloride solution as equal to the concentration of Tl^+? How could this question have been investigated by potentiometric methods? Compare the ease of carrying out the polarographic and potentiometric procedures.

Calculate the solubility product of thallous chloride from these data, and compare the resulting value with the values found in preceding experiments.

Why is direct polarography superior to direct potentiometry as an analytical method? Experimental errors aside, would you expect these two techniques to yield identical values for the concentration of a substance like mercuric chloride in a solution? Compare the utilities of direct polarography, direct potentiometry, and direct conductometry for the determination of the solubility of thallous chloride in 1 *F* potassium nitrate.

If the saturated thallous chloride solutions used in this experiment had contained small amounts of thallic chloride ($TlCl_3$), lead chloride, or hydrochloric acid, what would have been the effect on the calculated solubility product? After consulting a table of polarographic half-wave potentials (*e.g.*, in the 37th or any later edition of the "Handbook of Chemistry and Physics"), outline a method by which the concentration of thallic or lead ion in the saturated solution could be determined.

What electrode reaction is responsible for the wave obtained in this experiment? Is it reversible? How could you confirm your answers to these questions by further experiments? Why is the half-wave potential of the second wave on a polarogram of thallic ion identical with the half-wave potential of the thallous ion wave? Why is it impossible to secure an anodic diffusion current corresponding to the reaction $Tl^+ = Tl^{+++} + 2e$ at a dropping mercury electrode, whereas it is possible to secure a cathodic diffusion current corresponding to the reverse of this reaction?

Sketch the polarogram that would have been obtained if, under the conditions of one of your experiments, the dropping mercury electrode had been replaced by a dropping dilute thallium amalgam electrode.

The literature value for $E_{1/2}$ of thallous ion in 1 *F* potassium nitrate is -0.480 v. *vs.* S.C.E. How well does your value agree with this? Would you expect the diffusion current constants ($i_d/Cm^{2/3}t^{1/6}$) of thallous ion in 1 *F* potassium chloride and in 1 *F* potassium nitrate to differ appreciably? Why or why not?

In Expt. I-8 directions are given for the determination of the iodide concentration in a saturated lead iodide solution. What experimental difficulty would have been encountered in an attempt to carry out a similar determination of the chloride concentration in a saturated thallous chloride solution? What modification of the apparatus would be needed to render such a determination feasible?

EXPERIMENT II-9

Polarographic Study of the Complexation of Thallous Ion in Concentrated Chloride Solutions

Measure 10.0 ml. of 0.01 F thallous chloride[1] into each of two 100-ml. volumetric flasks, and to each add 2.0 ml. of 0.1 per cent Triton X-100. Add 8.6 ml. of concentrated (11.6 F) hydrochloric acid to one flask and dilute to the mark with water. This solution will be 1.00 F in hydrochloric acid. Dilute the other solution to the mark with concentrated hydrochloric acid; this solution will be 10.2 F in the acid.

Transfer exactly 25 ml.† of the 1 F acid solution to the cell, deaerate it, insert the capillary, and adjust the height of the mercury reservoir to give a drop time of 3 to 4 sec. Obtain the polarogram of the solution from −0.2 to −1.0 v. *vs.* S.C.E., using intervals of about 0.02 v. on the steeply rising portion of the wave.

Using a volumetric pipet, add exactly 5 ml. of the solution of thallous chloride in 10.2 F acid. Deaerate the mixture and, without changing the height of the mercury reservoir, secure the polarogram of this solution. Next obtain the polarogram of a solution prepared by adding another 5 ml. of the 10.2 F acid solution, and finally obtain the polarogram of the solution prepared by adding 10 ml. more of the 10.2 F acid. These solutions will contain constant concentrations of thallous ion and of the maximum suppressor, but will contain 2.5, 3.6, and 5.1 F hydrochloric acid, respectively.

Discard the final solution and repeat the above measurements, interchanging the 1 M and 10 M acid solutions. In this way you will secure polarograms of thallous ion in 10.2, 8.7, 7.6, and 6.1 F hydrochloric acid. Especially in the two most concentrated solutions, be sure to carry the measurements to potentials which are sufficiently negative to make you feel quite certain that the final current rise due to the reduction of hydrogen ion has begun.

Plot the polarograms obtained. On each curve extrapolate the flat portion of the residual current line preceding the wave, and measure the vertical distance between the extrapolated line and the experimental

[1] A saturated solution of thallous chloride will serve very well.

† Before doing this, make sure that both the capillary tip and the salt bridge will be covered when 25 ml. of solution is placed in the cell. If not, all of the volumes must be adjusted accordingly.

curve at a potential near the middle of the plateau of the wave. This is the diffusion current. Does it vary with the concentration of hydrochloric acid? If so, propose an explanation. (Look up the viscosities of concentrated hydrochloric acid solutions.)

For each polarogram plot $E_{d.e.}$ against log $i/(i_d - i)$. Are the waves reversible? (What assumption are you making about the value of n? What grounds do you have for this assumption?)

Construct a plot of $E_{1/2}$ *vs.* log $[Cl^-]$. What would you expect the slope of such a plot to be if thallous ion forms only a single complex with chloride and if the value of p in the formula of this complex $(TlCl_p^{1-p})$ is 1, 2, 3, or 4? Is your plot a straight line? What conclusion do you draw from this?

How do you expect the liquid-junction potential at the boundary $HCl(c)||KCl(4.2\ F)$ to vary as c is increased from 1 to 10 F? What effect does this have on your values for $E_{1/2}$? How could you modify the experiment to minimize the variation in E_j as the chloride concentration is changed?

Consult the literature to find values of the mean ionic activity coefficient of hydrochloric acid in concentrated solutions. Assuming[1] that $f_{Cl^-} = f_\pm$, plot $E_{1/2}$ against the activity of chloride ion. Is this more informative than the plot based on the concentration of chloride?

If the instructor so directs, consult the literature references listed in the last paragraph of Expt. I-9*a*.

EXPERIMENT II-10*a*

Amperometric Titration of Thallous Ion with Potassium Chromate

Pipet exactly 50 ml. of one of the saturated thallous chloride solutions into a polarographic cell, add 2 ml. of 0.1 per cent Triton X-100 and 5.0 g. (0.01 mole) of potassium nitrate, and deaerate the solution for several minutes. Insert a dropping electrode and adjust the height of the mercury reservoir to give a drop time of about 4 sec. Adjust the potential of the dropping electrode to roughly -1.0 v. *vs.* S.C.E. and, with the gas stream passing over the surface of the solution, measure the limiting current.

Add 0.5-ml. aliquots of standard 0.1 F potassium chromate from a 10-ml. microburet. After each addition deaerate the solution for a minute or two; then measure the limiting current. It is wise to repeat each measurement after a minute or two to ensure that solubility equilibrium has been reached.

[1] Are there any thermodynamic grounds for this assumption? *Cf.* G. P. Haight, Jr., *J. Am. Chem. Soc.*, **75**, 3848 (1953).

Repeat the titration with a 50-ml. aliquot of the other saturated thallous chloride solution. At the end of the second titration refer to the last paragraph of Expt. VI.

Plot the titration curves and calculate the concentrations of thallous ion in the two solutions. Compute the solubility product of thallous chloride.

How do these values of $[Tl^+]$ compare with the values obtained in Expt. II-7? Propose an explanation for any difference. Which technique would be better suited to the determination of thallous ion in a solution containing 1 *F* sodium chloride? Does amperometric titration appear to give results which are more or less reliable than those attainable by direct polarography? How accurately do you think that a potentiometric titration of thallous ion with chromate could be carried out? What could you use as an indicator electrode in such a titration?

Would an amperometric titration with a standard chloride solution be a feasible method of determining thallium?

What would have been the shape of the titration curve obtained if this experiment had been carried out at $E_{d.e.} = 0$ v. *vs.* S.C.E.? What would have been the practical advantage of carrying out the titration at that potential rather than at -1.0 v.?

EXPERIMENT II-10*b*

Amperometric Titration of Thallium with Potassium Ferricyanide

Pipet exactly 25 ml. of saturated thallous chloride into a polarographic cell and add 25 ml. of water and 10 ml. of 6 *F* sodium hydroxide. Secure a rotating platinum microelectrode and a synchronous rotator which will rotate the electrode at exactly 600 r.p.m.[1] Connect the rotating electrode to the positive terminal of the polarograph and adjust its potential to approximately $+0.2$ v. *vs.* a saturated calomel electrode contained in the other half of the H cell.

Titrate the solution with standard 0.1 *F* potassium ferricyanide from a 10-ml. microburet. Measure the current at the start of the titration and after the addition of each 0.5 ml. of ferricyanide. Wait for a minute or two after making each measurement; then check it to make sure that equilibrium has been attained before adding the next portion of ferricyanide.

Repeat the titration with a 25-ml. portion of the other saturated solution of thallous chloride. At the end of the second titration, clean the electrode by washing it thoroughly with dilute (1 *F*) nitric acid to remove any solid thallic oxide, Tl_2O_3. Let the electrode stand in 1 *F* nitric acid.

[1] These are commercially available from E. H. Sargent and Co., Chicago, Ill.

Plot the titration curve and locate the end point. Calculate the concentration of thallous ion in the thallous chloride solutions, and compute the solubility product of thallous chloride.

How do the results of this experiment agree with those of Expt. II-10*a*? On chemical grounds, which experiment seems likely to give the more reliable results?

How could the reaction between thallous ion and ferricyanide in alkaline solutions be employed in a potentiometric titration? Why would you not try to base a conductometric titration on this reaction?

Could thallous ion be determined by direct voltammetry with a rotating platinum microelectrode? How?[1] Might this have any advantages over the use of a dropping mercury electrode in analyzing mixtures of thallium with other elements?

Why is ferricyanide used in this titration instead of some other oxidizing agent?

EXPERIMENT II-12

Analysis of Thallous Chloride by Coulometry at Controlled Potential

Accurately weigh about 1.0 g. of the pure thallous chloride prepared in Expt. II-1 into a 125-ml. erlenmeyer flask; add exactly 1 ml. of a standard 0.1 *F* solution of zinc nitrate or chloride, 10 ml. of water, and 10 ml. of concentrated nitric acid; and evaporate to incipient crystallization. Cool, add 5 ml. of 72 per cent perchloric acid,[2] and evaporate again to copious fumes of the acid. Cool to room temperature; add 25 ml. of water, followed by 6 g. of sodium hydroxide. When this has dissolved (neglect any flocculent precipitate of hydrous thallic oxide), add 1 ml. of 85 per cent hydrazine and let stand for a minute or two until the precipitate redissolves completely.

Meanwhile prepare 100 ml. of 1 *F* sodium hydroxide and use this to fill the central and auxiliary electrode compartments of a cell for controlled-potential electrolysis. Quantitatively transfer the thallium solution to the working electrode compartment, using as little water as possible for rinsing out the flask. Proceed as directed in Expt. VII, carrying out the electrolysis at a working electrode potential of -0.65 ± 0.05 v. *vs.* S.C.E.[3] No harm will be done if the working electrode

[1] *Cf.* P. Delahay and G. L. Stiehl, *J. Am. Chem. Soc.*, **73**, 1755 (1951).

[2] Read the footnote on p. 437.

[3] The half-wave potentials of thallous and lead ions in 1 *F* sodium hydroxide are -0.48 and -0.76 v. *vs.* S.C.E., respectively. Consequently thallium will be reduced at -0.65 v., but lead will not. Advantage is taken of this fact in the following procedure for the estimation of the percentage of lead present in the thallous chloride. At the discretion of the instructor this latter portion of the experiment may be omitted,

potential momentarily becomes more negative than −0.7 v., provided that the potential is kept quite accurately constant at −0.65 v. for at least ten minutes at the end of the electrolysis. (Would this still be true if the reduction of lead were thermodynamically irreversible?)

When the electrolysis current has remained below 1 ma. for ten minutes or more, remove as much as possible of the electrolyzed solution from the cell without interrupting the electrolysis circuit. Transfer a suitable volume of this solution to the dropping electrode compartment of a polarographic H cell, add enough 0.1 per cent Triton X-100 to give a concentration of about 0.004 per cent, bubble nitrogen through it for a minute or two, and obtain its polarogram from about −0.5 to −1.8 v. *vs.* S.C.E.

Meanwhile adjust the leveling bulb of the coulometer so that the gas mixture in the buret is at atmospheric pressure, and record the volume of evolved gas, its temperature, and the barometer reading.

Discard the remaining liquid in the three compartments of the electrolysis cell, reserving the mercury for subsequent repurification.

Recalculate the volume of gas evolved in the coulometer to standard conditions of temperature and pressure, and compute the weight and percentage of thallium in the sample. Compare this result with the theoretical percentage of thallium in pure thallous chloride and also with the value found gravimetrically in Expt. II-2, suggesting possible reasons for any substantial discrepancies.

Plot the polarogram of the electrolyzed solution and measure the diffusion currents of lead and zinc. If the concentration of the standard zinc solution was exactly 0.1 M, and if the total volume of solution electrolyzed was V ml., the concentration of zinc was $100/V$ mM, while the concentration of lead corresponding to w mg. of this element in the sample was $1000w/207.2V$. The diffusion current constants of lead and zinc in 1 F sodium hydroxide are 3.40 and 3.14, respectively, so that

$$i_{d_{\text{Pb}}} = \frac{3.40 \times 10^3 w m^{2/3} t^{1/6}}{207.2V}$$

and

$$i_{d_{\text{Zn}}} = \frac{314 m^{2/3} t^{1/6}}{V}$$

Dividing one of these equations by the other eliminates V; and in addition $m^{2/3}t^{1/6}$ can be considered constant, since its variation is small compared to the relative errors in the measurement of the two diffusion currents, so that it disappears as well. Therefore the weight of lead in the sample can be calculated directly from the known weight of zinc added and the

---

and the thallium and lead may be determined together by measuring the quantity of electricity consumed in an electrolysis at −1.1 ± 0.2 v. *vs.* S.C.E. (On what grounds might one suspect the presence of lead as an impurity in the thallous chloride?)

ratio of the measured diffusion currents. This is a typical "pilot ion" method of analysis.

Report the percentage of lead in the sample. What is the probable uncertainty of this value? Would this percentage of lead have had any effect on the results of preceding experiments?

Why was the addition of hydrazine necessary? What would have happened if it had been omitted? Why could the same results not have been obtained by the controlled-potential electrolysis of a perchloric acid solution of thallous and lead ions? What other methods could be used for the estimation of lead in this sample?

How would you modify the experiment to permit the detection and determination of traces of both lead and zinc in a sample of thallous chloride?

The thallous wave in 1 *F* sodium hydroxide is thermodynamically reversible. What fraction of the thallous ion originally present would be deposited if 50 ml. of such a solution were electrolyzed with a mercury cathode having a volume of 25 ml. at a potential of −0.65 v. *vs.* S.C.E.?

Explain the abnormal shape of any small wave due to unreduced thallous ion remaining in the electrolyzed solution.

EXPERIMENT II-13

Determination of Thallium by Coulometry at Controlled Current

Transfer about 250 ml. of 2 *F* hydrochloric acid to the titration cell constructed in Expt. VIII. Insert the reference electrode and a helical platinum wire indicator electrode, and turn on the magnetic stirrer. Turn on the electrolysis current by means of switch *S* (Fig. 7-12) so that the current flows through R_3 instead of through the cell, and wait about 15 min. for the resistors to come to thermal equilibrium. R_1 should be adjusted to about 10000 ohms. Make sure that the working generator electrode is the anode by checking the polarity of the connections to the cell.

Meanwhile standardize the potentiometer against the standard cell, connect it across the standard resistor R_2, and balance it approximately. This will save time during the exact measurement of the voltage drop across R_2 while the electrolysis is taking place.

At the end of the 15-min. warm-up period add 3.5 g. (0.025 mole) of sodium bromide dihydrate and exactly 5 ml. of one of the saturated thallous chloride solutions. When the sodium bromide has completely dissolved, measure the potential of the platinum indicator electrode. Set the stop clock to zero and throw switch *S* in the opposite direction so that the current passes through the cell.

Immediately measure the voltage drop across the standard resistor R_2, working as quickly as possible. At the end of about 100 sec. reverse switch *S*, record the reading of the stop clock, and again measure the potential of the indicator electrode. Reconnect the electrolysis circuit, allow the reaction to proceed for about 50 sec. more, then again stop the electrolysis and measure the potential of the indicator electrode. When you think the end point is near, decrease the generation interval between successive potential measurements to 15 or 20 sec. Be sure that a steady value of the potential is obtained at each point.

When the end point has been passed, remeasure the voltage drop across R_2 with the electrolysis current flowing through the cell. Then discard the solution, rinse the cell thoroughly, and repeat the "titration" with a 5-ml. aliquot of the other saturated thallous chloride solution. If the current is allowed to flow through R_3 between the "titrations," it will be unnecessary to allow a second warm-up period.

Plot the potential of the indicator electrode against the generation time and locate the end points of the "titrations." Compute the quantity of electricity required to reach each end point, calculating the current from the mean of the initial and final voltage drops across R_2, which should be the same within ± 0.1 per cent. In the reaction which occurs at the generator anode the thallous ion is oxidized to the $+3$ state. Calculate the concentration of thallous ion in each of the saturated thallous chloride solutions, and compute the solubility product of thallous chloride. What do you think is the probable error of this value?

What would have happened if the generator cathode had not been isolated from the solution?

The formal potential of the half-reaction $Tl^{+++} + 2e = Tl^{+}$ is $+1.26$ v. *vs.* N.H.E. in 1 *F* perchloric acid, while the standard potential of the bromine–bromide couple is only $+1.07$ v. In view of these facts, how can you account for the observation that thallous ion is quantitatively oxidized by electrolytically generated bromine under the conditions of this experiment? From your data, what is the formal potential of the thallic–thallous couple in this medium? In the light of the results of Expt. II-9, do you feel that the complexation of thallous ion by bromide is likely to be significant under these conditions? Propose a method of securing information concerning the ionic state of thallium(III) in bromide solutions and estimating the dissociation constants of any complexes formed.

Do you think that the coulometric titration of thallous ion with electrolytically generated ferricyanide might be feasible? How would you suggest carrying out such a titration? In view of the results of Expt. II-2, what would probably be the chief deterrent to the success of such a procedure?

Compare the results of this experiment with those of Expt. II-4*b*.

Which of these procedures do you think is better suited to the analysis of a thallous chloride solution? How would you proceed to improve the other procedure? Which would be better suited to the analysis of a 10^{-5} M solution of thallous ion? How should the procedure be modified for the analysis of such a solution?

EXPERIMENT II-15

Spectrophotometric Determination of Thallium

Directions for the use of the spectrophotometer will be provided by the instructor. Turn the instrument on and allow it to warm up while the solutions are being prepared.

The following reagent solutions are required:

Bromine: dissolve 10 g. of sodium dihydrogen phosphate (NaH_2-$PO_4 \cdot H_2O$) in a mixture of 90 ml. of freshly prepared bromine water and 10 ml. of concentrated hydrochloric acid.

Phenol: dissolve 25 g. of phenol in 100 ml. of glacial acetic acid.

Potassium iodide: dissolve 0.2 g. of iodate-free potassium iodide in 100 ml. of water. This solution should be freshly prepared.

Starch: triturate 1 g. of soluble starch with 5 ml. of water, and pour the mixture slowly into 50 ml. of boiling water. Add 50 ml. of glycerol, boil 5 min., and cool before using.

Prepare a standard 1×10^{-4} F solution of thallous nitrate by appropriate dilution of an 0.1 F thallous nitrate solution used in previous experiments. Place 0.25 ml. of the 1×10^{-4} F thallous nitrate in one 50-ml. erlenmeyer flask, 0.50 ml. in a second, 0.75 ml. in a third, and so on up to 2.00 ml. Meanwhile dilute 10.0 ml. of each of the saturated thallous chloride solutions to 1000 ml. in a volumetric flask. Place 0.75 ml. of each of the resulting solutions in two more 50-ml. erlenmeyer flasks, and 1.00 ml. of each in yet another pair. In a final flask place 1 ml. of water; this will be carried through the same chemical manipulations as the other solutions, and will eventually serve as a reagent blank. Dilute each of the solutions to 10.0 ± 0.5 ml. and add 0.5 g. of solid ammonium chloride. When this has completely dissolved, add 10 ml. of the bromine reagent, heat just to boiling, and boil very gently until the solution is just decolorized. Cool rapidly under running water and transfer each solution quantitatively to a 50-ml. volumetric flask. (The solutions may now be allowed to stand for several days if the experiment must be interrupted. However, after the next step the experiment must be completed without delay.)

To each of the resulting mixtures of +3 thallium (probably $TlCl_6^{\equiv}$) and excess bromine add 0.25 ml. of phenol solution, mix, and let stand for 3 min. This serves to remove the small excess of bromine which remains after the boiling; the *s*-tribromophenol which is formed will not oxidize iodide in the next step.

Add 5 ml. of the potassium iodide solution to *one* of the flasks (preferably the one containing 1.00 ml. of the standard 1×10^{-4} *F* thallous solution), followed by 1 ml. of the starch solution. Dilute to volume, shake, and let stand for 5 min. Treat the blank solution in exactly the same way.

Transfer suitable aliquots of these solutions to spectrophotometer cells and measure the absorbance of the thallium solution as a function of wavelength between 400 and 700 mμ. Locate the wavelength of maximum absorption, and carry out the remaining measurements at that wavelength.

Now treat another of the thallium solutions with iodide and starch as directed above, and while this is standing for 5 min. discard the first thallium solution in the spectrophotometer cell and wash out the cell very thoroughly with distilled water. (Do not disturb the cell containing the reference blank.) Then measure the absorbance of the second thallium solution. When all of the known solutions have been measured in this way, proceed to make similar measurements with the four unknowns prepared from the thallous chloride solutions.

At the end of the experiment wash out both of the cells and *carefully* dry their outsides with lens paper or absorbent tissue, and put them away protected from dust.

Plot the absorbance against concentration for the known thallium solutions. Is Beer's law obeyed over this concentration range? Compute the molar absorptivity for each of the known solutions. What is the mean error of the data? Calculate the mean concentration of thallous ion in each of the thallous chloride solutions, and the solubility product of thallous chloride.

Write equations for the reactions which occur during the successive steps in the preparation of the final solutions. What other elements, if present, would interfere in the determination of thallium by this method? Do you think that water could be used as the reference solution in the spectrophotometric measurements, or is the reagent blank essential?

What are the principal sources of error in this experiment? How could they be eliminated?

EXPERIMENT II-16

Spectrophotometric Titration of Dichromate with Ferrous Sulfate

Pipet exactly 25 ml. of one of the saturated solutions of thallous chloride into a 100-ml. volumetric flask. Add *one drop* of 0.1 *F* sodium hydroxide, then exactly 15 ml. of a standard 0.01 *F* solution of potassium chromate.

Shake well, dilute to the mark with water, and let stand. Treat a portion of the other saturated thallous chloride solution in exactly the same way.

Assemble the apparatus for spectrophotometric titrations in accordance with the instructor's directions.[1] Turn on the spectrophotometer and allow it to warm up for at least 15 min. In the meantime pipet exactly 1 ml. of the 0.01 F potassium chromate into the clean titration cell; add 6 ml. of 85 per cent phosphoric acid, 15 ml. of dilute (3 F) sulfuric acid, and 25 ml. of water. Stir thoroughly and place the cell in the cell compartment of the spectrophotometer.[2]

Adjust the monochromator to 350 mμ, and set the dark current control so that the instrument indicates zero transmittance with the shutter closed. Set the slit control for minimum slit width and the sensitivity control for minimum sensitivity. Open the shutter and adjust the controls so that the instrument indicates an absorbance of 2.0 ± 0.1 with an effective slit width of 10 mμ or less. Record the absorbance indicated; then add 0.25 ml. of 0.1 F ferrous sulfate–1 F sulfuric acid from a 10-ml. microburet. Stir thoroughly and measure the absorbance again. Continue until at least a 50 per cent excess of the ferrous solution has been added. The absorbance will decrease linearly with increasing volume of ferrous solution up to the end point, and thereafter will remain practically constant.

Discard the solution and repeat the titration with a 2-ml. aliquot of the 0.01 F chromate solution.

Filter the thallous chromate suspensions through small medium- or fine-porosity sintered-Pyrex Buchner-type funnels into *clean dry* 125-ml. suction flasks. Use traps between these flasks and the aspirator. Discard the first 5 to 10 ml. of filtrate; then collect at least 50 ml. of each solution.

Pipet 50 ml. of one of these filtrates into the titration cell, add 7 ml. of 85 per cent phosphoric acid and 3 ml. of concentrated sulfuric acid, and titrate with the 0.1 F ferrous sulfate according to the above directions. Repeat with 50 ml. of the other filtrate.

Plot all of the titration curves. Is it necessary to apply a correction for dilution? Calculate the concentration of the ferrous sulfate solution,

[1] Instructions for this cannot be given here, because the mechanical details differ greatly from one spectrophotometer to another. In the following it is assumed that a 100-ml. titration cell is employed; if this is not the case, the volumes used must be adjusted accordingly.

[2] With some spectrophotometers it is possible to mount a magnetic stirrer under the cell so that the solution can be stirred throughout the titration. This is very convenient, but care must be taken to ensure that the stirring bar is well out of the light path, and also that the stirring is slow enough to prevent the formation of a vortex in the light path.

and compare the values secured with the two known chromate solutions. How does the precision of spectrophotometric titration compare with that of direct spectrophotometry? Why?

Calculate the concentration of chromate in each of the thallous chromate suspensions, and from this calculate the solubility product of thallous chloride.

Why was sodium hydroxide added to the thallous chloride before adding the potassium chromate? Justify the use of only a very small excess of chromate in the next step. Do you think that the spectrophotometric titration of thallous ion with standard chromate, at a wavelength where the excess chromate has a high absorptivity, would be a practical procedure? Why or why not?

Consult the paper by J. W. Miles and D. T. Englis, *Anal. Chem.*, **27**, 1996 (1955), for further details concerning this titration. On the basis of the spectra given in this paper, explain why the titration is carried out at 350 mμ rather than at a wavelength on the plateau of the dichromate curve in the ultraviolet.

Would any error result from the presence of a small amount of suspended silica in the thallous chromate suspension? Justify your answer by sketching the titration curve that would have been secured in the presence of the silica. Does spectrophotometric titration differ from direct spectrophotometry in this respect?

EXPERIMENT III-1

Preparation of Cupric Oxalate

In a 2000-ml. beaker dissolve 12.6 g. (0.1 mole) of oxalic acid dihydrate in 1000 ml. of water and add 7 ml. (0.1 mole) of concentrated ammonia. While stirring continuously, slowly add a solution of 25.0 g. (0.1 mole) of cupric sulfate pentahydrate in 1000 ml. of water. Cover with a watch glass and let stand for a day or two, stirring occasionally if possible.

Decant the supernatant liquid, wash the precipitate once or twice by decantation with distilled water, and finally add a solution of about 0.5 g. of oxalic acid in 1000 ml. of water. Heat just to boiling, cover, and let stand for a day or two more. This treatment should destroy any basic salt that may have been formed during the original precipitation.

Discard the supernatant liquid and wash the precipitate several times by decantation with distilled water. Finally filter it through a coarse-porosity sintered-Pyrex Buchner-type filtering funnel, wash several more times with water, suck as dry as possible, and spread the moist solid out on a large watch glass to dry. On this watch glass place a large glass triangle, and invert another watch glass of the same size on the triangle to protect the salt from dust.

Let stand for a week or so with occasional stirring.

EXPERIMENT III-2

Analysis of Cupric Oxalate

A. Determination of Water. Weigh two 1-g. portions of the air-dried salt into tared weighing bottles and determine the loss of weight on drying to constant weight at 105°C.

B. Determination of Copper. Weigh out two 0.5-g. portions of the air-dried salt into 250-ml. erlenmeyer flasks. To each add 10 ml. of dilute (3 *F*) sulfuric acid. Warm gently over a hot plate or a small bunsen flame until the evolution of gas bubbles ceases and the salt is completely dissolved; then increase the rate of heating and evaporate to dense fumes of sulfur trioxide. Let cool and *cautiously* add 25 ml. of water. Add 3 g. of sodium acetate trihydrate and 25 g. of potassium iodide, swirl for 10 to 15 sec., or until the potassium iodide is completely dissolved, then titrate with 0.1 *N* sodium thiosulfate (previously standardized against potassium iodate according to directions given in textbooks of elementary quantitative analysis). The end point may be taken as the point at which the last trace of yellow color disappears, or about 0.5 ml. of starch indicator may be added if desired.

For a detailed discussion of this titration consult L. Meites, *Anal. Chem.*, **24,** 1618 (1952).

C. Determination of Oxalate. Weigh out two 0.5-g. portions of the air-dried salt into 400-ml. beakers, add 200 ml. of 0.5 *F* sulfuric acid to each sample, and stir until the salt is completely dissolved. Add about 1 ml. of saturated manganous sulfate, heat to about 60°C., and titrate with an approximately 0.1 *N* (0.02 *F*) solution of potassium permanganate previously standardized against sodium oxalate by the same procedure.

Calculate the percentages of copper, oxalate, and water in the sample, and compute the formula of the salt. What is its purity on a dry (at 105°C.) basis? Is the mole ratio of copper to oxalate equal to 1 within experimental error? The literature contains a mention of the hemihydrate ($CuC_2O_4 \cdot \frac{1}{2}H_2O$); do your results confirm this formula?

In preparation for later experiments, place about 1900 ml. of conductivity water or the best distilled or deionized water available in each of two 2000-ml. glass-stoppered Pyrex erlenmeyer flasks, and add about 2 g. of the air-dried cupric oxalate to each. Warm one of the flasks to 40 to 50°C. for about an hour with frequent shaking, and then let both of the flasks stand at room temperature or, if one is available, in a water thermostat at 25°C. Label the flasks clearly so that they can be told apart later on, and shake them occasionally during the first week or so.

EXPERIMENT III-3

Analysis of a Saturated Cupric Oxalate Solution by Direct Potentiometry

Wind a length of heavy (16-gauge) pure copper wire into a helix as described in Sec. 3-8. Immerse the coiled portion in 6 *F* nitric acid for approximately 15 sec. or until the surface is etched slightly; then rinse it thoroughly and immerse it in a solution containing 0.1 *F* mercurous nitrate and 0.5 *F* nitric acid. When enough mercury has been deposited to give the surface of the electrode a uniform silvery color, wash the electrode thoroughly with distilled water. Let it stand in 0.1 *F* nitric acid except when it is actually in use. When it is first prepared, the electrode will tend to give unsteady and meaningless potentials, because the activity of copper at its surface is undefined. It must therefore be allowed to stand until the film of mercury becomes saturated with copper; this will require 15 to 30 min., and the surface of the electrode will then appear golden in color.

Place a reference electrode[1] in a clean dry 150-ml. beaker in such a way that its tip nearly touches the bottom of the beaker. Connect the reference electrode to the potentiometer (will the copper electrode be the negative electrode or the positive electrode?). Add 50.0 ml. of 0.1 *F* cupric sulfate and 1.0 ml. of 0.1 *F* perchloric acid to the beaker.

Standardize the potentiometer against the standard cell. Rinse the copper electrode thoroughly with distilled water, then immerse it in the cupric sulfate solution, connect it to the potentiometer, and measure the potential of the cell. When the potentiometer has been balanced, stir the solution briefly and then measure the potential again after changing the potentiometer setting by about 0.05 v. This will help to ensure that reproducible and meaningful values are being secured.

Remove the copper electrode from the solution and replace it in the 0.1 *F* nitric acid. Discard the solution and substitute a solution prepared by mixing 30.0 ml. of 0.1 *F* cupric sulfate and 2.0 ml. of 0.1 *F* perchloric acid in a 100-ml. volumetric flask and diluting to the mark. Measure the potential of the copper electrode in this solution; then proceed in the same way to make measurements with solutions containing 0.01, 0.003, 0.001, 0.0003, and 0.0001 *M* cupric ion.

[1] Directions for the preparation of one type of silver–silver chloride–saturated potassium chloride reference electrode are given in Expt. V. The instructor may prefer to use a commercial saturated calomel electrode in this experiment, and he should be consulted for directions.

Finally measure the potential of the copper electrode in the solutions prepared by mixing 1.0 ml. of 0.1 *F* perchloric acid with 50.0 ml. of each of the saturated cupric oxalate solutions from Expt. III-2. Filter the cupric oxalate solutions if necessary before adding the acid.

Disassemble the cell and store the reference electrode according to the instructor's directions.

Plot the potential of the indicator electrode against log $[Cu^{++}]$. How well does the slope of the best straight line through the experimental points agree with the theoretical value, 0.02957 v. at 25°C.? Extrapolate the data to log $[Cu^{++}] = 0$, and from the resulting value of E_{ind} and the known potential of the reference electrode calculate the formal potential of the cupric ion–copper couple. Compare your value with the literature value, +0.345 v. Account for any difference between these values. Can a good value of the standard potential be secured from these data by extrapolating a plot of $E_{ind} - 0.02957 \log [Cu^{++}]$ *vs.* the square root of the ionic strength to infinite dilution? Why or why not?

From the plot of E_{ind} *vs.* log $[Cu^{++}]$, find the concentration of cupric ion in each of the cupric oxalate solutions, and calculate the solubility product of cupric oxalate. What do you think is a fair estimate of the uncertainty in the value of K obtained from the data?

EXPERIMENT III-4

Potentiometric Titration of Copper with Vanadous Chloride

Dissolve 6.0 g. (0.05 mole) of ammonium metavanadate, NH_4VO_3, in 100 ml. of 1 *F* sodium hydroxide, and pour this solution slowly and with constant stirring into 100 ml. of 5 *F* hydrochloric acid in a 500-ml. erlenmeyer flask. Meanwhile dissolve 3 g. of mercurous nitrate in 100 ml. of 0.5 *F* nitric acid and add 50 g. of pure 20-mesh metallic zinc. Stir for several minutes; then pour off the solution and wash the zinc five times by decantation with 0.1 *F* nitric or perchloric acid and twice more with 0.1 *F* hydrochloric acid. After discarding the second portion of the hydrochloric acid, transfer about three-fourths of the amalgamated zinc to the flask containing the vanadate solution. Cover the remainder of the zinc with 0.1 *F* hydrochloric acid and reserve it for later use.

Loosely stopper the flask containing the vanadate and zinc and swirl it until the color of the solution has changed from the orange of vanadium(V) through the blue of VO^{++} and the green of V^{+++} (vanadic ion) to a clear purple (V^{++}, vanadous ion). Then add 300 ml. of water, mix thoroughly, and let the solution stand, loosely stoppered, until it is needed.

Secure a magnetic stirrer, if one is available, and attach it to a sturdy ring stand. On it place a 400-ml. beaker, preferably of the so-called "electrolytic" (lipless) type, and center this on the top of the stirrer. Place a magnetic stirring bar in the beaker and make sure that it rotates freely without striking the sides of the beaker. Select a solid rubber stopper of the appropriate size, and bore four holes in it. One of these should be of the proper size to admit and hold in place a saturated calomel or silver–silver chloride reference electrode. Another should be of the proper size to admit the tube of a coarse-porosity sintered-Pyrex gas-dispersion cylinder. The third should be large enough to admit the tip of a 10-ml. microburet and leave an annular ring about 2 mm. wide to permit the easy escape of gas. The fourth hole should be just large enough for a No. 00 rubber stopper.

Insert the reference electrode and the gas-dispersion cylinder in the holes prepared for them; then clamp the stopper in place so that the beaker can be removed from the assembly (by swinging the stirrer to one side) without disturbing the stopper. Clean a fairly heavy platinum wire indicator electrode by immersing it in 6 *F* nitric acid for about a minute, rinsing with distilled water, and heating to red heat in the *oxidizing* cone of a good Meker flame. Insert this in the fourth hole of the stopper and wedge it in place with a solid No. 00 rubber stopper. The tip of the coiled portion of this electrode must just clear the stirring bar, and the other end must project an inch or so above the top of the large stopper.

Connect the indicator electrode to one terminal of the potentiometer circuit, and the reference electrode to the other terminal. Connect the gas-dispersion cylinder to a nitrogen supply line.

The remainder of the experiment should be carried out in one laboratory period.

Place 200 ml. of 2 *F* hydrochloric acid and 2.00 ml. of standard 0.1 *F* cupric sulfate in the cell and mount it in position underneath the stopper. Bubble a fairly rapid stream of nitrogen through the solution for about 15 min. to remove dissolved air. Meanwhile fill a 10-ml. microburet to just below the 10-ml. graduation with the amalgamated zinc previously set aside. Pour several successive 2-ml. portions of the vanadous solution into this buret, draining the solution to just above the top of the zinc after each addition. Then fill the buret nearly to the top—well above the zero mark—with the vanadous solution, but do not mount the buret in position above the cell yet.

Standardize the potentiometer against the standard cell. Tap the column of zinc in the microburet sharply to dislodge any bubbles of gas trapped between the pieces of zinc. Drain the vanadous solution down to the zero mark, place the buret in position above the cell, and barely open the stopcock so that about 0.2 ml. of solution flows slowly into the cell. (This is important because any +3 vanadium present in the solution will not be completely reduced if the rate of flow past the zinc is too high.)

Measure the potential of the indicator electrode, wait a minute or two, and repeat the measurement to make sure that a steady potential is secured. Continue adding 0.2-ml. aliquots of the vanadous solution until the slope of the curve begins to increase; then add 0.05-ml. aliquots in the vicinity of the equivalence point. Continue the titration until at least 7.5 ml. of the vanadous solution has been added.

Discard the solution and titrate another 2.00-ml. portion of standard 0.1 F copper sulfate in exactly the same way. The positions of the end points should agree to within 0.01 ml. of vanadous solution.

Now titrate an aliquot of each of the saturated cupric oxalate solutions in the same way. Use 200 ml. of the cupric oxalate solution and 40 ml. of concentrated hydrochloric acid.

Interpret the shapes of the titration curves secured. What reactions take place when a chloride solution of +2 copper is titrated with a strong reducing agent? To what is the vanadous ion oxidized? From your data estimate the formal potential in 2 F hydrochloric acid of each of the three couples involved in this titration.

Why might it be desirable to use a mercury indicator electrode in this titration? Why did dissolved air have to be removed? Could the vanadous chloride have been replaced by any other reagent? What would have happened if the hydrochloric acid used in the titration had been replaced by sulfuric acid?

Calculate the normality of the vanadous chloride, the concentration of cupric ion in each of the saturated cupric oxalate solutions, and the solubility product of cupric oxalate. Do you feel that this solubility product is more or less reliable than the value calculated in Expt. III-3? Why?

EXPERIMENT III-5*a*

The Hydrolysis of Cupric Ion

Prepare 100 ml. of an 0.10 F solution of cupric sulfate in the best water available. The concentration of this solution should be known to within about ±1 per cent. Dilute 25.0 ml. of this solution to 100.0 ml. to prepare an 0.025 F solution; dilute 25.0 ml. of this in turn to prepare an 0.00625 F solution; and so on until a series of solutions is available whose concentrations range from 0.1 to 2.4×10^{-5} F in successive factors of four.

Set the pH meter to read 3.57 with the electrodes immersed in saturated potassium hydrogen tartrate. Rinse the electrodes thoroughly and immerse them in a portion of the 0.1 F cupric sulfate. When a steady reading is secured, discard the solution and replace it with another portion of the 0.1 F solution. If the pH now measured differs from the value finally secured with the first portion, continue in the same way until a constant pH is obtained. Then discard the cupric sulfate solution, rinse the electrodes thoroughly with dilute nitric acid and then with water, and measure the pH of the saturated potassium hydrogen tartrate solution. If the value secured lies between 3.55 and 3.59, reset the meter to 3.57 and proceed to measure the pH of the 0.025 F cupric sulfate solution. If not, reset the meter and repeat the measurements with the 0.1 F solution.

When the pH of each of the copper solutions has been measured, proceed to measure the pH of each of the saturated cupric oxalate solutions prepared in Expt. III-2.

For each of the cupric sulfate solutions compute the equilibrium constant K_a of the reaction $Cu^{++} + H_2O = Cu(OH)^+ + H^+$. What is the mean acid dissociation constant of cupric ion? How does the effect of ionic strength on the various activity coefficients enter into this experiment? How could this be eliminated?

Taking your experimental value for this constant and the literature value of K_2 for oxalic acid (6.1×10^{-5}), calculate the solubility product of cupric oxalate. The conservation equation here is essentially (since $[H_2C_2O_4]$ is negligible)

$$[Cu^{++}] + [CuOH^+] = [HC_2O_4^-] + [C_2O_4^=]$$

By using K_a, K_2, and the measured pH, this can be transformed into an equation involving only $[Cu^{++}]$ and $[C_2O_4^=]$. Another equation involving these quantities can be gotten in a similar fashion by starting with the electroneutrality equation

$$[H^+] + 2[Cu^{++}] + [CuOH^+] = [HC_2O_4^-] + 2[C_2O_4^=] + [OH^-]$$

Solving these equations simultaneously will provide the information needed for the calculation of the solubility product.

How would the formation of complex ions such as $Cu(C_2O_4)_2^=$ (see Expt. III-9) affect this calculation?

EXPERIMENT III-5*b*

Alkalimetric Titration of Cupric Ion

Prepare and standardize an approximately 0.1 F solution of sodium hydroxide according to the directions in Expt. I-5*b*.

Accurately weigh about 0.25 g. (1 millimole) of clear uneffloresced crystals of reagent grade cupric sulfate pentahydrate into a 250-ml. beaker and add exactly 100 ml. of water. Set the pH meter to read 3.57 with the glass and calomel electrodes immersed in a saturated solution of potassium hydrogen tartrate; then rinse the electrodes thoroughly with water and immerse them in the cupric sulfate solution.

Measure the pH of this solution (how does the result compare with the data obtained in Expt. III-5*a*?); then add 0.50 ml. of the standard 0.1 *F* sodium hydroxide solution from a 50-ml. buret, stir thoroughly, and measure the pH again. Continue the titration until 50 ml. of sodium hydroxide has been added, being sure to secure enough data to define the entire course of the titration curve. Note the volume of base required to produce the first permanent precipitate. Is the final solution perfectly clear? Discard the solution and rinse the beaker and the electrodes thoroughly with dilute nitric acid followed by water. Leave the electrodes standing in distilled water.

Plot the titration curve and locate the end point of the reaction $Cu^{++} + 2OH^- = Cu(OH)_2$. How does the volume of base required to reach this point compare with the expected value? How is this comparison affected by the adsorptive properties of hydrous cupric oxide? Do you think this titration could be developed into a useful method for the determination of copper? What does the composition of the precipitate appear to be?

Is there any indication of an end point corresponding to the reaction $Cu^{++} + OH^- = CuOH^+$? Would a failure to find such an end point prove that $CuOH^+$ is not formed during the titration? What does a failure to observe two inflection points during the titration of a weak dibasic acid indicate about the value of K_1/K_2?

At each experimental point preceding the end point, calculate the number of millimoles of sodium hydroxide added to the solution. Divide this by the number of millimoles of base needed to reach the end point, and take the resulting ratio as the fraction of the copper precipitated. Calculate the concentration of copper left in solution; this is $[Cu^{++}] + [CuOH^+]$. From this and the result of Expt. III-5*a* calculate $[Cu^{++}]$, and from this and the measured pH calculate the solubility product of $Cu(OH)_2$. Look up the literature value of this constant. Suggest an explanation for any difference between the average of your values and the literature value. Do your values seem to vary systematically as more of the copper is precipitated? What assumption in the prescribed method for estimating $[Cu^{++}] + [CuOH^+]$ might be responsible for such behavior?

EXPERIMENT III-5*c*

Analysis of a Saturated Cupric Oxalate Solution by Potentiometric Titration with Sodium Hydroxide

Pipet exactly 100 ml. of each of the saturated cupric oxalate solutions prepared in Expt. III-2 into a 250-ml. beaker. Titrate with 0.01 *F* sodium hydroxide prepared by diluting exactly 10 ml. of the 0.1 *F* solution of Expt. III-5*b* to 100 ml. with freshly boiled distilled water. The titrations should be made according to the directions of Expt. III-5*b*, keeping in mind the fact that only a few milliliters of base will be consumed in the titration.

Locate the end points of the titration curves. They will be very poorly defined, and you may wish to ask the instructor to confirm your judgment. From the volumes of sodium hydroxide used here and in Expt. III-5*b*, calculate the amount of cupric ion in each of the samples of saturated cupric oxalate solution, and from these values compute the solubility product of cupric oxalate. Why is the use of a standardization against cupric sulfate likely to give more accurate results than that of a standardization against, say, potassium hydrogen phthalate?

Compare the results of Expts. III-5*a* and III-5*c*. Which method seems to you to be likely to give the more reliable value? In the light of the results of Expt. III-3, does this expectation appear to agree with fact? Propose reasons for any discrepancies among the results of these three experiments.

EXPERIMENT III-5*d*

The Dissociation Constants of Oxalic Acid

Prepare and standardize an approximately 0.1 *F* solution of sodium hydroxide according to the directions given in Expt. I-5*b*.

Accurately weigh about 0.25 g. (2 millimoles) of oxalic acid dihydrate into a 250-ml. beaker and add exactly 100 ml. of water. Set the pH meter to read 3.57 with the electrodes immersed in saturated potassium hydrogen tartrate; then rinse the electrodes thoroughly and immerse them in the oxalic acid solution. Record the pH of this solution, then add 0.5 ml. of the sodium hydroxide solution from a 50-ml. buret and measure the pH again. Continue the titration until the entire 50 ml. of base has been used, being sure to make enough measurements to define the entire titration curve completely. Finally, rinse the electrodes thoroughly and leave them standing in distilled water.

Locate the second equivalence point and compute the weight of oxalic acid in the solution. How does this compare with the weight taken? What do you conclude about the purity of the oxalic acid?

Let V^* be the volume of sodium hydroxide required to reach the second equivalence point. How does the volume of base consumed at the first equivalence point compare with $V^*/2$? How closely do you think the first equivalence point can be located? Would a titration with sodium hydroxide be a practical method of analyzing a mixture of oxalic acid and sodium hydrogen oxalate?

For each of a number of volumes of base, V_b, such that $0 < V_b < V^*/2$, calculate the value of K_1 found from the measured pH. Is the equation $K_1 = [H^+]\left(\frac{V_b}{V^*/2 - V_b}\right)$, which is the ordinary buffered-solution equation, applicable in this case? Since K_1 is fairly large, what corrections should be made? [Consult Eq. (4-16).] Compare the value of K_1 calculated from these points with the value found from the pH of the original oxalic acid solution by using the equation $K_1 = [H^+]^2/(C_a - [H^+])$. Which is more reliable? Why?

From the measured pH values at each of a number of points for which $V^*/2 < V_b < V^*$, calculate K_2. By interpolation in the data find the pH at the point where $V_b = V^*$, and use this pH to calculate K_2. How do these values compare?

Compare your values of K_1 and K_2 with those in the literature. Usually these values of K_2 are much more nearly equal than the values of K_1. Suggest an explanation of this fact.

What do you think of the feasibility of determining the concentration of oxalate ion in the saturated cupric oxalate solution by a potentiometric titration with standard acid?

EXPERIMENT III-6

Determination of the Solubility of Cupric Oxalate by Direct Conductometry

Secure a dip-type conductance cell and a direct-reading a-c operated conductance bridge. If necessary, platinize the electrodes of the cell according to the directions in Secs. 4-3 and 5-13.

Place exactly 500 ml. of water in a clean dry 600-ml. beaker and measure its conductance. From a 10-ml. microburet add 0.10 ml. of standard 0.2 N (0.1 F) cupric sulfate, stir thoroughly, and measure the conductance again. Add 0.10 ml. more of the cupric sulfate and measure the conductance again. Then add successively 0.20, 0.20, 0.40, 1.00, 3.00, and 5.00 ml. of the cupric sulfate (a total of 10.00 ml.), measuring the conductance after each addition.

Discard the solution, rinse the cell and the beaker very thoroughly, and repeat the above measurements (including the measurement of the conductance of the water), using first 0.1 F potassium sulfate and then 0.1 F potassium oxalate in place of the cupric sulfate. Then measure the conductance of each of the saturated cupric oxalate solutions.

Finally measure the conductance of a standard potassium chloride solution supplied by the instructor. Then rinse the cell very thoroughly and leave it standing in a clean beaker with the electrodes covered with water.

Consult Expt. I-6 for a discussion of the treatment of the data. It will be found to be essential to apply solvent corrections in this experiment. Do you have any information concerning the conductance of the water used to prepare the saturated cupric oxalate solutions? What relative error in the solubility product of cupric oxalate would result from a difference of 10 per cent between the mean conductance of the distilled water on the day this experiment was performed and the conductance of the water used in making up the cupric oxalate solutions?

How could you use direct conductometry to check on the possibility that the solid cupric oxalate contains ammonium oxalate as a result of insufficient washing in Expt. III-1?

EXPERIMENT III-7

Determination of the Solubility of Cupric Oxalate by Conductometric Titration

Prepare a standard 0.005 F solution of salicylaldoxime by dissolving the required amount of reagent in 10 ml. of 95 per cent ethanol and diluting to 100 ml. with water. To a 100-ml. aliquot of one of the saturated cupric oxalate solutions add just enough water to cover the electrodes of a dip-type conductance cell, and titrate the resulting solution with the salicylaldoxime solution, using a 10-ml. microburet for the titration. Add the salicylaldoxime in 0.50-ml. aliquots, and continue the titration until at least a 50 per cent excess of the reagent has been added. Repeat the titration with a portion of the other saturated cupric oxalate solution.

Carry out similar titrations with 0.002 F lead nitrate as the reagent.

Plot the titration curves and locate the end points of the titrations. Calculate the concentrations of cupric and oxalate ions in the saturated cupric oxalate solutions, and the solubility product of cupric oxalate. Are the concentrations of cupric and oxalate ions in the saturated cupric oxalate solution equal within experimental error? If not, propose an explanation.

Explain the shapes of the titration curves. What is the apparent precision of your titrations? How could this precision be improved? Do you think that the solubility product of cupric oxalate obtained in this experiment is more or less reliable than the value obtained from Expt. III-6?

Consult the literature for information on the use of salicylaldoxime as a reagent for metal ions. Could copper be determined in the presence of zinc by a conductometric titration with this reagent? Outline a method for assaying salicylaldoxime by conductometric titration. Is salicylaldoxime a useful reagent for potentiometric titrations of metal ions?

EXPERIMENT III-8

Determination of the Solubility of Cupric Oxalate by Direct Polarography

Deaerate exactly 50 ml. of 1 F hydrochloric acid in the solution compartment of an H cell, insert the dropping electrode, and adjust and check the apparatus according to the directions given in the fourth section of Expt. VI.

Adjust the "bridge" H (Expt. VI, Fig. 1) to apply about 0.9 v. across the cell circuit. Make sure that the d.e. is the *negative* electrode. Measure $E_{\text{d.e.}}$ and E_R; then adjust the bridge to deliver about 1.0 v. and repeat the measurements. Continue in this way until E_R has begun to increase rapidly because of the onset of hydrogen evolution. This should be at $E_{\text{d.e.}} = ca. - 1.2$ v.

Then adjust the bridge to deliver 0.8 v. and carry out the measurements at successively less negative values of $E_{\text{d.e.}}$. The current (and with it E_R) will decrease as $E_{\text{d.e.}}$ becomes less negative, and will become equal to zero when $E_{\text{d.e.}}$ is about -0.3 v. At still less negative values of $E_{\text{d.e.}}$, a negative current will flow through the cell circuit.

The anodic oxidation of mercury does not begin until positive values of $E_{\text{d.e.}}$ are reached. In order to measure positive values of $E_{\text{d.e.}}$, it will be necessary to throw switch S_2 to the right rather than to the left. Continue the measurements out to successively more positive potentials until the (negative) current begins to increase rapidly because of the oxidation of mercury.

After the entire residual current curve has been secured, add exactly 1.00 ml. of a standard 0.0200 F copper sulfate solution. Bubble nitrogen through the solution for 3 to 5 min., then divert the gas stream over the surface of the solution, remove any gas bubbles from the tip of the capillary, and repeat the measurements. Pay special attention to the steeply rising portions of the waves in order to make sure that enough points are secured.

Repeat the measurements after adding successively 1.00 ml. of 0.1 per cent Triton X-100 and two more 1.00-ml. portions of 0.0200 F cupric sulfate. Then discard the solution and thoroughly rinse both the cell and the capillary.

To exactly 100 ml. of one of the saturated solutions of cupric oxalate add 10.0 ml. of concentrated hydrochloric acid and 2.00 ml. of 0.1 per cent Triton X-100. Rinse the cell with one or two small portions of this solution, then transfer about 50 ml. of the solution to the cell, deaerate it, and secure its polarogram. *The height of the column of mercury above the capillary must be exactly the same here as in the measure-*

ments with the known solutions. (Why?) Measure and record this height so that it can be reproduced in Expt. III-9.

Discard this solution and replace it with a similar one prepared from the other saturated cupric oxalate solution. Deaerate this and measure the limiting current at one or two potentials on the plateau of the second wave.

At the completion of the experiment refer to the last paragraph of Expt. VI.

Plot the residual current curve and the polarograms obtained with the stock copper sulfate solution. What effects does the addition of the maximum suppressor have on the heights and half-wave potentials of the waves? Calculate the concentration of cupric ion present in each of these solutions; measure the diffusion current ($= i_l - i_r$) of the total double wave on each curve. From these values calculate i_d/C. Measure i_d for each of the solutions prepared from the cupric oxalate, and from these and the mean value of i_d/C calculate the concentration of dissolved copper[1] in a saturated cupric oxalate solution. How well does this value agree with the values secured by other techniques? What are the principal sources of error in this experiment, and how could they be avoided?

What is the smallest diffusion current that can be detected with this apparatus, assuming that the students' potentiometer is accurate and precise to ± 0.5 mv.? To what concentration of cupric ion does this correspond under the conditions of this experiment? How could the circuit be modified to allow smaller currents to be detected? Is this a practical method of increasing the sensitivity of polarographic analysis?

What electrode reactions are responsible for the two waves? If your answer to the preceding question is correct, are the waves reversible? How could you confirm your answers to these questions by further experiments?

The literature values for the half-wave potentials of cupric copper in 1 F hydrochloric acid are $+0.04$ and -0.22 v. *vs.* S.C.E. How well do your values compare with these? If the total diffusion current constant of cupric copper in 1 F hydrochloric acid is 3.39, what is the value of $m^{2/3}t^{1/6}$ for your capillary? (To what value of $E_{d.e.}$ does this refer?)

EXPERIMENT III-9

Polarographic Study of the Copper(II) Oxalate Complex

Dissolve 18.4 g. (0.10 mole) of potassium oxalate monohydrate in 75 ml. of water and add exactly 1.00 ml. of 0.1 F cupric sulfate. Slowly

[1] In view of the formation of a complex copper oxalate ion, as demonstrated in Expt. III-9, what is actually being determined here?

add 0.02 F potassium hydroxide, stirring continuously, until the pH of the mixture becomes equal to 8.0 $\pm$ 0.2 as indicated by a glass electrode pH meter. Transfer the solution to a 100-ml. volumetric flask and dilute to the mark. In the same way prepare solutions containing 0.3 and 0.1 F oxalate.

Obtain the polarogram of each of these solutions, following the directions given in Expts. VI and III-8. The height of the mercury column above the capillary should be as nearly as possible the same as that in the measurements of Expt. III-8.

Plot the current–voltage curves. Compare the diffusion currents obtained here with the value of i_d/C secured for the 2-electron reduction of copper(II) in 1 F hydrochloric acid (Expt. III-8). What does this comparison reveal concerning (*a*) the value of n for the reduction of copper(II) from oxalate solutions, and (*b*) the relative mobilities of the chloride and oxalate complex ions? What purely chemical reasons are there for rejecting copper(I) as a possible product of the electrode reaction? Does the height of the wave secured with a fixed concentration of copper change as the potassium oxalate concentration is changed? If so, why?

Test each of the waves for reversibility by measuring $E_{3/4} - E_{1/4}$ and comparing with the theoretical value $\left(-\frac{0.05915}{2}\log 9\right)$ for a reversible 2-electron reduction. Plot $E_{1/2}$ against the logarithm of the oxalate ion concentration; if the complex is $Cu(C_2O_4)_p^{2-2p}$, to what value of p does the slope of the line most nearly correspond? What would be the effect of taking the change of the activity coefficient of oxalate ion into account? From the resulting plot estimate the value of $E_{1/2}$ at $\log [C_2O_4^{=}] = 0$. (Account for any difference between this value and the value of $E_{1/2}$ actually measured in 1 F oxalate.) Taking $E_{1/2}$ for cupric ion in 1 F potassium nitrate as +0.006 v. *vs.* S.C.E., calculate the dissociation constant of the cupric oxalate complex. Does this comparison provide an accurate compensation of the liquid-junction potential? If not, suggest a better procedure. How accurate do you think that this dissociation constant is likely to be?

From the total concentration of copper in a saturated cupric oxalate solution (as measured by Expts. III-5, III-7, and III-8) and the dissociation constant just calculated, compute the fraction of the dissolved copper which is present as the oxalate complex in such a solution. Does the resulting value seem to warrant any modification of the calculations of Expt. III-6? How could you estimate the value of λ^0 for the cupric oxalate complex from the data secured in Expts. III-8 and III-9?

From the results of all of the experiments thus far performed, estimate the solubility product of cupric oxalate. Combine this with the dissociation constant of the complex ion measured in this experiment to secure a value for the total concentration of dissolved copper in a solution containing 0.1 *F* oxalate (*i.e.*, $[C_2O_4^{=}] + p[Cu(C_2O_4)_p^{2-2p}] = 0.1$) which is just saturated with solid cupric oxalate. Devise a method by which this value could be measured experimentally with the aid of a polarograph.

EXPERIMENT III-10

Analyses of Saturated Cupric Oxalate Solutions by Amperometric Titrations

With a volumetric pipet transfer exactly 50 ml. of one of the saturated solutions of cupric oxalate to a polarographic cell, and add 2.0 ml. of 0.1 per cent Triton X-100, 5.0 g. (0.05 mole) of solid potassium nitrate, and 1.0 ml. of glacial acetic acid. Bubble nitrogen through the solution for about 5 min., then divert the nitrogen stream over the surface of the solution, insert the dropping electrode, and adjust the height of the mercury reservoir to give a drop time of about 3 sec.

Adjust the potential of the dropping electrode to approximately −0.5 v. *vs.* S.C.E. and measure the current at this potential. From a 10-ml. microburet add 0.25 ml. of 0.005 *F* salicylaldoxime (*cf.* Expt. III-7), bubble nitrogen through the solution for about a minute, then divert the nitrogen stream over the surface of the solution and measure the current again. Continue in this way until at least a 50 per cent excess of the reagent has been added, then wash the cell out with 6 *F* hydrochloric acid followed by distilled water, and repeat the titration with a 50-ml. aliquot of the other cupric oxalate solution.

By the same procedure, but omitting the acetic acid and using an applied potential of −1.0 v. instead of −0.5 v., titrate a 50-ml. aliquot of each cupric oxalate solution with 0.002 *F* lead nitrate (*cf.* Expt. III-7).

Prepare a standard solution of hydrazine dihydrochloride by dissolving 1.050 g. (accurately weighed) of the pure salt in water and diluting to exactly 100 ml., then diluting a 5.00-ml. aliquot of this solution to exactly 500 ml. The resulting solution will be 1.00×10^{-3} *F*. Pipet exactly 50 ml. of one of the cupric oxalate solutions into the H cell; add 2.5 g. (0.05 mole) of ammonium chloride, 4 ml. (0.05 mole) of concentrated ammonia, and 2.0 ml. of 0.1 per cent Triton X-100. Deaerate *very* thoroughly[1] and measure the current at −1.0 v. *vs.* S.C.E. While bubbling a rapid stream of nitrogen through the solution, add 0.20 ml. of the dilute hydrazine solution from a 5- or 10-ml. microburet. Divert the nitrogen over the surface of the solution and again measure the cur-

[1] The success of this titration depends primarily on the care with which oxygen is removed from the solutions. In ammoniacal or alkaline solutions hydrazine is an extremely powerful reducing agent, and it is very rapidly oxidized by oxygen as well as by the tetramminocupric ion.

rent at -1.0 v. *vs.* S.C.E. Continue until 2.0 ml. of hydrazine has been added; then add several 1-ml. aliquots of the hydrazine solution as above, measuring the current after each addition. Observe the appearance of the final solution. To what is the copper reduced? What is the oxidation product of the hydrazine? Discard the solution and titrate a 50-ml. aliquot of the other saturated cupric oxalate solution in the same way. After completing the second titration, secure the complete polarogram of the final solution.

Plot the titration curves and locate the end points. Calculate the concentrations of cupric and oxalate ions in the saturated cupric oxalate solution from the results of the titrations with salicylaldoxime and with lead nitrate, and compare the results with the values found conductometrically in Expt. III-7. Compare amperometric and conductometric titrations with respect to precision and ease of execution. Could the amperometric titration described here be used to determine copper in the presence of zinc? Interpret the titration curves secured in the titrations with salicylaldoxime and with lead nitrate. Which of these titrations gives the more reliable results? Why?

Compare the results secured in the titration with hydrazine with those obtained by the use of salicylaldoxime. In the former titration, what do you think was the fate of the oxygen which was dissolved in the reagent solution? If the solubility of oxygen in water at room temperature is 2.5×10^{-4} *M*, what relative error could it have caused? How could this error be eliminated? Interpret the polarogram secured with the solution containing excess hydrazine. What would have been the shape of the titration curve secured if the titration with hydrazine had been carried out at $E_{d.e.} = -0.35$ v. *vs.* S.C.E.? How would you locate the end point of such a titration?

Do you think that cupric copper could be titrated potentiometrically with hydrazine in an ammoniacal medium? If so, devise a brief set of instructions for carrying out such a titration. If not, explain the reason for your belief.

EXPERIMENT III-11

Analysis of Cupric Oxalate by Electrolysis at Constant Current

Weigh out two 0.5-g. portions of the air-dried cupric oxalate prepared in Expt. III-1 into 250-ml. erlenmeyer flasks. To each add 10 ml. of 3 *F* sulfuric acid. Warm gently over a hot plate or a small bunsen flame until the evolution of gas bubbles ceases and the salt is completely dissolved; then increase the rate of heating and evaporate to fumes of sulfur trioxide. Cool the solution, *cautiously* add 25 ml. of water, and quantitatively transfer the contents of the flask to a 250-ml. beaker,

Add about 5 ml. of dilute (3 *F*) sulfuric acid and 2 ml. of dilute (6 *F*) nitric acid, and dilute to about 150 ml.

Clean a platinum gauze cathode by immersing it in dilute (6 *F*) nitric acid, warming if necessary to remove the last traces of metallic deposit from a previous experiment. Rinse thoroughly with distilled water and heat the electrode just to a dull red heat in the *oxidizing* cone of a good Meker flame. Let cool to room temperature and weigh.

Immerse a platinum anode and the tared cathode in the copper solution and connect them to the terminals of a d-c power supply similar to that shown in Fig. 7-1. Make sure that the cathode is connected to the negative terminal. In order that the solution can be easily tested for completion of deposition near the end of the experiment, not more than about three-quarters of the cathode should be covered by the solution at this point. The anode and cathode should be as nearly concentric as possible. Cover the beaker with a split watch glass if one is available.

Start the stirrer, if one is provided with the apparatus, and set the rheostat R to zero. Close the electrolysis circuit and read the current and the applied voltage, both of which should be zero. Increase the applied voltage in 0.2-v. increments up to about 1 v., then in 0.1-v. increments up to about 2.5 v. Measure the current at each applied voltage, and record the time at which the applied voltage is set at 2.5 v.

Now allow the electrolysis at an applied voltage of 2.5 v. to proceed until the solution appears completely colorless; then add enough water to cover an additional 5 to 10 mm. of the cathode. If no visible deposit of copper has appeared on the freshly covered portion of the cathode after 30 min., proceed to the next paragraph. If such a deposit does appear, repeat the above test after another hour or so. The complete deposition of copper may require as little as 20 min. (if the solution is very efficiently stirred and if the current is high, so that the solution becomes hot because of the i^2R power dissipation) or as much as 24 hours (if the solution is unstirred and the current is fairly small). Note the time at which the electrolysis is discontinued.

Observe the current flowing through the cell at the end of the electrolysis. Stop the electrolysis by lowering the solution away from the electrodes without opening the electrical circuit. Immediately wash the cathode with distilled water, place it on a clean watch glass, and dry it at 110°C. for 10 to 15 min. before weighing.

Dissolve the deposited copper by immersing the cathodes in 6 *F* nitric acid; then rinse the cathodes with water and return them and the anodes to the instructor.

Plot the current–voltage curve and find the decomposition potentials with respect to copper deposition and hydrogen evolution. Calculate the concentrations of cupric ion in the original solutions, and compare the decomposition potentials found experimentally with the expected values. Take the overpotentials for the deposition of hydrogen on copper and for the deposition of oxygen on platinum as 0.60 and 0.40 v., respectively. Explain any difference between your curve and Fig. 7-2*a*.

From the mean of the initial and final currents at an applied voltage of 2.5 v., calculate the average current efficiency for the reduction of cupric ion. What would the current efficiency have been if the electrolysis had been allowed to proceed for two hours more? What would it have been if the entire electrolysis had been carried out with an applied voltage midway between the two decomposition potentials? What sort of current–time curve would you expect to find in an electrolysis at the latter voltage?

What is the significance of the slope of the current–voltage curve of the original solution at applied voltages higher than the decomposition potential with respect to hydrogen evolution?

Calculate the percentage of copper in the copper oxalate, and compare the resulting value with the data secured in Expt. III-2.

EXPERIMENT III-12

Determination of Copper by Coulometry at Controlled Potential

Accurately weigh about 0.6 g. of solid cupric oxalate into a 125-ml. erlenmeyer flask and add 40 ml. of a solution containing 1 *F* ammonia, 2 *F* ammonium chloride, and 1 *F* hydrazine. When the salt has completely dissolved and the solution is colorless, transfer it to the working electrode compartment of a controlled-potential electrolysis cell whose central and auxiliary electrode compartments are filled with the ammonia–ammonium chloride–hydrazine solution, and pass a fairly rapid stream of nitrogen into the solution for a few minutes. Keep the nitrogen stream bubbling through the solution during the entire electrolysis. *Cf.* the footnote in Expt. III-10.

Proceed as directed in Expt. VII, carrying out the electrolysis at a working electrode potential of -0.9 ± 0.2 v. *vs.* S.C.E. When the electrolysis current has remained smaller than 1 ma. for several minutes, remove most of the electrolyzed solution from the cell without interrupting the electrolysis current. Adjust the leveling bulb of the coulometer so that the gas mixture in the buret is at atmospheric pressure, and record the volume of evolved gas, its temperature, and the barometer reading.

Discard the remaining contents of the cell, reserving the mercury for repurification and reuse.

At the option of the instructor, the percentage of zinc present in the cupric oxalate may be determined by the following procedure, provided that the total volume of solution used in dissolving the salt was measured exactly. Transfer exactly 25 ml. of the electrolyzed solution to the dropping electrode compartment of a polarographic H cell, add 1.00 ml. of 0.1 per cent Triton X-100, deaerate the mixture briefly, and obtain its polarogram, starting at about -1.0 v. *vs.* S.C.E. Without changing the height of the column of mercury above the capillary, add exactly 1 ml. of a standard 0.1 F solution of zinc nitrate or chloride, deaerate again for a minute or two, and obtain the polarogram of the resulting solution.

Recalculate the volume of gas evolved in the coulometer to standard temperature and pressure, find the quantity of electricity to which this corresponds, and compute the percentage of copper in the cupric oxalate. Compare this with the values obtained in Expts. III-2 and III-11.

The concentration of zinc in the electrolyzed solution is related to the measured diffusion current by the equation

$$(i_d)_1 = kC^0_{\text{Zn}} \tag{1}$$

The total concentration of zinc in the solution after the addition of v ml. of the standard C_s F zinc solution to V ml. of the electrolyzed solution will be

$$C_{\text{Zn}} = \frac{VC^0_{\text{Zn}^{++}} + vC_s}{V + v} \tag{2}$$

and the diffusion current of zinc in the mixture will be given by

$$(i_d)_2 = kC_{\text{Zn}} \tag{3}$$

By combining Eqs. (1) to (3), secure an equation for C^0_{Zn} which does not involve k (which need only be the same throughout the experiment), and from this find the percentage of zinc in the cupric oxalate.

The hydrazine in this experiment serves to reduce any $Cu(NH_3)_4^{++}$ ions, which would be formed at the surface of the solution if traces of air were present in the cell, before they can reach the cathode and undergo electrochemical reduction. How would the results of the experiment have been affected by omitting the hydrazine?

Where do you think the zinc in the cupric oxalate might have come from? How might you proceed to prepare a sample of cupric oxalate completely free from zinc?

Compare the half-wave potential of zinc in this medium with the literature value, -1.35 v. *vs.* S.C.E., in a 1 F ammonia–1 F ammonium chloride solution. Explain any difference between these values.

Would the concentration of zinc finally found have been affected if the potential of the working electrode had erroneously been maintained at -1.9 v. *vs.* S.C.E. during the first minute of the electrolysis? Would this have affected the quantity of copper found? Explain.

EXPERIMENT III-13

Coulometric Titration of Copper with Electrolytically Generated Ethylenediaminetetraacetate

Turn on the electrical circuit constructed in Expt. VIII and allow the current to pass through R_3 for at least 15 min. to allow the resistors to come to thermal equilibrium. R_2 should be adjusted to about 20000 ohms.

Meanwhile secure or prepare 100 ml. of an approximately 0.1 F solution of mercuric ethylenediaminetetraacetate. This can be prepared by dissolving 3.36 g. of mercuric nitrate and 3.72 g. of reagent grade disodium dihydrogen ethylenediaminetetraacetate in 100 ml. of water.[1]

Thoroughly clean a cell such as that shown in Fig. 7-8 and used in Expt. III-11. Fill the central compartment with 0.1 F ammonium nitrate, stopper it tightly, and half fill the auxiliary electrode compartment with 0.1 F ammonium nitrate. Pipet exactly 25 ml. of one of the saturated cupric oxalate solutions into the working electrode compartment of the cell and add exactly 25 ml. of the mercuric ethylenediaminetetraacetate solution, roughly 25 ml. of water, and 0.6 g. (0.075 mole) of ammonium nitrate. Turn on the stirrer and pass a moderately rapid stream of nitrogen into the solution *via* the sintered-Pyrex gas-dispersion cylinder. Place a fiber-type saturated calomel electrode in the appropriate side tube of the working electrode compartment, and immerse a helical platinum wire electrode in the auxiliary electrode compartment to serve as the auxiliary generator anode.

While the solution is being deaerated, standardize the potentiometer against the standard cell, connect it across R_2, and balance it approximately. This will expedite the exact measurement of the voltage drop across R_2 during the electrolysis.

The potentiometer should be so arranged that it can be connected either across R_2 or across the cell comprised of a mercury pool at the bottom of the working electrode compartment and the saturated calomel reference electrode. The same mercury pool will be used as the working generator cathode, and it and the platinum generator anode should be appropriately connected to the electrolysis circuit assembled in Expt. VIII.

When the deaeration has proceeded for 10 min. or so, add about 25 ml. of mercury to the working electrode compartment. Connect the positive terminal of the potentiometer circuit to this mercury pool and

[1] In order to simplify the determination of the correction necessitated by the inevitable presence of a small excess of one of these reagents in the mixture, it is highly desirable to have mercuric ion rather than ethylenediaminetetraacetate in excess. The subsequent directions are based on the assumption that this is the case.

the negative terminal to the S.C.E., and slowly add concentrated ammonia to the solution until the potential of the mercury pool becomes $+0.11 \pm 0.01$ v. *vs.* S.C.E. Record the actual value secured; this is the first point on the "titration" curve. Then connect the potentiometer across R_2 (it will be found convenient to insert a double-pole–double-throw switch in the circuit to facilitate using the potentiometer for these two different purposes), set the stop clock to zero, and throw switch S (Fig. 7-12) in the opposite direction so that the current passes through the cell.

Immediately measure the voltage drop across the standard resistor R_2, working as rapidly as possible. At the end of about 100 sec. reverse switch S, record the reading of the stop clock, connect the potentiometer across the mercury pool and the S.C.E., and measure the potential. Disconnect the potentiometer from the cell and reverse switch S so that the electrolysis current flows through the cell for another 100 sec.; then measure the potential of the mercury pool again. Continue in this fashion until the potential decreases to about $+0.08$ v. *vs.* S.C.E.; then begin using smaller generating intervals between successive potential measurements. The end point will occur at a potential of approximately $+0.03$ v. *vs.* S.C.E., and the generation intervals around the end point should be about 20 sec.

When the end point has definitely been passed, measure the voltage drop across R_2 again with the electrolysis current flowing through the cell. Discard the solution in the working electrode compartment (but not the solutions in the other compartments) and repeat the "titration" with a 50-ml. aliquot of the other saturated cupric oxalate solution. If the current is allowed to flow through R_3 between the "titrations," it will be unnecessary to allow a second warm-up period.

The stock mercuric ethylenediaminetetraacetate solution will contain a small excess of mercuric ion, and this must be determined by a separate "blank" experiment. After the completion of the "titration" of the second sample of cupric oxalate, carry out an exactly similar "titration," using exactly 25 ml. of the mercuric ethylenediaminetetraacetate solution, but with 50 ml. of water instead of the saturated cupric oxalate.

Finally empty all three compartments of the cell, rinse them thoroughly, and leave them half filled with distilled water.

Plot the potential of the mercury pool against the generation time and locate the end points of the three "titrations." Subtract the quantity of electricity required to reach the end point of the blank "titration" from each of the other end-point values to secure the quantities of electricity equivalent to the copper contents of the cupric oxalate solutions. In computing the quantity of electricity consumed in each "titration," assume that the current is given by the mean of the initial and final voltage drops across R_2, which should agree within ± 0.1 per cent for each "titration."

The reactions which occur at the generator cathode may be represented by the following equations:

$Hg^{++} + 2e = Hg$ (reduction of excess mercuric ion)

$Hg(NH_3)Y^{=} + Cu(NH_3)_4^{++} + 2e = CuY^{=} + Hg + 5NH_3$ ("titration" of copper)

$Hg(NH_3)Y^{=} + nNH_4^{+} + 2e = Hg + (n + 1)NH_3 + H_nY^{n-4}$ (beyond the end point)

where Y^{-4} represents the ethylenediaminetetraacetate radical, $(OOCCH_2)_2$-$NCH_2CH_2N(CH_2COO)_2^{-4}$. Consequently 2 faradays of electricity is consumed by each mole of cupric ion. The potential of the indicator electrode may be calculated from the equation

$$E = E^{0'}_{Hg^{++},Hg} - \frac{0.05915}{2} \log \frac{1}{[Hg^{++}]}$$

where the concentration of mercuric ion is governed by the equilibrium

$$Hg(NH_3)Y^{=} + nNH_4^{+} = Hg^{++} + (n + 1)NH_3 + H_nY^{n-4}$$

Under the conditions of this experiment $[Hg(NH_3)Y^{=}]$, $[NH_4^{+}]$, and $[NH_3]$ all remain virtually constant, and so the rapid change of the potential around the end point reflects a rapid increase in the concentration of free ethylenediaminetetraacetate in the solution.[1]

Compute the concentration of cupric ion in each of the cupric oxalate solutions, and report the solubility product of cupric oxalate.

Why is it necessary to isolate the generator anode in this experiment? (Look up a polarogram of oxygen.) Would any error be caused by the reduction of some cupric ion to metallic copper? Could the end point of this "titration" have been located by amperometric measurements? How would you modify the procedure used in this experiment if you had to determine a much smaller amount of copper? What would have happened if the sample had contained equal concentrations of copper and zinc?

Could these same reactions be used in the determination of copper by a potentiometric titration? Outline a method by which such a titration could be carried out. Do you think it would be suitable for the analysis of a saturated cupric oxalate solution?

What would be the best way to proceed in this experiment if the mercuric ethylenediaminetetraacetate solution had been accidentally made up with excess ethylenediaminetetraacetate rather than with excess mercuric ion?

[1] A more detailed discussion of the rather intricate chemistry of this experiment is given by C. N. Reilley and W. W. Porterfield, *Anal. Chem.*, **28**, 443 (1956).

EXPERIMENT III-14

Determination of Copper by Visual Colorimetry

Evaporate exactly 250 ml. of each of the saturated cupric oxalate solutions nearly to dryness on a hot plate or steam bath. Cool, add 10.0 ± 0.5 ml. of 6 *F* ammonia, transfer to a 25-ml. volumetric flask, and dilute to the mark with water. Prepare an 0.001 *F* solution of the ammino-cupric ion for use as a reference by diluting 1.00 ml. of standard 0.1 *F* cupric sulfate and 40 ± 2 ml. of ammonia to 100 ml.

Clean the colorimeter cells, fill them with the known copper solution, and place them in position on the colorimeter. Raise the cells until their bottoms touch the plungers and read the two verniers. Both will read 0.0 if the instrument is properly adjusted; otherwise consult the instructor. Turn on the lamp (if one is used) and adjust the mirror so that the two halves of the field of view are brightly and equally illuminated.

Set the right-hand cell to a vernier reading of 30.0 (mm.) and lower the left-hand cell until the two halves of the field of view become indistinguishable. Record the reading of the left-hand vernier at this point; then lower the left-hand cell several millimeters more and raise it until balance is again secured. Repeat several times; then lower the right-hand cell until its vernier reads 45.0 and repeat the measurements again.

Empty the left-hand cell, rinse it once or twice with one of the ammoniacal solutions prepared from the cupric oxalate, and fill it with this solution. Repeat the above procedure with the right-hand cell set first at 45.0 and then at 30.0 mm. Treat the other prepared cupric oxalate solution in exactly the same way.

Calculate the concentrations of copper in the solutions prepared from the saturated cupric oxalate, consulting Sec. 8-5 for a discussion of the method of treating the data. Compute the solubility product of cupric oxalate, and compare the result with the values obtained in preceding experiments.

What is the apparent precision of the measurements? Would better results have been secured by using more concentrated solutions of the cupric–ammonia complex? (*Cf.* Table 8-2.) What would have been the effect of a difference between the concentrations of ammonia in the standard and unknown solutions? Do you think that better results could have been secured by using a hydrochloric acid medium rather than an ammoniacal one? What would have happened if the unknown solutions had contained cobalt as well as the copper? How would you proceed to determine the concentration of cobalt in such a mixture without interference from the copper?

EXPERIMENT III-15

Spectrophotometric Determination of Copper

Directions for the use of the spectrophotometer will be provided by the instructor. Turn the instrument on and allow it to warm up while the solutions are being prepared for measurement.

Prepare an 0.1 per cent (1 g. per l.) solution of neo-cuproine (2,9-dimethyl-1,10-phenanthroline) in reagent grade absolute ethanol. Also prepare a 1×10^{-4} *F* solution of cupric sulfate by appropriate dilution of the standard solution used in previous experiments. By means of a microburet measure a series of samples of the 1×10^{-4} *F* copper solution into each of a series of 50-ml. separatory funnels. Take samples of 0.2, 0.4, 0.6, 0.8, 1.0, and 1.2 ml., containing amounts of copper varying from about 1 to 8 γ. Dilute to 10 ml. with water; add 5 ml. of a 10 per cent solution (100 g. per l.) of hydroxylamine hydrochloride and sufficient 6 *F* ammonium hydroxide solution to give a pH between 4 and 6. Then add 10 ml. of the reagent solution and 10 ml. of chloroform. Shake for about 30 sec., allow the layers to separate, and draw off the chloroform layer into a 25-ml. volumetric flask containing about 3 to 4 ml. of absolute ethanol. Repeat the extraction with an additional 5 ml. of chloroform, combine with the first portion in the volumetric flask, and dilute to the mark with absolute ethanol.

Measure the absorbance of the solution containing about 5 γ of copper as a function of the wavelength between 400 and 600 mμ. Use as a reference solution 15 ml. of chloroform shaken with a blank of the reagents and diluted to 25 ml. with ethanol. Similarly, measure the absorbance of each of the remaining solutions at the wavelength of maximum absorption determined from the absorption spectrum obtained for the 5-γ sample. Make a plot of absorbance against concentration. Is Beer's law obeyed? Finally, determine the concentration of copper in each of the saturated solutions of cupric oxalate, using the same procedure by which the calibration curve was obtained. Take an aliquot of the cupric oxalate solution that will permit the measurement of absorbance in the region of maximum accuracy. Use data from previous experiments to estimate the concentration of copper in the saturated solution.

Calculate the solubility and solubility product of cupric oxalate.

Why are the absorbance measurements made at the wavelength of maximum absorbance rather than at some other wavelength? Do you think that a reference solution of a chloroform–ethanol blank is really necessary, or would water be satisfactory? How would you justify your conclusion? If copper were to be determined in a solution containing other metal ions such as Ni^{++}, Co^{++}, or Fe^{+++}, and if you wished to avoid a preliminary separation, what facts would you want to know about the behaviors of the possible interferences? Do you think cyanide ion would interfere? How? Why should the absorption of the copper–ammonia complex not be used for the precise spectrophotometric determination of copper?

EXPERIMENT III-16*a*

Determination of Copper by Spectrophotometric Titration with Sodium Ethylenediaminetetraacetate

Assemble the apparatus for spectrophotometric titrations in accordance with the instructor's directions.[1] Turn on the spectrophotometer and allow it to warm up for at least 15 min.

Meanwhile pipet a 50-ml. aliquot of one of the saturated cupric oxalate solutions into the titration cell and add 20 ml. of a buffer solution prepared by adding 85 per cent phosphoric acid to 0.5 F sodium dihydrogen phosphate until the pH is 2.5 ± 0.1. Place the cell in position in the cell compartment of a spectrophotometer (consult footnote 2, p. 463, if a magnetic stirrer is used in the assembly).

Adjust the monochromator to 745 mμ and set the dark current control so that the instrument indicates zero transmittance with the shutter closed. Set the slit control to minimum slit width and the sensitivity control to minimum sensitivity. Open the shutter and adjust the controls so that the instrument indicates an absorbance of 1.0 or higher with an effective band width of 10 mμ or less.[2] Record the absorbance indicated, add 0.50 ml. of a standard 0.002 F solution of disodium dihydrogen ethylenediaminetetraacetate (prepared determinately from the reagent grade salt) from a 5- or 10-ml. microburet, stir thoroughly, and measure the absorbance again without changing any of the instrument settings.

Continue in this fashion until at least 5 ml. of the reagent has been added. Discard the solution and carry out a similar titration with an aliquot of the other saturated cupric oxalate solution.

Plot the titration curves and locate the end points. Is it necessary to apply a correction for dilution? Calculate the concentrations of cupric ion in the saturated cupric oxalate solutions and the solubility product of cupric oxalate. Compare the results secured with those obtained in Expts. III-7 and III-10.

[1] The mechanical details of this apparatus differ so greatly from one spectrophotometer to another that instructions for its assembly cannot be given here. In the following it is assumed that a 100-ml. titration cell is employed; if this is not the case, the volumes used must be adjusted accordingly.

[2] This may not be possible with some spectrophotometers. In that event set the slit control to give a band width of 10 mμ and set the sensitivity control to maximum sensitivity.

What is the structure of the cupric ethylenediaminetetraacetate complex? Why does the titration of cupric ion with ethylenediaminetetraacetate give a sharp end point, whereas that of cupric ion with a monodentate ligand such as ammonia or thiocyanate does not? (*Cf.* A. E. Martell and M. Calvin, "Chemistry of the Metal Chelate Compounds," Prentice-Hall, Inc., Englewood Cliffs, N.J., 1952, pp. 473–479.)

EXPERIMENT III-16*b*

Spectrophotometric Titration of Oxalate with Standard Permanganate

Transfer exactly 50 ml. of one of the saturated cupric oxalate solutions to a 100-ml. beaker or other vessel suitable for use as a titration cell with the spectrophotometer employed. Add 5 ml. of 3 *F* sulfuric acid and place the cell in position in the cell compartment of the spectrophotometer.

Adjust the monochromator to 525 mμ and adjust the instrument to indicate zero absorbance with the light beam passing through the titration cell. Add a freshly prepared 0.0004 *F* solution of potassium permanganate (2.00 ml. of standard 0.02 *F* permanganate diluted to 100.0 ml. with water) from a 10-ml. microburet, stirring continuously, until the absorbance becomes slightly but definitely greater than zero. Record the buret reading and the absorbance; then add several successive 0.05-ml. portions of permanganate, measuring the absorbance after each addition.

Repeat the titration with a 50-ml. aliquot of the other saturated cupric oxalate solution. At the end of the experiment clean the titration cell thoroughly by rinsing it with 0.1 *F* sulfuric acid containing a little hydrogen peroxide and then with water.

Plot the titration curves and locate the end points. Calculate the concentrations of oxalate ion in the saturated cupric oxalate solutions, and compare these values with the results of Expt. III-16*a*.

Why must the dilute permanganate solution be freshly prepared? What would be observed if the permanganate solution were allowed to stand for several days before being used? Is this a peculiarity of permanganate solutions, or do other reagents behave in the same way? Of what significance is this fact in the analytical importance of coulometry at controlled current?

Would this titration have been successful if the solution had contained 1 *F* cupric ion and 10^{-4} *F* oxalate? (What does the spectrum of cupric ion look like?) What is the criterion for interference in a spectrophotometric titration by a substance which does not interfere chemically with the reaction that takes place?

How do the accuracy and sensitivity of spectrophotometric titrations appear to compare with those of potentiometric, amperometric, and conductometric titrations? What are the chief limitations of each of these techniques?

EXPERIMENT IV-1

Preparation of Silver Acetate

Dissolve 68.0 g. (0.5 mole) of sodium acetate trihydrate in 1000 ml. of water contained in a 2000-ml. beaker. Add 6 ml. (0.1 mole) of glacial acetic acid; then, while stirring continuously, slowly add a solution of 85.0 g. (0.5 mole) of silver nitrate in 1000 ml. of water.

Let the precipitate settle and pour off the supernatant liquid. Wash the precipitate two or three times by decantation with a 1 per cent solution of acetic acid, then once or twice with water. Finally, let the precipitate settle, decant as much of the water as possible, and filter the remaining suspension through a coarse-porosity sintered-Pyrex Buchner-type filtering funnel. Suck the salt as dry as possible, spread it out on a large watch glass to dry, and cover the watch glass with another of equal size resting on a large glass triangle. Let stand for a week, with occasional stirring if possible.

EXPERIMENT IV-2

Analysis of Silver Acetate

Because the accurate determination of acetate by classical procedures is not simple experimentally, two methods are given for the determination of silver. If time permits, both of these should be done to establish the purity of the salt as certainly as possible.

1. Volumetric. Following the directions given in any textbook of elementary quantitative analysis, standardize an 0.1 F potassium thiocyanate solution by titrating weighed portions of pure dry silver nitrate by the Volhard method.

Dissolve accurately weighed 0.5- to 0.7-g. samples of the air-dried silver acetate in 100-ml. portions of 1 F nitric acid. Add 1 ml. of saturated ferric ammonium sulfate to each and titrate with the standard thiocyanate.

2. Gravimetric. Dissolve 1-g. samples of the silver acetate in 250 ml. of water containing 2 ml. of concentrated nitric acid. Slowly and with continuous stirring add a small excess of an 0.1 F solution of sodium bromide. Heat nearly to boiling and let stand for a few minutes until the precipitate coagulates completely; then test for complete precipitation by adding a few drops of the sodium bromide solution to the supernatant liquid. Wash the precipitate three or four times by decantation with

0.01 *F* nitric acid, then collect it in a weighed medium-porosity sintered-glass filtering crucible. Wash thoroughly with 0.01 *F* nitric acid, and dry to constant weight at 105 to 120°C.

Compare the analytical data with the theoretical composition of silver acetate. Assuming that water is the only impurity present, calculate the percentage purity of the salt. Is there any difference between the volumetric and gravimetric results? To what might this be due?

Place about 25 g. of the air-dried silver acetate and 1900 ml. of water in each of two 2000-ml. glass-stoppered Pyrex erlenmeyer flasks. Warm one to about 35°C. and keep it at that temperature for about an hour with frequent shaking; add more of the salt if necessary to ensure that an excess is present. Label the flasks clearly so that they can be told apart; then let them stand at room temperature (or, if possible, in a water thermostat at 25°C.) until equilibrium is attained.

EXPERIMENT IV-3

Analysis of a Saturated Silver Acetate Solution by Direct Potentiometry

Wind a length of pure silver wire into a helix as described in Sec. 3-8. Immerse the coiled portion in 6 *F* nitric acid for about 15 sec.; then rinse it very thoroughly with distilled water.

Secure a reference electrode and a potassium nitrate salt bridge from the instructor, or prepare them in accordance with his directions.[1] Place two 150-ml. beakers side by side on the laboratory bench. Fill one about half full with saturated potassium nitrate, and immerse the tip of the reference electrode in this solution. In the other beaker place 50 ml. of 0.1 *F* silver nitrate and immerse the silver electrode in this solution. Connect the silver electrode and the reference electrode to the potentiometer (which of these will be the positive electrode?), and connect the two solutions by means of the salt bridge.

Standardize the potentiometer against the standard cell; then measure the potential of the cell. Stir the silver nitrate solution briefly; then measure the potential again. The two values should agree to within ± 0.5 mv.; if they do not, continue until a steady value is secured. Then discard the silver nitrate solution and replace it with another prepared by diluting 30.0 ml. of 0.1 *F* silver nitrate to 100.0 ml. Measure the potential of the silver electrode in this solution, and proceed in the same way to solutions containing 0.01, 0.003, 0.001, 0.0003, and 0.0001 *M* silver ion.

Discard the last silver nitrate solution and place 50 ml. of one of the silver acetate solutions prepared in Expt. IV-2 in the cell. Measure the

[1] Directions for the preparation of one type of silver–silver chloride–saturated potassium chloride reference electrode may be found in Expt. V.

potential of the silver electrode in this solution, discard it, and repeat the measurement with a portion of the other silver acetate solution. Finally repeat these two measurements as a check.

Disassemble the cell and store the reference electrode and salt bridge according to the instructor's directions. Let the silver electrode stand in a small beaker containing 1 F nitric acid.

Plot the potential of the indicator electrode against log $[Ag^+]$. How well does the slope of the resulting straight line agree with the theoretical value, 0.05915 v.? Construct another plot of $E_{ind} - 0.05915 \log [Ag^+]$ against $\sqrt{[Ag^+]}$ (which is, of course, the square root of the ionic strength). What does this plot indicate about the reason why the slope of the first plot deviates from the theoretical value? Extrapolate the data secured at low silver ion concentrations to $E_{ind} = 0$, and calculate the solubility product of silver chloride. Explain any difference between this value and the literature value (1.77×10^{-10} at 25°C.).

The cell used in this experiment is

$$Ag|AgCl(s), KCl(s)||KNO_3(s)||Ag^+|Ag$$

Would you expect the potential of the cell to be changed if the potassium nitrate in the salt bridge were replaced by some other salt? Explain.

Use the plot of E_{ind} *vs.* log $[Ag^+]$ to find the concentration of silver ion in each of the silver acetate solutions. Judging from the reproducibility of the measurements, how accurate do you think the resulting solubility product of silver acetate is likely to be? To what can you attribute any differences between the values of K obtained for the two silver acetate solutions?

EXPERIMENT IV-4

Analysis of a Saturated Silver Acetate Solution by Potentiometric Titration with Potassium Thiocyanate

Using a volumetric pipet, transfer exactly 50 ml. of one of the saturated silver acetate solutions prepared in Expt. IV-2 to a 250-ml. beaker and add exactly 25 ml. of 1 F nitric acid. Construct a potentiometric titration cell according to the directions given in the first paragraph of Expt. II-4*a*.

Measure the potential of the cell. If the potentiometer cannot be balanced, it may be necessary to interchange the wires leading from the indicator and reference electrodes to the potentiometric assembly. From a 50-ml. buret add 1 ml. of standard 0.1 F potassium thiocyanate (*e.g.*, that used in Expt. IV-2), and record the potential as soon as it becomes steady. Continue adding measured volumes of the thiocyanate solution until about a 100 per cent excess has been added. Near the beginning of the titration, where the slope of the titration curve is small, a milliliter or two of the thiocyanate may be added at a time. As the end point is approached, however, the thiocyanate should be added in successively smaller portions; around the end point the thiocyanate should be added in small (*e.g.*, 0.1-ml.) uniform portions, as shown in Table 3-2.

Discard the solution and repeat the experiment with a 50-ml. portion of the other saturated silver acetate solution prepared in Expt. IV-2.

Plot the titration curve and locate the end point. Calculate the molarities of silver in the solutions and (assuming that the concentrations of the silver and acetate ions are equal in the saturated solution) the solubility product of silver acetate. Since actually

$$[Ag^+] = [OAc^-] + [HOAc]$$

what error is introduced by this assumption if K_a for acetic acid is 1.8×10^{-5}? How does this compare with your estimate of the experimental errors involved?

Look up the potential of the silver–silver chloride–saturated potassium chloride reference electrode. If the standard potential of the half-reaction $Ag^+ + e = Ag$ is +0.7991 v., what is the activity of silver ion in a solution saturated with both silver and potassium chlorides? The solubility of potassium chloride at 25°C. is 4.2 F, and its mean ionic activity coefficient at this concentration is about 0.595. From these data calculate the solubility product of silver chloride. How does this compare with the literature value?

From the calculated activity of silver ion in the reference electrode, together with the measured potential of the titration cell before any thiocyanate had been added (take the average of the values secured in the two titrations), compute the activity of silver ion in the silver acetate–nitric acid mixture. Taking into account the fact that this solution was prepared by diluting 50 ml. of the saturated silver acetate solution to 75 ml., calculate the concentration of silver ion in the saturated solution; assume that the activity coefficient of silver ion in the silver acetate–nitric acid mixture is 0.70. How does this compare with the silver ion concentration found by the titration? How do you account for the difference between these values?

Indicate how, from the data available, the solubility product of silver thiocyanate could be calculated. Carry out the computation, using the potential measured in one titration at a point at which about a 50 per cent excess of thiocyanate had been added. (Could the solubility product of silver thiocyanate also be calculated from the interpolated potential at the equivalence point? Why would this be less accurate than basing the calculation on a measurement made in the presence of a considerable excess of thiocyanate?) Suggest an explanation for any difference between the value found and the accepted value.

EXPERIMENT IV-5*a*

Measurement of the Formal Dissociation Constant of Acetic Acid

A. By Potentiometric Titration. Weigh out approximately 1 g. (7 millimoles) of sodium acetate trihydrate and dissolve it in 100 ml. of water. Also prepare an 0.1 *F* solution of perchloric acid by diluting 8.2 ml. of the 72 per cent (12.2 *F*) or 10.8 ml. of the 60 per cent (9.2 *F*) reagent grade acid to 1 l.

Transfer about 50 ml. of the sodium acetate solution to a 150- or 250-ml. beaker. Adjust a stirring motor or a magnetic stirrer, if one is available, to keep the solution in slow but constant motion.

Turn on the pH meter, let it warm up for about 5 min., and make whatever electrical adjustments are prescribed in the instruction manual for the instrument used. Then immerse the glass and calomel electrodes in a saturated solution of potassium hydrogen tartrate, and adjust the meter to read pH 3.57. Wash the electrodes thoroughly with distilled water and immerse them in the sodium acetate. Make sure that the electrodes are well clear of the stirrer.

Measure the pH, wait a minute or two, then measure it again. Repeat until a steady value is obtained; this may require several minutes, for the solution is quite poorly buffered. Then add 0.5 ml. of the 0.1 *F* perchloric acid from a 50-ml. buret and measure the pH again. Continue adding successive portions of perchloric acid until at least a 10 per cent excess has been added. It may be convenient to plot the data roughly during the titration to aid in locating the end point approximately. Try to make each portion of perchloric acid just large enough to lower the pH by about 0.25 unit.

Especially during humid weather, you may find that the operation of a magnetic stirrer influences the reading of the pH meter. This is due to the electrical current induced by the rotating magnetic field of the stirring bar. To check this, turn the stirrer off when a steady reading has been obtained with the initial sodium acetate solution and observe whether this changes the reading. If it does, the stirrer must be turned off while each successive reading is made, then turned on again to stir in the next portion of perchloric acid, and so on.

Discard the solution, rinse the electrodes thoroughly, and let them stand in a small beaker of distilled water.

Plot the data and locate the end point as accurately as possible. Call the volume of acid added at this point V^*. For each of the experimental points preceding the end point compute the value of f, the fraction of acetate ion neutralized, by means of the equation

$$f = \frac{V}{V^*}$$

where V is the volume of acid corresponding to the point in question. Also compute the value of $[H^+]$ at each point from the measured pH and the equation pH $= -\log [H^+]$. From these calculate K_a by means of the equation

$$K_a = [H^+] \frac{1 - f}{f}$$

How well does the average value agree with the literature value? (*Cf.* the Appendix of "Quantitative Analysis," by W. Rieman, III, J. D. Neuss, and B. Naiman, 3d ed., McGraw-Hill Book Company, Inc., New York, 1951.) To what is the difference due? Account for any systematic variation of the values of K_a as the value of f changes from nearly 0 to nearly 1.

B. By Direct Potentiometry. Prepare about 250 ml. of an approximately 1 F solution of acetic acid in such a way that its concentration is known within about ±2 per cent.

Turn on the pH meter, let it warm up for about 5 min., and make whatever electrical adjustments are prescribed in the instruction manual for the instrument used. Set the meter to read pH 3.57 with the glass and calomel electrodes immersed in a saturated solution of potassium hydrogen tartrate. Wash the electrodes thoroughly and immerse them in a portion of the 1 F acetic acid solution.

Measure the pH, wait a minute or two while stirring the solution occasionally, then measure the pH again. Continue until a steady value is obtained; then discard the solution and replace it with a fresh portion of 1 F acetic acid. If the pH measured with this solution is not the same as the value finally secured with the first portion, the process must be repeated until a reading is obtained which does not change when the solution is replaced by a fresh one.

Meanwhile prepare an 0.4 F solution of acetic acid by diluting 100 ml. of the 1 F solution to 250 ml. Measure its pH in the same fashion, and proceed in the same way to measure the pH of 0.16, 0.064, 0.0256, 0.0102, 0.0041, and 0.0016 F acetic acid solutions.

At the end of the experiment rinse the electrodes thoroughly with distilled water and let them stand in a small beaker of water.

From each of the measured pH values compute the concentration of hydrogen ion, assuming that $\log [H^+] = -$pH. Use these values to calculate the dissociation constant of acetic acid from each of the measured pH values. Plot these values of K_a against the square root of the ionic strength (remember that the undissociated acetic acid does not contribute to the ionic strength of the solution), and extrapolate to infinite dilution to secure the thermodynamic value of this quantity. The literature value of K_a at infinite dilution is 1.754×10^{-5}. Account for any difference between this and your value.

Which of these two methods gives the more precise value of K_a? In view of the fact that K_a varies with changing ionic strength, how do you account for this?

EXPERIMENT IV-5*b*

Measurement of the Hydrolysis Constant of Acetate Ion

Prepare about 250 ml. of an approximately 1 *F* solution of sodium acetate in such a way that its concentration is known within about ±2 per cent.

Turn on the pH meter, let it warm up for about 5 min., and make whatever electrical adjustments are prescribed in the instruction manual for the instrument used. Immerse the glass and calomel electrodes in a saturated solution of potassium hydrogen tartrate, and adjust the meter to read pH 3.57. Wash the electrodes thoroughly with distilled water and immerse them in a portion of the 1 *F* sodium acetate solution.

Measure the pH, wait a minute or two, then measure it again. Repeat until a steady value is obtained; this may require several minutes, for the solution is quite poorly buffered. Equilibrium will be attained more rapidly if the solution is stirred gently between measurements. When the meter reading becomes steady, discard the solution and replace it with a fresh portion of the 1 *F* sodium acetate. Remeasure the pH, which obviously should be the same as the final value secured with the first portion of the solution. If it is not, continue in the same way until a reading is obtained which does not change when a new portion of the solution is used. (Why is this procedure necessary when working with poorly buffered solutions?)

Meanwhile prepare an 0.4 *F* solution of sodium acetate by diluting 100 ml. of the 1 *F* solution to 250 ml. Measure the pH of this solution in the same way. Then prepare an 0.16 *F* solution by diluting 100 ml. of the 0.4 *F* solution to 250 ml. and measure its pH. Continue in the same fashion to measure the pH of 0.064, 0.0256, 0.0102, 0.0041, and 0.0016 *F* acetate solutions.

At the end of the experiment rinse the electrodes thoroughly and let them stand in a small beaker of distilled water.

From each of the measured pH values compute the concentration of hydroxyl ion in the solution, assuming that $\log [OH^-] = pH - 14$. Use these values to calculate the hydrolysis constant of acetate ion; this is the equilibrium constant of the reaction $OAc^- + H_2O = HOAc + OH^-$. Plot the resulting values of K_h against the square root of the concentration of sodium acetate and extrapolate to infinite dilution. If the thermodynamic dissociation constant of acetic acid is 1.754×10^{-5} at 25°C., what is the predicted value of the hydrolysis constant of acetate ion? Explain any difference between your value and the predicted value. Calculate the effect on the measured hydrolysis constant of an error of ± 0.05 unit in a pH measurement. What error in the concentration of sodium acetate would be required to produce the same effect on the hydrolysis constant?

What is the effect of atmospheric carbon dioxide on the results secured in this experiment? Of a trace of acetic acid present as an impurity in the sodium acetate?

EXPERIMENT IV-5*c*

Analysis of a Saturated Silver Acetate Solution by Direct Potentiometry with a Glass Electrode

This experiment should be preceded by either Expt. IV-5*a* or Expt. IV-5*b*.

Adjust the pH meter to read 3.57 in a saturated solution of potassium hydrogen tartrate. Wash off the glass and calomel electrodes thoroughly and immerse them in a portion of one of the two saturated silver acetate solutions prepared in Expt. IV-2. Measure the pH of this solution by the procedure described in Expts. IV-5*a* and *b*. When a steady value is obtained, discard the solution, wash the electrodes thoroughly, and proceed to measure the pH of the other saturated silver acetate solution.

From the pH values of these solutions, together with the experimentally measured value of K_a, calculate the concentration of silver acetate in a saturated solution. Compare the result with those secured in Expts. IV-3 and IV-4. Does the potentiometric titration or the direct potentiometric procedure give the more reliable result? Why? Do you see any reason for preferring one of the direct potentiometric methods (one with a glass electrode, the other with a silver electrode) to the other? What results would you expect to obtain if the pH measurements were made with saturated solutions of two other silver salts, AgX and AgY, if each were just as soluble as silver acetate while the values of K_a for HX and HY were 10^{-2} and 10^{-10}?

EXPERIMENT IV-5*d*

Analysis of a Saturated Silver Acetate Solution by Potentiometric Titration with Perchloric Acid

Adjust the pH meter to read 3.57 with the glass and calomel electrodes immersed in a saturated solution of potassium hydrogen tartrate.

Transfer exactly 50 ml. of one of the saturated silver acetate solutions prepared in Expt. IV-2 to a 150- or 250-ml. beaker. Adjust a stirring motor or a magnetic stirrer, if one is available, to keep the solution in slow but constant motion. Immerse the glass and calomel electrodes in this solution and measure its pH. From a 50-ml. buret add a standard 0.1 *F* perchloric acid solution (*e.g.*, standardize the solution prepared in Expt. IV-5*a* by either visual or potentiometric titration of weighed samples of borax, sodium bicarbonate, or some other standard substance provided by the instructor) in approximately 5-ml. portions, measuring the pH after the addition of each portion. When the rate of change of the pH begins to increase, start adding smaller portions of the acid; around the end point this should be added in uniform 0.5- or 1.0-ml. portions. Continue until at least a 10 per cent excess of the acid has been added.

Wash the electrodes off thoroughly and titrate a 50-ml. sample of the other saturated silver acetate solution in exactly the same way.

Plot the titration curves and locate the end point on each. Calculate the concentration of acetate in each of the saturated solutions, and from these values compute the solubility product of silver acetate. Compare the average result with the result of Expt. IV-4. Which of these do you feel to be more reliable? Why?

Following the instructions given in Expt. IV-5*a*, method A, calculate the dissociation constant of acetic acid from the data secured in both titrations. How do these values compare with those secured by some other method?

EXPERIMENT IV-5*e*

Measurement of the Solubility Product of Silver Acetate by Potentiometric "Titration" with Acetic Acid

From the results of Expts. IV-3, IV-4, and IV-5*c* or *d*, estimate the concentration of a saturated silver acetate solution as accurately as possible, and prepare about 250 ml. of a sodium acetate solution of the same molarity (within about ± 2 per cent).

Into a 100-ml. volumetric flask measure accurately a quantity of glacial acetic acid which will give an 0.1 *F* solution after dilution to the mark.

Assume that the glacial acid is 99.5 per cent pure and contains 1046 g. of acetic acid per liter. Dilute to the mark with the sodium acetate solution just prepared. Fill a 50-ml. buret with this mixture and place exactly 50 ml. of the sodium acetate solution in a 150- or 250-ml. beaker.

Adjust the pH meter to read pH 3.57 with the glass and calomel electrodes immersed in a saturated solution of potassium hydrogen tartrate. Wash the electrodes and immerse them in the beaker containing the sodium acetate solution. Measure the pH of this solution; then add 0.50 ml. of the acetic acid–sodium acetate mixture and measure the pH again. Continue this "titration" until 50 ml. of the mixture has been added. Each addition of solution should be large enough to change the pH by about 0.2 unit.

Discard the solution and rinse the electrodes; then immerse them in a mixture of exactly 50 ml. of water and 1 to 1.5 g. of pure solid silver acetate. Let this mixture stand, with frequent stirring, until no further change of pH is observed; this will probably require about 10 or 15 min.

Meanwhile calculate the value of K_a from the data just secured. The concentration of acetic acid in the solution at each point is known, the concentration of acetate ion is constant, and the hydrogen ion concentration can be calculated from the pH meter reading and the equation $pH = -\log [H^+]$. (If the liquid-junction potential is neglected—*i.e.*, if the pH meter actually gives $\log 1/a_{H^+}$—what is the relationship between the K_a thus calculated and the thermodynamic dissociation constant of the acid? This procedure actually amounts to an empirical calibration of the pH meter scale in terms of the concentration of hydrogen ion.) This gives an effective dissociation constant of acetic acid at the ionic strength of a saturated silver acetate solution.

Now "titrate" the saturated silver acetate solution with an approximately 0.1 F (known to ± 2 per cent or better) solution of acetic acid in water, using the same procedure as that employed for "titrating" the sodium acetate solution with the acetic acid–sodium acetate mixture. Continue the "titration" until 50 ml. of the acetic acid has been added. Be sure to allow enough time for solubility equilibrium to be attained after each addition of the acid.

From these data calculate the concentrations of acetic acid and hydrogen ion at each point. Then, making use of the previously measured value of K_a, calculate the concentration of acetate ion at each point. These values should naturally all be equal.

How does the solubility product of silver acetate secured in this experiment compare with the values secured by other methods? How could this experiment have been carried out if you had had no previous knowledge of the solubility of silver acetate?

EXPERIMENT IV-5*f*

Measurement of the Solubility Product of Silver Acetate by Potentiometric "Titration" with Perchloric Acid

Adjust the pH meter to read 3.57 with the glass and calomel electrodes immersed in a saturated solution of potassium hydrogen tartrate.

Place about 2 g. of pure solid silver acetate in a 250-ml. beaker and add exactly 100 ml. of water. Stir the solution until its pH becomes constant, then for a few minutes longer to make sure that it is saturated. Record the equilibrium pH; then add 0.50 ml. of standard 0.1 F perchloric acid (*cf.* Expt. IV-5*d*) from a 50-ml. buret. Stir the mixture for a minute or two and record its pH; then stir it for several minutes more and record its pH again. When a constant reading is obtained, add another 0.50 ml. of perchloric acid and measure the equilibrium pH again. Add 1.00 ml. of perchloric acid and repeat; then add 2.00, 4.00, 8.00, and 16.00 ml. of the acid, thus continuing the "titration" until a total of 32 ml. of acid has been added. Solid silver acetate must be present throughout the "titration"; more of the salt may be added if necessary.

Discard the solution, rinse the electrodes thoroughly, and let them stand in distilled water.

For each of the experimental points calculate $[H^+]$ from the pH meter reading, assuming that $pH = -\log [H^+]$; calculate $[ClO_4^-]$ from the concentration of the perchloric acid and the volume relationships involved; and calculate $[OH^-]$ from the equation $[OH^-] = 1 \times 10^{-14}/[H^+]$.

According to the electroneutrality rule, the composition of the solution at any point is defined by the equation

$$[Ag^+] + [H^+] = [OAc^-] + [ClO_4^-] + [OH^-]$$

Since $[H^+]$, $[ClO_4^-]$, and $[OH^-]$ at each point can be calculated from the experimental data, it is convenient to collect these into one term Q which is defined by the equation

$$Q = [ClO_4^-] + [OH^-] - [H^+]$$

so that

$$[Ag^+] = Q + [OAc^-]$$

Moreover, since the dissolution of silver acetate produces silver ions and acetate radicals in equal numbers,

$$[Ag^+] = [OAc^-] + [HOAc]$$

and therefore

$$Q = [HOAc]$$

By combining these equations with the expressions for K_a, the dissociation constant of acetic acid, and K_s, the solubility product of silver acetate, show that

$$K_s = \frac{Q^2 K_a}{[H^+]}\left(1 + \frac{K_a}{[H^+]}\right)$$

Using this equation, find the value of K_s corresponding to each of the points. How does the average of these values compare with the data obtained in earlier experiments? For one of the experimental points calculate the error in K_s which would result from an error of ± 0.05 pH unit in the measurement of the equilibrium pH. How would the reliability and precision of the value of K_s be affected by changes in K_a? Could the method be applied to the study of a salt like calcium carbonate?

EXPERIMENT IV-6

Determination of the Solubility of Silver Acetate by Direct Conductometry

Secure a dip-type conductance cell and a direct-reading a-c operated conductance bridge. If necessary, platinize the electrodes of the cell according to the directions in Secs. 4-3 and 5-13.

Place exactly 500 ml. of water in a clean dry 800-ml. beaker and measure its conductance. Add 1.00 ml. of 0.100 N (0.100 F) silver nitrate, stir thoroughly, and measure the conductance again. Then add successively 1.00, 3.00, 5.00, 10.0, 30.0, and 50.0 ml. of the silver nitrate, stirring thoroughly and measuring the conductance of the mixture after each addition.

Discard the solution and repeat the measurements, using first 0.1 F sodium nitrate and then 0.1 F sodium acetate in place of the silver nitrate. Then dilute exactly 100 ml. of each of the saturated silver acetate solutions prepared in Expt. IV-2 with exactly 400 ml. of water, and measure the conductances of these solutions.

Finally measure the conductance of a standard potassium chloride solution secured from the instructor. Then rinse the cell very thoroughly and leave it standing in a clean beaker with the electrodes immersed in distilled water.

Consult Expt. I-6 for a discussion of the treatment of the data.

EXPERIMENT IV-7

Analyses of Saturated Silver Acetate Solutions by Conductometric Titrations

Pipet 10.00 ml. of one of the saturated silver acetate solutions into a beaker of suitable size and add enough water to immerse completely the electrodes of a dip-type conductance cell. Measure the resistance of the cell. Add 1.00 ml. of standard 0.1 *F* lithium chloride from a 10-ml. microburet, mix thoroughly, and measure the resistance again. Continue until 10 ml. of the lithium chloride has been used.

Discard the solution and wash the cell very thoroughly, using a little ammonia to remove the last of the silver chloride. Titrate a 10.00-ml. portion of the other silver acetate solution with standard 0.1 *F* perchloric acid. Finally titrate 10.00 ml. of each solution with standard 0.1 *F* hydrochloric acid. At the end of the experiment leave the conductance cell standing with the electrodes covered with distilled water.

Construct the titration curves, locate the end points, and calculate the concentrations of the saturated silver acetate solutions. Calculate the solubility product of silver acetate, and compare this value with the results of Expt. IV-5. Why does conductometric titration give better results in this case? Under what conditions would a potentiometric titration give better results in the analysis of a silver solution than a conductometric titration?

What would have been the shape of each of the titration curves in this experiment if the silver acetate solution had contained some sodium acetate as an impurity? In the first titration, why is lithium chloride used instead of sodium or potassium chloride?

EXPERIMENT IV-8

Determination of the Solubility of Silver Acetate by Direct Polarography

Deaerate exactly 50 ml. of a saturated solution of hydrazine dihydrochloride in the solution compartment of an H cell, insert the dropping electrode, and adjust and check the apparatus according to the directions given in the fourth section of Expt. VI.

Obtain the residual current curve by following the directions in Expt. III-8. The attainable range of potentials in this supporting electrolyte extends from about $E_{d.e.} = -0.1$ v. (where mercury begins to be oxidized) to -1.0 v. (where hydrogen ion begins to be reduced). The measurements need not be made at intervals smaller than about 0.1 to 0.2 v. except near the two extremes of the attainable range.

Add three successive 1.00-ml. portions of 0.0200 *F* silver nitrate. Deaerate the solution for several minutes after each addition, then divert the gas stream over the surface of the solution, remove any gas bubbles from the tip of the capillary, and carry out the measurements as directed in the preceding paragraph. Discard the final solution, rinse the cell and capillary thoroughly, and pipet into the cell exactly 50 ml. of saturated hydrazine dihydrochloride and 1.00 ml. of one of the saturated silver acetate solutions prepared in Expt. IV-2. Deaerate the mixture and measure the limiting current at a single applied potential, say −0.5 v. Discard this solution and carry out the same measurement with the other silver acetate solution. *The height of the column of mercury above the solution must remain unchanged throughout this experiment.* (Why?)

When the measurements are completed, consult the last paragraph of Expt. VI.

At room temperature the solubility of hydrazine dihydrochloride is roughly 9 *F*, and the solubility product of silver chloride is 1.8×10^{-10}. Why does silver chloride not precipitate from these solutions? How would you proceed to secure data which you could use to describe this phenomenon quantitatively?

Plot the current–potential curves for the supporting electrolyte and for the three known silver solutions. Why is it impossible to measure the half-wave potential of silver in this supporting electrolyte? Explain this in terms of the formal potentials of the silver–silver amalgam and mercurous–mercury couples. Would you expect this to be true in other supporting electrolytes as well?

Calculate the concentration of dissolved silver in each of these solutions, measure the diffusion currents of the waves, and calculate the value of i_d/C. Calculate the diffusion current obtained with each of the silver acetate solutions, and from these and the value of i_d/C just found compute the concentration of silver ion in a saturated silver acetate solution.

What effect on the result of this experiment would you expect from the fact that the dissolved silver can oxidize the pool of mercury at the bottom of the cell? What do you think that the mercury is oxidized to?

EXPERIMENT IV-10

Amperometric Titrations of Silver with Chloride

A. At the Dropping Mercury Electrode. Using a volumetric pipet, transfer exactly 50 ml. of one of the saturated solutions of silver acetate to a polarographic H-type cell. Add 1 ml. of a 1 per cent gelatin solution and enough 6 *F* nitric acid to make the solution about 0.1 *F*. Deaerate the solution for about 5 min., then insert the dropping electrode, and adjust the mercury reservoir so that the drop time is about 4 sec. Connect the dropping electrode to the negative terminal of the polarograph (see Expt. VI) and set the potential to −0.5 v. *vs.* the saturated calomel electrode contained in the other half of the H cell.

Titrate with a standard 0.1 *F* potassium chloride solution, using a 10-ml. microburet. After each addition of reagent, deaerate the solution for about one minute and then divert the gas stream over the surface of the solution before measuring the current. Work as quickly as possible to minimize the extent of the reaction with the mercury which collects at the bottom of the cell.

B. At the Rotating Platinum Microelectrode. Repeat the titration as in part A but replace the dropping electrode with a rotating platinum electrode (see instructor for directions) at a potential of +0.1 v. *vs.* S.C.E. It is not necessary to deaerate the solution. Why?

Use each method for the titration of an aliquot of each of the saturated solutions of silver acetate.

Make plots of the titration data and explain the shape of each graph. Locate the end points of the titrations and calculate the concentrations of silver in the silver acetate solutions. Then calculate the solubility product of silver acetate. Compare with the values obtained by other methods. Which electrode gives the more satisfactory titration? What would be the effect of neglecting to apply the dilution correction? May the gelatin be omitted in either case? Explain. What would be the effect of adding acetone or alcohol to the titration medium? What effects would omission of the nitric acid have? Would it be preferable to titrate with bromide, iodide, or thiocyanate? Explain. How would you proceed to analyze a solution containing 0.1 m*M* silver ion by an amperometric titration?

EXPERIMENT IV-12

Determination of Silver by Coulometry at Controlled Potential

Following the directions given in Expt. VII, fill the central and auxiliary electrode compartments of a cell for controlled-potential electrolysis with 1 *F* perchloric acid. Pipet exactly 25 ml. of one of the saturated silver acetate solutions into the working electrode compartment, add 2 to 3 ml. of 72 per cent perchloric acid, and immerse a clean platinum gauze electrode in the solution. Add 1 *F* perchloric acid if necessary to cover the gauze surface of the electrode. Turn on the stirrer (which should be positioned coaxially with the working electrode), and pass a slow stream of nitrogen into the solution for 5 to 10 min.[1]

Proceed according to the directions given in Expt. VII, carrying out the electrolysis at a working electrode potential of +0.15 ±0.1 v. *vs.* S.C.E. Continue the electrolysis until the current has fallen to a value smaller than 1 ma. and remained there for 5 to 10 min. At the end of this time adjust the leveling bulb of the coulometer so that the evolved gas is just at atmospheric pressure, then drain the cell, open switch S', and wash

[1] Silver ion can be completely reduced at potentials much too positive to permit the reduction of oxygen. Deaerating the solution here is merely a precaution to avoid the error which would otherwise be incurred if the working electrode potential happened to drift to a value as negative as +0.05 v. at any time during the experiment.

the electrodes and cell thoroughly with distilled water. Place the working electrode in a beaker containing warm dilute nitric acid to dissolve any silver which still adheres to it.[1]

Now read the volume of gas evolved in the coulometer, its temperature, and the barometric pressure. Finish stripping the deposit off the working electrode, and return the latter to the instructor.

Correct the volume of gas liberated to standard temperature and pressure, and compute the concentration of silver ion in the saturated silver acetate solution. Calculate the solubility product of silver acetate, and compare the results with those of preceding experiments.

From the potentials of the half-reactions involved estimate the fraction of the silver which would remain undeposited at the end of a sufficiently prolonged electrolysis under the conditions of this experiment. What would be the highest concentration of copper that could have been present at the start of the electrolysis without causing any co-deposition of copper with the silver? If the silver acetate solution were suspected of being contaminated with copper, how would you proceed to estimate the concentration of copper present after removing the silver ion?

What are the smallest amounts of silver that could be determined with an accuracy of ± 1 per cent by polarography and by coulometry at controlled potential, using the procedures described here and in Expt. IV-8? What modifications would you make in each of these procedures if it were necessary to determine one-tenth of this quantity of silver without any sacrifice of accuracy?

EXPERIMENT IV-15

Determination of Silver by Direct Spectrophotometry

Prepare a 1.00×10^{-3} M solution of silver ion by diluting exactly 1.00 ml. of standard 0.1 F silver nitrate to 100 ml. with 0.05 F nitric acid. Similarly dilute 1.00 ml. of each of the saturated silver acetate solutions to 100 ml. with 0.05 F nitric acid.

Thoroughly clean seven 100-ml. volumetric flasks (preferably Pyrex). Into four of these measure 0.300, 0.600, 1.000, and 1.500 ml. of the 1×10^{-3} F silver nitrate, using a 2- or 5-ml. microburet if one is available. Measure exactly 1 ml. of each of the prepared silver acetate solutions into

[1] The silver deposited from this solution is very poorly adherent, and can hardly be collected for weighing without serious danger of mechanical loss. If it is desired to weigh the silver deposited in order to compare its weight with the coulometer reading, the perchloric acid may be replaced with 1 F ammonia–1 F ammonium nitrate. Consult the instructor for directions. CAUTION: Ammoniacal silver solutions decompose slowly and may eventually explode with shattering violence. An ammoniacal silver solution which is as much as two or three hours old should be drowned in a large excess of water without delay.

the fifth and sixth flasks. The seventh flask will be used to prepare a blank solution for making the 100 per cent transmittance settings on the spectrophotometer. Add 75 ± 5 ml. of 0.05 *F* nitric acid to each of the flasks, followed by 2.00 ml. of 0.05 per cent rhodanine solution prepared by dissolving 0.05 g. of the reagent in 100 ml. of 95 per cent ethanol and filtering through a fine-porosity sintered-glass filter. Immediately dilute to the mark with 0.05 *F* nitric acid and mix thoroughly.

Rinse and then fill one of a pair of matched absorption cells with the blank solution, and rinse and fill the other cell with the most concentrated of the known silver solutions. Place the cells in the cell compartment of the spectrophotometer and, using the blank solution to make the 100 per cent transmittance setting at each wavelength, measure the absorbance of the known solution as a function of wavelength between 400 and 750 mμ. The measurements may be made at widely spaced wavelengths in the regions where the absorbance is small, but should be made at 10-mμ intervals in the vicinity of the absorbance maximum.

Discard the known solution and treat the other three knowns in exactly the same way. Finally measure the absorbance of each of the solutions prepared from the silver acetate; this need be done only at the wavelength of maximum absorption. When the measurements have been completed, rinse the cells thoroughly with water and then with 95 per cent ethanol, dry them carefully with lens tissue, and return them to the instructor.

The reagent used in this experiment is

HN——C=O
S=C C=CH⟨benzene⟩NR_2
S

where R may be either methyl or ethyl. This reacts with silver ion according to the equation

HN——C=O
S=C C=CH⟨benzene⟩$NR_2 + Ag^+ =$
S

AgN——C=O
S=C C=CH⟨benzene⟩$NR_2 + H^+$
S

and it is the resulting slightly soluble compound which is responsible for the color observed.

From the data secured with the four known solutions construct a plot of absorbance *vs.* concentration at the wavelength of maximum absorption. Does the system obey Beer's law? What is the most probable explanation of its failure to do so at high silver concentrations? On the graph locate the absorbances measured with the two silver acetate solutions, and compute the solubility product of silver acetate from the corresponding concentrations. What is your estimate of the error of the resulting value? Compare the probable accuracy which could be secured by this method with that obtainable by such other methods as direct conductometry if they were applied to the analysis of a saturated solution of silver chloride.

Judging from your knowledge of the conditions required to ensure success in the Fajans titration of chloride with silver, what do you think would have been the effect on this experiment of the presence of a high concentration of potassium nitrate in each solution? How do you think this analysis could be successfully carried out in the presence of a high concentration of potassium nitrate? Do you consider this system to be suitable for accurate analyses using a filter photometer and a 20-mμ band-width filter?

From the data secured with the most dilute known solutions, calculate the mean absorptivity of silver rhodaninate at the wavelength of maximum absorption. What are the principal factors which affect the accuracy of this value?

EXPERIMENT IV-16

Spectrophotometric Titration of Acetate with Perchloric Acid

Secure a stock 0.1 per cent solution of either methyl orange, bromphenol blue, or (preferably) methyl yellow. By dilution prepare two 0.001 per cent solutions of the indicator: one in 0.1 *F* perchloric acid, the other in a buffer of pH 6 to 7. Transfer one of the solutions to an absorption cell having a path length of 5 cm., and fill a similar cell with water. Measure the absorbance of the indicator solution as a function of wavelength between about 400 and 750 mμ; then replace this indicator solution with the other one and repeat the measurements.

Locate the wavelength at which the difference $A_a - A_b$ between the absorbances of the acidic and basic forms of the indicator is greatest,[1] and carry out the remainder of the experiment at this wavelength.

Assemble the apparatus for spectrophotometric titrations in accordance with the instructor's directions. In the following it is assumed that the titration cell has a capacity of about 100 ml.; if this is not the case, the volumes used must be adjusted accordingly.

[1] The titration can be carried out about equally well at a wavelength where $A_b - A_a$ is a maximum. Consult the instructor for directions.

Pipet exactly 50 ml. of one of the saturated silver acetate solutions into the cell and add 0.5 ml. of the stock indicator solution. Adjust the controls of the spectrophotometer so that it reads zero absorbance at the wavelength selected. From a 10-ml. microburet add 0.250 ml. of standard 1 F perchloric acid, stir thoroughly, and measure the absorbance of the mixture. Continue in the same way until a constant absorbance is reached which does not change on addition of more acid. Then discard the solution and repeat the titration with a 50-ml. aliquot of the other saturated silver acetate solution.

Correct the measured absorbances for dilution by multiplying them by $(V + v)/V$, where V is the volume of silver acetate solution used and v is the volume of perchloric acid added. If the corrected absorbance at any volume before the equivalence point is A_v, and if the corrected constant absorbance observed in the presence of excess acid is A_c, show that

$$\frac{[\mathrm{In}^-]}{[\mathrm{HIn}]} = \frac{K_{\mathrm{HIn}}}{[\mathrm{H}^+]} = \frac{A_c - A_v}{A_v} = \frac{K_{\mathrm{HIn}}[\mathrm{OAc}^-]}{K_{\mathrm{HOAc}}[\mathrm{HOAc}]} \tag{1}$$

Further, show that, except for values of v very close to either 0 or V^*, where V^* is the volume of acid required to reach the equivalence point,

$$\frac{[\mathrm{OAc}^-]}{[\mathrm{HOAc}]} = \frac{V^* - v}{v} = \frac{V^*}{v} - 1 \tag{2}$$

and consequently that a plot of $(A_c - A_v)/A_v$ *vs.* $1/v$ will be a straight line whose zero intercept is V^*.

Construct such a plot for each of the titrations. Extrapolate the two straight-line segments of each plot to their point of intersection to find V^* (why is this procedure preferable to a simple extrapolation of the data secured prior to the equivalence point to zero?), and from these values compute the concentrations of acetate ion in the silver acetate solutions and the solubility product of silver acetate. Compare these data with those obtained in preceding experiments, especially Expt. IV-5*d*. Explain why the spectrophotometric titration permits the end point to be more accurately located than does the potentiometric procedure.

For further details concerning the use of indicators in spectrophotometric acid-base titrations, consult T. Higuchi, C. Rehm, and C. Barnstein, *Anal. Chem.*, **28**, 1506 (1956).

EXPERIMENT V

Assembly of Potentiometric Apparatus

1. Electrical Circuit. The following apparatus is required:

P, Leeds and Northrup student-type potentiometer
R, four-dial decade resistance box (999.9 ohms $\times$ 0.1 ohm steps)

SC, unsaturated Weston standard cell
S, knife-type double-pole–double-throw (DPDT) switch
G, enclosed lamp-and-scale type a-c operated galvanometer
E, two 1.5-v. dry cells
K, tapping key
D, 1000-ohm linear-taper radio potentiometer

The apparatus is connected together as shown in Fig. 3-8 (page 42).

Before testing the assembly of the apparatus, make sure that it is conveniently arranged on the bench. The potentiometer and key should be toward the front since adjustments of these will be made frequently. The galvanometer should be in a position where it can be read easily. The resistance box and switch may be placed to one side, since these are less frequently adjusted, and the batteries and standard cell may be placed in the rear of the assembly.

To test the setup, set the potentiometer to the potential of the standard cell, close the DPDT switch to connect the standard cell into the circuit, set the resistance box to 999.9 ohms, and very rapidly close the tapping key. (Important! Do not hold the key in closed position as this will cause a large deflection in the galvanometer when the imbalance of the circuit is large. Permanent damage may then occur to the galvanometer.) Note the direction of deflection of the light spot on the galvanometer scale when the key is closed. Change the setting on the resistance box to zero and again note the direction of deflection of the galvanometer when the key is tapped. The deflection should be in the opposite direction. If not, reverse the polarity of the standard cell and repeat the test. If you still do not secure opposite deflections of the galvanometer with resistance settings of zero and 999.9 ohms, either the wiring is incorrect or one of the components of the circuit is defective.

Now standardize the potentiometer by adjusting the external resistance until no deflection is observed on tapping the key, leaving the potentiometer set to the potential of the standard cell. A resistance of about 150 ohms is usually required. As the batteries age the setting will have to be changed slightly to compensate for the drain on the batteries.

A fully charged battery changes in voltage rather rapidly during the first few per cent of its life; thereafter its voltage remains more nearly constant until fully discharged. To avoid errors which may result from this initial relatively rapid change, it is necessary to repeat the standardization at frequent intervals during the first few hours of use.

2. Preparation of Silver–Silver Chloride Reference Electrode. Add 3.5 g. of powdered agar to 90 ml. of water in a 125-ml. erlenmeyer flask. Place the flask in a large beaker of water or on a steam bath and heat until the agar is completely dissolved. This will probably not require more than 15 or 20 min. Do not mistake air bubbles for particles of undissolved agar. Then add 30 g. of reagent grade potassium chloride and stir gently until this is dissolved.

Meanwhile secure a 10-mm. Pyrex "sealing tube" with a medium-porosity fritted disc (Corning No. 39570). Cut off one end of the tube as close to the disc as possible and *gently* "sand" away the remainder of that end, using a clean wire gauze, so that it is flush with the surface of the fritted disc. Wash thoroughly and remove any excess water from the inside of the tube with a pipe cleaner or a small piece of rolled-up filter paper. Using a length of scrap glass tubing as a pipet, add enough of the molten agar gel to form a 5- to 10-mm. layer above the fritted disc. Let stand until the gel has completely solidified. Meanwhile remove the flask containing the rest of the gel from the heat. When it has cooled, stopper it tightly and reserve it for later use.

Fill the tube above the solidified gel with solid potassium chloride; then slowly add water until the excess salt is covered by about 5 mm. of solution. Dissolve a few crystals of silver nitrate in a few drops of water, and add this to the saturated potassium chloride until a permanent precipitate of silver chloride is formed.

Wind an 8-in. length of 16-gauge or heavier pure silver wire into a helix about an inch long and not more than 3/8 in. in diameter. Leave about 2 in. of the wire uncoiled at one end. Immerse the coiled portion of the wire in 6 F nitric acid for a minute; then rinse it thoroughly with water. Do not touch the cleaned portion of the electrode with your fingers for fear of forming a trace of silver sulfide on its surface.

Mount the wire in the electrode, with the entire helical portion covered with solution. Bore a very small hole in a No. 00 neoprene or gum rubber stopper, and force this down over the straight portion of the silver wire and into the tube. Apply a thin film of petroleum jelly or stopcock grease to the entire exposed portion of the stopper to prevent "creeping" of the potassium chloride.

Always keep the electrode in a small beaker of saturated potassium chloride when it is not in use. Allow a day or two for equilibrium to be attained before using the electrode for the first time. The electrode may be used immediately if an anodic current of a few milliamperes is first passed through it for a minute or two. Consult the instructor for directions.

3. Testing the Apparatus. Secure a saturated calomel electrode from the instructor and place it and the silver–silver chloride–saturated potassium chloride electrode in a small beaker of saturated potassium chloride. Connect the electrodes to the potentiometer circuit at the points indicated in Fig. 3-8, consulting Table B of the Appendix in order to find which will be the positive electrode.

Standardize the potentiometer scale against the standard cell as directed in part 1 above. Then throw switch S in the opposite direction, set the potentiometer to read zero, tap the key, and note the direction in which the galvanometer deflects. Repeat with the potentiometer set to read 0.100 v. (If the galvanometer deflects in the same direction as before, interchange the wires leading to the two electrodes.) Finally adjust the potentiometer until no galvanometer deflection can be detected when the tapping key is depressed. Record the reading of the potentiometer and the polarities of the electrodes.

Discard the beaker of saturated potassium chloride, replace it with a beaker of saturated potassium nitrate, and repeat the measurement.

The standard potentials of the half-reactions involved in this experiment are

$$AgCl + e = Ag + Cl^- \qquad E^0 = +0.2222 \text{ v.}$$
$$Hg_2Cl_2 + 2e = 2Hg + 2Cl^- \qquad E^0 = +0.2676 \text{ v.}$$

What would be the potential of the cell

$$Ag|AgCl(s),\ KCl(c),\ Hg_2Cl_2(s)|Hg$$

at infinite dilution? Which electrode should be the positive electrode?

What is your value for the potential of this cell with $c = 4.2\ F$ (the solubility of potassium chloride at 25°C.)? Which did you find to be the positive electrode? Would you expect the potentials at $c = 0$ and $c = 4.2\ F$ to be the same? Why or why not? (Silver chloride is appreciably soluble in concentrated chloride solutions, giving $AgCl_n^{-(n-1)}$ ions, where $n = 2$, 3, and 4. What effect does this have on the potential of the silver electrode?)

Is there a liquid-junction potential in any of the following cells?

$$Ag|AgCl(s),\ KCl(c),\ Hg_2Cl_2(s)|Hg$$
$$Ag|AgCl(s),\ KCl(c)\|KCl(c)\|KCl(c),\ Hg_2Cl_2(s)|Hg$$
$$Ag|AgCl(s),\ KCl(c)\|KNO_3(c)\|KCl(c),\ Hg_2Cl_2(s)|Hg$$

Did replacing the potassium chloride "bridge" with potassium nitrate change the measured potential of the cell? Explain.

Write the equation for the cell reaction. What is its equilibrium constant? Attempt to calculate the equilibrium constant at infinite dilution from the standard potentials given above. How do you explain this contradiction? If two moles each of silver, silver chloride, and mercury are brought into contact with one mole of mercurous chloride, what spontaneous reaction will occur? When will it stop?

EXPERIMENT VI

Assembly of Apparatus for Polarographic Measurements

1. Electrical Circuit. The following apparatus is required:

H, 100-ohm linear voltage divider (*e.g.*, Model T-10A Helipot)
R, 10,000-ohm precision resistor (*e.g.*, General Radio Co. Type 500-J)
S_1, S_2, S_3, knife-type DPDT switches
E, two 1.5-v. dry cells

Construct the electrical circuit shown in Fig. 1, making the appropriate connections to the potentiometer assembly set up in Expt. V. Make

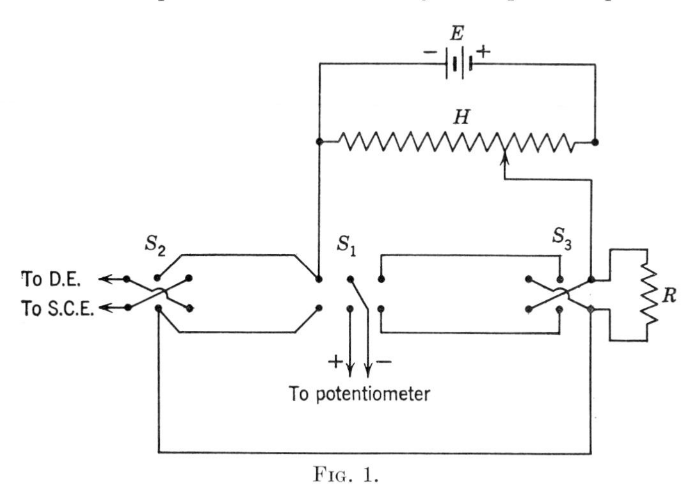

FIG. 1.

sure that you thoroughly understand the function of each component of this circuit before you attempt to put it to use.

2. Cell and Reference Electrode. Thoroughly clean an H-type cell and support it in a clamp with the cross-member vertical and the dropping electrode compartment down.

Prepare an agar–potassium chloride gel according to the directions in part 2 of Expt. V, or simply reheat the gel prepared in that experiment over a water or steam bath until it has liquefied. Using a piece of fairly large (*ca.* 12 mm. i.d.) glass tubing as a pipet, fill the cross-member of the cell with the molten gel. The remainder of the gel may be kept indefinitely in the tightly stoppered flask.

When the gel in the cross-member of the cell has congealed (it will appear perfectly white), remove the cell from the clamp and, while holding it in the same position, let cold water run over the outside of the cross-member until the gel is at or near room temperature.

Now support the cell in an upright position and pour enough pure mercury into the reference electrode compartment to give a pool about 10 mm. deep. Select a one-hole rubber stopper which fits the reference electrode compartment of the cell and into this insert a length of glass tubing of convenient size. This should be long enough so that, when the stopper is in place, the bottom of the tubing will be covered by the mercury, while its top will project 5 to 10 mm. above the top of the stopper. Fire polish the bottom end of the tube; then lay a 2-cm. length of 18- or 20-gauge platinum wire inside this end so that about half of the wire protrudes from the glass. Heat this end of the tube strongly until it melts around the platinum; make sure that no tiny hole remains. Let the tube cool to room temperature, insert it in the stopper, fill it to the level of the top of the stopper with mercury, and lay it aside.

To the reference electrode compartment of the cell add enough mercurous chloride to form a 3- to 5-mm. layer on top of the mercury, then fill this compartment to the top with solid potassium chloride. Slowly add water until the salt which remains undissolved is just covered by the solution. Insert the stopper tightly and apply a thin film of stopcock grease or petroleum jelly to the entire exposed portion of the stopper. (This prevents "creeping" of potassium chloride.)

Select a solid stopper which just fits the solution compartment of the cell. In the center of this stopper bore a hole large enough to permit the capillary to be inserted and removed without touching the stopper. Bore another hole near the rim of the stopper and in this insert an L-shaped piece of glass tubing about 1 in. on each side. This should barely protrude below the bottom of the stopper. When measurements are being made, an inert gas such as nitrogen or hydrogen will be passed through this tube to form a protective blanket over the solution in order to exclude dissolved air.

Add just 50 ml. of saturated potassium chloride solution to the dropping electrode compartment of the cell (this should completely cover the sintered disc), and let the cell stand for a day or two, if possible, before use.

3. Dropping Electrode Assembly. Attach a 3-in. ring clamp to a 36-in. tripod stand and place a 125-ml. leveling bulb in the ring. Insert a two-hole rubber stopper in the neck of the bulb. In one of its holes mount a glass tube filled with mercury and having a short length of platinum wire sealed through the bottom end. This should be prepared according to the directions given in the fourth paragraph of part 2 above. The tube should extend nearly to the bottom of the leveling bulb. In the other hole in the stopper place a short piece of glass tubing which has been bent into a U shape; this will help to keep dust out of the leveling bulb.

Attach a 24-in. length of sulfur-free tubing (Tygon, neoprene, Scimatco, etc.) to the leveling bulb. Secure a freshly cut 4- to 6-in. length of marine barometer tubing and attach this to the other end of the tubing. The newly cut end of the capillary should be placed at the bottom. Wind some fairly stout copper or stainless-steel wire around the tubing outside the capillary, and twist this tight so that no mercury can flow out around the capillary. Slit a 2-in. length of heavy pressure tubing lengthwise and slip this over the tubing around the wire. Arrange a clamp (on the same tripod stand that carries the leveling bulb) in such a way that it grips the pressure tubing and holds the capillary in a perfectly vertical position. Set the cell in its holder on the bench beside the capillary and move the latter up or down until its tip is about 1 cm. below the level of the 50 ml. of solution in the dropping electrode compartment of the cell. Do not put the tip of the capillary into the solution. Mark the position of the clamp on the tripod stand so that it can be reproduced at will.

Place a beaker on the bench underneath the capillary. Raise the leveling bulb as high as it will go, so that the tubing is perfectly vertical, and slowly pour pure mercury into the bulb until the tubing is entirely full of mercury and the leveling bulb is about half full. Make sure that there are no air bubbles anywhere in the tube.

Insert the stopper in the leveling bulb. When mercury begins to form in droplets on the tip of the capillary, lower the leveling bulb slowly until the flow of mercury just stops. Raise the leveling bulb 1 cm. above this point and mark the position of the ring clamp. Slip a screw-type pinchcock over the tubing just above the pressure tubing. Whenever the assembly has to be left for more than a few minutes, the capillary should first be washed very thoroughly with mercury flowing through it. Then the leveling bulb should be lowered to the point you have just marked, and finally the stopcock should be tightly closed.

Connect the mercury in the leveling bulb to the point in Fig. 1 marked "to D.E." Be sure that there is enough slack in this wire to permit the leveling bulb to be moved up and down freely. Similarly connect the reference electrode to the point in Fig. 1 marked "to S.C.E."

4. Testing the Apparatus. Empty the solution compartment of the cell (preferably by sucking the liquid out with an aspirator tube, rather than by turning the cell upside down) and rinse it thoroughly with water. Then rinse it with a small portion of a 1 *F* potassium chloride solution containing 1 m*M* cadmium ion, suck this out, and add 50 ml. of this solution to the cell. Insert the stopper into the solution compartment of the cell and bubble a slow stream of nitrogen or hydrogen through the solution.

Meanwhile raise the leveling bulb to a height of about 50 cm. above the tip of the capillary. Release the pinchcock and make sure that the entire dropping electrode assembly is free from air bubbles. Insert the dropping electrode in the solution, adjusting the clamp which holds the capillary to the position previously marked. The capillary must be accurately vertical.

While nitrogen or hydrogen continues to bubble through the solution, throw switch S (Expt. V) to the SC position and standardize the potentiometer scale against the standard cell. Return S to the e.m.f. position, and (referring to Fig. 1 above) set H at about 0.30 and throw S_1, S_2, and S_3 to the left. This impresses a voltage of about 0.9 v. across the series combination of the cell and the resistance R, and connects the potentiometer across the cell for measurement of $E_{d.e.}$. Note that this makes the dropping electrode the *negative* electrode of the cell. In all of the later work be careful *never* to subject the dropping electrode to any large *positive* potential.

Disconnect the gas delivery tube from the side arm of the cell and attach it to the L-shaped tube in the stopper of the solution compartment. If any gas bubbles are present on the capillary tip, tap the capillary gently until they are dislodged. Check to make sure that the tip of the capillary is immersed in the solution, that the mercury column in the dropping electrode assembly is continuous and free from air bubbles, that the platinum wires in the reference electrode and in the leveling bulb make contact with the mercury, and that all of the other connections are also satisfactory.

Now adjust the radio potentiometer across the galvanometer terminals so that the galvanometer is nearly but not quite shorted out. If the connections are made as indicated in Fig. 3-8, this will involve turning the shaft of the radio potentiometer counterclockwise nearly to the stop. Set the potentiometer (the students' potentiometer, not the radio potentiometer) to 0.70 v. and depress the tapping key. Note the direction in which the galvanometer spot deflects. Then repeat this operation with the potentiometer set to read 1.10 v. The galvanometer spot should move in the opposite direction. If it does not, interchange the two wires leading from switch S_1 to the potentiometer circuit. When this test shows that these wires are correctly connected, adjust the potentiometer so that the galvanometer spot oscillates back and forth around zero. It may be helpful in making the final adjustment to turn the shaft of the radio potentiometer slightly clockwise to decrease the galvanometer damping and thus increase the amplitude of the oscillations. Note that the key K must be held down and not merely tapped.

These oscillations reflect the fact that the potential of the dropping electrode changes continuously during the life of each drop. The total voltage E_H derived from the voltage divider H is distributed across both the cell and the series resistance R:

$$E_H = E_{\text{cell}} + E_R \tag{1}$$

where E_{cell} is the voltage drop across the cell and E_R is the voltage drop across the resistance R. By Ohm's law the latter is equal to the product of the current i and the resistance R of the resistor. If the cell is properly constructed and if the current is not too large we can neglect the iR drop through the cell itself and say that the total voltage drop across the cell is equal to the potential difference between its two electrodes. Then

$$E_H = E_{\text{d.e.}} + iR \tag{2}$$

Since both E_H and R are constant, any variation of i must be accompanied by a variation of $E_{\text{d.e.}}$. At the instant when a drop has just begun to form on the tip of the capillary, i is zero [Eq. (6-8)], and $E_{\text{d.e.}}$ is equal to E_H. However, as the drop grows, i increases, and consequently $E_{\text{d.e.}}$ becomes smaller until the instant at which the drop falls. Then i decreases abruptly to zero, and $E_{\text{d.e.}}$ again becomes equal to E_H. So the value of $E_{\text{d.e.}}$ oscillates between two extremes. The experimental technique described above gives an average value of $E_{\text{d.e.}}$ during the life of the drop.

Now that $E_{\text{d.e.}}$ has been measured, throw switch S_1 to the right, so that the potentiometer is connected across the resistance R, and set the potentiometer to zero. Depress the tapping key and observe the direction in which the galvanometer spot deflects. Set the potentiometer to read 0.2 v. and repeat the above operation. The spot should now move in the other direction. Proceed to measure E_R by the same method previously used to measure $E_{\text{d.e.}}$. The resulting value of E_R is the average value of the iR drop across R during the life of a drop, and so it is equal to $10^4\bar{i}$ (since R is 10000 ohms), where $\bar{i}$ is the average current (in amperes).

You have now measured the average current at one value of $E_{\text{d.e.}}$; this gives you one point on the current–voltage curve or polarogram. Repeat the measurements at a number of other settings of the voltage divider H (record the exact setting of H at each point). It is convenient to plot the data roughly as you go. Where the slope of the curve is small, the measurements can be made at fairly widely spaced values of $E_{\text{d.e.}}$. Especially near the half-wave potential (-0.648 v. *vs.* S.C.E., according to the literature), however, the points should be spaced much more closely in order that no important feature of the curve will be overlooked.

Since the potentiometer scale extends only up to 1.6 v., this is the highest value of $E_{\text{d.e.}}$ accessible by the procedure just described. However, it is frequently desirable to secure data at still higher potentials. One way to do this involves doubling the range of the potentiometer by setting it to read one-half the potential of the standard cell, adjusting the resistance R (Fig. 3-8, p. 42) until the potentiometer is balanced at this setting. (It will be necessary to use three dry cells rather than two as shown in the figure.) Now the reading of the potentiometer will be exactly half of the actual value of any unknown potential. In this way the instrument can be used for the measurement of potentials up to 3.2 v. When H is a precision voltage divider, however, a still simpler method is available.

Suppose, for example, that when H is set to 0.30 the values of E_{cell} and E_R are measured directly and found to be 0.845 v. and 0.058 v., respectively. Then E_H is 0.903 v.; since this is just $3/10$ of the voltage impressed across H by the two dry cells, the latter must be 3.01 v. Now suppose that H is set at 0.60 and that E_R is found by direct measurement to be 0.085 v. It is easy to compute E_{cell} from the equation

$$E_{cell} = (0.60)(3.01) - 0.085 = 1.721 \text{ v.}$$

with an accuracy entirely sufficient for any practical purpose.

The half-wave potential of the cadmium wave on your polarogram should be very close to -0.648 v., and its diffusion current should be of the order of 10 microamp. Show your polarogram to the instructor for approval.

Remove the capillary from the cell. Holding it over a beaker, wash it very thoroughly with distilled water. Then lower the leveling bulb to the position marked according to the directions in the fourth paragraph of part 3 above, and close the pinchcock tightly. Thoroughly rinse the solution compartment of the cell and fill it with saturated potassium chloride. Disconnect the two battery circuits and turn off the nitrogen or hydrogen stream before leaving the laboratory.

EXPERIMENT VII

Assembly and Use of Equipment for Controlled-potential Electrolysis and Coulometry at Controlled Potential

Secure a cell such as that shown in Fig. 7-8,[1] and mount it securely on a sturdy ring stand. Insert a commercial fibre-type saturated calomel electrode in the appropriate side tube of the conical solution compartment. In a solid rubber stopper selected to fit this compartment, bore a hole large enough to leave a 3-mm.-wide annular ring around the shaft of a glass paddle-type stirrer.[2] Insert the stirrer into the cell

[1] These cells are available from Analytical Instruments, Inc., Bristol, Conn., Catalog C-2. For use in Expt. IV-12 the same company's Catalog C-3 cell is preferred. The working electrode compartment of this cell is cylindrical, rather than conical, and this greatly facilitates the use of an ordinary platinum gauze cathode as the working electrode.

[2] Convenient stirrers are available from E. H. Sargent and Co., Chicago, Ill., Catalog S-76667.

and mount the stopper in place. Attach the stirrer to a stirring motor[1] and check to make sure that the stirrer will rotate freely, striking neither the walls of the cell nor the sealed-in gas-dispersion cylinder. The bottom of the stirrer blades should be within 5 mm. of the bottom of the cell, so that when mercury is added to the cell the blades of the stirrer will trail in the mercury. Only in this way can efficient stirring of the mercury–solution interface be secured.

Connect the tube leading to the gas-dispersion cylinder to a source of nitrogen, preferably prepurified.[2] Place a helical platinum wire electrode in the auxiliary electrode compartment, and keep the cell about half filled with water when it is not in use.

Assemble the electrical circuit shown in Fig. 1. The wires from the switch S'' and the potentiometer to the negative (working) electrode of the cell should be connected by a heavy battery connector to a straightened paper clip or other stout wire insoluble in mercury. One of the side tubes of the cell carries a heavy platinum wire which makes contact with the mercury working electrode. Half fill this tube with mercury and immerse the ends of the wires in the mercury. Connect the lead from the positive terminal of the potentiometer to the reference electrode. The remaining lead from the electrolysis circuit should be connected to the platinum auxiliary electrode by means of a battery connector.

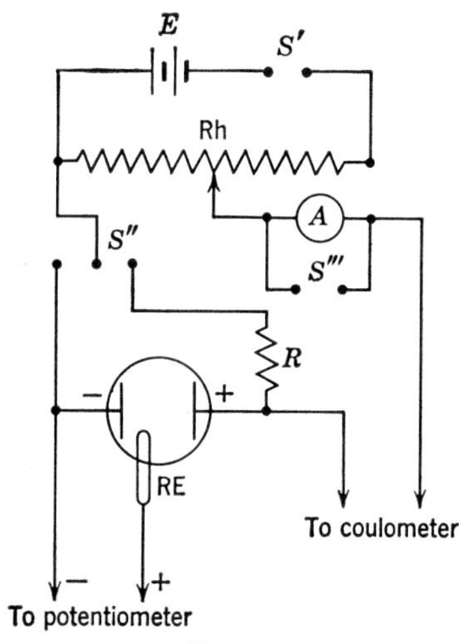

FIG. 1.

The potentiometer may be replaced by a vacuum-tube voltmeter if one is available.

Obtain a hydrogen-oxygen coulometer from the instructor. A drawing of a typical hydrogen-oxygen coulometer is shown in Fig. 7-10. Connect this in series with the cell as shown in Fig. 1 above. Unless it is already filled, raise the reservoir to the top of its travel and, with the buret stopcock open, pour freshly prepared 0.5 F potassium sulfate into the reservoir until the meniscus is just below the stopcock of the buret and just above the bottom of the leveling bulb. At the same time fill the

[1] It is convenient to interpose a flexible stirring shaft between the stirrer and the motor, because this makes it possible to mount the motor some distance away, where it will not interfere with access to the top of the cell. Suitable flexible shafts are available from E. H. Sargent and Co., Chicago, Ill., Catalog S-76655.

[2] Prepurified nitrogen has been specially purified by the supplier to reduce its oxygen content to 0.002 per cent or less.

water jacket with tap water and let the coulometer stand at least until it has reached room temperature.

The controlled-potential electrolyses and coulometric measurements required in Expts. I-12, II-12, III-12, and IV-12 are to be carried out by the following general procedure.

Empty all three compartments of the cell and rinse them thoroughly with distilled water. Fill the central compartment to the bottom of the ground joint with the supporting electrolyte being used, and stopper tightly. Fill the auxiliary electrode compartment about half full of the same solution. Pour the solution to be electrolyzed into the working electrode compartment and turn on a fairly rapid stream of nitrogen. Make sure that the level of the liquid in the auxiliary electrode compartment is at least 2 cm. above the level of the liquid in the working electrode compartment. Turn on the stirring motor and let the deaeration proceed for at least 10 min.

Meanwhile, with switch S'' open and S''' closed (to protect the milliammeter A) set the slider of the rheostat Rh to a position at which about one-fourth of the voltage of the storage batteries will be impressed across the cell circuit. Throw switch S'' to the *right*, open the stopcock of the buret in the coulometer, then close S'. This will cause a current of about 200 ma. to flow through the coulometer, resulting in fairly vigorous evolution of gas at the two electrodes. This ensures that the solution in the coulometer will be saturated with the hydrogen-oxygen mixture at atmospheric pressure when the electrolysis is begun. No less than 5 min. should be allowed for this presaturation, and it is better to allow 10 min. unless the instrument has been used within the preceding day or two with the same filling of electrolyte.

When the solution in the working electrode compartment of the cell is thoroughly deaerated, add 25 to 30 ml. of pure mercury.[1] Check to make sure that the stirrer blades are trailing in the mercury–solution interface, and that the tip of the reference electrode is within a few millimeters of the surface of the mercury.

Open switch S'' and *immediately* close the stopcock on the coulometer buret. Adjust the reservoir so that the levels of the solution in the buret and in the leveling bulb are nearly the same,[2] and read the thermometer and the level of the liquid in the buret.

[1] This is evidently unnecessary in Expt. IV-12, where a platinum working electrode is used. The addition of the mercury is delayed until this point in order to prevent oxidation of the mercury by dissolved air, which would produce mercurous or mercuric salts which would be rereduced during the subsequent electrolysis.

[2] This setting need not be very exact, for an error of 1 mm. corresponds to a difference of only 0.07 mm. between atmospheric pressure and the pressure on the confined gas.

Adjust the potentiometer to indicate the desired potential of the working electrode and set the variable damping resistor across the galvanometer terminals to a very small value. Lock down the tapping key, close switch S'' by throwing it to the *left* (thus connecting both the cell and the coulometer into the electrolysis circuit), and *immediately* adjust the slider of the rheostat until the galvanometer shows practically zero deflection. Continue to adjust the setting of the rheostat to keep the potentiometer balanced as the electrolysis proceeds. Adjustment will be required almost continuously at the start of the electrolysis, but will be needed less and less frequently as time goes on. The entire electrolysis will ordinarily consume 45 min. or less, depending primarily on the efficiency of the stirring. It is desirable to adjust the coulometer leveling bulb occasionally during the electrolysis to keep the confined gas at very nearly atmospheric pressure.

When the rheostat setting has remained constant for several minutes, open switch S'''. The electrolysis current will now be indicated by the milliammeter A. (It may be necessary to readjust the rheostat slightly immediately after opening S''' to compensate for the additional resistance thus introduced into the circuit.) Continue the electrolysis as before until the current has fallen to a value smaller than 1 ma.

Readjust the coulometer leveling bulb so that the solutions in the buret and in the leveling bulb are at the same height, and let the coulometer stand for a few minutes to ensure that equilibrium is reached. Meanwhile open switches S' and S'' and close S'''. Empty the cell, rinse it thoroughly with distilled water, and leave it half full of water. Consult the instructor for directions concerning the treatment of the used mercury. Do *not* pour mercury into the laboratory sink.

Read the meniscus of the electrolyte in the coulometer buret and the thermometer in the water jacket. Measure the atmospheric pressure and, taking the final temperature as the temperature of the evolved gas, calculate the volume of gas which would have been obtained at standard temperature and pressure. (Assume that the hydrogen-oxygen mixture is an ideal gas.) The partial pressure of water vapor above 0.5 F potassium sulfate is 18.4 mm. at 21°C., 20.6 mm. at 23°C., 23.3 mm. at 25°C., and 26.2 mm. at 27°C. Sufficiently accurate values at intermediate temperatures can be found by linear interpolation in these data. Careful experimental measurements have shown that 1 ml. of the hydrogen-oxygen mixture is produced by the flow of 0.0596 millifaraday of electricity through the coulometer.[1] This corresponds to the electrolytic

[1] This is about 0.1 per cent larger than the theoretical value; the source of the discrepancy is unknown. Recent research by J. A. Page and J. J. Lingane at Harvard University has revealed errors of as much as 20 per cent (the number of millifaradays required to produce one milliliter of gas becomes larger than the value just given)

reduction or oxidation of 0.0596 milliequivalent of material in the electrolysis cell.

EXPERIMENT VIII

Assembly of Equipment for Coulometry at Controlled Current

1. Electrical Circuit. The following apparatus is required:

E, two 45-v. heavy-duty B batteries (Burgess Type 5308 or equivalent)
R_1, 20000-ohm variable resistance
R_2, 100-ohm precision resistor (*e.g.*, General Radio Co. Type 500-D)
R_3, carbon or wire-wound resistor, 18 or 22 ohms ± 10 per cent, ½ watt
S, double-pole–double-throw switch
C, electric stop clock

Assemble these as shown in Fig. 7-12. The wires labeled "to potentiometer" are to be connected to the leads marked "to unknown e.m.f." in the potentiometer circuit shown in Fig. 3-8.

2. Cell.[1] Mount a 400-ml. beaker on a magnetic stirrer. Clamp a coarse-porosity sintered-Pyrex gas-dispersion cylinder (Corning 39533), which has been cut so that it is just 1 or 2 cm. higher than the beaker, in such a position that its tip is as close to the bottom of the beaker as possible without interfering with the rotation of the stirring bar. Secure about a foot of platinum wire and coil about 6 in. of it into a tightly wound helix which will just reach the bottom of the gas-dispersion cylinder. Insert this into the gas-dispersion cylinder and attach the free end to the electrical circuit, taking care to observe the polarity appropriate to the experiment which will be carried out. This auxiliary generator electrode must be connected to the positive terminal of the battery (*via* either R_1 and R_2 or S, depending on the connections which have already been made) if the working electrode is to be the cathode, and to the negative terminal of E if the working electrode is to be the anode.

Similarly coil another piece of platinum wire for the working generator electrode. Mount this in position by bending it over the rim of the beaker, and connect it to the other lead from the electrical circuit.

---

when the current density at the electrodes of the coulometer becomes very small; the reason for this is also unknown. In student work the error thus caused is negligible, because it affects only the last few per cent of the total quantity of current consumed, but it does greatly decrease the utility of the hydrogen-oxygen coulometer in work where the highest possible accuracy is essential.

[1] In Expt. III-13 it is more convenient to use a three-compartment cell, such as that shown in Fig. 7-8 and described in Expt. VII, in order to facilitate the necessary deaeration of the solution. Consequently this section of Expt. VIII may be omitted in the work with cupric oxalate.

Mount a rotating platinum microelectrode over the beaker if the "titration" is to be carried out amperometrically, or prepare another platinum wire helix to be used as an indicator electrode if the "titration" is to be followed potentiometrically. In the latter case it will be convenient to insert another double-pole–double-throw switch in the circuit to facilitate connecting the potentiometer across either R_2 or the indicator and reference electrodes.

Finally arrange a clamp to hold the silver–silver chloride–saturated potassium chloride reference electrode prepared in Expt. V. Like the other electrodes, this must be arranged so that it does not interfere with the rotation of the stirring bar.

EXPERIMENT IX-1

Characteristics of a Geiger Counter

Obtain from your instructor detailed directions for the operation of the electronic equipment available for this work.

Arrange a dipping or a jacketed counter in a fixed position near a source of radioactivity. Slowly increase the voltage on the counter until it just starts to register counts, then increase the voltage 20 to 40 v. more. Change the position of the source until the counting rate is about 100 per sec.; then fix the source and counter rigidly with respect to each other. Decrease the voltage until the counter stops.

Now increase the voltage slowly and note the value at which counting just starts. Continue to increase the voltage in steps of 20 v., and record the time for about 10000 counts at each setting. Plot the counting rate *vs.* voltage as the experiment progresses. Do not increase the voltage beyond the region in which the curve begins to turn up sharply; the counter tube may be irreparably damaged if a continuous discharge is started in it. Turn the voltage back to a point near the middle of the plateau of the curve.

Remove the source and determine the background counting rate of the counter in the same location.

Replace the source and record times for 30 successive counts of about 10000. Calculate the individual deviations from the average, and construct a plot showing the frequency distribution of the deviations. Plot on the same sheet the corresponding Gaussian distribution. Calculate and plot a few points of the Poisson distribution near the maximum of the curve. Calculate and give a precise statement of the accuracy with which you know the average counting rate for the series of thirty measurements.

Prepare a solution of some long-lived radioactive nuclide (perhaps Cs^{137}, Na^{22}, or Co^{60}) which in your counting arrangement registers about

200 per sec. Dilute a portion of this solution to precisely twice its volume; mix thoroughly. In this operation use volumetric flasks, *not a mouth-operated pipet.*

Dry the counter carefully, check its background count, and fill it with the more dilute solution. Determine the time for about 200000 counts. Remove the solution, clean and dry the counter, and check background. Determine the time for about 200000 counts with the more active solution. Calculate the resolving time of the counter and the probable accuracy of this figure.

If possible reserve your now well-characterized counter for future experiments.

EXPERIMENT IX-2

Adsorption of Traces: The Growth and Decay of a Daughter Activity

In this experiment the genetic relation Cs^{137}(30 yr.) = Ba^{137}(2.60 min.) is used to illustrate the specific adsorptive properties of a freshly prepared precipitate. Barium carbonate is prepared under conditions which lead to primarily adsorbed barium ion by carrying out the precipitation in the presence of a small excess of Ba^{++}. A solution of cesium chloride containing Cs^{137} in secular equilibrium with Ba^{137} is treated with this precipitate; the barium, being present in very minute quantity, is selectively adsorbed and nearly completely removed from the solution. The decay of the radioactivity in the precipitate and the growth of activity in the solution are observed.

In discussing the results of his experiment the student must remember that his observations of the activity of Ba^{137} are necessarily made over periods of time comparable to the half-life of the nuclide; he should use appropriately exact formulas in treating his results.

Prepare a solution of 0.6 g. of barium chloride dihydrate in 100 ml. of water. Heat nearly to boiling and add slowly with stirring a solution of 0.2 g. of ammonium carbonate in 50 ml. of water. Filter off the precipitate on loose textured paper and wash two or three times with hot water. Preserve the precipitate wet in its paper.

From this point on it will be necessary to work rapidly. Every detail of the experiment should be rehearsed, with nonradioactive material, to make certain that there will be no slip-up in the "hot" run. If duplicate sets of counting equipment are not available, it will be better to carry through the experiment twice, measuring the decay curve and the growth curve in separate experiments.

Obtain from the instructor a solution of cesium chloride spiked with Cs^{137}. Test this solution with bits of pH paper and, if necessary, add a drop or two of dilute ammonia to make sure that the solution is very faintly basic. Scrape your precipitate of barium carbonate into the cesium solution and stir thoroughly.

Arrange a small Buchner funnel with a receiver in the filter flask to collect the cesium solution. Filter off the precipitate, quickly remove and reserve the filtrate, and wash the precipitate with several portions of water.

For the measurements of radioactivity it is advantageous to have at hand two complete sets of counting equipment. One of these should be equipped with an end-window counter and the other with a dipping or jacketed cylindrical counter. Watch the decay of the barium activity adsorbed in the precipitate using the end-window counter, carefully recording times of counts and counting intervals. These intervals may be made longer and longer as the activity decays. The initial length of the intervals will, of course, be adjusted to the amount of activity taken. It is desirable to start with sufficient material to give several thousand counts a minute at the beginning of the observations. For weak preparations it may be desirable to let the counter run continuously and record times and total counts at frequent intervals.

Using the jacketed or dipping Geiger counter, observe the growth of the barium activity in the solution initially stripped of its barium.

Construct appropriate plots of the data and compare these with theoretical curves calculated for the known half-life of Ba^{137}, 2.60 min.

The interested student with laboratory time available will find it highly instructive to design and carry out an original investigation of the adsorptive properties of barium carbonate as they are affected by the conditions of the precipitation, age of the precipitate, and composition of the solution from which the adsorption takes place.

EXPERIMENT IX-3

Isotopic Dilution

The following experiment serves to illustrate the principles of the method of isotopic dilution. If one wishes to determine the equilibrium amount of the cation, say, in an ion-exchanging material, the principal difficulty is almost always the separation of the ion exchanger from the solution with which it is in equilibrium without in any way altering the exchanger itself. This is particularly difficult with gelatinous and otherwise ill-defined materials. Many natural organic and inorganic materials are of this character, *viz.*, the proteins and the clay minerals. The determination can, however, be made by measuring the distribution of radioactivity between exchanger and solution after carrying out only a partial separation. In the following experiment the sodium content of a bentonite clay is measured. Evidently many variants of the experiment are possible.

Prepare a solution of sodium chloride of accurately known strength, about 0.01 *F*, containing Na^{22} as a tracer. Obtain a sample of bentonite known to be almost entirely in the sodium form. Weigh accurately about 1 g. of the clay into a 100-ml. wide-mouth bottle, and measure in just 50 ml. of the labeled sodium chloride solution. At once stopper the bottle securely and shake violently to disperse the clay. Shaking at intervals, keep the clay suspended in the solution for about an hour. Now divide the clay suspension between two 50-ml. plastic centrifuge tubes and centrifuge, preferably at 4000 to 4500 r.p.m. in a 15-cm.-radius angle head, until a clear supernatant solution results. Carefully pipet off samples of the clear solution. Measure the activity of these and of the original solution at nearly the same time with identical counting arrangements. The activity of the solution should be reduced by about two-thirds by its contact with the clay.

Calculate the exchangeable sodium content of the clay mineral. Discuss the errors of the experiment and estimate the accuracy of the result. Suppose the clay contains a small amount of cation other than sodium (Ca^{++} or K^{+}) which is replaceable by sodium. How would your results be affected? Suggest means for determining the true total cation capacity of the mineral.

EXPERIMENT IX-4

The Coprecipitation of Cobalt and Iron

In this experiment a tracer of Co^{60} is employed to study the extent of coprecipitation of cobaltous hydroxide with ferric hydroxide under different conditions of forming the precipitates. The results obtained in precipitations with large excesses of ammonia are compared with that obtained in the method of Lingane and Kerlinger[1] in which the cobalt is caused to form a soluble complex with pyridine. In the latter method the cobalt is held in solution at pH 5.0 to 5.5, corresponding to a solution containing about equal concentrations of pyridine and pyridinium chloride. The student will be able to assess the efficacy of the separation.

Prepare or obtain from the instructor 500 ml. of a solution 0.1 *F* in ferric chloride and 0.01 *F* in cobaltous chloride. Add $Co^{60}Cl_2$ (preferably of high specific activity) to give a convenient counting rate in the solution, about 100 per sec.

A. Measure into two 600-ml. beakers 50-ml. portions of the Fe-Co solution and precipitate by adding 300 ml. of concentrated ammonia. By a separate experiment convince yourself that no precipitate containing cobalt is obtained from a similar treatment of an iron-free solution. Filter on porous paper and wash the precipitate with 20 to 25 small portions of water. Dissolve the precipitate into a small beaker with

[1] J. J. Lingane and H. Kerlinger, *Ind. Eng. Chem., Anal. Ed.*, **13,** 77 (1941).

several small portions of hot 6 *F* hydrochloric acid, and wash the paper with hot water. Transfer the solution of the precipitate to a 100-ml. volumetric flask and dilute to the mark.

B. Repeat as in A, but this time add the Fe-Co solution slowly with constant stirring to 300 ml. of warm concentrated ammonia (hood!).

C. Measure 50 ml. of the Fe-Co solution into each of two 100-ml. volumetric flasks. Add to each 2 ml. of concentrated hydrochloric acid and 5 ml. of pyridine. Dilute to the mark with water. Mix thoroughly and then allow the precipitate to settle. Filter and dissolve the precipitate as before.

Measure the activity of the original solution and each of the six solutions of precipitates, using the same counter for each measurement. Make sure the counting rate returns to background between each pair of measurements. Remember to halve the measured counting rate of the stock solution in comparisons with the rates for the solutions of the precipitates.

What is the stable oxidation state of cobalt in an ammoniacal solution exposed to air? What are the formulas of the complex ions formed? Does a similar reaction occur in the pyridine medium? What is the formula of the cobalt–pyridine complex? Which of these complex ions would you expect to be most strongly adsorbed onto a precipitate of hydrous ferric oxide? Why? For what other reason would you expect the pyridine separation to be the most efficient of the three separation procedures investigated in this experiment?

EXPERIMENT X-1

Preparation and Properties of an Ion-exchange Resin

The copolymerization of methacrylic acid and glycol *bis*-methacrylate is a free-radical reaction and quite temperature-dependent. It will be carried out in this experiment by mixing the monomers and warming them to a temperature at which a homogeneous solution is just formed, then adding a small amount of ammonium persulfate to initiate the reaction, and heating the resulting mixture in molds until polymerization is complete.

For the experiments here described the resin is conveniently prepared in the form of rods. For use as molds prepare six 10-mm. glass tubes about 6 in. long. These should have test-tube bottoms and should be beaded for corks at the top.

Mount a variable-speed stirring motor over a 2-l. beaker of water on a tripod. Arrange a shaft equipped with a large notched stopper and a heavy rubber band to carry three of the molds and rotate them within the beaker. By thus arranging to stir the water bath with the molds themselves, uniform heating of the mixtures during polymerization is assured.

Calculate the weights of glycol *bis*-methacrylate necessary for 1 per cent and 5 per cent cross-linking with 25 g. of methacrylic acid. Prepare small amounts of these mixtures and determine the temperatures at which clearing occurs.

Bring the bath to the temperature thus found for the 1 per cent cross-linked mixture. In it place three molds attached to the stopper on the stirring shaft, and a small flask containing 25 ml. of a 1 per cent aqueous solution of ammonium persulfate. Into another small flask weigh 25 g. of methacrylic acid and the amount of glycol *bis*-methacrylate calculated for 1 per cent cross-linking. Place this flask in the bath and let stand until its contents become homogeneous. Add the ammonium persulfate solution, warming slightly if necessary to obtain a clear solution. Quickly fill the molds, stopper them, and start the stirrer. Heat the bath slowly, noting the temperature (80 to 90°C.) at which gelling starts. After gelling has started the resin can be seen pulling away from the sides of the molds. Finally bring the bath to a boil and maintain this temperature for an hour to complete the polymerization. Turn off the heat and allow the bath to cool slowly to room temperature before removing the molds.

The rods of polymer are usually quite easily removed from the molds, although it is sometimes necessary to break the end of a mold in order to get a grip on the resin. Soak the rods in several successive portions of 0.01 F hydrochloric acid before use.

Prepare rods of 5 per cent cross-linked resin by an exactly similar procedure.

The rods of resin may be cut into sections for the following observations.

A. Cut 2-mm. slices and 1-cm. sections of 1X and 5X resin and place them in dilute (0.01 F) sodium hydroxide solution. Note the swelling characteristics of the two materials. Interpret any peculiar mechanical behavior of the thick pieces. Transfer the swollen resins to 0.1 F hydrochloric acid and record your observations.

B. Immerse 2-cm. sections of the two resins in 1 F cupric sulfate for some hours. At intervals remove one rod of each resin from the solution, rinse briefly with water, and cut each rod in half perpendicular to its axis. By inspecting the planes thus exposed, observe the progress of the diffusion into the material.

C. The resins may be converted to the sodium salts without too serious disintegration by soaking in half-saturated sodium chloride solution which is 0.1 F in NaOH. Prepare a piece of the sodium form and immerse it in silver nitrate solution. Observe the progress of the diffusion of silver ion into the material.

D. The acid character of the resin and the diffusion of long molecules into it may be illustrated as follows. Insert a one-inch piece of the 5X resin into a short piece of rubber tubing. Fill the tube with a neutral aqueous solution of methyl red, and let stand overnight. Pour off the indicator solution and immerse the resin in water to watch the progress of the diffusion.

EXPERIMENT X-2

Determination of Sodium *via* Ion-exchange Replacement

Prepare a column of Amberlite IR-120 using a tube as illustrated in Fig. 11-9*b*. The tube should be approximately 20 cm. long over-all and the inside diameter of its narrower portion should be about 12 mm. Place a thin mat of Pyrex glass wool in the bottom, and fill with water well into the enlarged portion. Sprinkle in 22 g. of the resin in the hydrogen form. Condition the column by passing through it 50 ml. of 1 F hydrochloric acid followed by a washing with 200 to 300 ml. of water. Make sure that the final effluent is neutral. Leave the column full of water to just above the bed of resin.

Prepare a standard solution of sodium chloride, using enough *dry* salt to make 250 ml. of approximately 0.08 F solution. Pipet into the column 25 ml. of this solution and allow it to pass slowly through, receiving it in a 500-ml. erlenmeyer flask. Wash the column with water, using five portions of about 25 ml., receiving the washings in the erlenmeyer flask.

Titrate the effluent with 0.1 F sodium hydroxide.

Regenerate the column with 50 ml. of 1 F hydrochloric acid, washing thoroughly with 200 ml. of water. Pass another 150 ml. of water through the column into a clean receiver; titrate with 0.1 F sodium hydroxide. Use this result as a blank correction.

Repeat the replacement reaction with the standard sodium chloride solution. Compare the result of the titration with the known amount of sodium chloride taken.

EXPERIMENT X-3

Separation of Iron, Cobalt, and Nickel on an Anion Exchanger

In the following experiment a chromatographic separation will be carried out which depends on the relative stability of the chloro-complex ions of the transition elements iron, cobalt, and nickel. Many metals form negatively charged complexes in concentrated hydrochloric acid. These vary markedly in the strength of adsorption by the strongly basic anion exchangers. Apparently, the singly charged complexes are the most strongly adsorbed. Thus Dowex 1 strongly adsorbs $FeCl_4^-$ from 12 F HCl. At this acid concentration cobalt is probably present in solution largely as $CoCl_4^{=}$; cobalt is retained on the resin much less strongly than is iron. Nickel is not at all adsorbed by the resin from concentrated HCl because it forms inappreciable concentrations of chloro anions.

In the following experiment, the separation will be made with only qualitative observations regarding nickel, while the character of the separation of cobalt from iron will be carefully followed using a radioactive tracer for cobalt and a spectrophotometric method for iron.

Prepare a column of Dowex 1 in a 15-mm. tube about 8 in. long equipped with a stopcock at the bottom. The stopcock plug should be grooved on opposite sides to make easy a close regulation of the flow at low rates. Put a pad of Pyrex glass wool in the bottom of the tube and fill it with water. Slowly sprinkle the resin into the tube, tapping it briskly all the while. When a column 6 or 7 in. long is obtained, place a thin pad of glass wool over the resin. Equip the column with a one-hole stopper carrying a funnel. The tip of the funnel should project just enough to be buried in the glass wool over the resin.

Adjust the level of water in the column so that it comes within the upper pad of glass wool. Now drain the column completely into a graduated cylinder in order to get a rough idea of its total free volume.

Connect a funnel to the exit tube of the column, using a piece of tubing about 2 ft. long, and pass water upflow through the column until all the air bubbles are removed. This operation may be greatly facilitated by first displacing the air with a stream of carbon dioxide; bubbles of the latter are rapidly dissolved away by the water wash.

Now pass downflow through the column at least 20 column volumes of concentrated (12 *F*) hydrochloric acid. Make very sure that air is at no time allowed to suck into the resin bed. Finally adjust the level of the acid to just above the tip of the funnel.

Prepare 250 ml. of a solution containing 0.1 *F* nickel chloride, 0.1 *F* cobalt chloride, and 0.1 *F* ferric chloride in 12 *F* hydrochloric acid. To make easier the analysis of this solution for iron, prepare separately 100 ml. of a solution containing about 0.5 *F* ferric chloride in 12 *F* hydrochloric acid. Use 50 ml. of this in the preparation of the mixed stock and reserve the remainder for the spectrophotometric analysis. Add sufficient $Co^{60}Cl_2$ (before bringing the solution to its final volume) so that the solution will give approximately 100 counts per sec. in the Geiger tube available for use in the experiment.[1]

[1] NOTE: Cobalt 60 is a long-lived gamma emitter. Handle the solutions with great care to avoid any spillage. All vessels containing the radioactive cobalt must be carefully rinsed into a single large container and turned over to your instructor for proper disposal. All the handling of the radioactive solutions should be done over large photographic trays, and the desk top should be covered with absorbent paper.

With the amounts of radioactive material used in this experiment, there is no

In transferring radioactive solutions a pipet should never be manipulated with the mouth. A simple expedient is to equip a pipet with the glass parts of a large hypodermic syringe, the plunger of which has been very lightly greased with petroleum jelly.

Pipet 20 ml. of the Ni-Co-Fe solution into the funnel over the column, making sure that no bubble of air is caught in the stem. Barely open the stopcock and very slowly, dropwise, allow 20 ml. of acid to flow from the column, closing the stopcock when the solution has reached the bottom of the stem of the funnel.

Catch the final 5 ml. of effluent in a test tube and test for the presence of nickel with dimethylglyoxime.

Rinse the funnel with two 5-ml. portions of 12 *F* HCl, withdrawing 5 ml. from the bottom of the column after each rinse. Test the effluent qualitatively for nickel.

Measure 70 ml. of 12 *F* HCl in a graduated cylinder and fill the funnel from this. Adjust the flow from the column to a rate of about 0.5 ml. per min. Collect samples in 10-ml. volumetric flasks or in test tubes previously marked at the 10-ml. level. If test tubes are used, these must be weighed in advance (with their stoppers) so that eventually sample sizes will be known. Note the appearance of the effluent during this initial operation when the nickel is being removed from the column. Roughly standardize your qualitative test for nickel and note semiquantitatively the manner in which the nickel appears in the effluent. (If desired, of course, quantitative determinations may be made.)

Next, remove the cobalt from the column by eluting with 100 ml. of 3 *F* HCl, collecting samples as before. Each of these samples is to be counted in the same manner as was the original solution. A convenient procedure is to use an all-glass jacketed Geiger counter tube holding somewhat less than 25 ml. of solution, and prepare each solution merely by diluting the 10-ml. sample to 25 ml.

Finally, elute the iron from the column with 100 ml. of 0.5 *F* HCl, again collecting 10-ml. samples. These, and the original stock solution of iron, are to be analyzed spectrophotometrically for their iron content by the thiocyanate method; see, *e.g.*, I. M. Kolthoff and E. B. Sandell, "Textbook of Quantitative Inorganic Analysis," 3d ed. The Macmillan Company, New York, 1952, p. 635. The absorbance measurements should be made near 365 mμ.

Calculate the ratio of effluent to input concentrations for cobalt and for iron and plot these against the effluent volume. Note on the graph the points at which eluent concentrations were changed.

---

hazard from radiation, but as a matter of good general practice, all radioactive solutions should be kept as far away from the body as possible.

A beta-gamma survey meter should be available and the radiation level should be frequently checked.

INDEX